Lecture Notes in Physics

Springer
Berlin
Heidelberg
New York
Barcelona
Hong Kong
London
Milan
Paris
Singapore
Tokyo

Physics and Astronomy ONLINE LIBRARY

http://www.springer.de/phys/

The Editorial Policy for Proceedings

The series Lecture Notes in Physics reports new developments in physical research and teaching – quickly, informally, and at a high level. The proceedings to be considered for publication in this series should be limited to only a few areas of research, and these should be closely related to each other. The contributions should be of a high standard and should avoid lengthy redraftings of papers already published or about to be published elsewhere. As a whole, the proceedings should aim for a balanced presentation of the theme of the conference including a description of the techniques used and enough motivation for a broad readership. It should not be assumed that the published proceedings must reflect the conference in its entirety. (A listing or abstracts of papers presented at the meeting but not included in the proceedings could be added as an appendix.)
When applying for publication in the series Lecture Notes in Physics the volume's editor(s) should submit sufficient material to enable the series editors and their referees to make a fairly accurate evaluation (e.g. a complete list of speakers and titles of papers to be presented and abstracts). If, based on this information, the proceedings are (tentatively) accepted, the volume's editor(s), whose name(s) will appear on the title pages, should select the papers suitable for publication and have them refereed (as for a journal) when appropriate. As a rule discussions will not be accepted. The series editors and Springer-Verlag will normally not interfere with the detailed editing except in fairly obvious cases or on technical matters.
Final acceptance is expressed by the series editor in charge, in consultation with Springer-Verlag only after receiving the complete manuscript. It might help to send a copy of the authors' manuscripts in advance to the editor in charge to discuss possible revisions with him. As a general rule, the series editor will confirm his tentative acceptance if the final manuscript corresponds to the original concept discussed, if the quality of the contribution meets the requirements of the series, and if the final size of the manuscript does not greatly exceed the number of pages originally agreed upon. The manuscript should be forwarded to Springer-Verlag shortly after the meeting. In cases of extreme delay (more than six months after the conference) the series editors will check once more the timeliness of the papers. Therefore, the volume's editor(s) should establish strict deadlines, or collect the articles during the conference and have them revised on the spot. If a delay is unavoidable, one should encourage the authors to update their contributions if appropriate. The editors of proceedings are strongly advised to inform contributors about these points at an early stage.
The final manuscript should contain a table of contents and an informative introduction accessible also to readers not particularly familiar with the topic of the conference. The contributions should be in English. The volume's editor(s) should check the contributions for the correct use of language. At Springer-Verlag only the prefaces will be checked by a copy-editor for language and style. Grave linguistic or technical shortcomings may lead to the rejection of contributions by the series editors. A conference report should not exceed a total of 500 pages. Keeping the size within this bound should be achieved by a stricter selection of articles and not by imposing an upper limit to the length of the individual papers. Editors receive jointly 30 complimentary copies of their book. They are entitled to purchase further copies of their book at a reduced rate. As a rule no reprints of individual contributions can be supplied. No royalty is paid on Lecture Notes in Physics volumes. Commitment to publish is made by letter of interest rather than by signing a formal contract. Springer-Verlag secures the copyright for each volume.

The Production Process

The books are hardbound, and the publisher will select quality paper appropriate to the needs of the author(s). Publication time is about ten weeks. More than twenty years of experience guarantee authors the best possible service. To reach the goal of rapid publication at a low price the technique of photographic reproduction from a camera-ready manuscript was chosen. This process shifts the main responsibility for the technical quality considerably from the publisher to the authors. We therefore urge all authors and editors of proceedings to observe very carefully the essentials for the preparation of camera-ready manuscripts, which we will supply on request. This applies especially to the quality of figures and halftones submitted for publication. In addition, it might be useful to look at some of the volumes already published. As a special service, we offer free of charge LaTeX and TeX macro packages to format the text according to Springer-Verlag's quality requirements. We strongly recommend that you make use of this offer, since the result will be a book of considerably improved technical quality. To avoid mistakes and time-consuming correspondence during the production period the conference editors should request special instructions from the publisher well before the beginning of the conference. Manuscripts not meeting the technical standard of the series will have to be returned for improvement.
For further information please contact Springer-Verlag, Physics Editorial Department II, Tiergartenstrasse 17, D-69121 Heidelberg, Germany

Series homepage – http://www.springer.de/phys/books/lnpp

Dany Page Jorge G. Hirsch (Eds.)

From the Sun to the Great Attractor

1999 Guanajuato Lectures on Astrophysics

Springer

Editors

Dany Page
Instituto de Astronomia, UNAM
Apdo. Postal 70-264, Cd. Universitaria
04510 México, D.F., México

Jorge G. Hirsch
Instituto de Ciencias Nucleares, UNAM
Apdo. Postal 70-543, Cd. Universitaria
04510 México, D.F., México

Library of Congress Cataloging-in-Publication Data applied for.

Die Deutsche Bibliothek - CIP-Einheitsaufnahme

From the sun to the great attractor : 1999 Guanajuato lectures on astrophysics / Dany Page ; Jorge G. Hirsch (ed.). - Berlin ; Heidelberg ; New York ; Barcelona ; Hong Kong ; London ; Milan ; Paris ; Singapore ; Tokyo : Springer, 2000
(Lecture notes in physics ; Vol. 556)
(Physics and astronomy online library)
ISBN 3-540-41064-3

ISSN 0075-8450
ISBN 3-540-41064-3 Springer-Verlag Berlin Heidelberg New York

Springer-Verlag Berlin Heidelberg New York
a member of BertelsmannSpringer Science+Business Media GmbH

Typesetting: Camera-ready by the authors/editor
Cover design: *design & production*, Heidelberg

Printed on acid-free paper
SPIN: 10780741 55/3141/du - 5 4 3 2 1 0

Preface

The Mexican School of Astrophysics (*Escuela Mexicana de Astrofísica 1999: EMA99*) was held in the city of Guanajuato on August 4 – 11, 1999. It was the second of its kind and marked the beginning of a hopefully long series of such events in the future. Both the quality of the lectures and the enthusiasm of the participants made it a very fruitful event. Moreover, the beauty of the colorful city of Guanajuato, as well as its sparkling life, made a wonderful setting for the school.

In keeping with the spirit of the previous school, the goal was to present a small set of topics of high current interest to advanced students and researchers in physics and astrophysics, The school consisted of eight courses which are presented here as the eight chapters of this book. A few short conferences and a poster session allowed the participants to present their own work. Each lecturer was set the difficult task of starting from the basics and culminate by bringing the audience to the forefront of her/his field. As the reader will see, the written texts of these lectures successfully fulfill this double challenge.

Mexico City,
July 2000

Dany Page
Jorge G. Hirsch

Supporting Institutions of EMA99

The organizers of the *Escuela Mexicana de Astrofísica 1999* aknowledge financial support from the following institutions:

Academia Mexicana de Ciencias
Centro Latinoamericano de Física (CLAF)
Consejo Nacional de Ciencia y Tecnología (CONACyT)
Departamento de Astronomía, Universidad de Guanajuato
Departamento de Física, CINVESTAV del IPN
Deutsche Akademischer Austauschdienst
Fundación México-Estados Unidos para la Ciencia
Instituto de Astronomía, UNAM
Instituto de Ciencias Nucleares, UNAM
Instituto de Geofísica, UNAM
Instituto de Física, UNAM
Instituto de Física y Matemáticas, Universidad Michoacana de SNH
Instituto Nacional de Astrofísica, Optica y Electrónica (INAOE)
Programa de Posgrado en Astronomía, UNAM
Programa de Posgrado en Ciencias Físicas, UNAM
Programa de Posgrado en Ciencias de la Tierra, UNAM

Organizing Committee of EMA99

Dany Page, Instituto de Astronomía, UNAM
Armando Arellano Ferro, Instituto de Astronomía, UNAM
Alberto Carramiñana, Departamento de Astrofísica, INAOE
Peter Hess, Instituto de Ciencias Nucleares, UNAM
Jorge Hirsch, Instituto de Ciencias Nucleares, UNAM
Tonatiuh Matos, Departamento de Física, CINVESTAV
Victor Migenes, Departamento de Astronomía, U. de Guanajuato
Lukas Nellen, Instituto de Ciencias Nucleares, UNAM
Maria Esther Ortiz, Instituto de Física, UNAM
Jose Francisco Valdes, Instituto de Geofísica, UNAM
Thomas Zannias, Instituto de Física y Matemáticas, U. Michoacana de SNH
Arnulfo Zepeda, Departamento de Física, CINVESTAV

Contents

List of Contributors

Dermott J. Mullan
Bartol Research Institute
University of Delaware
Newark, DE 19716, USA
mullan@brivs2.bartol.udel.edu

Moshe Gai
Laboratory for Nuclear Science
Department of Physics, U3046
University of Connecticut
2152 Hillside Rd.
Storrs, CT 06269-3046, USA
gai@uconn.edu

Karl-Heinz Rädler
Astrophysikalisches Institut Potsdam
An der Sternwarte 16
D-14482 Potsdam, Germany
khraedler@aip.de

Włodek Kluźniak
Copernicus Astronomical Center
ul. Bartycka 18
00-716 Warszawa, Poland
wlodek@camk.edu.pl

Trevor C. Weekes
Harvard-Smithsonian Center
for Astrophysics
Whipple Observatory
P.O. Box 97
Amado, AZ 85645-0097, USA
weekes@egret.sao.arizona.edu

Esteban Roulet
Departamento de Física
Universidad Nacional de La Plata
CC67, 1900, La Plata, Argentina
roulet@venus.fisica.unlp.edu.ar

Günter Sigl
DARC
Observatoire de Paris-Meudon
F-92195 Meudon Cédex, France
sigl@indigo.obspm.fr
and
Department of Astronomy
& Astrophysics
Enrico Fermi Institute
The University of Chicago
Chicago, IL 60637-1433, USA
sigl@humboldt.uchicago.edu

Renée C. Kraan-Korteweg
Departamento de Astronomía
Universidad de Guanajuato
Apartado Postal 144
36000 Guanajuato, GTO, Mexico
kraan@norma.astro.ugto.mx

Solar Physics: From the Deep Interior to the Hot Corona

Dermott J. Mullan

Bartol Research Institute, University of Delaware, Newark DE 19716, USA

Abstract. We present an overview of the thermal properties of the Sun from the hot interior to the hot corona. For pedagogical reasons, we confine the discussion to certain relevant solutions of the energy conservation equation. In the interior, quantitative information can be obtained by using a polytropic equation of state: internal temperatures obtained in this way are found to be reliable to about 10%, and we can obtain a good estimate of the depth of the convection zone. In the chromosphere, acoustic waves originating in the convection zone do work on the gas: as the gas heats up, the atomic energy levels of many elements (especially hydrogen) exert a strong thermostatic control so that the temperature is confined to a steady value in the range 5000–10^4 K. In long-lived coronal loops, a steady state balance between thermal conduction and radiative losses causes the temperature of the electrons to lie in the range (1–2) million K. Coronal ions are heated to greater temperatures than electrons. In flares, processes of heating and cooling are explicitly non-steady, and short-lived excursions to temperatures as high as 25 million K (or more) are observed in the largest flares.

1 Internal Structure of the Sun

The most important quantity in determining stellar structure and evolution is the **TEMPERATURE** inside the star: this determines thermonuclear reaction rates at the center, and it also determines how the energy is transported. So in order to understand anything about the Sun and its operation, we need to determine T and how it varies as a function of radial distance from the center.

Three conservation laws in general are needed in order to determine how the fluid in or near a star behaves. These are the conservation of (i) mass, (ii) momentum, and (iii) energy. The *mechanical* properties of the material inside the star can be determined if we solve only (i) and (ii). But the *thermodynamic* properties of the material in general require us also to solve (iii). If we can solve all three equations, then we obtain the desired model of the star, i.e. we obtain radial profiles of density, pressure, and temperature. Now, a full solution of (iii) can be a difficult process. However, from a pedagogical standpoint, it is fortunate that valuable information can be obtained about stellar structure without solving (iii) in detail. Let us see how far we can go.

1.1 Mechanical Equilibrium

Consider the mechanical properties. In a spherical shell at radius r and thickness dr, the mass contained in the shell is $\mathrm{d}M(r) = 4\pi r^2 \rho(r)\mathrm{d}r$. This allows us to write (i) as

$$\frac{\mathrm{d}M(r)}{\mathrm{d}r} = 4\pi r^2 \rho(r). \tag{1}$$

We write (ii) conservation of momentum as:

$$\frac{\partial \mathbf{v}}{\partial t} + \mathbf{v} \cdot \nabla \mathbf{v} =$$

$$= -\frac{\nabla p}{\rho} + g. \tag{2}$$

A static solution of this equation (i.e. $\mathbf{v} = 0$) is possible if right-hand side equals zero, i.e. if the pressure gradient balances gravity:

$$\frac{\mathrm{d}p(r)}{\mathrm{d}r} = -\rho g. \tag{3}$$

This particular equation describes hydro*static* equilibrium (HSE). Let us look at how we need to treat g in different regions in the Sun.

First, in the layers of the Sun near the visible surface (i.e. near the photosphere), g can be taken as constant: $g(surface) = -GM_{\text{sun}}/R_{\text{sun}}^2$. Inserting values appropriate for the Sun, i.e. $M_{\text{sun}} \approx 2 \times 10^{33}$ gm, and $R_{\text{sun}} \approx 7 \times 10^{10}$ cm, we find $g_{\text{surface}} \approx 2.7 \times 10^4$ cm sec^{-2}. This leads to a simple solution if we are dealing with an isothermal perfect gas: $p(z) = p(0)\mathrm{e}^{-z/H}$. Here, $z = r - r_0$, where r_0 is a reference location at which the pressure has the value $p(0)$, and $H = R_{\text{gas}}T/\mu g$ is the "pressure scale height". (The local temperature and molecular weight are T and μ; R_{gas} is the gas constant.)

Second, outside the Sun, $g = -GM_{\text{sun}}/r^2$. When we discuss the corona in Sect. 5 below, we will use this to arrive at a *non*-static solution of (3), i.e. one in which $\mathbf{v} = v(r)$ is non-zero.

Third, inside the Sun, $g(r) = -GM(r)/r^2 \to 0$ as $r \to 0$. In order to model the interior of the Sun, we need to use this radially-dependent expression for $g(r)$. Rewriting HSE with this choice of $g(r)$, we see that

$$M(r) = -\frac{r^2}{G\rho}\frac{\mathrm{d}p}{\mathrm{d}r}. \tag{4}$$

Now we differentiate (4) with respect to r and use (1):

$$\frac{1}{r^2}\frac{\mathrm{d}}{\mathrm{d}r}\left(\frac{r^2}{\rho}\frac{\mathrm{d}p}{\mathrm{d}r}\right) = -4\pi G\rho \tag{5}$$

The interior of the Sun (and any other stable spherically symmetric object) obeys this equation. However, we cannot yet solve it: there are TWO unknowns ($p(r)$, $\rho(r)$) but only one equation. To proceed, we need more information: in principle, the solution of the full energy equation would give us the information. But we can get an overview of the internal structure without going so far.

1.2 Polytropes

The trick is to adopt a particular solution of (iii) and then solve (i) and (ii). We consider a special class of solutions where we can decouple thermodynamics from mechanics. In this class, pressure and density are assumed to be related by a power law:

$$p = K\rho^{\delta}. \tag{6}$$

A particular case where such a relation exists is well known from studies of thermodynamics: when a parcel of gas behaves adiabatically, p and ρ are related by $p \sim \rho^{\gamma}$ where $\gamma = C_{\rm p}/C_{\rm v}$ is the ratio of specific heats at constant pressure and at constant volume. In a monatomic gas, where $C_{\rm p} = (5/2)\ R_{\rm gas}/\mu$ and $C_{\rm v} = (3/2)\ R_{\rm gas}/\mu$, the value of γ is 5/3. Adiabatic behavior is one particular solution of the energy equation. But (5) is more general than (6): it describes how the pressure is related to the density in situations which need not be adiabatic.

It is customary to write (6) in a slightly different form. We introduce a parameter n (the **polytropic index**) such that $\delta = 1 + 1/n$. Then

$$p = K\rho^{1 + 1/n}. \tag{7}$$

If the gas pressure and density can be related in this way, then the gas is said to behave like a polytrope. If the perfect gas law ($p \sim \rho T$) is also obeyed, the density and pressure in the polytrope satisfy $\rho \sim T^n$ and $p \sim T^{n+1}$. For future reference, we note that this implies d(log p)/d(log T) $= n + 1$ in a polytrope.

Let us look at two regions of the solar interior to see whether it is reasonable to rely on a polytrope. We shall see below (Sect. 1.13) that the solar interior consists of two major regions: a "radiative zone" between the center and about $0.7R_{\rm sun}$, and a "convective envelope" between $0.7R_{\rm sun}$ and the surface. In the convection zone (see Sect. 1.12), gas moves around in such a way that the motions are close to adiabatic, i.e. $\gamma = 5/3$ in (6). This corresponds to the polytrope $n = 3/2$. The $n = 3/2$ polytrope actually does a good job of describing the structure of the convection zone in the Sun.

But in the radiative zone in the interior of the Sun, photons do the energy transport. Modelers who solve for all the details about energy conservation in this region find that p and T have certain radial profiles. How close are these profiles to polytropic? To answer this, we refer to the recent solar model of Christensen & Dalsgaard [1] (hereafter JCD): using this model, we can construct numerically the gradient d(log p)/d(log T). Since this gradient should have the value $n + 1$ if the medium behaved exactly like a polytrope, we can define an "effective polytropic index" by setting $n_{\rm eff}$ = d(log p)/d(log T) $- 1$. Values of $n_{\rm eff}$ are shown in Fig. 1.

We see that, near the solar surface, between radii of 0.7 and 0.95 solar radii, $n_{\rm eff}$ has a value which turns out to be remarkably constant, just as a polytrope would have. In this region, a polytropic model with $n = 1.5$ is an excellent approximation to the radial variation of pressure, density, and temperature. Why

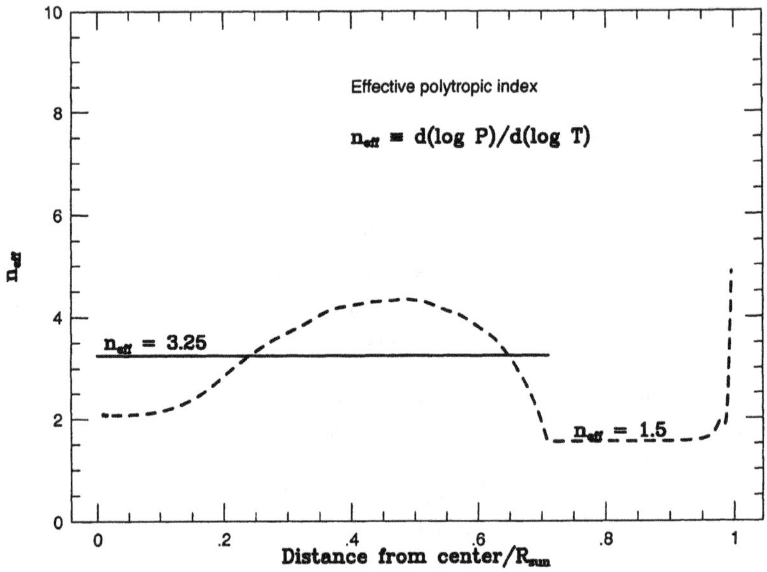

Fig. 1. Effective polytropic index as a function of radial distance in the solar model of JCD

is the $n = 1.5$ polytrope such a good approximation for this region of the Sun? We shall discuss the answer to this question in Sect. 1.12.

Deep in the solar interior, at $r \leq 0.7R_{\rm sun}$, $n_{\rm eff}$ is no longer strictly constant, but shows some variations with radius. Therefore, we do not expect that a polytropic model will be quite as successful in describing the structure of the deep interior as it is in the outer region ($0.7 \leq r \leq 0.95\ R_{\rm sun}$). Nevertheless, we note that the variations of $n_{\rm eff}$ in Fig. 1 do not extend over an arbitrarily wide range, but are mainly confined between 2 and 4. A value $n_{\rm eff} = 3.25$ is actually a fair approximation to a mean value in the radiative interior. (Reasons why $n_{\rm eff}$ $= 3.25$ is a plausible value for the solar interior will be discussed in Sect. 1.11 below.) The variations in $n_{\rm eff}$ are small enough that, we might be able to obtain a good zeroth order approximation to conditions inside the radiative interior of the Sun by assuming a polytropic equation of state.

1.3 A Temperature Variable

The advantage of using a polytrope is that we can now solve for the radial profiles of density and pressure. To do this, we introduce a dimensionless function $\theta(r)$ according to

$$\frac{\rho(r)}{\rho_c} = \theta^n \tag{8}$$

where subscript c denotes values at the center of the star. Combining (8) and (7), we find that the pressure obeys

$$\frac{p(r)}{p_c} = \theta^{n+1}. \tag{9}$$

The advantage of using the function θ can be seen when we compare eqs. (8) and (9) in order to obtain the ratio of $p(r)/p_c$ to $\rho(r)/\rho_c$: this ratio is simply θ. Now, if the material which makes up the star consists of a perfect gas (with $T \sim p/\rho$), then $\theta(r) = T(r)/T_c$ where T_c is the central temperature. So θ is simply the scaled temperature inside the star. Once we solve for $\theta(r)$ as a function of radius, we will then also have the information we set out to acquire: the radial profiles of p, ρ, and T.

1.4 Solving the Polytrope

To solve the polytrope, define a new radial variable: $\xi = r/r_n$, where the so-called Emden unit of length r_n is defined by $r_n^2 = (n+1)p_c/4\pi G\rho_c^2$. With this new variable, the polytrope equation becomes

$$\frac{1}{\xi^2}\frac{\mathrm{d}}{\mathrm{d}\xi}\left(\xi^2\frac{\mathrm{d}\theta}{\mathrm{d}\xi}\right) = -\,\theta^n. \tag{10}$$

This is an ordinary differential equation in one unknown. With suitable boundary conditions, (1) $\theta = 1$ at $\xi = 0$, and (2) $\mathrm{d}\theta/\mathrm{d}\xi = 0$ at $\xi = 0$ (because $g = 0$ at center), eq. (10) has a unique solution θ_n once n is specified. The function $\theta_n(\xi)$ decreases monotonically with ξ. When θ_n reaches zero for the first time at $\xi = \xi_1$, the values of T, p, and ρ fall to zero. This indicates that at $\xi = \xi_1$, we have reached the "surface" of the star.

Analytic solutions exist for $n = 0$, 1, and 5. For example, with $n = 0$ (constant density), the solution is $\theta_0(\xi) = 1 - \xi^2/6$. The first zero occurs at $\xi_1 = \sqrt{6}$. Converting to dimensional units, the radius r_1 corresponding to ξ_1 is $R_0 = \sqrt{(6p_c/4\pi G\rho^2)}$ for the $n = 0$ polytrope. How accurate is this result? Let us apply it to a nearly incompressible body: the Earth. With mean density $\rho \approx 5.5$ gm cm^{-3} and radius $R_0 = 6371$ km, the polytrope solution predicts a central pressure p_c of about 2×10^{12} dyn cm^{-2}. This is within a factor of 5 of the pressure predicted by the most detailed model of the Earth.

1.5 Central Condensation

The temperature function $\theta(\xi)$ is a maximum at the center of the Sun and falls off with increasing radius. Since density and pressure scale as the nth and $(n+1)^{th}$ powers of θ, the density is peaked more sharply than temperature towards the center. And the pressure is peaked more sharply still. In order to show that this behavior of the polytrope solution is relevant to a "real" solar model, we show in Fig. 2 how p and T behave in the JCD model. It is apparent from Fig. 2 that

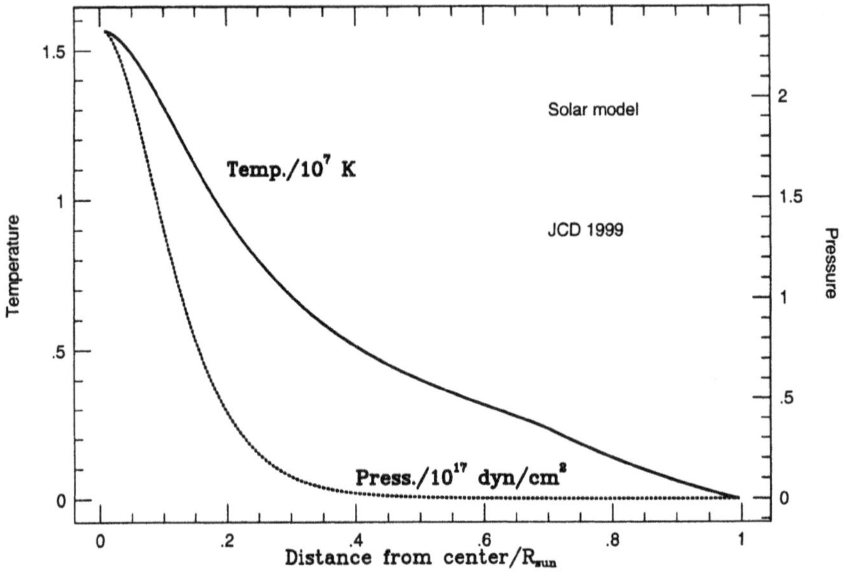

Fig. 2. Radial variation of temperature and pressure in a solar model

the model results are entirely consistent with the above comments about the centrally peaked nature of the various parameters.

As a measure of how sharply peaked the density is, we refer to the "central condensation" (CC), which is the ratio of central density to mean density. Each polytrope has a unique value of CC. To evaluate CC, we first estimate the total mass M_n of polytrope n: in dimensional units, we do this by integrating $\mathrm{d}M(r)$ ($\sim \rho\, r^2\, \mathrm{d}r$) from center to surface. In terms of θ, this means integrating $\theta^n\ \xi^2\ d\xi$ from $\xi = 0$ to ξ_1. We find that M_n depends on ξ_1 and on the numerical value of the radial gradient of θ at the surface, $(\theta')_1$. Once we know M_n, we can evaluate the mean density $\rho_{\rm av}$ in terms of central density. This leads to an expression for the CENTRAL CONDENSATION

$$CC \ \equiv\ \rho_{\rm c}/\rho_{\rm av} \ =\ -\,\xi_1/3\theta_1'. \tag{11}$$

For n=1.5, the value of CC is 5.99071. For a star like the Sun, where $n \approx$ 3.25 in the radiative interior, the central density must exceed the mean density by a factor of about 88. Since the mean density of the Sun $M_{\rm sun}/(4\pi R_{\rm sun}^3/3)$ is 1.4 gm cm^{-3}, the polytrope model with n = 3.25 predicts that the density at the center of the Sun should be about 123 gm cm^{-3}. It is remarkable that, with such a simple approach, we have obtained a central density which is within 25% of the value obtained in sophisticated modern models.

Moreover, integration also leads to a precise prediction of the central pressure for polytrope n:

$$p_{\rm c} = \frac{M^2}{R^4}\frac{G}{4\pi}\frac{1}{(n+1)(\theta_1')^2}. \tag{12}$$

Converting to solar units of mass ($M_s = M/M_{sun}$) and radius ($R_s = R/R_{sun}$), we find:

$$p_c = 8.947 \times 10^{14} \, (M_s^2/R_s^4) \, [1/(n+1)(\theta_1')^2]$$

where the units are dyn cm^{-2}. For the case $n = 3.25$ (appropriate to the solar interior), numerical integration gives $\theta_1' = -0.03032$. This yields $p_c \approx 2.29 \times 10^{17}$ dyn cm^{-2}. With $\rho_{av} = 123$ gm cm^{-3}, and assuming a perfect gas, we find that the central temperature T_c is $\approx$ 13.5 million K, within (10–15)% of the best model predictions (JCD).

Thus, without doing thermodynamics explicitly, but simply from mechanical arguments, we have arrived at a rough working estimate of the central temperature in the Sun. There is a sound physical reason why a mechanical approach might be expected to work rather well for a star in equilibrium: gravity has the effect that the weight of the entire Sun wants to fall into the core, but pressure wants to make the gas expand to infinity. Therefore, when equilibrium is achieved between these two opposing tendencies, the central temperature (which is related to the central pressure) must be related to M_{sun} and R_{sun} in terms of natural constants. From dimensional arguments we see that $T_c \sim p_c/\rho_c \sim (M^2/R^4)/(M/R^3) \sim M/R$. More precisely, from the polytrope solution, we see that $R_{gas}T_c/\mu = (GM/R) \times 1/(n+1)|\theta_1'|$. This shows that the mean thermal speed at the center of the Sun ($v_{th,c} \sim \sqrt{T_c/\mu}$) is proportional to the escape speed from the surface ($v_{esc} \sim \sqrt{GM/R}$). For the polytropes which are relevant to us here, the constant of proportionality between $v_{th,c}$ and v_{esc} does not differ from unity by orders of magnitude. Thus, a global property of the Sun (its escape speed) determines the physical conditions at the center of the Sun.

1.6 Waves in the Sun: Relevant Time-Scales

How can we test our solution for $T(r)$ inside the Sun? One answer is: by studying the propagation of waves whose properties depend on $T(r)$. For example, acoustic waves travel at the speed of sound $c_s = \sqrt{\gamma R_{gas} T/\mu}$. Therefore, empirical quantities which pertain to acoustic propagation inside the Sun permit us to test (to some extent) the temperature inside the Sun, and its radial variation.

Helioseismology provides a powerful tool for studying waves inside the Sun. There are two major classes of waves: p-modes rely on pressure for their restoring force, while g-modes rely on gravity. Both classes of waves occur in many modes: each mode has an eigenfunction characterized by three integers n_r, n_L, and n_m, representing the number of nodes in radial, latitudinal, and longitudinal directions. Because of the rough equality between $v_{th,c}$ (which is important for p-modes) and v_{esc} (which is important for g-modes), there is a rough equality between certain asymptotic periods of p and g modes in the Sun.

For p-modes, the relevant asymptotic period is the time required for sound to travel from the center of the Sun to the Surface:

$$t_{sound} = \int_0^{R_{sun}} dr/c_s(r) \sim \int_0^{\xi_1} d\xi/\sqrt{\theta}. \qquad (13)$$

All p-modes in the solar polytrope have periods shorter than $t_{\rm sound}$. Since $\theta(\xi)$ is a fairly gradual function, there is a large part of the interior of the Sun where θ is almost constant. If T were equal to T_c at all r, then we would use the above scalings to find that $t_{\rm sound} \sim \sqrt{R_{\rm sun}^3/M_{\rm sun}}$.

For the g modes, the relevant asymptotic period is that of a pendulum with length L equal to $R_{\rm sun}$. This leads to $t_{\rm gravity} \sim \sqrt{R_{\rm sun}/g}$. Inserting the expression for g, we find that $t_{\rm gravity} \sim \sqrt{R_{\rm sun}^3/M_{\rm sun}}$ All g-modes have periods longer than (roughly) $t_{\rm gravity}$. We see that both $t_{\rm sound}$ and $t_{\rm gravity}$ depend on identical functions of stellar mass and radius.

How do the numerical values of $t_{\rm sound}$ and $t_{\rm gravity}$ compare to each other? The integration in (13) can be performed accurately for any particular polytrope [2,3]. For a polytrope with $n = 3$ and solar mass and radius, $t_{\rm sound}$ is found to be 4049 seconds, while $t_{\rm gravity}$ is found to be 3497 seconds. The fact that these two time-scales, which depend on entirely different physical processes, are within 15% of each other in a model of the Sun is striking. Since both time-scales depend similarly on stellar parameters, the rough equality between $t_{\rm sound}$ and $t_{\rm gravity}$ will also be true in other stars. But this is not an accident: it is simply another indication that a star in equilibrium has a structure which is determined by a balance between gravity and pressure.

Empirically, it is certainly true that all p-modes detected so far in the Sun have periods less than $t_{\rm sound}$: no g-modes have been detected so far, so we cannot test the prediction for $t_{\rm gravity}$.

There is another test we can apply to our model: helioseismology predicts that at high frequencies, two neighboring p-modes with equal n_L and equal n_m, but with n_r differing by unity, should be separated in frequency $\delta f_{n_r,n_{r+1}}$ by a constant spacing: the interval should be $1/(2t_{\rm sound})$. For the $n = 3.25$ polytrope with solar mass and radius, $\delta f_{n_r,n_{r+1}}$ is predicted to be 120.88 μHz [2]: empirically, $\delta f_{n_r,n_{r+1}}$ in the Sun is observed to be about 135 μHz (see, e.g. [4]). It is remarkable that a model as simple as a polytrope predicts a value for $\delta f_{n_r,n_{r+1}}$ which is within (10–15)% of the observed value.

Of course, we should not expect to reproduce the Sun's properties precisely by means of a single polytrope: it is clear from Fig. 1 that a single polytropic index is not appropriate for the entire Sun. If we wanted to obtain more accurate results, we might attempt to model the Sun as composed of two polytropes: an outer shell with $n = 1.5$, and an inner sphere with $n \approx 3.25$, with appropriate matching at the interface. But such an attempt would take us far beyond the simplified approach that we use here. The point is this: when it comes to obtaining rather reliable information about the radial profile of the speed of sound (i.e. the temperature) in the interior of the Sun, we can do quite well by using a single polytrope, i.e. without having to solve the energy equation in complete detail.

1.7 The Existence of a Star: A Competition Between Atomic Constants

The Sun (and any star) depends for energy generation on having T_c high enough to drive thermonuclear reactions at sufficiently rapid rates. The rates at which nuclear reactions occur are extremely sensitive to the local temperature: the reason is that none of the gas particles in the solar interior can participate in a nuclear reaction unless it first tunnels through the Coulomb barrier into a target nucleus. The only particles which are fast enough to do this tunneling lie on the far tail of the thermal Maxwellian distribution. Such particles therefore represent an exponentially small fraction of the total population, and yet they are essential for thermonuclear reactions to occur. Now, it is formally true to say that some thermonuclear reactions would occur even at low temperatures: but at such temperatures, the rates of reaction approach zero exponentially rapidly. In order to be occurring at a fast enough rate to be useful in the context of stellar power generation, the temperature must exceed a threshold value $T_{pp} \approx$ 5 million K for proton-proton reactions [5]. As we have seen, the numerical value of T_c in our polytropic model of the Sun ($\approx$ 13.5 million K) is certainly large enough to exceed T_{pp}. We conclude that, at least in the center, our model of the Sun is hot enough to drive nuclear reactions.

This is an important consistency check to see that we have in fact modeled an object which we may fairly refer to as a "star". Moreover, nuclear reactions do not occur only at the center of the Sun. In the polytrope solution, $\theta(\xi)$ is rather flat-topped near the center, and falls off rather gradually with radius. As a result, the temperature inside the Sun remains higher than T_{pp} out to a radial distance of about $R_{sun}/3$: therefore, some (3–4)% of the Sun's *volume* is involved in generating the Sun's power. We refer to this volume as the "energy-generating core". Of course the density is much higher near the center: so the fraction of the Sun's *mass* which resides in the energy-generating core is large, some (60–80)%.

The most important parameter as far as nuclear reactions are concerned is T_c. Now, the value of T_c depends on two constants of nature (R_{gas} and G), and on M/R. In view of the M/R dependence, there is a lower limit of M/R below which T_c falls below T_{pp} . In this case, nuclear reactions are simply too slow, and the object would not qualify for the title of "star" at all. It could be at most a brown dwarf or a planet.

However, as we consider stars where T_c increases more and more above the threshold T_{pp}, the result is not simply an increase in nuclear reactions rates: another effect also begins to have an effect. Radiation pressure builds up according to the formula $p_r = (1/3)aT^4$ where $a = 7.5634 \times 10^{-15}$ ergs cm^{-3} K^{-4} is the radiation density constant. With a large enough M/R, radiation pressure eventually exceeds gas pressure in the process of supporting the star. Gravity has a harder time holding onto photons than onto material particles: as a result, if radiation pressure becomes too large, the star is no longer stable.

The competition between getting T_c large enough to drive reactions, but not so large as to destabilize the star is a close one: it depends on the relative magnitudes of R_{gas}, G, and a. There is actually only a relatively narrow range of masses in which stable stars can exist. Although there are 80 orders of magnitude between the mass of a proton and the mass of the universe, there are only 2 orders of magnitude in that enormous range of masses in which essentially all stable stars occur: the range of stable stars extends from roughly 10^{32} to 10^{34} gm. The Sun lies close to the middle of this range.

1.8 Luminosity and Energy Flux

Now that we know that temperatures near the center of the Sun are hot enough for thermonuclear reactions to occur, we might in principle consider how the reactions operate. But this would take us too far afield. (For details on nuclear reactions inside the Sun, the reader should refer to the article by M. Gai in this volume.) Here, we simply assume that energy is generated in the core, and then consider some of the consequences.

In astronomical parlance, we refer to the total power generated inside a sphere of radius r as the local "luminosity" $L(r)$ ergs/sec. Outside the energy-generating core, there are no further additions of energy, and as a result, $L(r)$ remains constant, and equal to the observed power output $L_{\text{sun}} = 4 \times 10^{33}$ ergs sec^{-1}.

Clearly, in order to generate the power L_{sun}, the amount of mass which must be converted into energy every second (via nuclear reactions) must be $\dot{M}_{\text{nuc}} = L_{\text{sun}}/c^2$ where c is the speed of light. This indicates that the Sun transforms 4–5 tons of mass every second into energy via nuclear reactions. We shall find below that the Sun also loses mass at a comparable rate via the solar wind.

Once the luminosity reaches its constant value (i.e. once we are at radial distances of $0.3R_{\text{sun}}$ and larger), the expression for the flux of energy $F(r)$ which must be transported across a sphere at radius r becomes simple:

$$F(r) = L_{\text{sun}}/4\pi r^2. \tag{14}$$

1.9 Heat Transport

Given that energy is generated inside the core, we now ask: how does this energy make its way through the Sun and eventually escape from the surface? In order to answer this question, we need to study how energy is transported from one point in the Sun to another.

The three standard methods to transport heat are conduction, radiation, and convection. In the Sun, all three play a role in one way or another. We now turn to how heat is transported in the *interior* of the Sun. Later (in Sects. 4 and 6), we shall discuss how heat is transported in the hot outer atmosphere.

1.10 Transport of Heat by Photons

When a diffusive process such as conduction is at work, the simple and well known formula of Fick's law states that energy flows down the temperature gradient, and that the magnitude of the energy flux is proportional to the local magnitude of the temperature gradient:

$$F(r) = -k_{\rm th}({\rm d}T/{\rm d}r) \tag{15}$$

where $k_{\rm th}$ is the thermal conductivity. In the context of the Sun or stars, where a certain flux of energy is supplied from the core (see (14)) we view (15) from the following perspective: it tells us what value dT/dr must adopt in order to transport the flux which is supplied.

The concept of a thermal conduction coefficient $k_{\rm th}$ will appear not only in the context of the deep interior of the Sun, but also in the context of the solar corona (Sect. 4). However, the "particles" which do the conducting are quite distinct in these two contexts, and this gives rise to quite different dependences of $k_{\rm th}$ on local parameters.

In the kinetic theory of gases, conduction of heat occurs when hot particles collide with cooler particles. In this theory, a general formula can readily be derived for $k_{\rm th}$ (e.g. [6]):

$$k_{\rm th} = \frac{1}{3}\, \lambda\, \rho\, C_{\rm v}\, u. \tag{16}$$

Here, λ is the mean free path between collisions, ρ is the mass density of the material, $C_{\rm v}$ is the specific heat per gram at constant volume, and u is the speed of the particles which are transporting heat.

Now, we need to ask: what is it that transports heat in the Sun's interior? Is it particles or photons? Conditions deep inside the Sun are such that *particles* are not very efficient at transporting heat: the density of material in the core of the Sun is so large that λ is very short, and $k_{\rm th}$ is small. It is only in the very dense interior of certain stars (including white dwarfs and red giant cores), that thermal conduction of the usual kind (involving degenerate electrons) becomes dominant in stellar interiors. This process is of no relevance in the *interior* of the Sun in its present evolutionary state.

It turns out that, in the solar interior, photons are much better than particles at transporting heat. For this reason, the interior of the Sun is referred to as a "radiative zone". So let us consider how heat is transported through a mixture of particles and photons, each of which contributes a different component to the process. We can use the general result of kinetic theory (eq. (15)) to guide us here. There are 4 quantities required to evaluate $k_{\rm th}$ according to (15). Particles provide the mass density, while photons provide the transport. This allows us to write down two of the quantities in (15) directly: ρ can be equated to the local mass density, and u can be equated to c, the speed of light.

As regards the third quantity required in (15), we note that $C_{\rm v}$ is defined by $(\partial U/\partial T)_{\rm v}$, where U is the internal energy density per unit *mass*. In a medium where the photons are serving as transporters, we note that the energy density of photons per unit *volume* is given by $E_{\rm ph} = aT^4$ per cm^3 where a is the radiation density constant mentioned above. Since photons provide energy for transport while particles provide mass we can regard the energy density *per gram* of the particle-photon transporter mixture as $U = aT^4/\rho$. Using this, we find $C_{\rm v} = 4aT^3/\rho$ erg gm^{-1} K^{-1}.

In order to evaluate the fourth quantity in (15) (the mean free path), we need to introduce the concept of optical depth, τ. As photons travel through a medium which does not have perfect transparency, the photon flux decreases according to $F = F_o\,\mathrm{e}^{-\tau}$, where τ is defined as follows. In a medium with opacity κ cm^2/gm and density ρ, photons traveling through an increment of distance dx experience an increment in optical depth $d\tau$ defined by $\kappa\ \rho\ dx$. With the above definitions, the mean free path of the photon (which appears in (15)) is $\lambda = 1/(\kappa\rho)$.

Combining the four quantities, we arrive at an expression for the "thermal conductivity" which is relevant for photon-mediated transport:

$$k_{\mathrm{th}} = 16\sigma_{\mathrm{SB}}T^3/3\kappa\rho. \tag{17}$$

In deriving (17), we have converted from the radiation density constant a to the Stefan-Boltzmann constant $\sigma_{\mathrm{SB}} = ac/4$.

Combining eqs. (14), (15), and (17), we can now evaluate the magnitude of the temperature gradient. With local luminosity L, we find

$$\left|\frac{\mathrm{d}T}{\mathrm{d}r}\right| = \frac{3\kappa\rho L}{64\pi\sigma_{\mathrm{SB}}r^2T^3}. \tag{18}$$

This is what the gradient must be in the radiative interior of the Sun, where photons diffuse outward.

1.11 Effective Polytropic Index in the Radiative Zone

Now that we know how the temperature must vary with radial distance in the radiative interior, we can estimate an effective polytropic index n_{eff}. Recalling that $n_{\mathrm{eff}} = \mathrm{d}(\log\ p)/\mathrm{d}(\log\ T) - 1$, we can estimate n_{eff} by comparing $\mathrm{d}p/\mathrm{d}r$ with $\mathrm{d}T/\mathrm{d}r$. We already know $\mathrm{d}p/\mathrm{d}r$ from HSE (eq. (3)): applying (3) to the part of the star where $M(r)$ has reached most of its final value M, we find $\mathrm{d}p/\mathrm{d}r = -GM\rho/r^2$. Interestingly, $\mathrm{d}p/\mathrm{d}r$ depends on the same combination of ρ/r^2 as appears in $\mathrm{d}T/\mathrm{d}r$ (eq. (18)). Therefore, when we take the ratio of $\mathrm{d}p/\mathrm{d}r$ to $\mathrm{d}T/\mathrm{d}r$, the radial dependence, and the dependence on density, disappear. We find that

$$\frac{\mathrm{d}p}{\mathrm{d}T} \sim \frac{T^3}{\kappa} \tag{19}$$

in the radiative zone.

In order to proceed further, we need to know how the opacity κ behaves as a function of the physical variables. Now, the value of κ is very complicated to calculate in detail: because there are ions of many species and many stages of ionization in the solar material, light can be absorbed by literally millions of different transitions between numerous bound atomic levels and continua. There is no simple way to estimate κ reliably. One example of a set of calculations of opacities (taken from [7]) is shown in Fig. 3.

Each curve shows the opacity (averaged over all wavelengths in a manner which leads to the so-called Rosseland mean) as a function of temperature T for a series of constant pressures p. Two obvious features of the opacity curves in

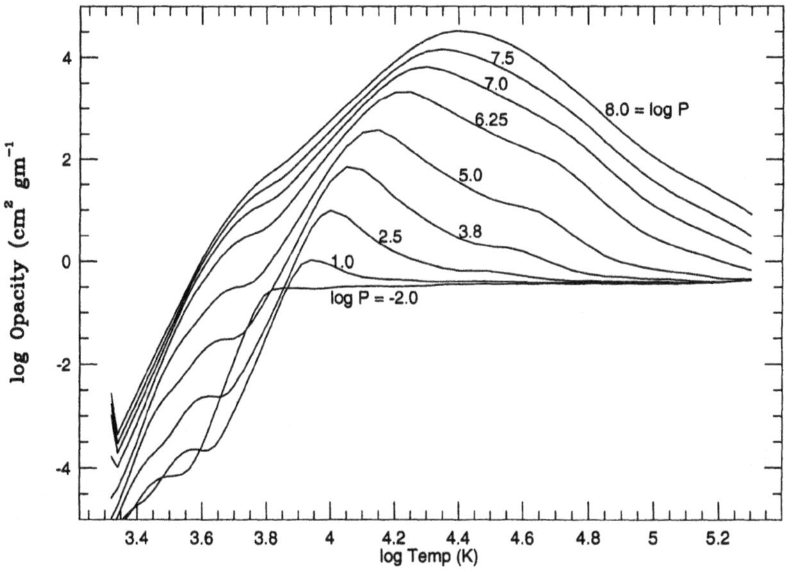

Fig. 3. Opacities for the solar mix of elements as a function of temperature at constant pressure. The plotted quantities are Rosseland mean values

Fig. 3 are particularly relevant to us. First, at high temperatures, the opacity *decreases* as T increases, i.e. $\kappa \sim T^{-\beta}$ where β is a positive number. Second, at low temperatures, the opacity *increases* rapidly as T increases. In the present section, we are primarily interested in the first of these. We shall return to the second in Sect. 2, when we discuss the chromosphere.

As regard the opacity at high temperatures, we note that in a gas where T exceeds $10^{4-4.5}$ K, atoms become progressively stripped of more and more electrons as $\log T$ increases. Now, stripped atoms are less capable of absorbing photons, and so κ declines with increasing T (at a given pressure). But the higher the density is, the more particles there are per cm^3 to do the absorbing, and therefore the larger the opacity. Thus, $\kappa \sim \rho^{+\alpha}$ where α is a positive number. To be sure, in the "real Sun", pressures extend to much higher values than those which are plotted in Fig. 3: but the range of pressures plotted in Fig. 3 (and provided by Kurucz) allow us to see the principal features which are of interest.

A useful approximation exists to describe the behavior of opacity at the high temperatures which are characteristic of the deep interior of the Sun and stars, the so-called Kramers opacity law (see, e.g., p. 62 – 73 of [5]):

$$\kappa \sim \rho\, T^{-3.5}. \tag{20}$$

The Kramers opacity has functional dependences on T and ρ which are consistent with those mentioned above, i.e. $\beta = +3.5$ is positive, and $\alpha = +1$ is also positive. It is important not to try to apply (20) outside the regimes of parameter space for which it was derived: it applies to the deep interior of the Sun, but it cannot be applied to the surface layers. Note also that the decline in κ with increasing

T cannot be extrapolated indefinitely: there is a strict asymptote on κ (≈ 0.4 cm^2 gm^{-1}) at high T due to electron scattering (see Fig. 3). This asymptotic behavior is not of great interest in a star such as the Sun: solar conditions are such that the opacities in the interior lie on the declining slopes in the upper right hand corner of Fig. 3.

Combining eqs. (20) and (19), we find $\mathrm{d}p/\mathrm{d}T \sim T^{6.5}/\rho$. Using the perfect gas equation of state ($\rho \sim p/T$), we can eliminate ρ to find

$$p\,\mathrm{d}p \sim T^{7.5}\mathrm{d}T. \tag{21}$$

Integrating this, we find

$$p \sim T^{4.25}. \tag{22}$$

Now, for a polytrope, we recall that p varies as T^{n+1}. Therefore, *the radiative interior of a star where Kramers opacity is at work is actually a polytrope with $n = 3.25$.* In fact, as we have already seen (Fig. 1), the interior of a sophisticated solar model can be described in terms of an effective polytropic index which on the average is not far from 3.25. This the reason why, although polytropes seem at first to be much too simplified to be of interest in learning about the "real Sun", nevertheless polytropic models *can* provide useful quantitative information about the structure inside a star such as the Sun. However, we should not push the polytrope approximation too far: in particular, as we can see from Fig. 1, the value $n = 3.25$ is *not* a good fit to the outer layers of the Sun. Photon conductivity with Kramers opacity must not be applicable to those layers: $n = 1.5$ obviously provides a much better fit. Why is $n = 1.5$ suitable for the outer layers of the Sun? To answer this, we now leave our discussion of photon transport and consider a very different method of heat transfer.

1.12 Transport of Heat by Convection

At the simplest level of approximation, it is worthwhile to estimate the mean temperature gradient between the center of the Sun and the surface:

$$\left(\frac{\mathrm{d}T}{\mathrm{d}r}\right)_{\mathrm{mean}} = \frac{T_{\mathrm{c}}}{R_{\mathrm{sun}}} \approx 2 \times 10^{-4}\ \mathrm{deg\ cm}^{-1}. \tag{23}$$

Why is the temperature gradient of interest to us here? The reason is that there is a critical temperature gradient which enters into the process of heat transport: this is the so-called adiabatic gradient of temperature Γ_{ad}. To see the physical significance of the adiabatic gradient, we reason as follows.

Consider a region of the star where the temperature is falling off with increasing radius in such a way that the gradient has a certain absolute value $\Gamma = |\mathrm{d}T/\mathrm{d}r|$. We wish to know whether or not this region is stable or unstable to gas motion.

To evaluate the stability, we perform the following thought experiment. Consider an element of gas with a mass of one gram which initially lies at radial distance r_{s} with total energy E_{s}. Suppose that a thermal fluctuation raises the temperature of this element infinitesimally. When constant pressure is established, the element is infinitesimally less dense that the surroundings. Because of the reduced density, buoyancy forces cause the element to move upwards, eventually reaching a larger radial distance $r_\mathrm{f} = r_\mathrm{s} + \mathrm{d}h$. The question we ask is: how does the final total energy E_f of the element compare to the starting value E_s? Has the total energy increased, or decreased, or remained unchanged?

To answer that, we note that two components contribute to the total energy: potential and internal. The potential energy increases by $\Delta(PE) = +g\,\mathrm{d}h$. Suppose that the element moves slowly enough that it is continually able to adjust its temperature and pressure to be equal to the ambient temperature and pressure surrounding the element. Now, between r and $r + \mathrm{d}h$, the ambient temperature outside the element decreases by an amount $\Gamma\,\mathrm{d}h$. Therefore, the internal temperature of the element also decreases by $\Gamma\,\mathrm{d}h$. As a result, the internal energy changes by $\Delta U = -C_\mathrm{p}\Gamma\,\mathrm{d}h$. Adding the two contributions $\Delta(PE)$ and ΔU, we see that the final total energy E_f of the element differs from its original value E_s by

$$\Delta E \equiv E_\mathrm{f} - E_\mathrm{s} = +g\,\mathrm{d}h - C_\mathrm{p}\Gamma\,\mathrm{d}h. \tag{24}$$

Three cases can be considered. First, suppose ΔE turns out to be positive. In this case, the element has a higher energy at r_f than at r_s. Therefore, the element needs to have work done on it to raise it from r_s to r_f. Energetically, this is not favorable, so the element will tend to return at its initial position. In this case, the gas is stable.

Second, suppose ΔE turns out to be negative. In this case, the element actually *loses* energy by being raised to the higher level: the work done by buoyancy is more than offset by the cooling of the element. The total energy in this case can be driven to even lower values by having the element move to even greater heights. From the standpoint of energetics, it is favorable for the element to move to higher and higher levels in order to achieve lower and lower total energies. As a result, even a small initial perturbation is enough to start an upward motion which will continue. Analogously, if we start with a temperature fluctuation which is negative, downward motion will be initiated, and will also continue. Thus, the gas is unstable to upward and downward motion. Of course, these motions cannot continue indefinitely: the internal excess (or deficit) of heat will eventually be wiped out by some sort of energy exchange with the surroundings.

Third, ΔE is zero. In this case, the motion of the element involves no change in total energy. Such a change is adiabatic.

We see now the importance of estimating the temperature *gradient* in any particular model of a stellar interior: in the presence of gravity, there exists a critical *gradient* which determines whether the gas is stable or unstable. The critical gradient is

$$\Gamma_\mathrm{ad} = g/C_\mathrm{p}.$$

If the absolute temperature gradient equals g/C_p, then ΔE is zero, and the gas behaves in such a way that it neither gains nor loses energy in its motion. This is the definition of adiabaticity. As a result, g/C_p is referred to as the "adiabatic temperature gradient" $(\mathrm{d}T/\mathrm{d}r)_\mathrm{ad}$. If the absolute temperature gradient is steeper than g/C_p, it is energetically favorable for gas to move upwards. This motion provides a very efficient transfer of energy: heat is transported not by "hot photons" or by "hot particles", but by macroscopic "blobs" (or turbulent eddies) of material in large-scale flows. These flows give rise to thermal convection: heat transport is driven by buoyancy forces acting on thermal fluctuations. Thus, the criterion for the onset of convection is

$$|\mathrm{d}T/\mathrm{d}r| \geq g/C_p \ . \tag{25}$$

We can now evaluate the adiabatic temperature gradient in the Sun. Near the surface of the Sun, where $g = g_{\mathrm{surface}}$ and $C_p \approx 2.5 R_{\mathrm{gas}} \approx 2 \times 10^8$ ergs cm^{-1} K^{-1}, we find

$$\left(\frac{\mathrm{d}T}{\mathrm{d}r}\right)_{\mathrm{ad}} \approx 1.4 \times 10^{-4} \text{ deg cm}^{-1}. \tag{26}$$

The significance of the numerical value of $(\mathrm{d}T/\mathrm{d}r)_{\mathrm{ad}}$ can be appreciated when we compare eqs. (26) and (23): we see that the mean temperature gradient between the center of the Sun and the surface is comparable to the adiabatic temperature gradient near the surface. At first sight, this appears as a remarkable coincidence. After all, why should the processes which determine the temperature at the center of the Sun have anything to do with the processes which control the adiabatic gradient near the surface? But upon reflection, we see that the coincidence is less remarkable than it first seemed. Recall that conditions at the center of the Sun are determined by (among other things) GM/R (which is related to surface gravity) and the gas constant (which relates pressure and temperature). And these are precisely the variables which also enter into $(\mathrm{d}T/\mathrm{d}r)_{\mathrm{ad}}$. Once again, we encounter the fact that the global properties of the Sun are controlled by a balance between gravity and pressure.

In order to discuss convection, we can do better than simply using the mean temperature gradient between center and surface. In a polytrope, the θ vs. ξ curve (which is a proxy for temperature) is shallow near the center, and becomes steeper as we move away from the center. (Such behavior in the temperature profile is apparent in Fig. 2 above). As a result, it becomes easier to satisfy the convective criterion (eq. (25)) as we approach the surface of the Sun. Therefore, in the outer layers of the Sun, convection transports heat.

Is there evidence for convection in the outer layers of the Sun? Yes. Images of the solar surface with angular resolution of at least 1 arcsec or better reveal a "granular" pattern consisting of a multitude of short-lived cells with bright centers and dark edges: hot gas rises in the center of a cell, and cool gas sinks in the dark lanes, with velocities of order 1 km sec^{-1}. This pattern of gas motion on or near the surface of the Sun is characteristic of convection. Full modeling of the three-dimensional nature of convective motions is very complicated, but with the advent of large computers, this is an active area of modern solar research (see, e.g., [8,9]).

Deep inside the solar convection zone, the motion of macroscopic blobs of matter is so efficient at transporting heat that the energy transport through the solar material can be accomplished by having $|\mathrm{d}T/\mathrm{d}r|$ only slightly steeper than

g/C_p (see [5]). As a result, $|dT/dr|$ in the solar convection zone remains close to g/C_p, i.e., the convection zone is essentially adiabatic: $p \sim \rho^{5.3}$. This explains why a model of the convection zone can be fitted very well by a polytrope of index $n_{\rm eff} = 1.5$ (see Fig. 1). In the case of convection, $n_{\rm eff}$ remains essentially identical to 1.5 almost all the way through the convection zone. However, in the topmost layers (very close to the surface), it is no longer impossible for the gas to behave in a strictly adiabatic manner, and so $n_{\rm eff}$ departs from 1.5 there.

1.13 Transition Between Radiative Core and Convection Zone

The Sun consists of a radiative core, where energy is transported by photon "conduction", and an outer envelope, where convection does the energy transport. Therefore, when we consider the Sun, and imagine what it would be like to penetrate inwards from the surface, we would find ourselves at first in a convection zone. Eventually, we would reach the base of the convection zone, and beyond that, we would find ourselves in the radiative interior. We refer to the transition between radiative core and convective envelope as the radiative-convective boundary (RCB).

The question we ask here is: how far below the surface of the Sun does the RCB lie? To address this topic, we recall that the RCB is situated at the radial position where the (absolute) temperature gradient in the radiative interior rises to a value which is steeper than the adiabatic gradient. Why does the absolute value of dT/dr increase as we move outwards from the center of the Sun? Mainly because the gas cools, and this allows for more bound electrons to be retained by the atoms in the gas. The more bound electrons there are, the large the opacity will be. Let us refer to quantities at the RCB with a subscript b. Then the temperature gradient on the radiative side of the RCB is given by $|dT/dr|_r = 3F\kappa_b\rho_b/16\sigma_{\rm SB}T_b^3$. By definition of RCB, this absolute temperature gradient must equal the local value of the adiabatic gradient. Thus, the RCB occurs at the location where

$$3F\kappa_b\rho_b/16\sigma_{\rm SB}T_b^3 = g/C_p. \tag{27}$$

We recall that if κ is determined by Kramers opacity, κ will depend on density and temperature as $\kappa = C_K\rho/T^{3.5}$ where C_K is a constant. In order to proceed, we need to insert a numerical value for the constant of proportionality C_K. Referring to Schwarzschild ([5] eq. 9.16), we find that in a medium such as the Sun, where metal abundance Z is of order 0.02, the opacity is mainly determined by bound-free transitions. For these, $C_K \approx 10^{24}$ in c.g.s. units. The flux F at RCB is larger than the surface flux ($F_{\rm surf} = 6.4 \times 10^{10}$ ergs cm^{-2} sec^{-1}) by a factor of about 2, since RCB occurs at some depth below the surface. Also, g is lower than the surface value by a factor of about 2. Moreover, with ionization complete at the RCB, we set $C_p = 5R_{\rm gas}$. Inserting these in (27) and rearranging, we find $T_b^{6.5} \approx 10^{42.7}\rho_b^2$

We can eliminate ρ_b if we know how density and temperature are related in the convection zone. Since the latter is a polytrope of index $n = 1.5$, we know that in the convection zone, $\rho = K_{\rm ad} T^{1.5}$. Unfortunately, there is no simple way to estimate the value of $K_{\rm ad}$ from surface values. The difficulty is that there are superadiabatic regions just beneath the photosphere which control which adiabat the solution follows at great depths. A detailed model is required in order to determine the constant of proportionality $K_{\rm ad}$. Referring to the JCD model, we find that in the deep convection zone, the density and temperature are related roughly by $\rho \approx 10^{-10} T^{1.5}$.

Combining the above relations, we find $T_{\rm b}^{3.5} \approx 10^{22.7}$. We note that $T_{\rm b}$ is raised to a rather high power: as a result, our estimate of $T_{\rm b}$ is not too sensitive to the various parameters. Finally, we arrive at

$$T_{\rm b} \approx 3 \text{ million } K. \tag{28}$$

A full model (such as JCD) suggests that the value of $T_{\rm b}$ is about 2.3 million. Thus, even with the crude approximations we have made (especially the Kramers opacity assumption), it is possible to obtain an estimate for the temperature at the base of the convection zone which is reliable within ~25%.

Now that we know $T_{\rm b}$, we can now address the question: how deep does the base of convection zone lie? To answer this, we note that the temperature gradient in the convection zone is essentially adiabatic. Therefore, the base of the convection lies at a depth $z_{\rm b}$, where $z_{\rm b} \approx T_{\rm b} / |(\mathrm{d}T/\mathrm{d}r)_{\rm ad}|$. Inserting $T_{\rm b} \approx$ 3 million K, and $|(\mathrm{d}T/\mathrm{d}r)_{\rm ad}| \approx 1.4 \times 10^{-4}$, we find $z_{\rm b} \approx 2.1 \times 10^{10}$ cm, i.e. $\approx$ 0.3 solar radii below the surface. This estimate is quite close to the value $z_{\rm b} =$ 0.29215 solar radii which occurs in the detailed JCD model.

In summary, the Sun has an inner radiative core which extends from $r =$ 0 to $r \approx 0.7 R_{\rm sun}$ and an outer convective envelope which extends from from $r \approx 0.7 R_{\rm sun}$ to $R_{\rm sun}$.

One important consequence of the convective envelope concerns the abundances of certain elements in the surface layers of the Sun. The circulation of material which occurs in a convection zone has the effect that material is swept down to great depths on short time scales. The time required for this sweeping $t_{\rm sw}$ can be estimated from the ratio of the depth of the convective envelope to the mean convective speed. We find that $t_{\rm sw}$ may be as short as a few days or weeks. This means that material which we see at the surface of the Sun today will be swept quickly down to the base of the convection zone, where the temperatures reach 2–3 million K. Now, certain elements can be destroyed in thermonuclear reactions at temperatures of 3 million degrees or less: the elements which belong to this category include deuterium and lithium. Because of the properties of convection, therefore, it is expected that the mean abundances of D or Li at the solar surface are very small.

2 The Solar Atmosphere: Photosphere, Chromosphere, and Corona

We have seen that the outer envelope of the Sun is convective: that is, gas moves in bulk flows upwards and downwards through the atmosphere. However, as we approach close to the surface of the Sun, i.e. close to a region of "empty space", the temperature falls off to very low values. Therefore, the density ρ ($\sim T^{1.5}$ in the convection zone) also tends to zero. With less and less material available to do the transport, it eventually becomes impossible for moving gas to transport the required flux of energy. That flux, with its well-defined value of $F_{\text{surf}} = 6.4 \times 10^{10}$ ergs cm^2 sec^{-1}, must somehow still be transported out through the surface. Since this energy must leave the Sun, and propagate through empty space, it is clear that energy transport must eventually revert to the only form of energy which can propagate in free space: radiation. This reversal begins to reach significant proportions when the convective medium finds itself at a level where the overlying material has an optical depth τ of about 1: at this level, hot gas rising from below can lose energy by radiating into space. The level at which this happens is also just about the deepest level in the Sun which we can observe directly. Because we can see light coming from that level, it is referred to as the photosphere (or "light-sphere"). As we look in from outside the Sun, we can therefore peer into the Sun down to the level where convection is occurring. That is why we see evidence for convection (i.e. granules) when we look carefully at the solar surface.

Strictly speaking, therefore, although the gross structure of the Sun consists of only two main components (radiative core plus convective envelope) there is in fact a third: it is a thin "skin" right at the surface where radiation transports the energy.

2.1 Radiative Transfer in the Photosphere

So in order to consider the surface layers of the Sun, we turn again to radiative transfer. If the approximations of "photon conductivity" were applicable, we could combine eqs. (15) and (17) and obtain

$$F_{\text{surf}} = \frac{16\sigma_{\text{SB}}T^3}{3\kappa\rho}\frac{\mathrm{d}T}{\mathrm{d}r}. \tag{29}$$

Actually, the assumptions which went into deriving (15) break down as we approach the surface: diffusive processes simply do not work well in the rarefied gas close to the surface. However, we can use (29) to estimate roughly some quantities which are of interest. Near the surface, as the temperature falls well below 10^4 K, κ falls rapidly towards very small values (see Fig. 3). The reason for this behavior is straightforward: in cool gas, bound electrons in the atoms of the dominant constituents of the solar atmosphere (hydrogen and helium) are predominantly in the ground state. Now, optical photons have energies of only a few eV, and these are certainly not enough to populate even the second energy level, let alone ionize the atoms. Therefore, there is little incentive for the optical photons (which are the predominant emission of the solar surface) to have any interaction with the gas. As a result, the photons stream almost freely through the gas with essentially no absorption. Thus, $\kappa \to 0$. In view of this, (29) suggests that the flux F_{surf} can be transported outwards even if $\mathrm{d}T/\mathrm{d}r \to 0$.

Therefore, with radiation performing the transfer of energy in the atmosphere, T tends to a constant value in the photosphere. The value is expected to be about 4000–4500 K.

2.2 Breakdown of Radiative Transfer

Empirically, there is evidence that indeed the temperature gradient tends to zero in the upper photosphere. But the predicted constancy in T does *not* persist throughout the atmosphere: as one moves upward from the level $\tau = 1$, T falls from about 6000 K, reaches a minimum value $T_{\rm min}$ of 4000–4500 K at a height $h_{\rm min}$ of a few hundred kilometers above the level $\tau = 1$, and then T begins to *increase* with increasing height. At first, the increase is rather modest: T rises by a few thousand K, and then remains almost constant (at about 6000 K) up to $h_{\rm c} \approx 2000$ km. The interval of the solar atmosphere between $h_{\rm min}$ and $h_{\rm c}$ is referred to as the *chromosphere*. Above $h_{\rm c}$, the value of T is observed to increase very rapidly to values of order 10^6 K: this super-hot gas is referred to as the *corona*.

Prior to the development of modern solar observatories, the only occasions on which observers could see the chromosphere and the corona was during a total eclipse of the Sun. On such an occasion, the chromosphere (meaning literally "sphere of color") appears as a narrow rose-colored aureole close to the Moon's limb, while the corona (literally: a "crown") appears as a diffuse pearly-white halo extending far from the Sun. Modern observations of the chromosphere indicate that the reddish color is due to a strong spectral line emitted at a wavelength of 6563Å. And modern observations of the corona indicate that the corona changes its shape over an 11-year cycle: this cycle is caused by a cyclic occurrence of a variety of magnetic phenomena in the solar atmosphere (including sunspots, active regions, prominences, etc.: see, e.g. [10]). When solar magnetic phenomena are most active, the corona is observed to be almost uniformly bright at all latitudes. But when magnetic activity is low, the corona is bright only in the equatorial regions, where so-called "streamers" of denser material point out into space. At the latter times the North and South poles of the Sun appear comparatively dark, and the term "coronal holes" has been coined to describe these dark regions. At all times, the brightest parts of the innermost corona are observed to have a brightness $I_{\rm cor}$ which is a few times 10^{-6} times the brightness of the visible disk of the Sun, $I_{\rm disk}$. We shall return to these observational results below.

The fact that $\mathrm{d}T/\mathrm{d}r$ actually becomes positive in the upper photosphere is remarkable. After all, heat is supposed to flow *down* a temperature gradient (see (15)), but in the chromosphere, the heat flows outward in the presence of an upward $\mathrm{d}T/\mathrm{d}r$. Clearly, the diffusion of heat according to Fick's law (eq. (15)) is irrelevant to the physics of the chromosphere. So what is happening in the chromosphere?

2.3 The Role of Mechanical Work

The answer is that internal energy ("heat") is not the only form of energy which is present in the solar photosphere. Thermodynamically, an increment of energy $dQ = dU + p\,dV$ can be provided to a gas in two forms, internal (dU) and work ($p\,dV$). In the solar chromosphere, there must be some agent which does work on the gas. What could this agent be?

We saw in Sect. 1.6 that the material of which the Sun is composed supports waves of various kinds. In particular, *acoustic waves* are present in the solar atmosphere. Such waves involve compressions and rarefactions which propagate at the speed of sound c_s through the ambient medium: the compressions associated with these waves provide us with an agent which can do work on the medium. Of course, the rarefactions tend to undo the work which is done by the compressions, but if there is an asymmetry in the wave (e.g. if it is steep enough to include a shock front), then the compressions can "win out" and do net work on the gas. It is widely believed that acoustic waves do indeed give rise to the increase of T in the chromosphere.

As far as the solar corona is concerned, it seems unlikely that acoustic waves can be at work: these waves deposit essentially all of their energy in the chromosphere. So what agent is available to do work on the coronal material? The fact that the corona changes its shape significantly during the 11-year magnetic cycle gives us a clue as to where we should look for an answer: the magnetic field. In magnetic regions of the solar atmosphere, the high electrical conductivity means that plasma and field are tightly coupled: the field and the plasma are "frozen" together. In such a medium, magnetohydrodynamic (MHD) wave modes of several kinds can be supported, of which the best known are Alfven waves.

To understand an Alfven wave, we note that a magnetic field of strength B in a plasma of mass density ρ behaves like an stretched string under tension $T_r = B^2/4\pi$. Because the field is tightly coupled ("frozen in") to the plasma, the density of the plasma effectively provides inertia to a field line. When such a field line is disturbed, it responds in the same way as a stretched string being plucked: a transverse wave propagates along the field line at a speed $V_A = \sqrt{T_r/\rho} = B/\sqrt{4\pi\rho}$. The speed V_A is referred to as the Alfven speed. Alfven waves are of particular interest in the corona: they can propagate into the upper regions of the solar atmosphere where acoustic waves do not survive.

Alternatively, because the corona is highly ionized, electric currents may also provide localized sources of energy deposition.

3 The Chromosphere

In order to discuss the energetics of the chromosphere, we need to address three issues: (i) how much acoustic energy is generated? (ii) how rapidly is this energy deposited in the atmosphere? (iii) how does the chromosphere respond to the deposited energy?

As regards (i), we note that acoustic waves are created basically by motions of compressible gas. Therefore, we first need to evaluate the physical characteristics of these motions, i.e. the characteristics of convective flows. As regards (ii), we need to know the density gradient in the atmosphere. And as regards (iii), we need to know how effective the gas is at radiating away excess energy.

3.1 Granulation

Turning first to (i), let us consider the properties of the convective motions which are known to exist in the Sun, i.e. the granules. The mean granule diameter D is of order 1200–1400 km [11]. The depth of a granule H cannot be measured directly: the simplest model of convective instability in a laboratory setting [12] suggests that maximum instability occurs when $H \approx D/2 \approx$ 600–700 km. Other models suggest a value of a few times the local pressure scale height H_{p}, i.e. a few hundred km.

The gas flows in granules have speeds v_{conv} which have a range of values: they can be as large as 6 km sec^{-1} [13], with a mean of about 1–2 km sec^{-1}. Because the solar convection zone is a highly turbulent medium, granule evolution is very complicated. When movies of the solar surface are viewed, a trained eye can identify an individual granule for a certain length of time: but as time goes by, the granule becomes more and more difficult to distinguish as an identifiable entity. It appears to "dissolve" gradually into the background, or explode, or fade out, or some combination of these. In any case, the original granule eventually loses its identity, and other granules become identifiable. Amidst this complexity, quantitative studies of correlations between images taken at different times suggests that measurable correlations persist for a finite time, and then go to zero. From the correlation plots, it is possible to speak roughly about an average e-folding time, or "lifetime", of a granule. The best estimates of these lifetimes (after allowing for effects of acoustic waves) are in the range 10–15 minutes [14].

It is instructive to compare this mean lifetime with the "turnover time" t_{turn}, i.e. the time required for gas to circulate once around the granule. Since the circulation length once around the granule L_{circ} is of order $(D + 2H)$, we estimate

$$t_{\mathrm{turn}} \approx L_{\mathrm{circ}}/v_{\mathrm{conv}} \approx 10^3 \text{ sec.} \tag{30}$$

Comparing to the observed mean lifetime, it appears that granules survive for about one turnover time. This is an indication of how turbulent the convection in the Sun really is: conditions are very far removed from the long-lived hexagonal "Benard cells" which are the hall-mark of laminar convection in the laboratory [12]. Nevertheless, high resolution images of granules in the Sun do suggest that some granules have shapes which look like (irregular) polygons. This has led to the application of the term "convection cells" to granules on the Sun. On the other hand, because of the turbulent nature of the convection, the granules are also sometimes thought of as turbulent eddies.

Whatever the term we use, an essential aspect of convective energy transport is the fact that gas circulates in the cell or eddy. As a result, if something interferes with the circulation, then the efficiency of convective heat transport may be impeded. What can interfere with the circulation in a granule? A magnetic field can: because gas in the solar atmosphere is "frozen in" to the magnetic field, the gas is not free to move arbitrarily across field lines.

Magnetic flux is created deep inside the Sun by dynamo action: kinetic energy associated with vortical flows of electrically conducting material can (in certain conditions) be converted into magnetic energy. Newly created magnetic flux emerges from time to time at the surface. An erupting flux tube creates a magnetic bipole, i.e. two neighboring regions of opposite magnetic polarity. The magnetic field emerges from the solar surface nearly vertically from one of these regions (a "foot-point"), loops up into the overlying atmosphere, and then returns to enter the solar surface nearly vertically in the other foot-point. The spatial area $A_{\rm mag}$ of a foot-point depends on how much magnetic flux $F_{\rm mag}$ is present in the particular tube: $A_{\rm mag} = F_{\rm mag}/B_{\rm surf}$. The field strength at the surface $B_{\rm surf}$ is controlled by the local gas pressure $p_{\rm gas}$, and also by the ram pressure $p_{\rm ram} \approx \rho v^2$ of large-scale organized flows if such flows are present. When the combination of the confining pressures $p_{\rm gas} + p_{\rm ram}$ reaches rough equality with the horizontal magnetic pressure ($p_{\rm mag} = B_{\rm surf}^2/8\pi$), horizontal equilibrium becomes possible, and the foot-point of the bipole can survive as a well defined feature on the Sun's surface.

The diameter of a foot-point $D_{\rm mag} \sim \sqrt{A_{\rm mag}}$ is one of the factors which determines whether the foot-point is bright or dark. Thus, if the flux tube is smaller than a granule diameter, i.e. if $D_{\rm mag} \leq 1200$–1400 km, then the magnetic effects are confined to a small enough scale that they do not interfere seriously with the convective circulation. Thus, the normal upward convective transport of heat continues unabated. In fact, the circulation may be strong to push the flux tube around, and this gives rise to emission of MHD waves which can heat the overlying atmosphere.

On the other hand, if the foot-point is large, specifically, if $D_{\rm mag} \geq 1200$–1400 km, then the magnetic flux tube exceeds the diameter of a granule. In such a situation, with vertical magnetic field lines covering an entire convection cell, the field (to which the plasma is "frozen") is in a position to interfere with the horizontal motions of the circulation pattern inside the cell. The stronger the field, the more severe is the interference. In flux tubes where B is as large as 2–3 kilogauss, the horizontal flows can be stopped altogether, and the usual convective circulation is effectively "switched off". Vertical motions are not affected, but such up-and-down oscillatory motions are a poor substitute for the normal convective heat transfer. As a result, a dark spot appears on the solar surface. In such a sunspot, the emergent flux of energy is only 10–20% of the normal value. Such a spot will survive as long as the vertical flux tube (a) retains a horizontal dimension in excess of a granule, and (b) retains a field strength of 2–3 kG.

The dynamo which is at work inside the Sun continually ejects new flux into the atmosphere. As new flux emerges, it interacts with old flux in a variety of ways. The most energetic of these interactions gives rise to the phenomenon of "solar flares". In a flare, the dynamo process is reversed: magnetic energy is re-converted into kinetic energy. We shall discuss flares in Sect. 6.

3.2 Wave Modes in a Compressible Gas

The natural modes of a compressible gas (in the absence of magnetic fields) include gravity waves and acoustic waves. Gravity modes occur in two categories: stable and unstable. The stable gravity modes are oscillatory, with gravity as the restoring force: there is as yet no convincing evidence for such waves in the Sun. Unstable gravity modes give rise to the sort of "run-away" motions that we discussed above in connection with convection . Thus, convective circulation, driven as it is by buoyancy forces (in which gravity plays an essential role), is readily identifiable as a gravity mode.

However, in the course of circulating, the gas also varies in *density*: it is, after all, buoyancy forces acting on density fluctuations which drive thermal convection in the first place. As a result, the circulation of gas in a convection cell inevitably contains spatial fluctuations in density or pressure, i.e. rarefactions and condensations. These are the precisely the phenomena which, if they also have appropriate temporal behavior, constitute an acoustic wave. The question we would like to address in this context is: how much energy flux is in acoustic form in the solar granulation? The answer to this question will help us to determine the properties of the solar chromosphere.

3.3 Flux of Acoustic Waves

We note first that any acoustic power which is present in the convection derives ultimately from the convective motions themselves: therefore, the kinetic energy density E_k of the convective motions is the source of acoustic power. Now, in the convection flows, we have that $E_k \approx \rho v_{conv}^2$ ergs cm^{-3}. Since the convective eddy lasts for only a finite time ($\approx t_{turn}$), the energy E_k of an eddy survives for only a short time, and then "dissolves" back into the background medium on a time-scale t_{turn}. The rate R_c at which kinetic energy is converted from convective form back into the medium is therefore of order $R_c \approx E_k/t_{turn}$. Inserting quantities from above, we find

$$R_c \approx \frac{\rho v_{conv}^2}{t_{turn}} \approx \frac{\rho v_{conv}^3}{L_{circ}}. \tag{31}$$

As the eddy dissolves, a fraction η_{ac} of the original kinetic energy of the eddy is converted into acoustic power. Based on dimensional arguments, the efficiency η_{ac} of conversion into a wave of wavelength λ_w is expected to scale as

$$\eta_{ac} \sim \left(\frac{L_{circ}}{\lambda_w}\right)^{2m+1}$$

where m is the multipole term which contributes to acoustic power generation. In the Sun, quadrupole terms are dominant: $m = 2$. The efficiency of acoustic power generation is maximum when the turnover time t_{turn} equals the period of the acoustic wave t_{wave}. For a wave of wavelength λ in a medium with sound speed c_s, the value of t_{wave} equals λ/c_s. Equating t_{turn} to t_{wave}, we find

$$\frac{L_{circ}}{\lambda} \approx \frac{v_{conv}}{c_s} \equiv M_{conv} \tag{32}$$

where $M_{\rm conv}$ is the Mach number of the convective flow. Thus, $\eta_{\rm ac} \sim M^5_{\rm conv}$.

Combining this with (31), we find that the rate of acoustic power generation per unit volume is

$$P_{\rm ac} = \eta_{\rm ac} R_{\rm c} \sim \rho\, v^3_{\rm conv}\, M^5_{\rm conv}/L_{\rm circ} \text{ ergs cm}^{-3} \text{ sec}^{-1}. \tag{33}$$

We note that $P_{\rm ac}$ is very sensitive to the local velocity. As a result, the source of acoustic emission is extremely peaked in the region of maximum $v_{\rm conv}$. To obtain the *flux* of acoustic power $F_{\rm ac}$, we integrate the power $P_{\rm ac}$ over the depth of the region of peak power emission, which is of order $L_{\rm circ}$. This finally leads to an estimate for the flux of acoustic power:

$$F_{\rm ac} \sim \rho v^3_{\rm conv} M^5_{\rm conv} \text{ ergs cm}^{-2} \text{ sec}^{-1}. \tag{34}$$

Estimates suggest that the constant of proportionality in (34) is about 20 [15].

To estimate some numerical values for the Sun, we note that in the photosphere, $\rho \approx 3 \times 10^{-7}$ gm cm^{-3}, and $c_{\rm s} \approx 8$ km sec^{-1}. With mean $v_{\rm conv}$ values of 1–2 km sec^{-1}, $M_{\rm conv}$ has a maximum value of 0.25. Combining these numbers, we find

$$F_{\rm ac} \leq 5 \times 10^7 \text{ergs cm}^{-2} \text{ sec}^{-1}. \tag{35}$$

Because of uncertainties in various parameters, the above quantitative estimates of $F_{\rm ac}$ are subject to considerable uncertainty. Nevertheless, the upper limit cited in (35) agrees well with the results of a detailed calculation of acoustic emission from solar convection [16].

From a qualitative point of view, it is important to note that acoustic power emission is *inevitable* when flows are present in a compressible medium. Convection *always* generates acoustic power. In particular, the extreme sensitivity of $F_{\rm ac}$ to $v_{\rm conv}$ (essentially to the 8th power) means that local regions of faster than average flow act as strong localized sources of acoustic emission. The *p*-modes which allow us to probe the interior of the Sun in such detail (see Sect. 1.6 above) represent the low-frequency end of the spectrum of acoustic modes which is emitted by solar convection. At higher frequencies, the waves in the spectrum can propagate upwards and these are responsible for heating the chromosphere.

3.4 The Rate of Mechanical Energy Deposition

We turn now to item (ii) in the list which appears in the opening paragraph of this section. To estimate the rate of energy deposition, we note that the solar atmosphere is stratified by gravity (see Sect. 1.1 above): the density ρ falls of with height as $e^{-z/H}$. With values appropriate to the solar surface, the scale height H has a numerical value of 100–200 km. Therefore, as acoustic waves emerging from the convection zone propagate up into the solar atmosphere, they encounter gas whose density ρ is becoming progressively smaller. Now the energy flux associated with a sound wave with velocity amplitude $v_{\rm w}$ in a medium where the sound speed is $c_{\rm s}$ is $F_{\rm w} \sim \rho v^2_{\rm w} c_{\rm s}$. In order to conserve energy flux, when the acoustic wave propagates in a medium where ρ is declining with increasing height, the amplitude $v_{\rm w}$ must increase as $\rho^{-0.5}$, i.e. $v_{\rm w}$ increases exponentially with height ($\sim e^{z/2H}$). Therefore, even though the initial amplitude of the wave may have been small compared to $c_{\rm s}$ (cf. $M \equiv v_{\rm w}/c_{\rm s} \approx v_{\rm conv}/c_{\rm s}$ has a maximum value of 0.25 in the convection zone), the occurrence of exponential growth has the effect that at heights of several hundred kilometers above the photosphere, $v_{\rm w}$ reaches values of order $c_{\rm s}$. At this point, the sound wave steepens to form a shock, and the shock does work on the gas. Shock dissipation is efficient enough that we may consider that the acoustic energy is deposited in the gas as soon as the shock forms.

Therefore, the convection zone provides a flux of acoustic power $F_{\rm ac}$ at the base of the solar atmosphere, and this flux is deposited over a typical length scale of $2H$. This allows to estimate the rate at which acoustic energy is deposited into the atmosphere per unit volume: $\dot{E}_{\rm mech} \approx F_{\rm ac}/2H$ ergs cm^{-3} sec^{-1}. Inserting values as given above, we find that the rate at which work is done on the chromospheric gas per unit volume is of order

$$\dot{E}_{\rm mech} \approx 1 \text{ erg cm}^{-3}\text{ sec}^{-1}. \tag{36}$$

Of course, the estimates of the various factors which enter into our evaluation of $\dot{E}_{\rm mech}$ are quite crude. As a result, the numerical value of $\dot{E}_{\rm mech}$ is subject to considerable uncertainty, perhaps by an order of magnitude or more. Fortunately, we shall find that our estimates of chromospheric temperature increases are quite insensitive to these uncertainties.

3.5 Increase of Temperature in the Chromosphere

Turning now to item (iii), we ask: by how much does the temperature of the gas increase when energy is deposited in it at the rate $\dot{E}_{\rm mech}$? To answer this, we note that as the gas heats up, it will lose energy at an increased rate. If this increased rate of energy loss can be made equal to $\dot{E}_{\rm mech}$, then the gas will find equilibrium. Let us search for this equilibrium in terms of the properties of gas in the solar atmosphere.

How fast can a gas lose energy? For gas in the solar chromosphere at levels where the optical depth is small ($\tau \leq 1$), the fastest means of losing energy is to radiate it away. (Conduction and convection are not important in the chromosphere as far as energy loss is concerned: but they will become important when we consider the corona.) The time-scale on which gas cools in the chromosphere is the radiative cooling time-scale $t_{\rm cool}$.

To estimate $t_{\rm cool}$, we consider an element of gas of volume $\mathrm{d}V$ and surface area $\mathrm{d}A$ in the solar atmosphere. The internal energy of the element is $\mathrm{d}E_{\rm int} = c_{\rm v} T \rho\, \mathrm{d}V$. If the element were optically thick, then energy would be radiated from the surface at a rate given by the black body law: $(\mathrm{d}E/\mathrm{d}t)_{\rm bb} = \mathrm{d}A\ 4\pi\sigma_{\rm SB}T^4$. However, in the region of the solar atmosphere which we are considering, the element will *not* be optically thick: its optical depth $d\tau$ is in general less than unity. In this condition, the rate at which energy is radiated away is $(\mathrm{d}E/\mathrm{d}t)_{\rm rad} = (\mathrm{d}E/\mathrm{d}t)_{\rm bb} \times \mathrm{d}\tau = \mathrm{d}A\ \mathrm{d}\tau\ 4\pi\sigma_{\rm SB}T^4$

The cooling time-scale $t_{\rm cool} \approx {\rm d}E_{\rm int}/({\rm d}E/{\rm d}t)_{\rm rad}$ therefore can be written as

$$t_{\rm cool} = \frac{C_{\rm v}\rho}{4\pi\sigma_{\rm SB}T^3}\frac{{\rm d}V}{{\rm d}A\,{\rm d}\tau}. \tag{37}$$

Now the ratio ${\rm d}V/{\rm d}A$ is of order ${\rm d}s$, the linear dimension of the element. The value of ${\rm d}s$ is related to ${\rm d}\tau$ by the standard definition of optical depth and opacity: ${\rm d}\tau = \kappa\ \rho\ {\rm d}s$. Therefore

$$t_{\rm cool} = \frac{C_{\rm v}}{4\pi\kappa\sigma_{\rm SB}T^3}. \tag{38}$$

Now we know the cooling time-scale, let us ask: can an equilibrium be obtained? To answer this, note that deposition of energy makes the gas heat up somewhat: let us say that the increase in temperature is ΔT. The excess thermal energy per unit volume $\Delta E_{\rm m}$ therefore equals $C_{\rm v}\rho\Delta T$ ergs cm^{-3}. Equilibrium occurs if the gas radiates the excess $\Delta E_{\rm m}$ away on a time-scale $t_{\rm cool}$, such that the rate of cooling equals the rate at which the shock heating is depositing energy: $\Delta E_{\rm m}/t_{\rm cool} = \dot{E}_{\rm mech}$. In view of (36), this means that

$$\frac{C_{\rm v}\rho\Delta T}{t_{\rm cool}} = 1 \text{ ergs cm}^{-2}\text{ sec}^{-1}. \tag{39}$$

Solving for ΔT, we find

$$\Delta T^4 \approx \frac{10^4\ {\rm K}^4}{\rho\kappa}. \tag{40}$$

To proceed further, we now need to know the details of κ. Recall that we are dealing with gas in the upper solar photosphere which starts off with a temperature of 4000–4500 K. At such temperatures, κ is a rapidly increasing function of temperature (see Fig. 3). There is also a slight dependence on density. Fitting power law approximations to curves such as those in Fig. 3, we find that at temperatures below 10^4 K, the opacity can be written as $\kappa \approx 10^{-17}T^6\rho^{0.7}$ cm^2 gm^{-1}.

Inserting this into (40), we find $\Delta T^{10} \approx 10^{21}/\rho^{1.7}$ Taking the tenth root of both sides, we finally have an estimate of the temperature increase in equilibrium:

$$\Delta T \approx 100\ {\rm K}\ \rho^{-1/6}. \tag{41}$$

An attractive feature of (41) is that our estimate of ΔT is *very* insensitive to the parameters which went into the calculation. In particular, even if we are wrong in our estimate of acoustic flux $F_{\rm ac}$ by a factor of 1000, the final value of ΔT will be wrong by a factor of only 2!

Now, in the region of the solar atmosphere in which we are interested (at heights between 0 and 2000 km above the photosphere), models indicate that typical densities fall from 10^{-6} to 10^{-12} gm cm^{-3}, i.e. $\rho^{-1/6}$ increases from 10 to 100. Thus, ΔT increases with height, from a value of about 1000 deg K just above the photosphere to a value of order 10^4 at 2000 km.

This region of the solar atmosphere where local heating by a few thousand degrees enable radiative losses to balance shock heating is the CHROMOSPHERE.

Obviously, the remarkable insensitivity of the value of ΔT to input parameters is connected in part with the fact that κ increases very rapidly with increasing T. Why is κ so sensitive to T at temperatures below 10^4? The reason is that at such temperatures, the dominant constituent of the atmosphere (hydrogen) is neutral: if the bound atomic levels of hydrogen can be populated, they serve as effective absorbers of photons. Increasing temperature leads to exponentially increasing populations of the bound levels, at least up to temperatures where ionization is not rapid. The process of populating excited levels in hydrogen (and in other ions as well) therefore leads ultimately to the conclusion that ΔT varies very slowly, only as the 10th root of the input power. This very slow dependence results in rather small variations in ΔT. The bound energy levels of the atoms and ions in the solar chromosphere serve as a sort of "thermostat" to hold the temperature nearly constant. This "thermostat" allows the gas to deal with even relatively large rates of energy deposition by increasing its temperature only slightly. This gives rise to a temperature plateau in the chromosphere.

We conclude that the chromosphere in the Sun exists at a well defined temperature essentially because of the existence of bound atomic levels. Since hydrogen is the most abundant element in the Sun, the bound levels of hydrogen play a significant role in the chromospheric thermostat.

During an eclipse, the eye sees the chromosphere as a narrow colorful aureole around the dark moon. We can now understand why the chromosphere is narrow: it extends only to heights of 2000 km above the photosphere, corresponding to an angular thickness of only 2–3 arcsec at the distance of the Sun. We can also understand why the chromosphere is "rose-colored": the strongest radiative losses from the bound levels of hydrogen in visible light occur in the Balmer-α spectral line in the red part of the spectrum (at wavelength 6563 Å).

In summary, the chromosphere exists essentially as long as hydrogen remains mainly neutral, and the human eye can actually detect that hydrogen radiation is important in cooling the chromosphere.

4 Transition from Chromosphere to Corona

At the top of the chromosphere, ΔT rises to values of order 10^4. At such temperatures, hydrogen quickly begins to ionize. As a result, the most plentiful supply of bound atomic energy levels is no longer available, either to absorb radiation, or to emit spectral lines as coolants. The disappearance of absorbing power shows up in Fig. 3 as a decrease in hydrogen opacity with increasing temperature. Of course the total opacity contains contributions from more than hydrogen: all ions with at least one electron left in a bound state contribute to opacity. Therefore, even though hydrogen no longer contributes to opacity at temperatures above $\log T = 4$–4.3, other elements still contribute bound level opacity even up to temperatures of $\log T = 4.4$–4.5. Eventually, however, at high enough temperatures, every element loses the electrons which can absorb optical light, and the Rosseland mean opacity then begins to decrease with increasing T. In other words, the presence of a definite maximum in opacity at a certain temperature is inevitable: beyond that, κ decreases when the gas heats up. As a result, the cooling time (eq. (38)) becomes longer, i.e. cooling is less efficient.

This behavior has a de-stabilizing effect on the equilibrium we considered above: when energy is dumped into a volume element, and the gas heats up, there is no longer an accompanying increase in radiative efficiency to help the gas get rid of its excess energy. In fact, as the gas gets hotter, it becomes *less efficient* at cooling itself. As a result, the temperature undergoes "thermal runaway" to high temperatures.

We see, then, that once hydrogen ionizes at the top of the chromosphere, the temperature rapidly increases to much higher values. This runaway appears in the solar atmosphere as an abrupt "transition region" (TR) between the chromosphere (where $T \leq 10^4$) and the corona (where $T \gg 10^4$ K). The transition region is very narrow: some estimates put it at no more than 100 km thick.

Once the temperature runaway starts at the top of the chromosphere, can a new equilibrium be found at higher temperatures? If such an equilibrium exists, it must involve some process in addition to radiative cooling. The reason that radiative losses are no longer effective in controlling the temperature is that radiative cooling efficiency ($\sim \kappa$) *decreases* with increasing T above the TR. To help achieve energy balance, we need to find another process in which the cooling efficiency *increases* as T increases.

Reverting to the three standard processes of transferring heat (radiation, conduction, and convection), we ask: is conduction or convection at work above the TR? Convection seems unlikely in the quiet corona. So let us consider thermal conduction. A new equilibrium will occur if it is possible to satisfy

$$\dot{E}_{\text{mech}} = (\mathrm{d}E/\mathrm{d}t)_{\text{cond}} + (\mathrm{d}E/\mathrm{d}t)_{\text{rad}} . \tag{42}$$

4.1 Thermal Conduction

As was mentioned above (eq. (15)), the heat flux carried by thermal conduction is given by $F_{\text{cond}} = -k_{\text{th}} \nabla T$ ergs cm^{-2} sec^{-1}. The rate of energy loss per unit volume is obtained by taking the divergence of this equation:

$$(\mathrm{d}E/\mathrm{d}t)_{\text{cond}} = \nabla \cdot (k_{\text{th}} \nabla T). \tag{43}$$

To evaluate k_{th}, we return to the general expression in (16) in order to apply it to the corona. We recall that when we considered k_{th} in the interior of the star (Sect. 1.10), ρ was determined by the heaviest particles (protons), whereas λ, v, and C_{v} were determined by the fastest moving "particles" (photons). In the coronal plasma, protons and electrons are the dominant constituents, and we have analogous contributions to k_{th}: ρ is still determined by protons, while the other quantities are determined by fast moving electrons.

Let us see how the various quantities in k_{th} depend on temperature and density.

The value of ρ is simply $n_{\text{i}} m_{\text{p}}$, where $n_{\text{i}} = n_{\text{e}}$ is the number density of protons (or electrons), and m_{p} is the mass of the proton.

The value of λ is $1/n_i\sigma_{ie}$, where the cross-section for Coulomb collisions is σ_{ie}. In a gas where electrons have temperature T_e, the value of σ_{ie} is given by $\pi e^4 \ln \Lambda/(kT_e)^2$, where e is the electron charge, and k =1.38 × 10^{-16} ergs deg^{-1} is Boltzmann's constant. The term ln Λ is a slowly varying term which allows for distant encounters between charged particles [17]. In coronal conditions, ln Λ has a value of about 20.

The r.m.s. speed of the electrons v is $\sqrt{2kT_e/m_e}$.

The specific heat of the electrons per unit volume is $3kn_i/2$. Since the protons dominate the mass, the mass of a unit volume is $n_i m_p$. Therefore, the specific heat per gram C_v is $3k/2m_p$.

Combining the four factors together, we finally obtain

$$k_{th} = K_0 \, T_e^{2.5}. \tag{44}$$

where the constant of proportionality K_0 is related to several physical constants: $K_0 \sim k^{3.5}/\pi e^4 \sqrt{m_e} \ln \Lambda$. The numerical value in c.g.s. units is $K_0 \approx 10^{-6}$ [17]. The key point to recognize in (44) is that k_{th} increases rapidly with increasing temperature. It is this rapid increase of k_{th} which helps stop the runaway of temperature in the corona. Note also that it is the *electron* temperature which appears in (44): we shall return to this point below.

4.2 Radiative Losses in the Corona

The optical thickness of the corona is very small. Therefore, when we consider the radiative losses from a volume element in the corona, it is hardly appropriate to consider emission from the *surface* of the element, as if we were able to "see" only the material near that surface, Now, we can "see" essentially every particle in the volume element as it emits. It is therefore more convenient to express the loss rate in terms of how effective the gas is at emitting *from the entire volume.* The *emissivity* $\Phi(T_e)$ is the rate at which an ion in the gas emits energy when it is excited by a collision with an electron of temperature T_e. Since there are n_e electrons per unit volume to excite any given ion, and n_i (=n_e) ions per unit volume which can be excited, the total radiative loss rate per unit volume per sec is $(\mathrm{d}E/\mathrm{d}t)_{rad} = n_e^2 \Phi(T_e)$.

Now we ask: what determines the emissivity $\Phi(T_e)$? The answer is: it is determined by processes whereby free electrons in the plasma collide with bound electrons in atoms and ions, and excite these bound electrons to higher energy levels. The subsequent decay of the excited states leads to photon emission. Therefore, in order to calculate $\Phi(T_e)$, it is necessary to know first how many of each species of ion and atom are present in the plasma: this is typically calculated by assuming ionization equilibrium, where collisional ionizations by electrons are balanced by radiative recombination. Since *electrons* are responsible for determining both the ionization equilibrium and the excitation of bound levels, it is not surprising that $\Phi(T_e)$ is a function of the *electron* temperature.

Without going into the many details of how $\Phi(T_e)$ is calculated, we note that the very same bound atomic levels and continua which are involved in creating emissivity are also involved in creating opacity: therefore, arguments based on opacity and those based on emissivity must overlap to some extent. In fact, the curve $\Phi(T_e)$ as a function of temperature has a shape which is comparable to the topmost opacity curve in Fig. 3. That is, $\Phi(T_e)$ is very small at low T, rises steeply to a maximum value Φ_{max} at $T_e = 10^{4-5}$ K, and then falls off as T_e increases from 10^5 to 10^7 K (see, e.g. [10]). The overall shape of the $\Phi(T_e)$ curve is controlled by the combined effect of millions of individual transitions.

Over certain ranges of temperature, the temperature dependence of $\Phi(T_e)$ can be represented roughly as a power law: in particular, at temperatures which are appropriate for gas in the transition region and above (between, say, 10^5 and 10^7 K), we find that $\Phi(T_e)$ can be described within a factor of about 2 by the expression

$$\Phi(T_e) \approx 10^{-19}\, T_e^{-0.5} \text{ ergs cm}^{-3} \text{ sec}^{-1} . \qquad (45)$$

4.3 The Non-Flaring Corona: A Balance Between Conductive and Radiative Cooling

Magnetic sources of some kind supply mechanical energy to the corona. The supply is strongest in closed magnetic field regions, where the magnetic field lines emerge from one place on the solar surface, arch up to some finite height, and then loop back down to the surface. No wind escapes from these closed magnetic loops, so there are no convective losses involved. In such loops, we may consider energy losses in terms of conduction and radiation only.

Without specifying in detail the sources of coronal heating, we can proceed to discuss a steady state in the corona by noting the following. Radiative losses ($\sim T_e^{-0.5}$) tend to cause the coronal electrons to "run away" to high temperatures, whereas conductive processes tend to keep the corona cool. In view of these competing tendencies, it is plausible to suppose that the corona can find an equilibrium at the electron temperature T_{eb} where there is a rough balance between the magnitude of the radiative and conductive processes. That is, at $T = T_{eb}$, $|(\mathrm{d}E/\mathrm{d}t)_{cond}|$ should be roughly equal to $|(\mathrm{d}E/\mathrm{d}t)_{rad}|$.

It is important to be aware that the present discussion applies only to the temperature of the *electrons* in the coronal plasma. This point would be of no particular significance if we were dealing with a plasma in thermal equilibrium, such as in the deep interior of the Sun: in such a plasma, temperatures of ions and electrons are equal. (Therefore, the T which appears in eq. (17) applies equally to all particle species, and even photons also.) And in the densest regions of the corona (such as in streamers in the low corona), collisions may be rapid enough to keep $T_e = T_i$. However, in certain parts of the corona, thermal equilibrium does *not* exist. Thus, in regions of low density, as in coronal holes where fast wind originates, collision rates may be so small that there is no longer any physical reason why *electrons* and *ions* should have identical temperatures (see Sect. 5.3). Therefore, when we draw conclusions about an equilibrium state of the plasma based on using eqs. (44) and (45), we should remember that the results apply specifically to the temperatures of *electrons*.

4.4 Electron Temperatures in the Non-Flaring Corona

Let us estimate the temperature T_{eb}. Consider a closed magnetic loop of half-length L: the top of the loop is at coronal temperatures while the footpoint of the loop is at a much lower temperature. Heat is conducted along the loop, and so the spatial gradient of temperature can be approximated by $\nabla T_e \approx T_e/L$. The divergence operator can be approximated by $1/L$. Therefore the conductive loss rate $(\mathrm{d}E/\mathrm{d}t)_{\mathrm{cond}}$ can be written as $K_0 T_e^{3.5}/L^2$.

Assuming that the magnitude of conductive losses equals the magnitude of the radiative losses at temperature T_{eb}, we find

$$\frac{K_0 T_{eb}^{3.5}}{L^2} = n_e^2 \Phi(T_{eb}). \tag{46}$$

Using $\Phi \approx 10^{-19} T_{eb}^{-0.5}$, and $K_0 \approx 10^{-6}$, this leads to

$$T_{eb}^4 \approx 10^{-13}\, n_e^2 L^2. \tag{47}$$

The units in (47) are c.g.s.: i.e., with n_e in units of cm^{-3}, and L in cm, the units of T_{eb} are degrees K.

Now, when we study the transition region (TR) between chromosphere and corona, it is convenient to use the pressure p as a variable rather than density. The reason is that the TR thickness is much less than one pressure scale height: therefore, p remains constant across the TR. Now, with $p = 2n_e k T_e$, where k is Boltzmann's constant, eq. (47) can be re-arranged to give

$$T_{eb}^6 \approx (pL)^2 \times 10^{-13}/(4k^2)$$

Solving for T_{eb}, and noting that the high power of T_{eb} makes for a reliable solution, we find

$$T_{eb} \approx 1000 \times (pL)^{1/3} \quad \text{deg K}. \tag{48}$$

In the upper chromosphere and low corona, empirical estimates in active regions suggest that $p \approx 1$ dyn cm^{-2}. Therefore the electron temperature where the rate of conductive losses equals the rate of radiative losses is $T_{eb} \approx 1000 L^{1/3}$ deg K.

We now need to insert actual values of coronal loop lengths. Most loops in solar active regions are short compared to the solar radius: typical values of L are in the range 10^9 to 10^{10} cm. Inserting these, we finally find $T_{eb} \approx$ **(1–2) million K.**

Is there any empirical evidence that the solar corona indeed has electron temperatures of 1–2 million K? Yes: even in the optical spectrum, there are some lines which are created by highly ionized iron. These highly ionized ions are created when fast electrons in the ambient medium strip many bound electrons from the ion. In order to achieve the amount of stripping which is observed in coronal iron, the fast electrons must have temperatures of 1–2 million K. Moreover, images of the Sun in X-rays detect bremsstrahlung radiation emitted by electrons which accelerate in the vicinity of ions. The bremsstrahlung emission is controlled mainly by the electron temperature: the images show that active regions contain a copious supply of electrons at temperatures of several million K.

Thus, on the basis of the physical characteristics of energy *losses* via the channels of conduction and radiation, we have been able to obtain rather reliable estimates of the temperatures of coronal electrons in closed loops. At first sight, this seems curious: it seems that we have discussed losses of energy without discussing energy *supply*. Recall that, when we were considering the heating of the *chromosphere*, we first had to specify the rate at which energy is being supplied ($\dot{E}_{\rm mech}$) before we could evaluate the temperature in the chromosphere. But here, we have said nothing explicit about the rate of deposition of mechanical energy in the corona. So how have we managed to reach conclusions about temperature? The answer is: we have actually allowed for $\dot{E}_{\rm mech}$ implicitly when we assigned a numerical value to the pressure p in (48). In a region of the Sun where more mechanical energy is being deposited (such as in an active region, where MHD wave fluxes are higher), the local pressure p will be larger. Therefore, $T_{\rm eb}$ will also be larger in such a region.

In an open field region, where gas is free to escape from the Sun, some energy is carried away into the wind (see Sect. 5). This leaves less energy to be distributed among conduction and radiation. Therefore, we expect that open field regions contain cooler gas than closed loops. Empirically this is borne out: coronal holes, from which fast solar wind escapes, have $T_{\rm e}$ values which are about 0.8 million K.

Finally, we note that, although the temperature jumps almost discontinuously from chromospheric values ($\approx 10^4$ K) to coronal values ($\approx 10^6$ K) across the TR, the pressure remains practically constant across the TR. Therefore, across the TR the 100-fold jump in temperature is accompanied by a 100-fold drop in density. With densities at the top of the chromosphere of order $n_{\rm ct} = 10^{11-12}$ cm^{-3}, we see that the densities at the base of the corona $n_{\rm cb}$ must be in the range 10^{9-10} cm^{-3}.

5 Expansion of the Solar Corona

We have seen that $T_{\rm eb} = 1$–2 million K is a good estimate of an average steady state temperature in closed loops in the corona. The coronal temperature which is observed in the quiet sun is also close to the above value. There is one particularly important physical property of a corona which has a steady temperature as high as 1–2 million K. We turn to that property now.

5.1 Breakdown of Hydrostatic Equilibrium

Let us examine hydrostatic equilibrium (HSE) in the corona. Outside the Sun, g is not a constant, but varies with radius as $g = -GM_{\rm sun}/r^2$. The question is: with this choice of g, can HSE (i.e. (3)) be satisfied?

To answer this question, we need to know how to handle the energy equation. Just as we did for the interior of the Sun, we can simplify the problem by accepting a particular solution. In the solar corona, the thermal conductivity ($\sim T^{2.5}$) is so large that the gas is close to isothermal. Let us therefore set $T =$ constant. (We specify nothing about the source of this heating: we simply assume that something is available which heats the corona to the same T at all radii.) Since coronal material behaves as a perfect gas, we use $p = R_{\rm gas}\rho T/\mu$, and then (3) can be written as an ordinary differential equation for ρ as a function of radial distance. The solution of this equation is readily obtained:

$$\rho(r) = \rho_o \, \mathrm{e}^{[A(r_o/r - 1)]} \tag{49}$$

where ρ_o is the density at radial distance r_o. If the corona is in HSE, then the radial density profile must obey eq. (49).

There are two points to note about (49). First, the functional form is such that as $r \to \infty$, ρ does *not* vanish. This is in striking contrast to the solution for $g =$ constant in a plane-parallel atmosphere: there, the density approaches zero exponentially rapidly. (See discussion following (3) above.) In the spherical corona, on the other hand, eq. (49) indicates that as the radial distance increases, the density does *not* tend to zero. Instead, it approaches a constant value $\rho(\infty) = \rho_o \mathrm{e}^{-A}$. Second, the numerical value of the constant A plays a crucial role: $A = GM_{\rm sun}\mu/R_{\rm gas}Tr_o \sim v^2_{\rm esc}/v^2_{\rm th,cor}$ is a measure of how effectively the thermal pool of the coronal gas fills up the Sun's gravitational well.

Now let us test the above solution in coronal conditions. Setting $T = 10^6$ K, and using $\mu \approx 0.5$ as befits fully ionized hydrogen, we find $A \approx 12$. Recall that the gas at the base of the solar corona has a number density $n_{\rm cb}$ of 10^{9-10} protons cm^{-3}. With this as the inner boundary density, a corona *in hydrostatic equilibrium* would therefore have a number density at infinity n_∞ which is less than $n_{\rm cb}$ by a factor of e^{-12}. Thus, the number density of a *hydrostatic* solar corona with $T = 10^6$ K would be n_∞ 6 $\times 10^{3-4}$ protons cm^{-3}.

However, this is not an acceptable solution: the density of gas in the interstellar medium (ISM) is typically 1 proton cm^{-3}. Because the ISM has a density (and pressure) which is many times smaller than n_∞, it is impossible for the ISM to contain the corona: the latter has a pressure which is simply too high to be confined.

It is important to note that this conclusion depends sensitively on the value of the coronal temperature. If the coronal temperature were reduced by a factor of only 2, i.e. if T were 0.5 million K, then n_∞ would turn out to be less than 1 cm^{-3}: the ISM could contain such a gas. However, our estimate of coronal temperature $T_{\rm eb}$ in (48) is a robust one, and it is not easy to alter $T_{\rm eb}$ by a factor of 2: our estimate of coronal pressure p would have to be incorrectly high by almost an order of magnitude. Such large errors are unlikely: empirical values of p are known to much better than an order of magnitude.

5.2 A HydroDYNAMIC Solution of the Momentum Equation

Now that we know that the solar corona cannot be in HSE, we need to know: what happens when HSE breaks down? To answer that, we recall that the equation of HSE (eq. (3)) is itself only a special case of a more general equation (eq.

(2)) which describes the law of conservation of momentum. In cases where HSE is satisfied (see (3)), the right-hand side of (2) is exactly zero: $v = 0$ is then a valid solution. But since in the corona, pressure forces do NOT balance gravity, the right-hand side of (2) is non-zero. As a result of the unbalanced forces, the gas must accelerate. Instead of hydro-"static" conditions, we now have to deal with unbalanced forces (i.e. dynamics). The onset of acceleration in the radial direction means that the corona must expand. This expansion of the coronal gas gives rise to a flow which is named the "solar wind".

It is important to note that when we talk of the solar wind, we are *not* talking about evaporation, as if a small fraction of the coronal material were "boiling off": there is nothing evaporative about the process described by (2). The solar wind involves a truly hydrodynamic expansion *of the entire corona.*

The fact that an outflow of some sort from the Sun exists has been known for decades. The speed of the outflow can be measured *in situ* by spacecraft, or by remote sensing of distant radio sources. By studying "scintillations" of background radio sources which happen to pass close to the Sun at certain times of the year, it was known already in the 1970's that the fastest wind (with speeds of 700–800 km sec^{-1}) emerges from the coronal holes at the North and South poles of the Sun. In recent years, the Ulysses spacecraft has measured the speed *in situ* at almost all latitudes. A polar plot of the wind speeds obtained by Ulysses over a time interval of several years is shown in Fig. 4 (from [18]). The results are striking: there is confirmation of the scintillation results that fast flows do indeed emerge from the polar regions. However, it is not only from "polar regions" (as traditionally defined) that the fast wind emerges. Rather, fast wind is detectable at latitudes ranging all the way from 90 N to perhaps 20 N, and from 90 S to perhaps 20 S. Only within about 20 degrees of the equatorial plane do the wind speeds slow down, and even then, there are some high speed flows present occasionally.

A remarkable aspect of Fig. 4 is the near constancy of the solar wind speed at high latitudes in both hemispheres. With a mean value of $v_{\rm mean} \approx 750$ km sec^{-1}, we see that the fluctuations in speed above and below $v_{\rm mean}$ are at about the 10% level. This is particularly interesting because, in the course of the several years that elapsed between the earliest and latest measurements in the plot, the Sun was continually evolving through its 11-year cycle of magnetic activity. But despite these variations in magnetic activity, the solar wind speed remained essentially unchanged. Fig. 4 leaves one with the impression that the Sun has for the most part a spherically symmetric wind, on which certain slower disturbances are superposed at low latitudes.

Let us see if we can understand the observed flow speeds in terms of what we know about the corona.

In the simplest case of steady flow, the velocity of outflow v does not depend on the time, but varies with radial distance. In this case, eq. (2) can be written in the form

$$v\frac{\mathrm{d}v}{\mathrm{d}r} = -\frac{1}{\rho}\frac{\mathrm{d}p}{\mathrm{d}r} - \frac{GM_{\rm sun}}{r^2}. \tag{50}$$

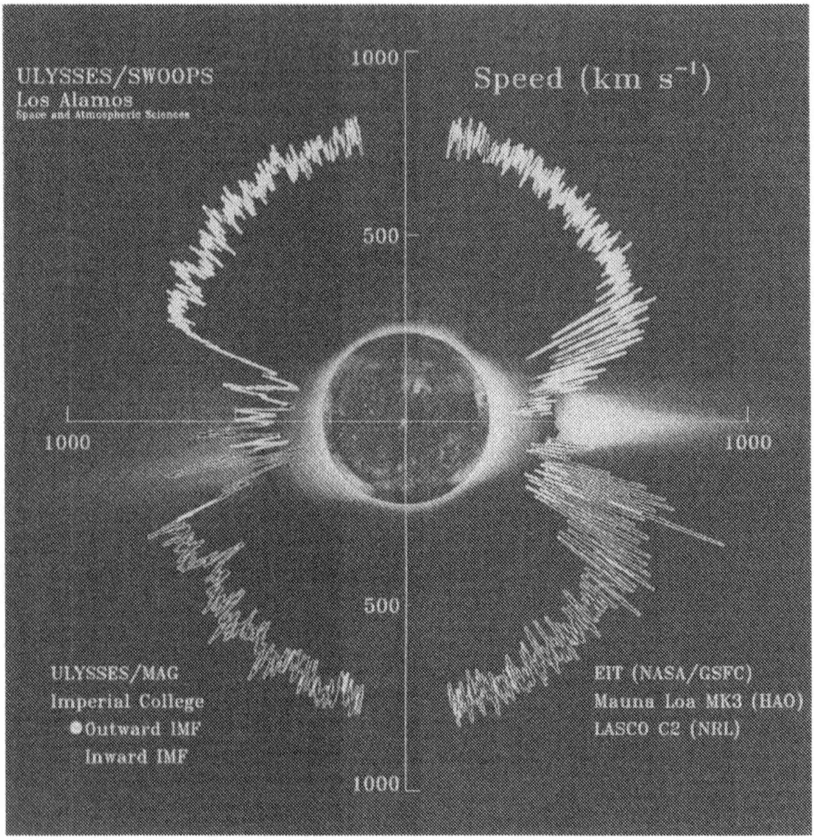

Fig. 4. Polar plot of solar wind speed as measured by Ulysses as it traversed latitudes both below and above the solar equatorial plane. In the center is a sample image of Sun in extreme ultraviolet light, plus a sample image of the corona obtained by combining two different images obtained by the C2 coronograph on SOHO and the MKIII instrument at Mauna Loa (Reprinted courtesy of McComas et al. [18] and Geophys. Res. Lett. ©*The American Geophysical Union.*)

In an isothermal corona, with $T = T_{\rm cor}$, this becomes

$$v\frac{\mathrm{d}v}{\mathrm{d}r} = -\frac{R_{\rm gas}T_{\rm cor}}{\mu}\frac{\mathrm{d}\log\rho}{\mathrm{d}r} - \frac{GM_{\rm sun}}{r^2}. \tag{51}$$

Invoking conservation of mass, we have that $4\pi r^2\rho v$ is a constant at all radial distances. Taking the radial derivative, we find $\mathrm{d}\log\rho/\mathrm{d}r = -\mathrm{d}\log v/\mathrm{d}r - 2/r$ Substituting in (51), we obtain the solar wind equation which was first discussed by Parker [19]:

$$\frac{\mathrm{d}\log v}{\mathrm{d}\log r} = \frac{2a^2 - GM/r}{v^2 - a^2}. \tag{52}$$

Here, $a = \sqrt{R_{\mathrm{gas}}T_{\mathrm{cor}}/\mu}$ is the isothermal sound speed in the corona. Parker obtained a solution for v as a function of r with the condition that the flow speed passes through the sound speed (i.e. $v = a$) at the so-called sonic point: this lies at a radial distance $r_s = GM_{\mathrm{sun}}/2a^2$. With T_{cor} = 1–2 million K, this leads to a sonic point distance of

$$r_s \approx (4-7)\, R_{\mathrm{sun}}. \tag{53}$$

An isothermal wind has the property that at great distances, the profile of velocity tends to the solution $v(r) \sim \sqrt{\log r}$. This is a very slowly increasing function of distance. Because of this slow variation, the velocity of the wind has become almost constant by the time the wind reaches a radial distance of 1 astronomical unit (AU), i.e. the Earth's orbit. The value of the velocity at Earth v_{E} increases with increasing coronal temperature. With $T_{\mathrm{cor}} = (1-2) \times 10^6$ K, $v_{\mathrm{E}} \approx$ 500–750 km sec^{-1}. These values are actually too large to be consistent with the solar wind observed near the Earth's orbital plane.

Of course, our assumption that the corona remains exactly isothermal at all radial distances is an extremely special solution of the energy equation: it implies that whatever the heating agent is, the agent must operate at all radial distances with precisely the strength required to make the local plasma temperature there equal to the T at the base of the corona. It is not clear precisely what agent would have such a remarkable property: more likely, the agent would be most effective at depositing heat closer to the Sun, but less effective far away from the Sun. Eventually, the supply of energy probably runs out, and beyond that distance, the wind should behave adiabatically, with $p \sim \rho^{5/3}$. Parker [19] suggests that rather than assuming isothermal conditions, it would be better to consider solutions of the energy equation of the form $p \sim \rho^{\delta}$, where δ varies with distance. Near the Sun, where energy is being supplied, the corona remains almost isothermal, and δ should be close to unity. But farther out, δ should approach 5/3.

We note that Parker's suggested p *versus* ρ relation for the solar wind is nothing other than the polytropic equation of state (see (6) above) which we found so helpful in studying the *interior* of the Sun. Now the polytropes make their appearance again in the corona, with the isothermal solution represented by the special case $\delta = 1$. Parker considers mixed solutions where isothermal conditions prevail inside $r = b$, while adiabatic conditions prevail at greater distances. In a corona with T_{cor} = 1 or 2 million K, the choice $b = 8R_{\mathrm{sun}}$ leads to v_{E} = 310 or 550 km sec^{-1} respectively. These are ~200 km sec^{-1} slower than the isothermal solution, and are more consistent with empirical speeds in the Sun's equator (see Fig. 4).

5.3 Unequal Temperatures of Ions and Electrons in the Corona

The solar wind is a plasma which contains both electrons and protons. Since the electrons are much less massive than the protons, one might imagine that the electrons could outrun the protons: but this does not happen. Electrostatic coupling between electrons and protons is so strong that ions and electrons expand away from the Sun with the same speed. As a result, the momentum is carried predominantly by the *ions*. Therefore, when we study the momentum equation for the solar wind, the temperature which enters into the equations is the *ion* temperature. This is in contrast to our earlier discussion of conduction where we obtained estimates of T_e, the *electron* temperature.

Are the ion and electron temperatures necessarily equal? In certain conditions the answer is yes: this is the case in the chromosphere, photosphere, and the densest parts of the corona (especially in streamers) where collisions are rapid enough to ensure close coupling between species. But as we move into the more rarefied parts of the corona, collisions become progressively rarer, and the temperatures of ions and electrons need not be equal.

A remarkable discovery of the SOHO satellite has been that ions in coronal holes are considerably hotter than electrons. Moreover, ions of greater mass are hotter than ions of lesser mass [20]. Thus, whereas electron temperature T_e are ≈ 1 million K, proton temperatures in the coronal hole wind are 2–3 million K, ions of magnesium with charge +9 have $T_{\rm Mg}$ of tens of millions K, and oxygen ions with charge +5 have $T_{\rm O}$ in excess of 100 million K. SOHO data also indicate that ions are heated preferentially in directions perpendicular to the magnetic field.

Why are the ions so much hotter than the electrons? Part of the answer is that in general, it is easier for an electron to lose energy than for an ion. For example, an electron of a given energy (say, 100 eV) can lose its energy by exciting bound electrons in the plasma, whereas an ion with an energy of order 100 eV is very inefficient at this process. Moreover, the electron is more efficient than the ion at thermal conduction. Therefore, even if energy is dumped at equal rates into electrons and ions, the asymptotic value of T_e will be less than T_i.

Another part of the answer is that heating processes in the corona may actually dump energy preferentially into ions rather than electrons. For example, energy deposition processes which increase with increasing mass of the particle would have this feature. The information contained in these SOHO results should eventually help to answer the question: are the ions in the corona being preferentially heated? Searches for an answer to this question are an active area of contemporary solar research. Among the various answers which have been developed recently, magnetic effects of various kinds play a central role. In this regard, models which have been developed in quantitative detail include dissipation of low frequency Alfven waves [21] and damping of high frequency waves [22].

The sound speed in the hot coronal hole protons, $c_{s,p}$ is large enough to drive a fairly fast proton wind. But it may not be altogether sufficient to explain the fast polar wind that is observed [20]. Thus, although we have been interested here mainly in the question of supplying *energy* to the corona, it appears likely that some process may also be supplying *momentum*, at least to the polar wind. There is widespread interest in Alfven waves in the solar corona because they have the ability to supply not only energy but also momentum to the solar wind (see e.g. [21]).

5.4 HydroSTATIC versus HydroDYNAMIC: Where Does the Change Occur?

Since the equation of HSE is a special solution of the more general momentum equation, we may wonder: where in the solar corona does the approximation of HSE change over to the hydrodynamic solution? The answer is: the transition occurs when the wind has a velocity v which is large enough to render the term $v\mathrm{d}v/\mathrm{d}r$ in the conservation of momentum equation comparable to the term $(1/\rho)\mathrm{d}p/\mathrm{d}r$. Close to the Sun, specifically, inside the sonic point, velocities of outflow are small compared to the sound speed: $v^2 \ll a^2$. In this limit, we can neglect $v\mathrm{d}v/\mathrm{d}r$ compared to $(1/\rho)\mathrm{d}p/\mathrm{d}r$, and the solution of the solar "wind" equations reverts essentially to that of HSE.

Therefore, in the innermost corona, where g is almost constant, the density has the following vertical profile: $n_{\mathrm{cor}}(z) \approx n_{\mathrm{cb}}\mathrm{e}^{-(z/H_{\mathrm{cor}})}$. Using typical coronal temperatures of 10^6 K, we find that the scale height H_{cor} is $\approx 5 \times 10^9$ cm.

However, as we move out from the solar surface, and the wind speed increases, the HSE approximation becomes less reliable. Certainly as we approach the sonic point, HSE must break down, and the full dynamics of solar wind acceleration come into play. According to the Parker solution, the sonic point is especially interesting because the wind is accelerating most rapidly at that radial distance. Thus, in order to study the physics of wind acceleration, the most valuable data are those which pertain to wind speeds in the transonic regime. Based on the estimates we made above of r_{s}, the sonic point radius, it seems that we should pay particular attention to the properties of the wind at radial distances of (say) $(5\text{–}20)R_{\mathrm{sun}}$. Unfortunately, this is precisely the range of radial distances where measurements of wind properties are most difficult to make. On the one hand, the coronal densities have fallen to such small values that spectroscopic or optical instruments (which give us information about the low corona) are not sensitive enough to detect any signal from plasma beyond (perhaps) $(3\text{–}4)R_{\mathrm{sun}}$. On the other hand, the closest approach that a spacecraft has made to the Sun is about $60R_{\mathrm{sun}}$ (Helios). The only way to study the transonic region of the solar wind is by remote sensing: using spacecraft beacons as sources, scintillations of intensity, phase, frequency, and line broadening can be used to derive various properties of the turbulent wind plasma. A great deal of information about properties of the solar wind can be obtained from studying scintillations of various kinds: for reviews, see Mullan and Yakovlev [23] and Yakovlev and Mullan [24].

5.5 The Corona: What Are the Densities?

Now that we know the scale height H_{cor} in the low corona (where HSE applies roughly), we can use the observed brightness of the corona ($I_{\mathrm{cor}}/I_{\mathrm{disk}} \approx 10^{-6}$)

to estimate roughly the density of the material at the base of the corona n_{cb} electrons cm^{-3}.

To do this, let us imagine what happens to photons which emerge from a surface element of area one sq. cm in the solar photosphere during an eclipse of the Sun. These photons stream away from the Sun, mostly along the radial direction, and by the time they have passed through the corona, the total column depth of electrons N_{col} which they have passed by in their 1 sq. cm. column is given by $N_{col} = n_{cb} \times H_{cor} \approx 5 \times 10^9 n_{cb}$.

There is a finite probability χ_{es} that when a photon passes by an electron, the photon will be scattered: the quantity which determines χ_{es} is the so-called Thomson cross-section, $\sigma_e \approx 0.665 \times 10^{-24}$ cm^2. The probability that a photon will scatter after passing through a column of N_{col} electrons cm^{-2} is $\chi_{es} \approx N_{col} \times \sigma_e$, i.e.

$$\chi_{es} \approx 3 \times 10^{-15} \times n_{cb}. \tag{54}$$

Therefore, of the photons which stream outward from the Sun during an eclipse, a fraction χ_{es} will be scattered into our line of sight.

Since the observed intensity of the corona I_{cor} is a few times 10^{-6} times the photospheric intensity I_{disk}, we conclude that χ_{es} has an empirical value of order 3×10^{-6}. Referring to (54) above, we see that n_{cb}, the electron number density at the coronal base, must be of order 10^9 cm^{-3}. This is consistent with the estimates at the end of Sect. 4.

The proton density in the solar wind at Earth orbit can be measured by spacecraft: it lies in the range 5–10 cm^{-3} on average. The mean speed of the solar wind at Earth orbit is also measurable: 300–400 km sec^{-1}. Using these numbers, we can estimate that the Sun loses mass at a rate $\dot{M}_{wind}$ of a few million tons per second as a result of the solar wind. It seems that the value of $\dot{M}_{wind}$ is comparable to $\dot{M}_{nuc}$, the rate at which mass is consumed in the core of the Sun by nuclear reactions. Thus, the corona and the core are both plasmas with temperatures of millions of degrees K, and both are responsible for loss of mass from the Sun. Of course, the physical origin of the loss of mass is completely different in the core of the Sun from what it is in the corona. It is not obvious whether there is a physical reason why $\dot{M}_{nuc}$ should be comparable to $\dot{M}_{wind}$. Whether this is a coincidence or not, mass loss has very little influence on solar evolution: the combined mass loss rates from these very different processes are such that, over the course of the Sun's evolutionary history (some 10^{10} years), the Sun's mass alters by no more than 1 percent.

6 Flares

So far, we have been considering the solar atmosphere in contexts where input of mechanical can be balanced by radiation and/or conduction. In locations where this balance can be achieved, it is meaningful to consider steady solutions for the chromosphere and corona. We have seen that certain properties of these steady states can be estimated with some degree of confidence.

However, radiation and conduction cannot dispose of arbitrarily large inputs of energy. There is a maximum rate of input $\dot{E}_{\rm max}$ which can be "handled" by the plasma in terms of radiation and conduction. For example, in optically thin gas with density $n_{\rm e}$, the maximum rate at which radiation can remove energy is $n_{\rm e}^2 \Phi_{\rm max}$. Now, the numerical value of $\Phi_{\rm max}$ is no more than 10^{-21} ergs cm^3 sec^{-1} (see, e.g. [10]). As a result, in a volume element in the upper chromosphere with $n_{\rm e} \approx 10^{12}$ cm^{-3}, radiation can remove energy at a rate which is at most $R_{\rm max,uc} = 10^3$ ergs cm^{-3} sec^{-1}. And for a volume element in the low corona, where $n_{\rm e}$ is at most 10^{10} cm^{-3}, radiation can dispose of energy at a rate which is at most $R_{\rm max,cor} = 0.1$ ergs cm^{-3} sec^{-1}.

Moreover, as far as conduction is concerned, the thermal conductivity $k_{\rm th} \sim T^{2.5}$ cannot increase indefinitely as a solar loop heats up. Once the temperature rises to a value where the mean free path of electrons λ ($\sim T^2$) becomes as long as the loop length L, the expression for classical conductivity becomes less reliable. If we want to use (16) in these "saturated" conditions, we should at least replace λ with L. This limits the conductivity to $k_{\rm sat} \approx L\, C_{\rm v}\, \rho\, u$.

Now, the fact that material in the solar atmosphere can dispose of only so much energy deposition by radiation and conduction is not "known" to the dynamo deep inside the Sun. That dynamo creates magnetic flux according to its own dynamics, and then leaves it to the atmosphere to dispose of the emergent magnetic energy as best it can. Because of this independence between source and sink, it is inevitable that from time to time, magnetic energy will be released into the solar atmosphere at a rate which exceeds $\dot{E}_{\rm max}$. What happens then? In such conditions, radiation and conduction are temporarily overwhelmed, and the pressure at the flare site rises (for at least a certain interval of time) to high values. Other channel(s) of energy loss must come into operation to relieve the excess pressure. Events in which localized energy releases overwhelm the usual coronal equilibrium ($\dot{E}_{\rm flare} \geq \dot{E}_{\rm max}$) are called "solar flares".

An energy loss channel which comes into play is kinetic energy: coronal gas begins to move in bulk in an attempt to transport energy away from the site of the flare. This bulk flow is a form of convection, although it is distinct from the convection which occurs inside the Sun: in the latter case, buoyancy forces drive the flow, but in the corona, gravity is not responsible for driving flare ejecta. The driving of coronal ejecta originates in the localized high pressure. The flows which develop around certain flares cause blast waves and shock fronts to propagate in the corona, sometimes with enough energy to survive into interplanetary space.

6.1 Flare as a "Reverse Dynamo"

The locations in which flares occur have a characteristic property: they are magnetically complex regions where flux loops of complicated topology are in close proximity to one another. As a result of motions of the foot-points in the sub-photospheric convection zone, there are times when one loop finds itself forced into contact with another one. The resulting magnetic gradients can give rise to very large electric current densities j in localized regions. When j exceeds a certain threshold $j_{\rm crit}$, the plasma quickly becomes turbulent, and the electrical conductivity σ_e suddenly falls by several orders of magnitude. The Joule dissipation rate ($\sim j^2/\sigma_e$) jumps to a high value, and the currents dissipate rapidly. Dissipation of currents corresponds to re-arrangements of the magnetic fields into a state of lower magnetic energy: this lower energy state has a simpler field configuration, and it looks as if the field lines have been "cut and re-connected". The label "reconnection" is given to this process of magnetic energy release. The magnetic energy which "disappears" in reconnection is converted into heat and kinetic energy at the heart of the flare: jets of material are ejected from the reconnection site at speeds of the order V_A, the Alfven speed.

Thus a flare, where magnetic energy is converted to kinetic energy, involves the opposite process to that in a dynamo.

6.2 Rate of Energy Deposition in Reconnection

At what rate does magnetic reconnection dump energy into the flaring solar atmosphere? To answer that, we need to know (i) how much energy is deposited per unit volume at the flare site, and (ii) how rapidly is it released?

As regards (i), we begin by noting that the strengths of magnetic fields in the *corona* can unfortunately not be measured directly. Indirectly, "coronal magnetic fields" of 1–10 G are often cited on the basis of extrapolating solar wind magnetic data back to the Sun. But these are irrelevant in the context of flares: they refer to conditions in coronal holes. Instead, we need to obtain field strengths in active regions. Here, we can rely on radio polarization data, and the coronal fields are strong: 30 to 600 G [25]. In a reconnection process, the magnetic energy density $B^2/8\pi$ ergs cm^{-3} is reduced by a factor ϕ. For order of magnitude purposes, let us suppose $\phi = 0.5$. Then the change in energy density of the fields ΔE_{mag} with $B = 30$–600 G ranges from about 10 ergs cm^{-3} to about 10^4 ergs cm^{-3}.

As regards (ii), when oppositely directed flux tubes are forced into contact over transverse length scales of L_B, the reconnection time-scale is of order $t_{\text{rec}} \approx L_B/v_{\text{rec}}$. The reconnection velocity v_{rec} is a fraction ψ of the local Alfven speed V_A. In the active region sample of Schmeltz et al. [25], the values of V_A range from 3.5×10^8 to 3.7×10^9 cm sec^{-1}. Even if ψ is as small as 0.1, the solar active corona provides us with v_{rec} of order $(0.3$–$3)\times 10^8$ cm sec^{-1}. Since granule motions are responsible for pushing the fields around, L_B may be comparable to granule dimensions ($\sim 10^8$ cm). With these values, we find that t_{rec} may be of order 0.3–3 seconds.

We see that the regions where most energy is released (i.e. where the coronal B is strongest) are also the regions where t_{rec} is shortest. That is, larger total energy releases go hand in hand with faster rates of energy conversion. The combination of large ΔE_{mag} and short t_{rec} in the strong field regions means that $\dot{E}_{\text{flare}}$ is maximum in large flares. Using the numbers given above, we find that magnetic reconnection in solar active regions leads to volumetric energy release rates in the range

$$\dot{E}_{\text{flare}} \approx 3 - 30000 \text{ ergs cm}^{-3} \text{ sec}^{-1}. \tag{55}$$

In both weak and strong field regions, the rate of energy dumping in the flare $\dot{E}_{\rm flare}$ clearly exceeds $R_{\rm max,cor}$, the upper limit on coronal radiative losses. In stronger field regions, the rate of flare dumping may even overwhelm the maximum radiative capacity $R_{\rm max,uc}$ in the upper chromosphere.

We conclude that reconnection of coronal magnetic fields is capable of dumping energy into the solar atmosphere at a rate that is so fast that equilibrium cannot be maintained. We therefore expect that many active regions will be easily able to satisfy the radiative runaway condition, at least in the corona.

6.3 The Transient Nature of a Flare

In a flare, steady state cannot be maintained: flares last only for a finite time. The length of time which a flare lasts depends on the region of the electromagnetic spectrum in which observations are made. In hard X-rays, some flares last for less than 1 second, while others last for several hundred seconds. But one does not observe bursts of hard X-rays lasting for (say) hours or days.

These data suggest that there are definite upper limits on flare durations. It is as if a certain amount of energy is "available" for the flare, and once that is expended, the flare comes to an end. In the context of a "reverse dynamo", one might even imagine that a built-in regulatory mechanism might cause a flare to quench itself after a finite time. For example, in the reconnection scenario, we note that the onset of turbulence depends on having the current density j exceed a threshold $j_{\rm crit}$: once that threshold is exceeded, the electrical conductivity $\sigma_{\rm e}$ becomes very low. This reduction in $\sigma_{\rm e}$ causes the rate of dissipation $(j^2/\sigma_{\rm e})$ to speed up by orders of magnitude. Dissipation causes j to decrease, and eventually, j falls below $j_{\rm crit}$. Then $\sigma_{\rm e}$ reverts to a large value, and dissipation falls essentially to zero. At this point, the flare event will cease. The next event will start at such times as the external forcing process (photospheric motions) set up the appropriate conditions.

6.4 Flare Temperatures

During a flare, with radiation and conduction overwhelmed, convection sets in. That is, material from the flare site begins to flow in bulk. Ejecta are seen to emerge at high speed from the flare site. One attractive aspect of the reconnection scenario of flares is that it provides a natural source for ejecta: models of reconnection predict that jets should emerge from the reconnection site with a speed of order the local Alfven speed. When these jets run into the ambient atmosphere, their kinetic energy is distributed as heat among the ambient ions.

How hot does the flare plasma become? Empirically, the maximum temperatures are found to be typically (2–3) $\times 10^7$ K (see, e.g. [26]). However, the values of $T_{\rm max}$ are not the same in all flares: there is a systematic trend with flare "size", i.e. with overall radiative power. A quantitative classification of flare "sizes" which is in widespread use has been developed based on the peak flux $F_{\rm X,max}$ of X-rays: flares are classified as A, B, C, M, and X (in order of increasing fluxes) based on 10-fold increases in $F_{\rm X,max}$ in the 1–8Å channel on the GOES satellite. A flare belonging to class A has $F_{\mathrm{X,max}} = 10^{-8}$ W m^{-2}, while a class X flare has $F_{\mathrm{X,max}}$ of 10^{-4} W m^{-2}. Empirically, larger flares are observed to contain hotter gas [26]: in a sample of almost 1000 flares, T_{max} was observed to increase linearly with the logarithm of $F_{\mathrm{X,max}}$. We have used the results of Feldman et al. to extract the following relationship:

$$T_{\mathrm{max},6} \approx 46.1 + 5.4 \log F_{\mathrm{X,max}} \tag{56}$$

over the range $-8 \leq \log F_{\mathrm{X,max}} \leq -4$. Here, $T_{\mathrm{max},6}$ is the maximum temperature in units of 10^6 K. (Note that the numerical coefficients in (56) do not agree with those given by Feldman et al. in their expression relating T_{nor} with x: the coefficients in (56) above are in agreement with the data plotted in their Fig. 6.) For the flares reported by Feldman et al., the values of $T_{\mathrm{max},6}$ range from about 5 to about 25. Feldman et al. point out that the relationship in (56) may not be applicable to the very largest flares: for the very largest flares in past solar cycles, where $\log F_{\mathrm{X,max}}$ may have been somewhat in excess of -3, $T_{\mathrm{max},6}$ may have risen to as large as 50.

The fact that flares of larger "size" contain distinctly hotter plasma than flares of smaller "size" means that an X-class flare does *not* consist simply of an agglomeration of many class A flares. This is consistent with the correlation mentioned above between ΔE_{mag} and $\dot{E}_{\mathrm{flare}}$: one expects that the faster the flare energy is released, the hotter the flare material will become.

The slow increase in T_{max} (by ≈ 5) reported by Feldman et al. [26] as $F_{\mathrm{X,max}}$ increases by a much larger factor (10^4) is noteworthy. It suggests that there exists some form of limiting process which imposes strict controls on temperature "runaway". Conduction may be this limiting process. Even in "saturated" conditions (i.e. $\lambda \approx L$), the volumetric rate of conductive energy losses $(\mathrm{d}E/\mathrm{d}t)_{\mathrm{cond}} \approx k_{\mathrm{sat}}\, T/L^2$ is still somewhat sensitive to temperature. Thus, consider a coronal loop of length $L = 10^9$ cm, where number densities are 10^{10} cm^{-3} and the temperature is T_6 million K. In such conditions, we find that the "saturated" $(\mathrm{d}E/\mathrm{d}t)_{\mathrm{cond}}$ has a value of about $C_{\mathrm{sat}}\, T_6^{1.5}$ ergs cm^{-3} sec^{-1}, where the coefficient C_{sat} has a value of order unity. We see that in a flare plasma with $T_6 = 20$–50, $(\mathrm{d}E/\mathrm{d}t)_{\mathrm{cond}}$ is in the range 10^{2-3} ergs cm^{-3} sec^{-1}. Comparing these with the rates of energy release in flares (eq. (55)), we see that conductive energy losses can indeed "handle" flare energy releases (at least up to the median of the range of $\dot{E}_{\mathrm{flare}}$ values in (55)) without allowing temperatures to rise above 20–50 million.

6.5 Flares: Storage of Energy and a Trigger

In the chromosphere and in the quiet corona, where equilibrium between heating and cooling can be maintained, it seems that energy is released from a volume element "immediately", i.e. as fast as it is deposited. But in a flare, it seems that something different may be happening: the energy which is added to the corona is *not* released immediately. Instead, this energy is stored in a reservoir of some kind for a finite time, and then, following the operation of a trigger, the stored energy is suddenly released. Any viable flare theory should account for both storage and trigger.

As regards the storage problem, electric currents provide one possible solution. The process of building up energy in a reservoir in the corona during the pre-flare time period presumably involves the magnetic field. The lowest energy state of the magnetic field (the so-called "potential" field) is one in which electric currents are completely absent. There may well be some fields of this kind in the solar atmosphere, but they surely do not survive for long. After all, the magnetic fields in the Sun have foot-points which are embedded in the convective motions near the photosphere. These turbulent motions act continually on the foot-points of the loops, causing the loop to be twisted and stretched in a complicated manner. As the fields in the loop become more and more stressed, the magnetic energy grows to values which are in excess of the energy of a potential field. In this sense, energy is being stored in the magnetic stresses. The stresses correspond to increasingly large electric currents in the loop.

As regards a trigger, electric currents also provide an attractive possibility: when a current flows in a plasma, it may be subject to a variety of instabilities [27] Each instability has its own threshold in terms of density and temperature: once a certain criterion is violated, the corresponding instability sets in abruptly and rapid current dissipation is the result.

From the point of view of a flare theory, one may think of the pre-flare phase as a time interval during which the currents are indeed growing, but have not yet reached the threshold for instability. Energy is being stored during this phase. The duration of this phase $t_{\mathbf{storage}}$ would depend on (a) how rapidly the stresses are being created, and (b) which instability threshold is the first one to be violated.

6.6 Coronal Heating and Flares: Are They Distinct?

Finally, the fact that $t_{\mathbf{storage}}$ varies from one flare to another leads us to raise the question: what happens in an event where $t_{\mathbf{storage}}$ is shorter than our time resolution? Then we would not classify such an event as a flare: instead, it would appear to us as if the energy were being released "immediately". This is reminiscent of what happens in the process of "heating" the chromosphere and quiet corona. Might the heating of the quiet corona actually consists of a large number of (very) small "flares" (i.e. microflares or nanoflares)? Or is there in fact some fundamental distinction between coronal heating and flaring? These questions have been in the literature since at least 1982 (see, e.g. [28]) but no definitive answer has yet been given.

The problem is partly empirical: if the corona is truly heated by many (very) small flare events, then these events must occur in large numbers. Formally, the requirement is that the number of flares $\mathrm{d}N$ per unit time with energies between E and $E + \mathrm{d}E$ must obey $\mathrm{d}N/\mathrm{d}E \sim E^{-\epsilon}$ where ϵ must be at least as large as 2. To test this requirement, it is crucial to evaluate not only the energies of the microflares (or nanoflares), but also the frequencies with which they occur.

Unfortunately, these tiny events are precisely the ones which are most difficult to observe reliably.

However, the work of Feldman et al. [26] suggests that there might be another way to approach the question. Let us take (56) above, which applies to what we might call *bona fide* flares, and attempt to extend it to microflares and nanoflares. If this is a permissible extrapolation, we should be able to predict what the temperature in the quiet corona might be. To see how well this works, we note that, by definition, a microflare (or nanoflare) is 6 (or 9) orders of magnitude smaller in total energy than the largest solar flare. Now, the largest solar flares reported by Feldman et al. had $F_{\mathrm{X,max}} \approx 10^{-3}$. We might therefore expect that a microflare (or nanoflare) would have $F_{\mathrm{X,max}} \approx 10^{-9}$ (or 10^{-12}). Inserting these into (56), we find that $T_{\mathrm{max,6}}$ is negative! Thus, our attempt to extrapolate the flare relationship (eq. (56)) to very small events leads to a meaningless result. This suggests, but does not prove, that flaring and coronal heating involve distinct processes.

Acknowledgements

I am very grateful to the organizers of the School, especially Dr. J. Valdes-Galicia and Dr. D. Page, for making it possible for me to visit the historic and beautiful city of Guanajuato. The United States-Mexico Foundation for Sciences in acknowledged for financial support during my stay in Mexico.

References

1. J. Christensen-Dalsgaard: Solar model with helium diffusion and settling (1999) (JCD): available online at website http://helios.tuc.noao.edu/teams/models/gong-www.l4b.d.18.html
2. D. J. Mullan and R. K. Ulrich: Ap. J. **331**, 1013 (1988)
3. D. J. Mullan: Ap. J. **337**, 1017 (1989)
4. G. Isaak et al.: in *Advances in Helio- and Asteroseismology*, ed. by J. Christensen-Dalsgaard (Reidel, Dordrecht 1986) pp. 53–57
5. M. Schwarzschild: *Structure and Evolution of the Stars*, (Princeton Univ. Press 1958)
6. F. W. Sears: *Thermodynamics*, 2nd edn. (Addison Wesley, Reading MA 1959), pp. 267-269
7. R. Kurucz: Rosseland mean opacities for solar composition (1992): available online at website http://cfaku5.harvard.edu/OPACITIES/ROSSELAND/kapp00.ross
8. M. Hossain and D. J. Mullan: Ap. J. **416**, 733 (1993)
9. R. F. Stein and A. Nordlund: Ap. J. **499**, 914 (1998)
10. P. Foukal: *Solar Astrophysics*, (Wiley Interscience, New York 1990), p. 116
11. O. Namba and W. E. Diemel: Solar Phys. **7**, 167 (1969)
12. Lord Rayleigh,: Phil. Mag., Ser. 6, **32**, 529 (1916)
13. J. M. Beckers: Solar Phys. **3**, 258 (1968)
14. A. M. Title et al.: Ap. J. **336**, 475 (1989)
15. L. Biermann and R. Lust: in *Stellar Atmospheres* ed. by J. L. Greenstein (Univ. of Chicago Press, Chicago 1960) p. 272

16. Z. Musielak et al.: Ap. J. **423**, 474 (1994)
17. L. Spitzer: *Physics of Fully Ionized Gases*, (Interscience, New York 1962), p. 128
18. D. J. McComas et al.: Geophys. Res. Lett. **25**, 1 (1998): the figure is available online at website http://swoops.lanl.gov/lasco_swoops.html
19. E. N. Parker: *Interplanetary Dynamical Processes*, (Interscience, New York 1963), p. 75
20. R. Esser et al.: Ap. J. Lett. **510**, L67 (1999)
21. I. Cuseri et al.: Ap. J. **514**, 989 (1999)
22. S. R. Cranmer et al.: Ap. J. **518**, 937 (1999)
23. D. J. Mullan and O. I. Yakovlev: Irish Astron. J. **22**, 119 (1995)
24. O. I. Yakovlev and D. J. Mullan: Irish Astron. J. **23**, 7 (1996)
25. J. T. Schmeltz et al.: Ap. J. **434**, 786 (1994)
26. U. Feldman et al.: Ap. J. **460**, 1034 (1996)
27. D. S. Spicer and J. C. Brown: in *The Sun as a Star* ed. by S. Jordan (NASA SP-450 1982), pp. 413–471
28. D. S. Spicer: in: *Activity in Red Dwarf Stars* ed. by P. B. Byrne and M. Rodono (Reidel, Dordrecht 1982), p. 560

Precision Laboratory Measurements in Nuclear Astrophysics

Moshe Gai

Laboratory for Nuclear Science, Department of Physics, U3046 University of Connecticut, 2152 Hillside Rd., Storrs, CT 06269-3046, USA (gai@uconn.edu; http://www.phys.uconn.edu)

Abstract. After reviewing some of the basic concepts, nomenclatures and parametrizations of Astronomy, Astrophysics, Cosmology, and Nuclear Physics, we introduce a few central problems in Nuclear Astrophysics, including the hot-CNO cycle, helium burning and solar neutrinos. We demonstrate that in this new era of Precision Nuclear Astrophysics *Secondary or Radioactive Nuclear Beams* allow for progress.

1 Introduction

In this lecture notes we discuss some aspects of Nuclear Astrophysics and Laboratory measurements of nuclear processes which are of central value for stellar evolution and models of cosmology. These reaction rates are important for several reason. At first they allow us to carry out a quantitative detailed estimate of the formation (and the origin) of the elements; e.g. the origin of ^{11}B *or* ^{19}F . In these cases the understanding of the nuclear processes involved is essential for understanding the origin of these elements. The understanding of the origin of these elements on the other hand, may teach us about exotic processes such as neutrino scattering that may occur in stars and are believed to produce the observed abundances of ^{11}B *and* ^{19}F. More importantly, in most cases details of many astronomical events, such as supernova, are hidden from the eyes of the observer (on earth). In most cases the event is shielded by a large mass and only telltales arrive on earth. Such telltales include neutrinos, or even some form of radiation. One of the most important telltale of an astronomical event are the elements produced by the thermonuclear nucleosynthesis. And in this case it is imperative that we completely understand the nuclear processes so that we can carry out an accurate test of the cosmological or stellar evolution models. In some cases, such as in the solar model, understanding of the nuclear processes in hydrogen burning allow for a test of the standard model of particle physics and a search for phenomena beyond the standard model, such as neutrino masses (neutrino magnetic moment?) or neutrino oscillations. Type 1a supernova on the other hand proved to be a very useful cosmological yard stick allowing for accurate measurements of some of the largest distances of the order of a few Billion Light Years (GLY). Such measurements gave evidence for an accelerating expanding Universe and appear to be one of the most disturbing discovery in Cosmology in recent times. In this case one needs to understand the process of helium burning in a type 1a supernova. In all cases one needs to understand

Nuclear reaction rates at energies which are considerably below where they can be measured in the laboratory, and one needs to develop reliable method(s) for extrapolation to low energies.

In spite of concentrated effort by Nuclear Astrophysicists on both experimental and theoretical sides a number of problems remain unsolved, including specific processes in helium and hydrogen burning. In contrast to many cases in Nuclear Astrophysics in the case of the solar neutrinos and type 1a supernovae, the processes of hydrogen burning and helium burning, respectively, must be measured with high precision of the oredr of 5-10%. These problems are in fact central to the field and must be addressed in order to allow for progress. In these lectures we will address these issues and suggest new experiments and new solutions.

Radioactive Nuclear Beams (RNB) now available at many laboratories around the world have already yielded some solutions to problems of current interest, e.g. in the Hot CNO cycle or hydrogen burning, and appear very promising for extending our knowledge to processes in exploding stars, such as the rp process. We will review in this lectures some of the current and future applications of such secondary (radioactive) beams.

In the first section we will define some scales, classifications of stars, nomenclatures, parameters and parametrization of relevance for nuclear astrophysics. We will then review some of the classical reaction chains in burning processes and discuss traditional laboratory measurements of the relevant nuclear reaction rates. In the later part of the lecture series we will develop new ideas for laboratory measurements of the required rates, mostly carried out in the time reversed fashion. We will demonstrate that by measuring the reaction rates in a time reversed fashion we construct a **"Narrow Band Width Hi Fi Amplifier"** that may allow for a measurement of the small cross sections involved. It is important to test whether in fact we construct a "Hi Fidelity Amplifier", so that we are indeed measuring rates relevant for nuclear astrophysics. These new techniques allow us to tackle some of the oldest open questions in Nuclear Astrophysics including the rate for the $^{12}C(\alpha,\gamma)^{16}O$ reaction of helium burning and the $^{7}Be(p,\gamma)^{8}B$ reaction of importance for the solar neutrino problem.

2 Scales and Classification of Stars

Most stars have been around for long time and thus have reached a state of statistical (hydro-dynamical) equilibrium. Indeed most properties of stars arise from simple hydrodynamical consideration or from the fact that stars are nearly (but not perfect) black body radiators. Some of the most obviously required observational parameters of a star are its distance from the earth and its spectrum of light emission and thus its color.

Early studies by Kepler and scientist of the Newtonian era allowed for accurate measurements of the radii and periods of orbital motion of the various planets, including the earth. In these measurements the appearance of comets were very pivotal and indeed the return of Halley's comet in April of 1759, as reported by Harvard astronomers, was announced as a confirmation of Newton's law of gravity. Ironically, when Halley's comet was late to return and did not show up between September 1758 and early April 1759, as predicted by Edmund Halley using Newton's $1/r^2$ law of gravity, Newton's law of gravity was (prematurely) declared wrong [1] by the "skeptics". It is also worth noting that while the earliest western record of Halley's comet is from AD 66 (that was linked to the destruction of Jerusalem), the Chinese records go back for another 679 years, as shown in Table 1 [2]. From these measurements of radii and periods, it was possible to determine the mass of the sun and planets with high precision; one solar mass $M_\odot = 1.989 \times 10^{30}$ Kg, and $M_E = 3\mu M_\odot$.

Some of the very early measurements (developed around 1838) of the distance of stars from the earth used the parallax method [3]. It was found that the nearest star, Alpha-Centauri visible in the southern hemisphere (a triple star system composed of Alpha-Centauri Proxima, A and B) produced (after corrections for its angle) 1.52 sec of arc of angular displacement, or a parallax of 0.76 arc sec. Knowing the earth average orbit radius = 149.6 MKm = 1 AU (Astronomical Unit), or approximately 8 light minutes, we calculate 1 parsec = 3.086×10^{16} meter, or 3.262 light years (LY). Indeed our closest neighbor is hopelessly far from us, at a distance of approximately 4.2 LY. Modern days (optical) telescopes have an accuracy of the order of 0.01 sec of an arc and with the use of interferometry one can improve the resolution to 0.001 sec of an arc. Hence, the parallax method has a limited use, for stars closer then 1 kpsc. In Fig. 1, taken from Donald Clayton's book [3], we show characteristic distances and structures in our galaxy. Note that the period of rotation of our galaxy is of the order of 100 million years.

Early measurements performed on stars also defined its color index [3], using the response of detectors (photographic plates) with band widths spanning the Ultraviolet, Blue and Visual spectra. The color index is defined as Blue magnitude minus the Visual magnitude. Note the magnitude is roughly proportional to -2.5 log(intensity). Hence, hot stars are characterized by small and in fact negative color index while cold stars have large color index. Astronomers are also able to correlate the color index with the (effective) surface temperature of a star, an extensively used parameter in stellar models. Stars are also characterized by their absorption spectra as O, B, A, F, G, K, and M stars (that can be memorized using a non quotable slogan.

2.1 Classification of Stars

Based on this color index one classify stars using a Hertzsprung-Russell Diagram (after the Danish and American astronomers that developed such diagrams around 1911-1913). In an H-R diagram one plots the Luminosity of a star or the bolometric magnitude (total energy emitted by a star) Vs the surface temperature, or the color index of a star. In Fig. 2 we show such an H-R diagram [3], for star clusters with approximately equal distance to the earth. These stars are believed to be formed within the same time period of approximately 100 million years, which allow for the classification.

Table 1. Chinese records of Halley's Comet [2]

Return	Date	Reign/year	Return	Date	Reign/year
-40	-1057- -1056	The Conquest of Zhou by Wu-Wang	-20	AD 451	Song Yuanjia 28
-39			-19	530	Liang Zhongdatong 2
-38			-18	607	Sui Daye 3
-37			-17	684	Tang Guangzhai 1
-36			-16	760	Qianyuan 3
-35			-15	837	Kaicheng 2
-34	BC 614	Zhou Qing Wang 5	-14	912	Liang Qianhua 2
-33			-13	989	Song Duangong 2
-32	465	Zhou Zhending Wang 3	-12	1066	Zhiping 3
-31			-11	1145	Shaoxing 15
-30			-10	1222	Jiading 15
-29	240	Qin Wang Zheng 7	-9	1301	Yuan Dade 5
-28	162	Han Wen Di Houyuan 2	-8	1378	Ming Hongwu 11
-27	86	Wu Di Houyan 2	-7	1456	Jingtai 7
-26	11	Yuanyan 2	-6	1531	Jiajing 10
-25	AD 65	Yongping 8	-5	1607	Wanli 35
-24	141	Yonghe 6	-4	1682	Qing Kangxi 21
-23	218	Jiunan 23	-3	1759	Qianlong 24
-22	295	Jin Yuankang 5	-2	1835	Daoguang 15
-21	374	Ningkang 2	-1	1910	Xuantong 2

Stars that reside on the heavy diagonal curve are referred to as main sequence stars [4]. For the main sequence stars we find the brightest star to be with highest surface temperature and of blue color. The main sequence stars spend most of their life burning hydrogen and acquire mass that is related to their luminosity: $L = const \times M^\nu,\ with\ \nu\ =\ 3.5\ to\ 4.0$. Stellar evolution is most adequately described on an H-R diagram, and for example the sun after consuming most of its hydrogen fuel will contract its core while expanding its outer layers (to a radius that will include the earth). The contraction at first raises the luminosity and then the sun will expand and redden, or move up and then to the right in an H-R diagram. At a later stage the helium fuel will ignited in the contracted core and the sun will move to the left on (an asymptotic branch on) the H-R diagram. At the end of helium burning the sun will further contract to a white dwarf, see below, and reside (forever) at the lower bottom left of the H-R diagram. For main sequence stars the luminosity is given by Planck's law $L = 4\pi R^2 \sigma T_e^4$, (we introduced here the effective temperature - T_e, since stars do not have a well defined surface and are not perfect black body radiators). Hence one can determine with limited accuracy the relative radii of main sequence stars. One common way of measuring the radii of stars is by using the interferometry method and the Hanbury-Brown Twiss (HBT) effect [5]. In this measurement one measures the pair correlation function (in momentum space) of two photons and

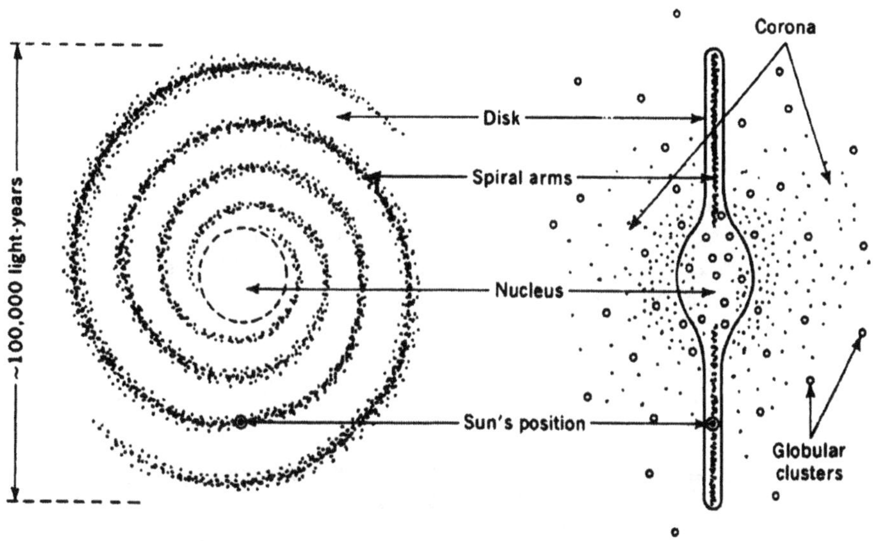

Fig. 1. Scales of our galaxy [3]

by using boson's statistics one relates the correlation width to the radius (of the source of incoherent photons). For example the sun's radius (not measured via the HBT effect) is $R_\odot = 6.9598 \times 10^8$ meters, or 0.69598 MKm, and $R_E = 1\% R_\odot$. While the average sun's density is $\rho_\odot = 1.4\ g/cm^3$ ($\rho_E = 5.5\ g/cm^3$), the central density of the sun is considerably larger, and it was determined (from stellar hydrodynamical models) to be $\rho = 158\ g/cm^3$ with a central temperature of 15.7 MK [8,7,6]. Indeed the gravitational contraction of the sun's central core allows for the heating of the core (from a surface temperature of approximately 6,000 K) and the ignition of the hydrogen burning that occurs at temperatures of a few MK. The convective zone of the sun terminates at a radius of approximately 74% at a temperature of approximately 2 MK and density of approximately 0.12 g/cm^3.

Above and to the right of the main sequence stars we find the **Red Giant** stars that are characterized by large luminosity and therefore they are easily seen in the sky. This class includes only a small number of stars, a few percent of the known stars. The redness of these stars arises from their large radii and they represent a star in its later stages of evolution, after it consumed its hydrogen fuel in the core and consist mainly of helium. The subgiant are believed to be stars that expand their outer envelope while contracting their helium cores, leading to the burning of helium. The horizontal branch stars, on the other hand, are believed to be at various stages of helium burning. The supergiant stars are believed to be stars at the advance stages of their stellar evolution and perhaps approaching the end of their energy-generating life.

In the lower left corner of the H-R diagram we find the **white dwarfs** representing approximately 10% of known stars, which are very dense stars of mass comparable to a solar mass, with considerably smaller radii, comparable to the earth radius. Due to the small surface area these stars have large surface temperature (blue color) in order to allow them to radiate their luminosity. These group composes of the universe's cemetery of stars that are inactive and simply radiate their pressure energy. The white dwarfs are so dense that the electron degeneracy keeps them from collapsing [9], hence can not have a mass larger then approximately $1.4 M_\odot$, the Chandrasekhar limit, beyond which the electron degeneracy can not overcome the gravitational collapse. Such massive stars (or cores of massive stars) collapse to a neutron star or a black hole under their own gravitational pressure.

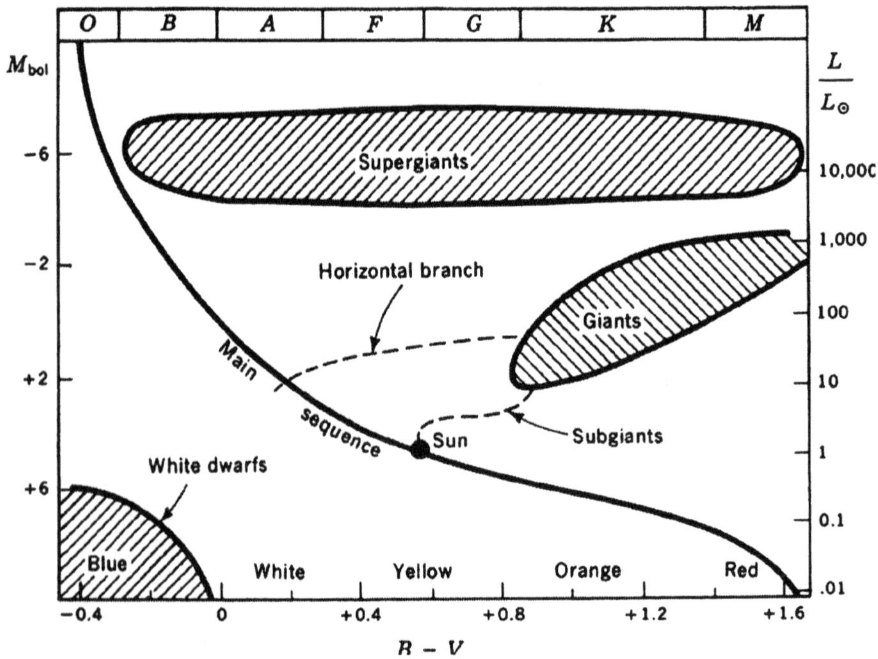

Fig. 2. Hertzsprung-Russell Diagram [3]

Cluster of stars are found very far from the sun, see Fig. 1, and they may contain as many as $10^5 - 10^7$ stars in spherical distribution with a radius in the range of approximately 10 parsec (globular cluster), other clusters include only a few stars. Based on the characteristics of these stars in an H-R diagram it is believed that the age of stars in the globular cluster is of the order of 14 ± 3 billion years (GY) [10], or as old as the universe itself (minus 1 GY). Within this cluster we find a relatively young class with blue giants as the most luminous, called population I, and an older class with red giants as the most luminous members, called population II. The galactic cluster Pleiades (or

Subaru in Japanese) includes its brightest star of blue color, and the M3 globular cluster that includes some 10^5 stars, include its brightest star of red color.

2.2 Age of Stars

First generation stars are stars that coalesced from the primordial dust that includes approximately 24% helium and 76% hydrogen with traces of lithium. Some of these stars are small enough, and have not evolved and are still burning hydrogen, others already converted to dwarfs. For example the sun (which is not a first generation star) has burned its hydrogen fuel for the last 4.6 Billion years and will do so for approximately 5 more Billion years. Such first generation stars are expected to have very small amount of elements heavier then carbon (some times generically referred to as metals). Thus one defines the metalicity of a star, to be the ratio of its iron (or some time oxygen) to hydrogen content, divided by the metalicity of the sun. This ratio (denoted by square brackets) is usually expressed in a log scale, typically varying between -4 and 0. Stars with metalicity of -3 to -4 are believed to be primordial with ages in the range of 10 to 15 Billion years. It should be emphasized that while the metalicity of a star is measured on its surface, one needs to know the core metalicity and hence one needs to introduce a stellar atmospheric model(s), and thus these data in some cases are model dependent.

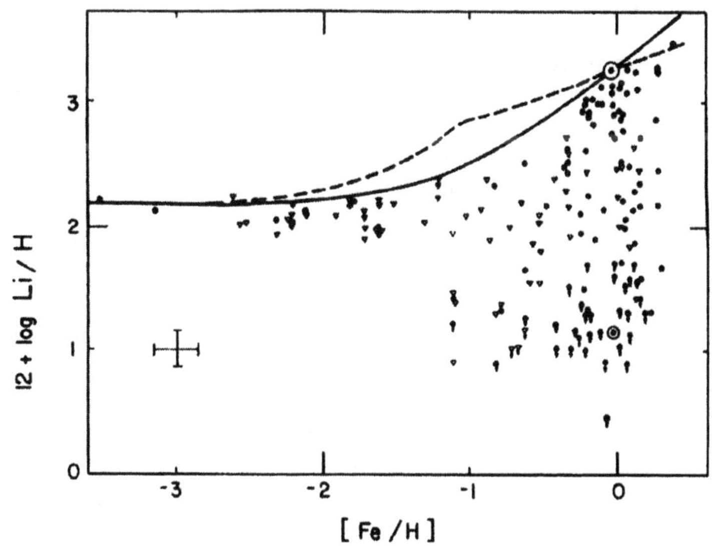

Fig. 3. Lithium abundance Vs metalicity [13]

One of the key questions in cosmology is the primordial abundance of the elements, produced during the epoch of primordial nucleosynthesis [11,12]. In Fig.

3 we show the abundance of Li Vs metalicity [13]. Lithium is a very volatile element, since it readily reacts with low energy protons via the $^7Li + p \to \alpha + \alpha$ reaction, that we depict as $^7Li(p,\alpha)\alpha$. Consequently younger stars show large fluctuations in Li abundance. Fig. 3 includes stars with metalicity as low as -3 and -3.5, and we extrapolate the Li primordial abundance in the range of 10^{-10} *to* 10^{-9}, relative to hydrogen. For younger stars we expect to have an additional lithium roughly proportional to the metalicity. This addition arise from the fact that the inter-stellar gas, from which younger stars coalesce, includes more produced lithium as it exist for longer times. The destruction of lithium in the stellar environment would yield to a depletion in younger stars. Indeed, the measurements of primordial lithium abundance and D and ^3He (first measured on the moon, with the Apollo mission [14]) were very pivotal for confirming Big Bang Nucleosynthesis [11,12]. In Fig. 4 we show the predicted primordial nucleosynthesis. In these calculations [11] one varies the ratio of photon density to baryon density to yield the observed primordial abundances. And with the knowledge of the photon density, from measurements of the cosmic microwave background, one deduces the baryon density that appears to be less then 10% of the (critical) density required to close the universe. Indeed if one assumes the universe is critically closed (as suggested in inflation models), big bang nucleosynthesis provides some of the strongest evidence for the existence of dark matter in the universe.

2.3 Distances to Far Away Stars and Galaxies

One of the most useful (optical) method to determine the distances of far away stars is with the use of Cepheid Variable stars [3]. These stars undergo periodic variations, which are not necessarily sinusoidal. Sir Edington demonstrated that the pulsation of the Cepheid Variables are due to the transfer of thermal energy of the star to mechanical energy that leads to pulsation [3]. As a consequence the star's period of pulsation is directly related to its mass and its luminosity. Hence, if one measures the apparent luminosity of a Cepheid Variable star (on earth) and its period of pulsation one can infer the distance to the Cepheid Variable and thus the distance of its galactic host.

Type 1a supernova proved to be a very useful and accurate tool in measuring large distances [15]. Type 1a supernova occur in a white dwarf Red Giant binary star system with the white dwarf accumulating hydrogen from the upper stratosphere of the Red Giant. When the white dwarf mass reaches the Chandrasekar limit of 1.4 solar mass, see below, it collapses under its own gravity. The time period of the buildup of light in the light curve of a type 1a supernova (see later Fig. 16), is directly related to its predicted luminosity, and thus measuring the shape of the light curve for type 1a supernova yield its expected luminosity that can be compared to the observed luminosity to yield the distance to the type 1a supernova and its host galaxy. Such modern measurements let us to conclude that the Universe expansion rate is accelerating in recent cosmological times.

One of the first uses of the Cepheid variable stars as an astronomical Yard Stick were carried out by Edwin Hubble with the 100 inch telescope at Mt.

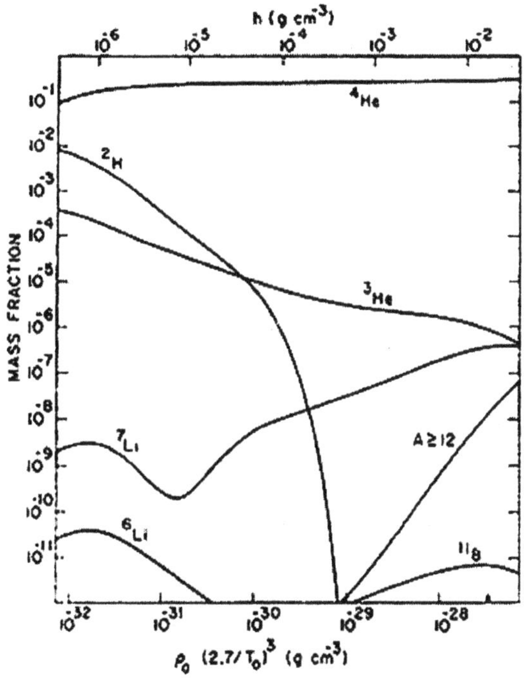

Fig. 4. Big Bang Nucleosynthesis [11]

Wilson observatory near Pasadena, California, during the 1920's [16]. Hubble was able to identify Cepheid Variable stars at a distance of 930,000 LY, and thus well outside our galaxy, of diameter of approximately 100,000 LY (see Fig. 1). Hubble was able to show that these "Faint Nebula" correspond to galaxies different then ours. These nebula were catalogued by Charles Messier in 1781 (with the Crab Nebula being M1) to allow observer to distinguish such objects from comets. Hubble's faint nebula are identified as the M31 (galaxy in Andromeda) and M33 spiral galaxies. Today the distance to the Andromeda nebula is estimated to be over 2 MLY.

Hubble later noticed that the known lines of emission from Hydrogen, Oxygen, Calcium, etc. from stars within the same galaxy are shifted toward the red, which he correctly interpreted as a Doppler shift. Hubble plotted the relative velocity (deduced from the accurate measurement of the redshift) Vs the distance, as he could best estimate using the Cepheid variable. Hubble's original discovery, see Fig. 5, was of a linear relationship between the velocities and distances $v = H \times R$, where H is Hubble's constant. Hubble's measurements of distance were less accurate then possible today, and they yielded the Hubble constant $H = 500\ Km/sec/Mpc$, as can be extracted from Fig. 5.

One of the immediate consequences of Hubble's observation was that it gave credence to the Big Bang hypothesis, developed as one possible solution to Ein-

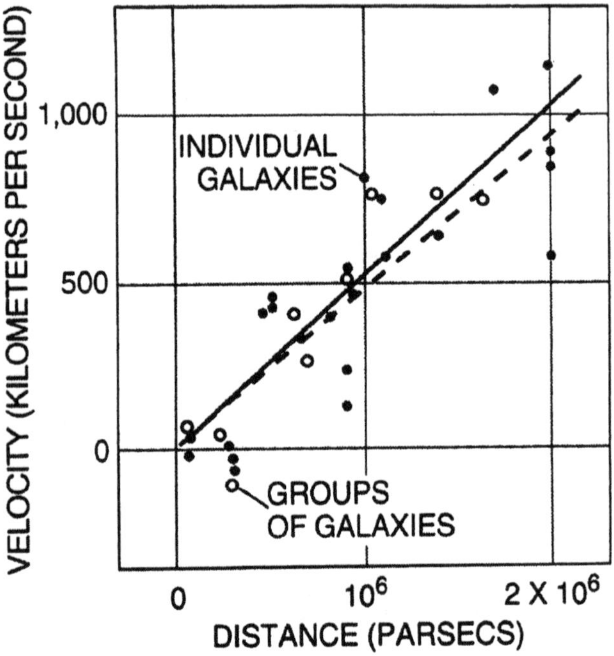

Fig. 5. Hubble's observation of $v = H \times R$ [16]

stein general relativity, in the early 20's by Alexandre Friedman in Russia, and George Lemaitre in Belgium. Details of Big Bang nucleosynthesis were later worked out by de-Sitter and Gamow in the 40's. Incidentally it is suggested that the name Big Bang was coined by Sir Fred Hoyle as a way of ridiculing suggestion of George Lemaitre who referred to his own theory as the theory of the primeval atom. It is ironic that Hoyle who to this date still prefers the steady state theory (and rejects the Big Bang theory), got to name the rival theory. Unfortunately Hubble's determination of H requires a universe that is only 2 Billion Years old. At that time one already knew that the earth and the solar system are much older, of the order of 4.6 Billion years, and the Big Bang theory was rejected. Today due to more accurate determinations of distances (e.g. a factor of 2 change for M31, see above), we believe that the Hubble constant is between 50 to 100 Km/sec/Mpc, with the most probable value at 65, corresponding to a universe between 20 to 10 Billion years old with the most probable age of approximately 14 GY.

The expanding universe allow us to define the Fractional Red-Shift, as the fractional stretching of wave length: $Z = \Delta\lambda/\lambda_0$, with the Doppler shift $\omega = \omega_0\gamma(1 + \beta cos\theta)$, and use it to parametrize distances to far away galaxies, radio galaxies, and quasars (young galaxies at the time of formation, mostly composed of gas with luminosity mostly composed of radio electromagnetic radiation). Measurements of these far away objects allow us to look back to the instant of

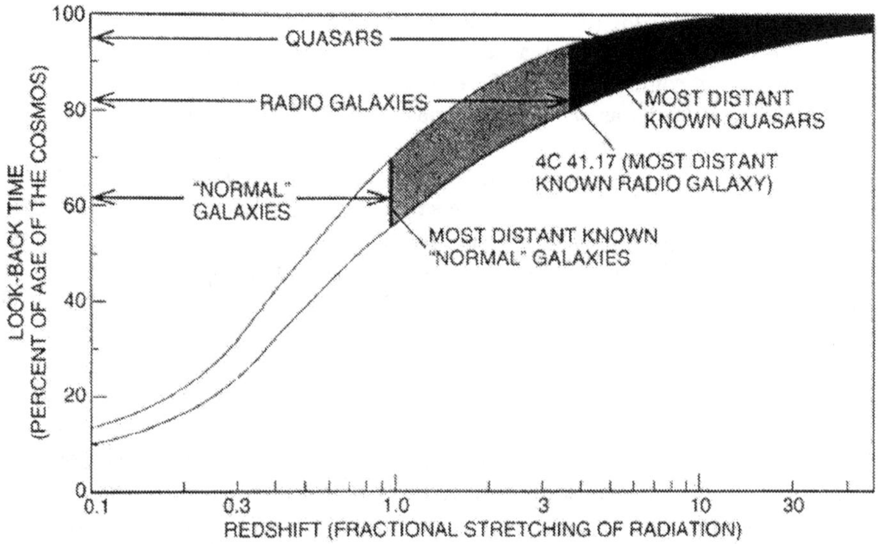

Fig. 6. Look back time VS Red Shift

the big bang as shown in Fig. 6, with the oldest known quasar at 5-10% of the age of the universe and the oldest radio galaxy (4C 41.17) at 10-15% of the age of the universe.

2.4 The Big Bang Theory

The big bang theory most vividly confirmed today by the COBE satellite mission, received one of its first strong confirmations in the work of Arno A. Penzias and Robert W. Wilson in 1964 [17], where they discovered the isotropic emission of microwave radiation from a (cosmological) source at a temperature of approximately 2.7 K. Penzias and Wilson were careful to characterize this thermal source, but did not point to its origin from the expanding universe of the big bang theory. This possibility was in fact pointed out by Peebles and Dicke. Indeed in a preceding paper [18] they demonstrated, that Penzias and Wilson measured the expected microwave remnants of the big bang. In fact Penzias and Wilson who originally only designed an antenna for microwave communication with satellites, first interpreted the continuous hum they detected from all directions of space as arising from pigeon dropping on their antenna.

According to the big bang theory when the Universe was just below 10 μsec, its temperature was approximately 200 MeV and hence the universe was composed of quarks and gluons solely. At that time a phase transition from the quark gluon plasma to hadron matter occurred. At the age of approximately 1 sec the universe had a temperature of approximately 1 MeV (approximately 10 GK) and then the inverse beta decay process of the neutron to the proton stopped, hence the ratio of neutrons to protons was fixed by the temperature and the mass difference following Boltzmann law. At approximately 100 sec after the big bang when the temperature was approximately 100 keV the epoch of big bang nucleosynthesis commenced [11,12] and it lasted for a few minutes. During big bang nucleosynthesis as we believe today all the available neutrons were captured to form helium, with a well understood helium fraction of $Y_p = 24\%$. At approximately 300,000 years when the temperature was approximately 10 eV, atoms emerged and accidentaly in the same time the universe became transparent to radiation (decoupling). At this point the universe changed its character from being radiation dominated to matter dominated. As the universe expands all characteristic dimensions expand and radiations from a source of 1 eV (10,000 K) temperature, were redshifted to larger wave lengths of today's observed microwave radiation, corresponding to a source at 2.7 K. Galaxies and stars we believe, first formed when the universe was approximately 1 Billion years old.

Recent speculations suggest that big bang nucleosynthesis may have in fact occurred in an inhomogeneous inflationary universe [19–25]. This model predicts a low but significantly different, abundance of heavy elements as for example produced in the rapid neutron capture process of supernova [26]. The observation of such heavy elements could test whether the quark-gluon to hadron phase transition is in fact first order. The nature of this phase transition is of great concern for lattice QCD calculations [27] and indeed for understanding QCD. Recent observation of the abundance of ^9Be [28] and ^{11}B, at first appeared promising for this model but subsequent analysis showed that the recently observed abundances (in particular the ratio $^{11}B/^9Be$) are consistent with spallation reaction [29] and no definitive evidence was found for these models of inhomogeneous big bang nucleosynthesis and the standard model of big bang nucleosynthesis prevails.

3 Reaction Theory, Methods and Applications

The gravitational pressure in a stellar environment leads to heating of the nuclear fuel. When hydrogen is heated to a temperature in excess of a few MK, it is ignited and nuclear fusion takes place. The fusion of light elements is the source of energy in stars and indeed the most readily available source of energy in the universe today. These fusion reactions aside from "driving stars" are also the origin of the elements heavier then helium. The understanding of thermonuclear processes entails a complete understanding of nuclear reactions as measured in the laboratory, as reviewed by Willie Fowler [30,31] and the seminal papers of FCZ I [33] and FCZ II [34]. A review of these reactions can also be found in Rolfs and Rodney's book [4]. Usually one would like to know if a reaction rate is sufficiently important to generate the energetic of a stellar environment, and whether it favorably competes with other possible reactions and decays. In this case one needs to define the reaction time scale, or the inverse of its rate, as we discuss below.

Consider two particles a and X, contained in a form of an ideal gas, interacting with each other. The reaction rate per unit volume (r) is given by:

$$r_{aX} = \sigma\, J_a\, N_X \tag{1}$$

where σ is the energy dependent cross section, N is the concentration of particles per unit volume and J is the flux, $J_a = v\, N_a$, hence:

$$r_{aX} = \sigma\, v\, N_a\, N_X \tag{2}$$

In a star the relative velocities of a and X are distributed in a Maxwell-Boltzmann distribution $\phi(v)$, with $\int \phi(v)dv = 1$, and the total thermonuclear reaction rate is given by:

$$r_{aX} = N_a N_X \int v\, \sigma(v)\, \phi(v)\, \mathrm{d}v = N_a N_X <\sigma\, v> \tag{3}$$

and for identical particle we need to introduce a further trivial correction (to avoid double counting):

$$N_a N_X \longrightarrow \frac{N_a N_X}{(1+\delta_{aX})}$$

We define $\lambda \equiv <\sigma v>$, the reaction per unit particle, and (3) becomes:

$$r_{aX} = \lambda_{aX}\, \frac{N_a N_X}{(1+\delta_{aX})} \tag{5}$$

We are usually interested in characteristic time scale for the reaction and the time that it takes to remove particle a from the stellar ensemble, which we may want to compare for example to the beta decay lifetime of that particle a, and we define:

$$\begin{aligned}\left(\frac{\partial N_X}{\partial t}\right)_a &= -\frac{N_X}{\tau_a(X)} \\ &= -r_{aX}\end{aligned} \tag{6a}$$

hence:

$$\tau_a(X) = \frac{1}{\lambda_{aX}\, N_a} = \frac{1}{<\sigma v>\, N_a} \tag{6}$$

with the correct units of inverse time. Note that the symmetry factor $(1+\delta_{aX})$ is now on both sides of (6a) and it drops out. In order to know if a reaction rate competes favorably with a decay rate, one needs to evaluate equ. 6 for that reaction. It is customary to include Avogadro's number, $\mathcal{N}_A = 6.023\times10^{23}$, in (6) and one usually quotes: $\mathcal{N}_A <\sigma v> N_a$ with N_a specified in units of moles/volume.

Inserting the Maxwellian into the integral in (6), we find:

$$<\sigma v> = 4\pi \left(\frac{\mu}{2\pi kT}\right)^{3/2} \int v^3\, \sigma(v)\, \mathrm{e}^{-\mu v^2/2kT}\, \mathrm{d}v \tag{7}$$

with μ the reduced mass.

Equations (6) and (7) include information from both nuclear physics (the cross section - σ) and stellar models (the stellar density and temperature). The integral is then the meeting ground for nuclear physics and stellar physics. Clearly the goal of nuclear astrophysics is to evaluate reactions rates relevant to stellar environments, by use of theoretical or experimental methods.

3.1 The S-Factor

The nuclear cross section (of s-wave interacting particles) was parametrized by Bethe and Gamow based on general principles of quantum mechanics, as:

$$\sigma(E) = \frac{S(E)}{E} \times \mathrm{e}^{-2\pi\eta} \tag{8}$$

where η is the Sommerfeld parameter

$$\eta = \frac{Z_1 Z_2 e^2}{\hbar v} \ .$$

It is immediately clear that 1/E originates from the π/k^2 that appears in the expression for the cross section in reaction theory, and the exponent accounts for the penetration factor of the two charged particles Z_1 and Z_2.

3.2 Non-resonant Reactions

The reaction cross section and S-factor for the $^{12}C(p,\gamma)^{13}N$ are shown in Fig. 7. The region of interest for stellar environment around 30 keV, (the CNO cycle, see below) is indicated in the figure, and it lies just beyond the region where experiments are still possible (i.e. cross section of 20 pbarns). It is clear that one needs to extrapolate to the energy region of stellar conditions and the extrapolation of the S-factor allows for additional confidence, since the S-factor varies more slowly. Inserting (8) to (7), we find:

$$\lambda = <\sigma v> = \left(\frac{8}{\mu\pi}\right)^{1/2} \times \frac{1}{(kT)^{3/2}} \int S(E) \times \mathrm{e}^{-\left[\frac{E}{kT} + \frac{b}{E^{1/2}}\right]} \mathrm{d}E \tag{9}$$

where we abbreviated $b = \pi Z_1 Z_2 \alpha (2\mu c^2)^{1/2}$, and $\alpha \equiv e^2/\hbar c$. And for a constant S-factor (S_0) we have:

$$\lambda = \left(\frac{8}{\mu\pi}\right)^{1/2} \times \frac{S_0}{(kT)^{3/2}} \int E^{-\left[\frac{E}{kT} + \frac{b}{E^{1/2}}\right]} \mathrm{d}E \tag{10}$$

In this case one finds that the convolution of the Maxwellian and cross section leads to a window of most efficient energy (E_0) for burning, the Gamow window, as shown in Fig. 8.

$$E_0 = \left(\frac{bkT}{2}\right)^{3/2} = 1.22\ (Z_1^2 Z_2^2 \times A \times T_6^2)^{1/3}\ \text{keV} \tag{11}$$

where T_6 is the temperature in million degrees Kelvin, and $A = A1A2/(A1+A2)$. For example helium burning in Red Giants occurs at 200 MK ($T_6 = 200$), hence the reaction $^{12}C(\alpha,\gamma)^{16}O$ needs to be measured at energies of approximately 315 keV where helium burning is most effective. As we shall see below this turned out to be a formidable task.

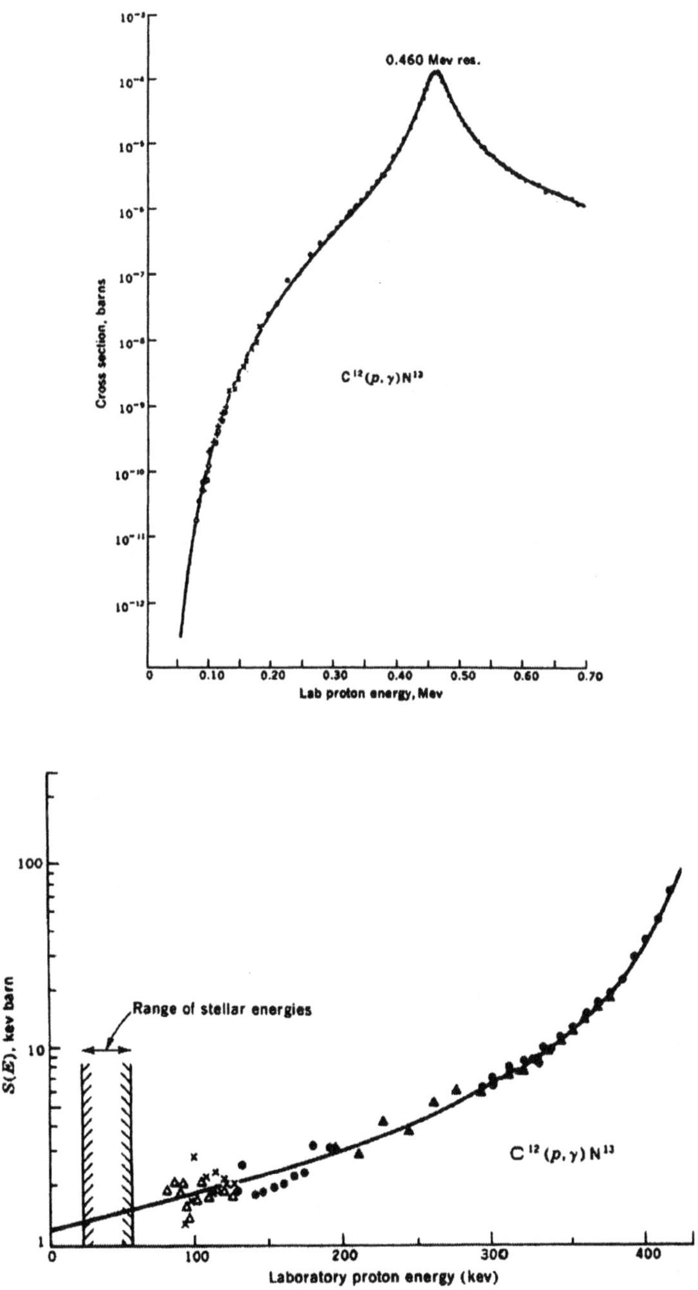

Fig. 7. Cross section (top) and S-factor (bottom) for the ${}^{12}C(p,\gamma){}^{13}N$ reaction [3]

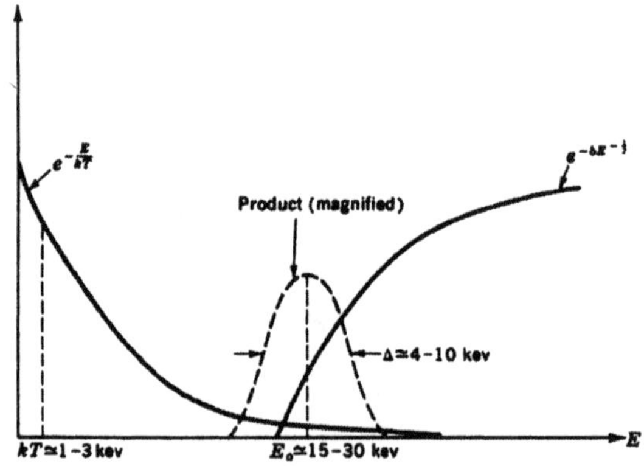

Fig. 8. The Gamow window predicted by Eqs. (10) and (11) [3] for the $^{12}C(p,\gamma)^{13}N$ reaction

3.3 Resonant Reactions

In many cases the relevant reaction rates are governed by a resonant nuclear state. Such states are either low lying and with narrow width, or higher lying but acquire large width ($\Gamma > 0.1E_r$), and can contribute significantly to the reaction rate at low energies. For narrow states the contribution to the thermonuclear rate arises from the tail (at higher temperatures) of the Boltzmann distribution and for the broad state the thermonuclear rate arises from the tail (at lower energies) of the resonant state.

The cross section for an interaction of particles $a+b$, of spins J_1 and J_2, in a relative angular momentum state ℓ via an isolated low lying (at E_r close to threshold) nuclear state, is given by the Breit-Wigner formula:

$$\sigma_{r,\ell}(a,b) = \frac{2\ell+1}{(2J_1+1)(2J_2+1)} \times \frac{\pi}{k^2} \times \frac{\Gamma_a \Gamma_b}{(E-E_r)^2 + (\frac{\Gamma}{2})^2} \tag{12}$$

with Γ_i the partial widths and the total width $\Gamma = \sum_i \Gamma_i$. The partial widths are given by, $\Gamma_i = 2P_\ell \gamma_i^2$, where γ_i^2 is the reduced width and P_ℓ the penetrability factor, e.g. the Coulomb penetrability:

$$P_\ell = \frac{kR}{G_\ell^2 + F_\ell^2}$$

Note that since the pentrability factor is a property of the exterior region (of the nuclear potential), the results are independent of the choice of the penetration factor (e.g. WKB penetration Vs. Coulomb penetration factor), but strongly depends on the choice for nuclear radii. One defines the statistical factor

$$\omega = \frac{(2J+1)}{(2J_1+1)(2J_2+1)} \, .$$

Note that for most reaction rates the total width are exhausted by one particle width (with other particle widths being energy forbidden), and the radiation width is much smaller. However the radiation width is the one that allows the resonant state to de-excite to the ground state and hence form the element of interest, as we illustrate in Fig. 9. Cross sections of astrophysical interest are small for energies near the resonant energy due to the smallness of the radiation width ($\Gamma_\gamma/\Gamma \approx 10^{-5} - 10^{-7}$), and at energies below resonance they are hindered by the penetrability. It is immediately clear that the cross section is most directly affected by the energy of the nuclear state, the lower the resonant energy the larger the cross section. And the width of the state is second in this hierarchy.

For a broad state we can write the S-factor:

$$S(E) = \frac{\pi\hbar^2}{2\mu}\,\omega\,\frac{\Gamma_1\Gamma_2}{(E-E_r)\,+\,\Gamma^2/2}\;\mathrm{e}^{2\pi\eta} \tag{13}$$

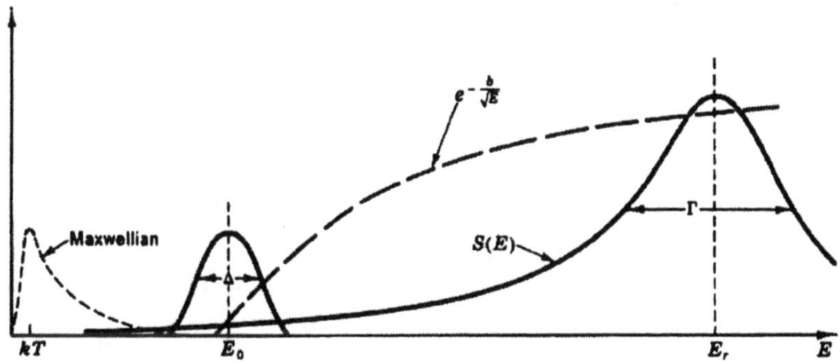

Fig. 9. Nuclear reaction governed by a (broad) nuclear state [3]

For computational purpose it is useful to remember that $\hbar c = 197.33$ MeV fm and $\alpha = 1/137.03$, hence $e^2 = 1.44$ MeV fm. In many cases the evaluation of thermonuclear reaction rates is reduced to accurate measurements of the partial widths that appear in (12) [35]. When measurements are not possible one attempts to calculate the S-factor with the use of standard nuclear models such as sum-rules [26,36], and the thermonuclear cross section could be calculated using (9) or (10). We see here that the investigation of the properties of nuclear states, i.e. Nuclear Structure Studies, are directly linked to Nuclear Astrophysics.

For a narrow state we drive the thermonuclear rate:

$$\lambda_i = \hbar^2 \left(\frac{2\pi}{\mu kT}\right)^{3/2} \omega_i\,\frac{\Gamma_1\Gamma_2}{\Gamma}\;\mathrm{e}^{-\frac{E_r}{kT}} \tag{14}$$

And it is immediately clear that the reaction is possible due to the tail of the Boltzmann factor, or the last term on the right hand side of (14).

In the following we shall use concepts that we developed in the above discussion of reaction theory to discuss particular processes in stars.

The PP Chain(s): Stars in the main sequence like our sun, spend most of their energy generating lifetime burning hydrogen. The burning of hydrogen occurs in several chains known as the PP chains [3,6], as we list below:

$$\begin{array}{ll} {}^1H + {}^1H \rightarrow {}^2D + e^+ + \nu_e & \\ {}^2D + {}^1H \rightarrow {}^3He + \gamma & \underline{\text{PPI}} \\ {}^3He + {}^3He \rightarrow {}^4He + 2\,{}^1H & \end{array}$$

$$\begin{array}{ll} {}^3He + {}^4He \rightarrow {}^7Be + \gamma & \\ {}^7Be + e^- \rightarrow {}^7Li + \nu_e & \underline{\text{PPII}} \\ {}^7Li + {}^1H \rightarrow 2\,{}^4He & \end{array}$$

$$\begin{array}{ll} {}^7Be + {}^1H \rightarrow {}^8B + \gamma & \\ {}^8B \rightarrow {}^8Be + e^+ + \nu_e & \underline{\text{PPIII}} \\ {}^8Be \rightarrow 2\,{}^4He & \end{array}$$

The PPI chain is the main source of energy in the sun. It amounts to the fusion of 4 protons to a helium nucleus with the release of approximately 25 MeV energy, and 95% of the photon luminosity is produced within 0.36 $M_\odot$ and $R < 0.21\ R_\odot$. The majority of the energy is released in a form of heat (kinetic energy of alpha-particles) and radiation (gamma rays), and some energy (2.3%) is released in the form of solar neutrino's. The reaction rate is dictated by the weak interaction process, the first process in the PPI chain, with a calculated S-factor $S(0) = 3.78 \pm 0.15 \times 10^{-22}$ keV-barn and linear term coefficient $\frac{dS}{dE} = 4.2 \times 10^{-24}$ barn. Inserting this S-factor and T = 15 MK, with the solar density of $\rho = 150\ \mathrm{g/cm}^3$ and $X_{He} = X_H = 0.5$, in (9) we derive a reaction time, $\lambda^{-1} = 10\ GY$, i.e. the expected lifetime of the sun. Using available luminosities (i.e. available beams and targets) we expect in the laboratory at energies of astrophysical interest, an approximate rate of one p + p interaction per year, which is clearly non measurable. However, this rate is considered to be reliable (within ±1%) as it is extracted from known weak interaction rates such as the neutron lifetime. We also note that the PPI neutrino luminosity (see above) is directly calculable from the total luminosity of the sun and thus the PPI neutrino flux is considered to be estimated with great certainty.

The burning of hydrogen release a large flux of neutrino's and with the knowledge of the various branching ratio's and reaction rates we derive [6,37] for the standard solar model the neutrino flux as shown in Fig. 10.

The Solar Neutrino Problem: Attempts to measure solar neutrino's were carried out over the last two decades [6]. The detection of solar neutrinos is expressed in terms of the SNU, the Solar Neutrino Unit, which is the product of the calculated characteristic solar neutrino flux (in units of $\mathrm{cm}^{-2}\ \mathrm{sec}^{-1}$) times the theoretical cross section for neutrino interaction in the detector (in units of

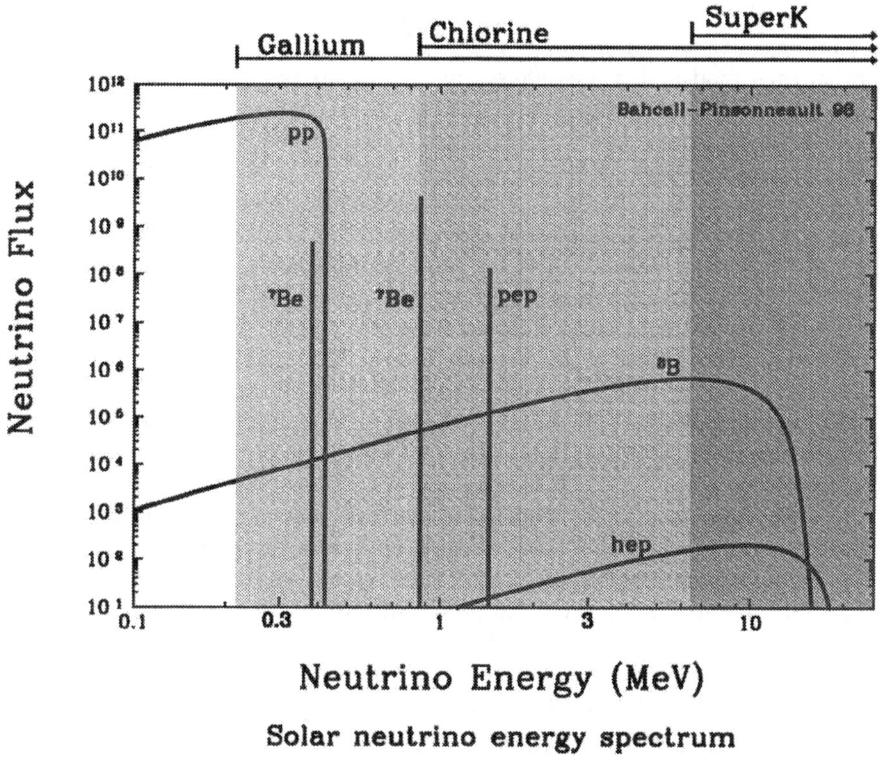

Fig. 10. Predicted Solar neutrinos fluxes [6]

cm^2). Hence the SNU is in units of rate, events per target atom, per second, and is chosen for convenience equal to 10^{-36} sec^{-1}. For a detector with 10^{31} atoms, one SNU yields one interaction per day. This counting rate is characteristic of solar neutrino detectors.

The first neutrino detector was constructed over three decades ago in the Homestake mine, by Raymond Davis Jr. [6] and it includes 10^5 gallons of the cleaning agent carbon tetra chloride. In this detector neutrino's with energies above 800 keV (threshold) yield the reaction:

$$\nu_e \ + \ ^{37}Cl \ \rightarrow \ e^- \ + \ ^{37}Ar \tag{15}$$

and the nobel gas argon is collected by bubbling helium through the tank and collecting it in chemical adsorbers. The decay products of the activity of ^{37}Ar are counted in a proportional counter in a low background environment. For this chlorine detector one predicts using Bahcall-Uhlrich Standard Solar Model and Bahcall-Pinsonneault SSM [37,38] 7.9 $\pm$ 2.6 $SNU's$. The observed rate of the Chlorine detector is averaged over the last three decades of counting to yield the quoted rate of: 2.2 $\pm$ 0.2 SNU, or for example 28% $\pm$ 3% of the rate predicted by Bahcall and Uhlrich [37]. The B-U model was later improved by Bahcall and Pinsonneault [38] and predict yet higher ^8B neutrino flux. As we discuss below other solar models that use different nuclear inputs (see below the S_{17} problem) predict a smaller neutrino fluxes [39–41].

The Kamiokande proton decay detector (Kamiokande I) was outfitted for a solar neutrino detector (Kamiokande II) and was used since January 1987. It detects the Cerenkov radiation of electrons elastically scattered by the neutrino's and it had at first a threshold of approximately 9.5, which was later improved to 7.5 MeV. This detector observed after approximately 1000 days of counting 46% $\pm$ 5%($stat$) $\pm$ 6%($syst$) of Bahcall's predicted flux [42]. Kamiokande III which consists of improved detection systems with larger efficiency for light collection using extensive mirrors and water considerably cleaner with less Rn contaminant(s) and hence smaller threshold (7 MeV), in operation since 1991 [43], reported 56% $\pm$ 6%($stat$) $\pm$ 6%($syst$) of Bahcall's predicted flux. The average of six years of counting with the Kamioka detector amounts to 50% $\pm$ 4%($stat$) $\pm$ 6($syst$) of the B-U Standard Solar Model [43] and 66% of the SSM of Turck-Chieze and Lopez [40,41]. For over two years a new SuperKamiokande detector came to operation and is taking data with threshold as low as 5 MeV and it quoted the rate [44] of 35.8% $+0.9-0.8$%($stat$) $+1.4-1.0$%($syst$) of the Bahcall and Pinsonneault [38] predicted rate.

More recently results from gallium detectors were reported. These detectors have a very low threshold, of 233 keV, and hence detect the neutrinos of the PPI chain, that extends to approximately 400 keV. In fact the detection of the PPI neutrino's constitute the first direct evidence that the sun burns hydrogen as its primary source of energy. The (updated) SAGE collaboration reported [45] 70 $\pm$ 20 SNU's and the GALLEX collaboration [46] (updated) rate is: 79$\pm$ 10($stat$) $\pm$ 7($syst$), compared to the expected rate of 132 $+20$ -17 $SNU's$. The PPI neutrino's contribute most of the predicted rate for Ga detectors (approximately 55%) and for PPI neutrino's all theoretical predictions are within a reasonable agreement of each other, and for example Turch-Chieze predicts 125 $\pm$ 7 SNU expected Ga detection rate.

The Sudbury Neutrino Observatory (SNO) detector [47,48] became operational in 1999 [49]. This detector uses 1000 tons of heavy water and is expected to have a much improved performance, as well as detect a variety of additional neutrino processes such as neutral current interactions, and would also serve as a neutrino spectrometer.

The most popular theoretical interpretation of the hindrance of the solar neutrino flux, by approximately a factor of 2, is the neutrino flavor oscillation induced by a density dependent resonance effect, known as the MSW effect [50,51]. We however note that in order to reconcile all the currently available data in one theoretical frame, one requires additional energy dependence of the resonance process with 1 MeV neutrinos maximaly oscilating.

The CNO Cycle: In 1939 in a seminal paper delivered in a meeting at Washington DC, Hans Bethe proposed that stars slightly more massive then the sun

($M > 2M_{\odot}$, but with temperatures smaller then 100 MK, may generate their energy more efficiently by burning hydrogen with the help of carbon (i.e. carbon is acting as a catalyst), now known as the CNO cycle. The main branch of the CNO cycle:

$$^{12}C(p,\gamma)^{13}N(\beta^+)^{13}C(P,\gamma)^{14}N(p,\gamma)^{15}O(\beta^+)^{15}N(p,\alpha)^{12}C \tag{16}$$

$$^{12}C(p,\gamma)^{13}N(p,\gamma)^{14}O(\beta^+)^{14}N(p,\gamma)^{15}O(\beta^+)^{15}N(p,\alpha)^{12}C \tag{16a}$$

We note that indeed in the CNO process (16), like in the PP chain, four protons were used to produce a helium nucleus, with the production of fusion energy and the emission of electron neutrino's. In addition the star will now have carbon and nitrogen isotopes at various concentrations due to this cycle. For stars of core temperature larger then 17 MK [7] the CNO cycle provides a more efficient energy source and indeed these stars generate a large fraction of their energy through the CNO cycle as shown in Fig. 11.

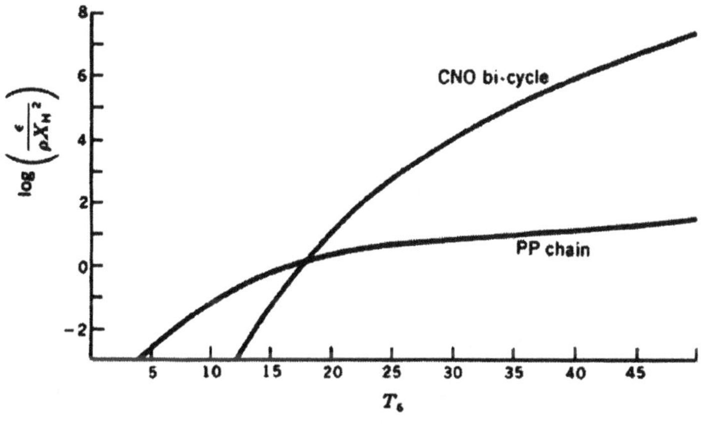

Fig. 11. The CNO - PP transition [3]

The Hot CNO Cycle: The beta decay lifetime of ^{13}N is 863 sec and of ^{15}O is 176.3 sec. The lifetime of ^{13}N is slow enough to allow for a different branch of the CNO cycle to develope, see equ. 16a. Clearly if the temperatures and densities rises, such as in explosive hydrogen stellar environments, it should be possible to reach a point where the $^{13}N(p,\gamma)^{14}O$ reaction rate is fast enough that it could favorably compete with the slow beta decay of ^{13}N, leading to the hot-CNO cycle (16a). This rate is given by (6),

$$\frac{1}{<\sigma v> N_{13}} < 863 \text{ sec} ,$$

and the conditions are related to the reaction cross section, density and temperatures. One then clearly needs to know the cross section for the reaction

$^{13}N(p,\gamma)^{14}O$ at low energies, in order to determine the stellar conditions (density and temperature) where stars may break into the hot CNO cycle. This reaction is governed by the 1^- state at 5.17 MeV in ^{14}O, as shown in Fig. 12.

The hot CNO cycle is found in hydrogen rich environments, at large temperatures and densities, usually involving a binary star system(s) such as Novae etc., hence further capture of protons and alpha-particles on elements from the hot CNO cycle may allow for break out of the hot CNO cycle and into the rp process [52]. In this case the production of ^{19}Ne via the $^{15}O(\alpha,\gamma)^{19}Ne$ reaction, and various related branches of the hot-CNO cycle, play a major role. These processes may in fact produce yet heavier elements, such as ^{22}Ne and elements as heavy as mass 60 nuclei, however we will not cover in this lecture notes these processes.

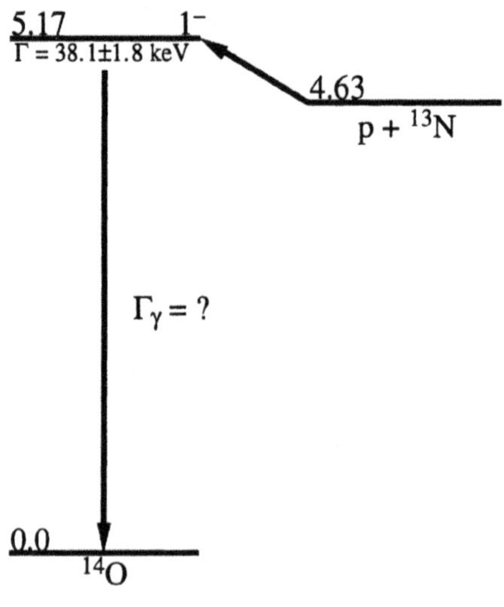

Fig. 12. Nuclear states in ^{14}O relevant for the hot-CNO cycle

Nucleosynthesis in Massive Stars: As stars consume their hydrogen fuel in the core, now composed mainly of helium, it contracts, raising its temperature and density. For example, in 25 solar masses stars the hydrogen burning last for 7 Million years. At temperatures of the order of 200 MK [4], the burning of helium sets in. The first reaction to occur is the $\alpha + \alpha \rightarrow {}^8Be$ due to the short lifetime of ^8Be this reaction yield a small concentration of ^8Be nuclei in the star. However, this reaction is very crucial as a stepping stone for the next reaction that is loosely described as the three alpha-capture process:

$$^8Be(\alpha,\gamma)^{12}C \tag{17}$$

The formation of small concentration of ^8Be, allows for a larger phase space for the triple alpha-capture reaction to occur. This reaction was originally proposed by Fred Hoyle, as a solution for bridging the gap over the mass 5 and 8, where no stable elements exist, and therefore the production of heavier elements. In fact the triple alpha capture reaction is governed by the excited 0^+ state in ^{12}C at 7.654 MeV, as shown in Fig. 13. This state was predicted by Fred Hoyle prior to its discovery (by Fred Hoyle and others) at the Kellog radiation lab [30]. One loosely refers to this 0^+ state as the reason for our existence, since without this state the universe will have a lot less carbon and indeed a lot less heavy elements, needed for life. Extensive studies of properties of this state by nuclear spectroscopist allow us to determine the triple alpha-capture rate using (14). The triple alpha process is in fact accurately known to better then 10%. A possible alternative to the formation of ^{12}C was suggested via the hot pp cycle [53]: the reaction chain $^7Be(\alpha,\gamma)^{11}C(p,\gamma)^{12}N(\beta^-)^{12}C$.

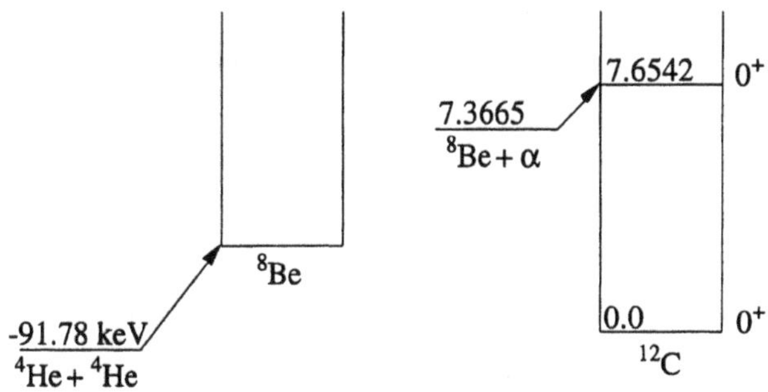

Fig. 13. Nuclear levels in ^{12}C and ^8Be, relevant for the triple alpha-particle capture reaction

At the same temperature range (200 MK), the produced ^{12}C nuclei can undergo subsequent alpha-particle capture to form ^{16}O:

$$^{12}C(\alpha,\gamma)^{16}O \tag{18}$$

Unlike the triple alpha-capture reaction this reaction occurs in the continuum, as shown in Fig. 14. This reaction is governed by the quantum mechanical interference of the tail of the bound 1^- state at 7.12 MeV (the ghost state) and the tail of the quasi-bound 1^- state at 9.63 MeV, in ^{16}O. As we shall see in section 4 of this lecture notes, these effects eluded measurements of the S-factor of $^{12}C(\alpha,\gamma)^{16}O$ reaction for the last two decades, in spite of repeated attempts. More recently great hopes were introduced for solving this problem [54] via beta-delayed alpha-particle emission of ^{16}N [55–57], but this hopes appear to have faded away [58–60], as we discuss below. Helium burning lasts for approximately 500,000 years in a 25 solar mass star [4], and occurs at temperatures of approximately 200 MK. As we shall see below the outcome of helium burning (i.e. the

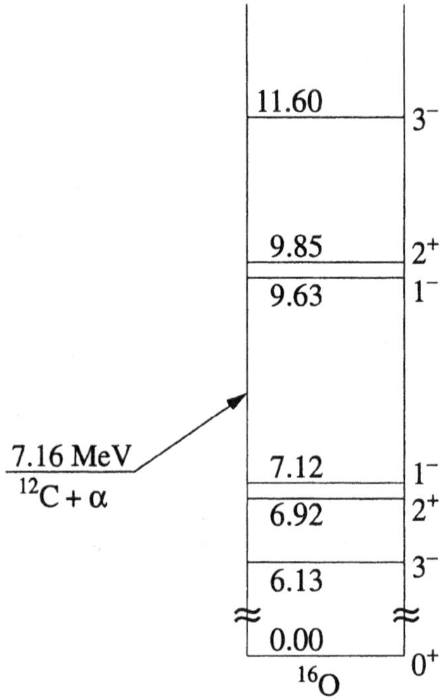

Fig. 14. Nuclear levels in ^{16}O relevant for helium burning

ratio between carbon and Oxygen) is very crucial for determining the final fate of a massive star prior to its supernova collapse.

Stars of masses smaller then approximately 8 solar masses will complete their energy generating life cycle at the helium burning cycle. They will be composed mainly of carbon and oxygen and contract to a dwarf lying forever on the left bottom corner of the H-R diagram. More massive stars at the end of helium burning, commence carbon burning at a temperature of approximately 600-900 MK. Carbon burning lasts for 600 years in 25 solar masses stars [4]. The main reaction process in carbon burning is the $^{12}C(^{12}C,\alpha)^{20}Ne$ reaction, but elements such as ^{23}Na, and some ^{24}Mg are also produced. At temperatures of approximately 1.5 BK (or approximately 150 keV) the tail of the Boltzmann distribution allows for the photo- disintegration of ^{20}Ne, with an alpha-particle threshold as low as 4.73 MeV. This reaction $^{20}Ne(\gamma,\alpha)^{16}O$ serves as a source of alpha-particle which are then captured on ^{20}Ne to form ^{24}Mg and ^{28}Si. The neon burning cycle lasts for 1 year in a 25 solar masses stars. These alpha-particles could also react with ^{22}Ne, as suggested by Icko Iben [61], to yield neutron flux via the $^{22}Ne(\alpha,n)^{25}Mg$ reaction and give rise to the slow capture of neutrons and the production of the heavy elements via the s-process. At this point the core is rich with oxygen, and it contracts further and the burning of oxygen commence at a temperature of 2 BK, mainly via the reaction $^{16}O(^{16}O,\alpha)^{28}Si$, with the additional production of elements of sulfur and potassium. The oxygen burning period lasts for approximately 6 months in a 25 solar masses star [62]. At temperatures of approximately 3 BK a very brief (one day or so) cycle of the burning of silicon commence. In this burning period elements in the iron group are produced. These elements can not be further burned as they are the most bound (with binding energy per nucleon of the order of 8 MeV), and they represent the ashes of the stellar fire. The star now resemble the onion like structure shown in Fig. 15.

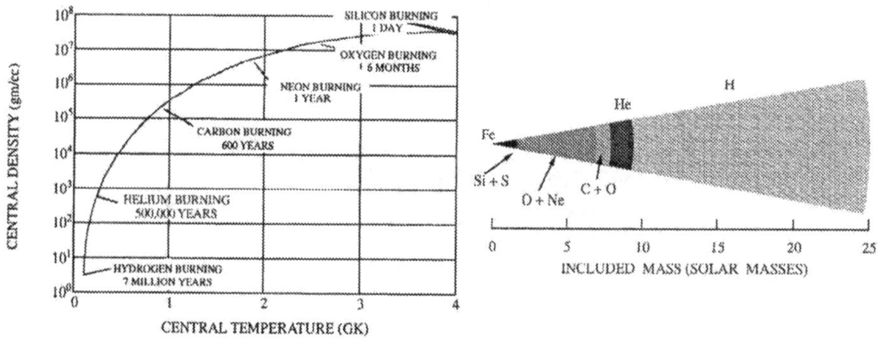

Fig. 15. Burning stages and onion-like structure of a $25M_\odot$ star prior to its supernova explosion [4,62]

As the inactive iron core aggregates mass it reaches the Chandrasekar limit (close to 1.4 solar mass) and it collapses under its own gravitational pressure, leading to the most spectacular event of a supernova. During a supernova the electrons are energetic enough to undergo electron capture by the nuclei and all protons are transposed to neutrons, releasing the gravitational binding energy (of the order of $\frac{3}{5}\frac{GM^2}{R} \approx 3 \times 10^{53}$ ergs) mostly in the form of neutrino's of approximately 10 MeV (and temperature of approximately 100 GK). As the core is now composed of compressed nuclear matter (several times denser then nuclei), it is black to neutrino's (i.e. absorbs the neutrino's) and a neutrino bubble is formed for approximately 10 sec, creating an outward push of the remnants of the star. This outward push is believed by some to create the explosion of a type II supernova. During this explosion many processes occur, including the rapid neutron capture (r process) that forms the heavier elements of total mass of approximately $M \approx 2\% M_\odot$.

The supernova explosion ejects into the inter-stellar medium its ashes from which at a later time "solar systems" are formed. Indeed the death of one star yields the birth of another. At the center of the explosion we find a remnant neutron star or a black hole, depending on the outcome of helium burning.

One of the early records of supernova was provided by Chinese astronomers from July 4th 1054 AD [4]. That explosion left behind a cloud known as the Crab Nebula. Additional observation were made by Ticho Braha and later by

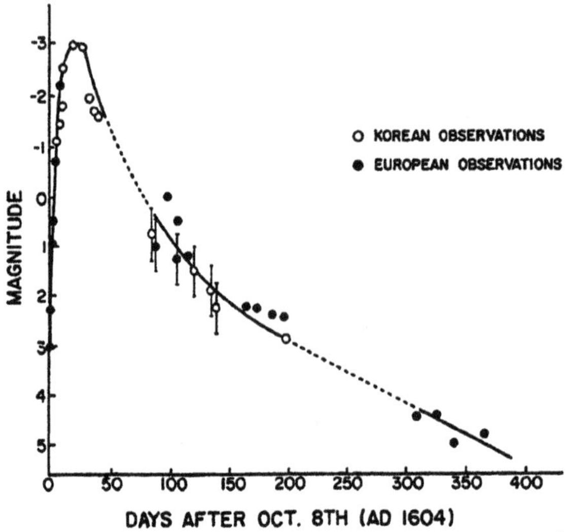

Fig. 16. Light curves obtained from western and eastern historical records, indicating a type I supernova [63,64]

his student Kepler. These include a supernova explosion on October 8, 1604 AD in the constellation Ophiuchus, shown in Fig. 16 [63,64] and one in 1667 AD in the constellation Cassiopeia A. Some speculate that the star of Beth-Lechem may correspond to a supernova explosion that occurred in the year 3 AD. More recent explosions, supernova 1987A and 1993J allowed for a more detailed examination of the nucleosynthesis as well as the observation(s) of neutrino's from such explosions.

It is clear from Fig. 15, that if in the process of helium burning mostly oxygen is formed, the star will be able to take a shorter route to the supernova explosion. In fact if the carbon to oxygen ratio at the end of helium burning in a 25 solar masses star, is smaller then approximately 15% [65], the star will skip the carbon and neon burning and directly proceed to the oxygen burning. In Fig. 17 we show the results of the neon burning as a function of the S-factor for the $^{12}C(\alpha,\gamma)^{16}O$ reaction [65], and clearly for a cross section of the $^{12}C(\alpha,\gamma)^{16}O$ reaction that is twice the accepted value [31,32] (but not 1.7 the accepted value), a 25 solar masses star will not produce ^{20}Ne, and the carbon burning is essentially turned off. This indeed will change the thermodynamics and structure of the core of the progenitor star and in fact such an oxygen rich star is more likely to collapse into a black hole [65] while carbon rich progenitor stars is more likely to leave behind a neutron star. Hence one needs to know the carbon to oxygen ratio at the end of helium burning (with an accuracy of the order of 15%) to understand the fate of a dying star and the heavy elements it produces.

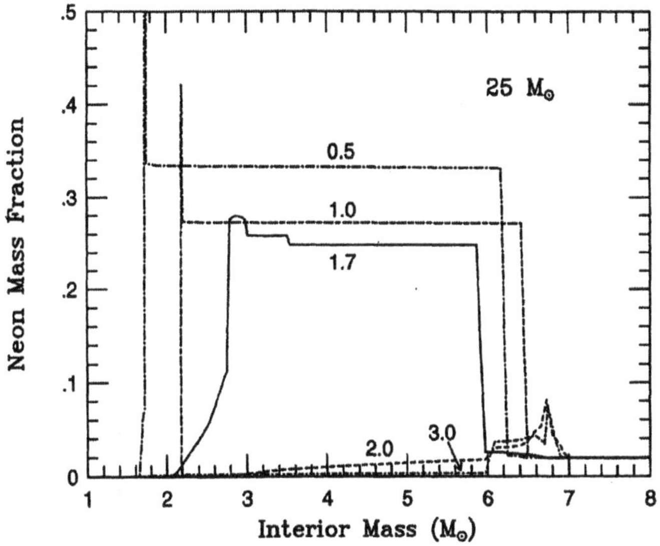

Fig. 17. Neon Formation; the turning off of carbon burning (at twice the [31] accepted value for the $^{12}C(\alpha,\gamma)^{16}O$ reaction), is evident by a small production of neon [65]

Since the triple alpha-particle capture reaction: $^{8}Be(\alpha,\gamma)^{12}C$ is very well understood, see above, one must measure the cross section of the $^{12}C(\alpha,\gamma)^{16}O$ reaction with high accuracy of the order of 15% or better. Unfortunately as we discuss in the next chapter this task was not possible over the last two decades using conventional techniques and initial hopes spured by the measurement of the beta-delayed alpha-particle emission of ^{16}N [55–57], did not materialize either [58–60].

4 Central Problems in Nuclear Astrophysics

In this chapter we review some of the central problems of nuclear astrophysics. We review the difficulties encountered and in some cases suggest that radioactive beams could be used to solve these critical problems of nuclear astrophysics.

4.1 The ^{8}B Solar Neutrino's and the $^{7}Be(p,\gamma)^{8}B$ Reaction

The predicted PPI solar neutrino flux is NOT sensitive to the details of the weak interaction nuclear process and only depends on knowledge of the luminosity of the sun, 1.37 kW/m^2 at 1 AU, and $L_\odot = 3.86 \times 10^{33}\ erg\ sec^{-1}$. This conclusion is due to the fact that the kinematics of hydrogen burning in the PPI chain requires that approximately 2.5% of the solar luminosity is radiated with neutrinos. The flux of the ^{8}B solar neutrino's, composing 75% of those detected by Ray Davis' chlorine detector, and 100% of the Kamiokande detector and also the SNO detector, on the other hand is very sensitive to the details of the nuclear inputs and in particular to the $^7Be(p,\gamma)^8B$ reaction, as well as the exact solar model including opacities and central temperatures.

The accepted value of the S-factor used by Bahcall and Uhlrich [37] for the $^7Be(p,\gamma)^8B$ reaction at zero energy is, $S_{17} = 24.3$ eV-barn. The more recent value adopted by Bahcall and Pinsonneault [38] is 22.4 eV-b. Turck-Chieze adopted the value measured by Filippone of 20.9 eV-b [39]. This small value is one of the most significant differences between her SSM and Bahcall's SSM. The value of S_{17} was studied in details by Barker and Spear [66] and Jonson, Kolbe, Koonin and Langanke [67]. Barker and Spear point out to problems in the value of normalization used for the $^7Be(p,\gamma)^8B$ studies, i.e. the $^7Li(d,p)^8Li$ reaction. They discuss the evolution of the value of the $^7Li(d,p)^8Li$ reaction cross section measured on the 770 keV resonance, as well as other factors and suggest the very low value of $S_{17} = 17$ eV-b, or approximately a 30% reduction in S_{17} as compared to the value adopted by Bahcall and Uhlrich, as shown in Fig. 18. This would imply a reduction of 30% in the expected ^8B solar neutrino flux, indeed a large decrease. Johnson et al. point out to some discrepancies between

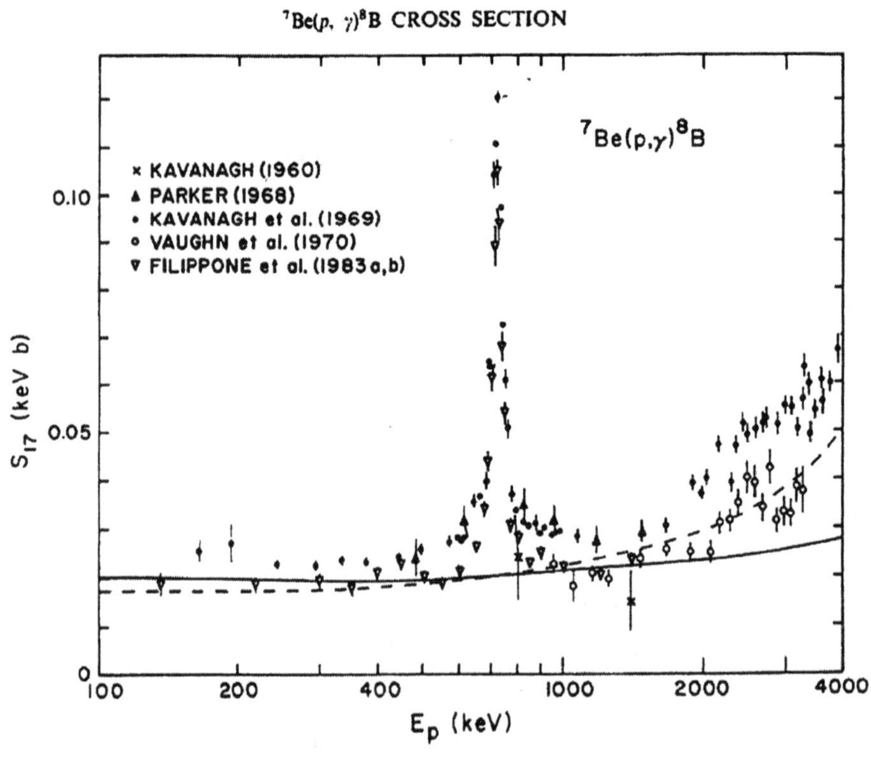

Fig. 18. The extrapolated S_{17} factor of Barker and Spear, who first suggested a low value of $S_{17}(0)$ of 17 eV-b [66]

data obtained by Filippone et al. [68] and the unpublished data of Kavanagh et al. [69]. Johnson et al. [67] adopt the value of $S_{17} = 22.4$ eV-b, as adopted by Bahcall and Pinsonneault but 8% below the value accepted by Bahcall and Uhlrich, as shown in Fig. 19.

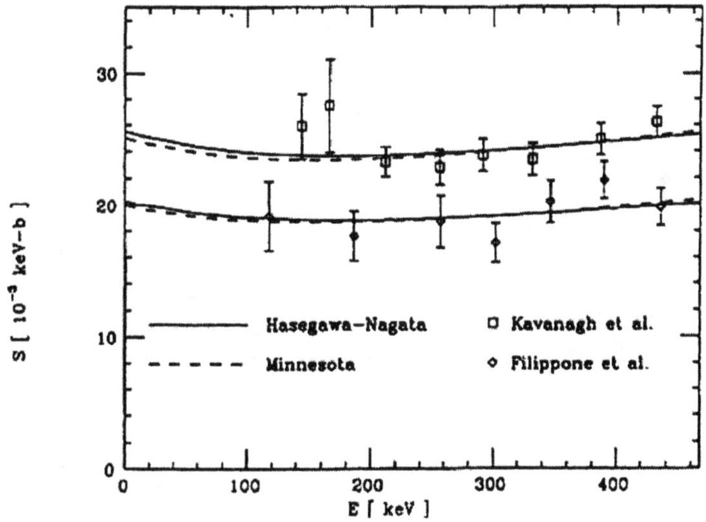

Fig. 19. Comparison of the measurement of Filippone [68] and Kavanagh [69]. And the S_{17} factor extracted by Johnson et al [67]

In Fig. 20 we show the ratio of the cross sections measured by Filippone et al. [68], Kavanagh et al. [69], Parker [70], and Vaughn et al. [71]. The data of Parker and Kavanagh et al. are in agreement with each other, as are the data of Filippone et al. and Vaughn et al. The two data sets are also in good agreement on the energy dependence of the two cross sections. However as shown in Fig. 20 the two data sets are in disagreement by approximately 35% on the absolute value of the cross section.

In a recent review of Solar fusion cross section [72] in a workshop in the INT at Seattle the cross section of the $^7Li(d,p)^8Li$ and S_{17} were reviewed with a re-eavluation of $\sigma_{dp} = 147 \pm 11$ [72–74] and $S_{17}(0) = 19 + 4 - 2$ eV-b. More recent direct measurements with a ^7Be radioactive [75,76] agree with the lower value adopted by the Seattle workshop [72]. A new ^7Be radioactive target produced at TRIUMF [77] allows for yet another mesuement with ^7Be radioactive target, and in the next chapter we discuss the most important experiment with accelerated ^7Be beams.

The importance of the $^7Be(p,\gamma)^8B$ reaction for the evaluation of the ^8B solar neutrino flux calls for a continued interest and additional accurate measurements of the $^7Be(p,\gamma)^8B$ reaction, and in particular measurements that can distinguish between the two absolute values of the cross sections, see Fig. 20, are very much needed. In the next chapter we discuss an interesting new approach with a measure of success success, at attacking this problem with ^8B radioactive beams and the use of a new technique involving the Coulomb Dissociation (Primakoff) process.

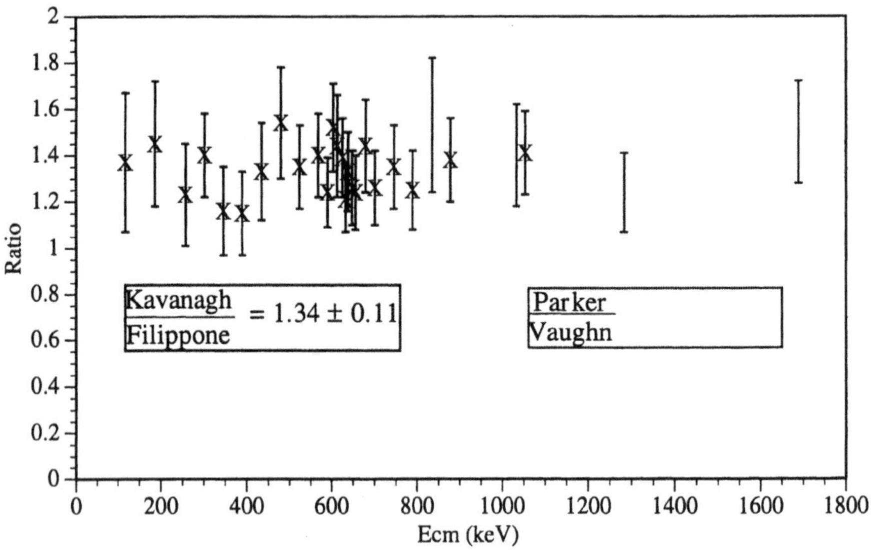

Fig. 20. The ratio of the cross sections for $^7Be(p,\gamma)^8B$ measured by Kavanagh et al. [69] and Parker [70] Vs Filippone et al. [68] and Vaughn et al. [71]

4.2 Extrapolation of S_{17} to Solar Energies

The discrepancy in the measured absolute value of the cross section of the $^7Be(p,\gamma)^8B$ reaction is clearly disturbing and as we show later it is quite possibly best addressed with a ^7Be radioactive beam and a hydrogen target, allowing for a direct measurement of the beam-target luminosity. However, additional uncertainty exists in the theoretical extrapolation of the measured cross section to solar energies (approximately 20 keV). A few theoretical studies suggest an extrapolation procedure that is accurate to approximately $\pm1\%$ [78]. Without discussing these rather strong statements we consider a similar situation that haunted Nuclear Astrophysics a few years back– the S-factor of the $d(d,\gamma)^4He$ reaction. It was assumed that in this case d-waves dominate and no nuclear structure effects should play a role at very low energy, as low as 100 keV. Much in the same way, it is stated today that s-waves dominate the $^7Be(p,\gamma)^8B$ reaction and we do not expect nuclear structure effects to play a role at low energies in the $^7Be(p,\gamma)^8B$ reaction. In Fig. 21 we show Fowler's extrapolated d-wave

S-factor that is a mere factor of 32 smaller than measured, due to a small non d-wave component in the d + d interaction [79]. A small nuclear structure effect, namely the d-wave component of the ground state of ^4He, gives rise to a change by a factor of 32 in the predicted astrophysical S-factor. Similarly we may ask whether a small non s-wave component in the low energy interaction of p + ^7Be could alter the extrapolated $S_{17}(0)$ value by more than one percent. A measurement of $S_{17}(0)$ with an accuracy of $\pm 5\%$ mandates that the cross section be measured at low energies, as low as possible, so as to also test the extrapolation procedures [78].

4.3 The Hot CNO Cycle and the $^{13}N(p,\gamma)^{14}O$ Reaction

As we discuss in section 3.3.4, the value of the cross section of the $^{13}N(p,\gamma)^{14}O$ reaction at low energies is governed by the 1^- state at 5.17 MeV in ^{14}O, see Fig. 12. Hence an indirect measurement of the cross section could be carried out by measuring its partial width. The knowledge of the energy of the state [80], its total width [81] and its partial radiative width, or branching ratio for gamma decay [35], should allow for determination of the cross section, see (12) and (14). This determination turned out to be a formidable task [82–84]. In Fig. 22 we show the radiative width extracted in these experiments [35,82–84] where it is deduced from a measurement of the branching ratio for the 5.17 MeV gamma decay and the total width of the state [81]. Only the measurement of Fernandez et al. appears useful for this study. As a comment in passing we note that the use of the Energy Weighted Dipole Sum Rule (EWDSR):

$$S_1(E1) = \sum E(1^-) \times B(E1 : 0^+ \rightarrow 1^-) = \frac{9}{4\pi}\frac{NZ}{A}\frac{e^2\hbar}{2m} \tag{19}$$

yield an upper limit on the radiative width of approximately 5 eV. In this case we assume that the $B(E1:\ 1^- \rightarrow 0^+)$ does not exhaust more then 1% of the EWDSR. Note that even the largest known B(E1)'s in ^{11}Be and ^{13}N exhaust 0.09% and 0.2% of the EWDSR, and based on our understanding of dipole electromagnetic decays, as first suggested by Gell-Mann and Telegdi [85] and Radicati [86] for self conjugate nuclei, and with advances made by theoretical and experimental studies of B(E1) in nuclei [36], we can estimate that the E1 decay should exhaust less then 1% of the EWDSR, as shown in Fig. 22. The sum rule model then allow us to place an upper limit on the value of the radiative width of the 1^- state. In spite of a concentrated effort and with the exclusion of the Seattle result of Fernandez et al., it is clear that an accurate determination of the partial widths of the 1^- state at 5.17 MeV in ^{14}O is needed. By way of introduction to the next chapter, we show in Fig. 22 the accurate results obtained (in experiments that lasted for only a few days each) with radioactive beams [87–89].

4.4 Helium Burning and the $^{12}C(\alpha,\gamma)^{16}O$ Reaction:

For understanding the process of helium burning and in particular the oxygen to carbon ratio at the end of helium burning we must understand the $^{12}C(\alpha,\gamma)^{16}O$

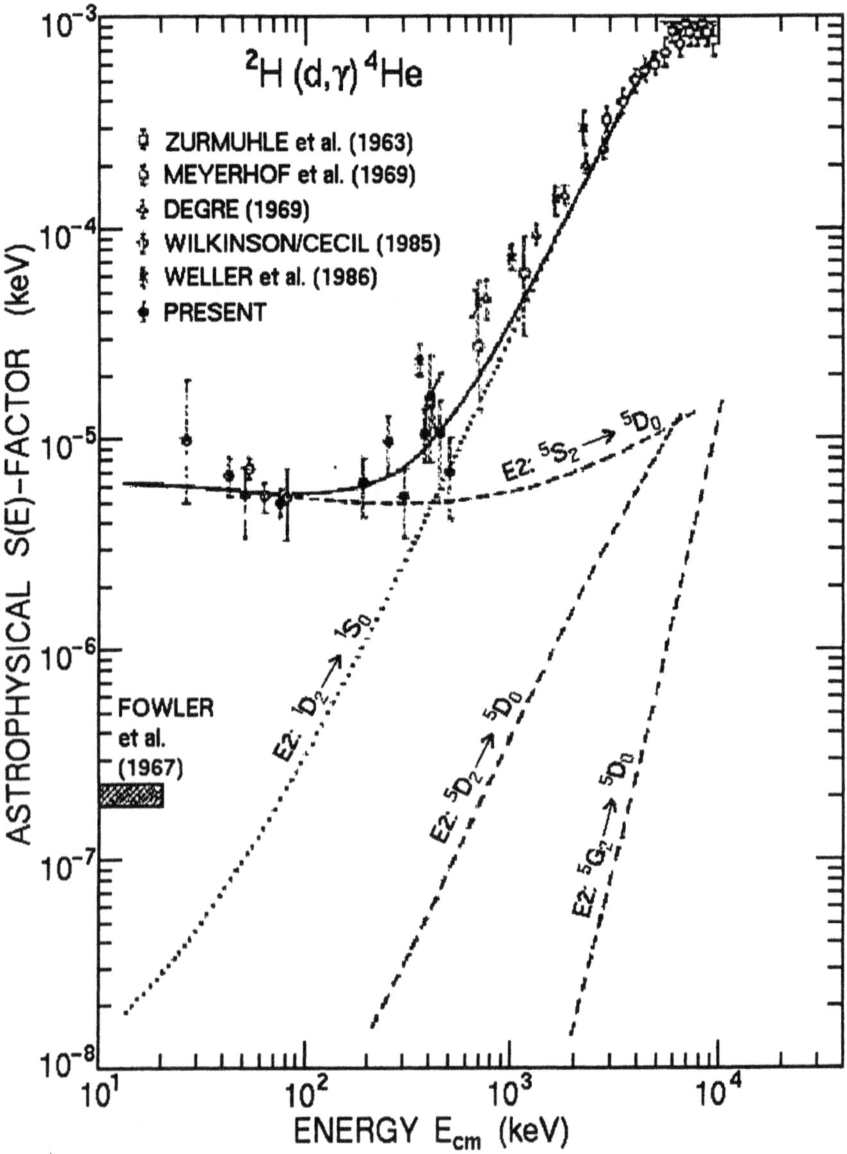

Fig. 21. Extrapolation of d-wave S-factor of the $d(d,\gamma)^4He$ reaction[79]. Note the presence of small non d-wave components that yield a discrepancy from Fowler's extracted S-factor by a factor of 32

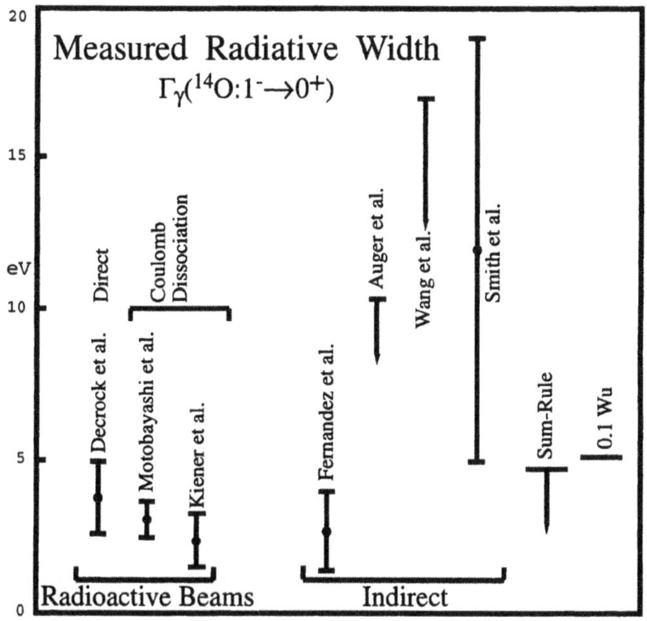

Fig. 22. Measured $\Gamma_\gamma(^{14}O\ :\ 1^- \ \rightarrow\ 0^+)$ using indirect and direct methods. Most indirect measurements, except for the Seattle one [35], yield results less sensitive than (even) the sum rule. The advent of radioactive beams is clear

reaction as in (18), at the most effective energy for helium burning of 300 keV, see (11). At this energy one may estimate [30] the cross section to be 10^{-8} nbarn, clearly non measurable in laboratory experiments. In fact the cross section could be measured down to approximately 1.2 MeV and one needs to extrapolate down to 300 keV, see Fig. 23. As we discuss below the extrapolation to low energies (300 keV) which in most other cases in nuclear astrophysics could be performed with certain reliability, is made difficult by a few effects.

The cross section at astrophysical energies has contribution from the p and d waves and is dominated by tails of the two bound states of ^{16}O, the 1^- at 7.12 MeV (p-wave) and the 2^+ at 6.92 MeV (d-wave), see Fig. 14. The p-wave contribution arises from a detailed interference of the tail of the bound 1^- state at 7.12 MeV and the broad 1^- state at 6.93 MeV, see Fig. 14. The contribution of the bound 1^- state arises from its virtual alpha-particle width, that could not be reliably measured or calculated. Furthermore, the tails of the quasi-bound and bound 1^- states interfere in the continuum and the phase can not be determined from existing data. Existing data could be measured only at higher energies and therefore it does not show sensitivity to the above questions. Hence, the cross section of the $^{12}C(\alpha,\gamma)^{16}O$ reaction could not be measured in a reliable way at 300 keV, and the p-wave S-factor at 300 keV, for example, was estimated to be between 0-500 keV barn with a compiled value of $S_{E1}\ =\ 60\ +60\ -30$

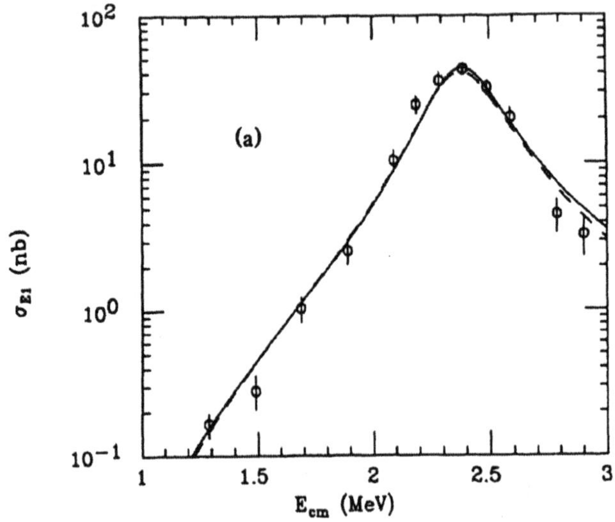

Fig. 23. The $^{12}C(\alpha,\gamma)^{16}O$ reaction cross section [30]

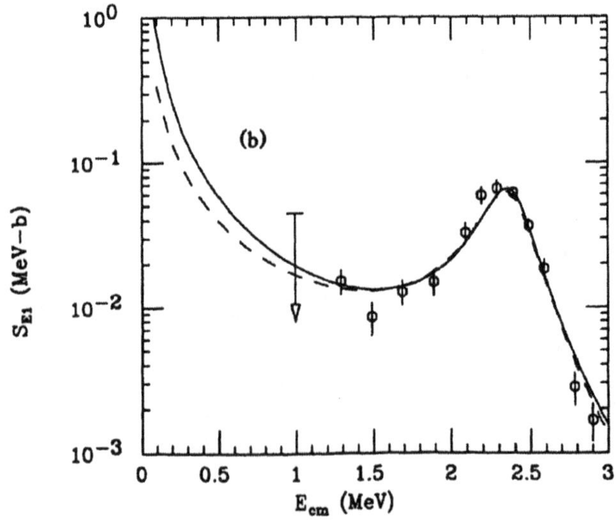

Fig. 24. Measured S - factor(s) for $^{12}C(\alpha,\gamma)^{16}O$ from [56]

keV-b [31,32] and $S_{E2}(300) = 40 + 40 - 20$ keV-b. This large uncertainty is contrasted by the need to know the S-factor with 15% accuracy, see chapter 3 and Fig. 17. In Fig. 24 we show the results obtained over two decades for the p-wave S-factor, with the most notable disagreement in the extracted results of the Munster group, that quoted a very large S-factor with a small error bar. We refer the reader to [55,56] for a complete reference list and review of the subject. The situation is best described as in Fig. 25 where a blind man attempts to find out wether the elephant trunk is up or down by holding its tail. He is clearly performing an experiment with small sensitivity to the question at hand. In the next section we will discuss new idea(s) for measuring this process (in the time reversed fashion with ^{16}O disintegrating to α + ^{12}C). Great hopes for measuring the p-wave S-factor in the beta-delayed alpha-particle emission of ^{16}N [54], turned out to be false and we propose a new experiment, the photodisintegration of ^{16}O, the $^{16}O(\gamma,\alpha)^{12}C$ to be performed at the Duke-HIGS facility.

In the previous chapter we have already described great advances made with the use of radioactive beams to study the $^{13}N(p,\gamma)^{14}O$ and the hot-CNO cycle, see Fig. 22. These studies were performed at the Louvain-La-Neuve (LLN) Radioactive beam facility with ^{13}N radioactive beams [87] and with ^{14}O radioactive beams at Riken [88] and at Ganil [89]. While the facility at LLN uses an ISOL type source and works at low energies, see Fig. 26, the facility at Riken, see Fig. 27, as well as that at Michigan State University, see Fig. 28, use high energy beams from fragmentation process.

4.5 The p-wave S-factor of $^{12}C(\alpha,\gamma)^{16}O$ from the Beta-Delayed Alpha-Particle Emission of ^{16}N, Facts and Fallacies

The beta-delayed alpha-particle emission of ^{16}N may allow us to study the $^{12}C(\alpha,\gamma)^{16}O$ reaction in its time reverse fashion, the disintegration of ^{16}O to α + ^{12}C, and it provides a high sensitivity for measuring low energy alpha-particles and the reduced (virtual) alpha-particle width of the bound 1^- state in ^{16}O at 7.12 MeV, see Fig. 14. As shown in Fig. 29, low energy alpha-particle emitted from ^{16}N correspond to high energy beta's and thus to a larger phase space and enhancement proportional to the total energy to approximately the fifth power. In addition the apparent larger matrix element of the beta decay to the bound 1^- state provides further sensitivity to that state.

5 Possible Solutions (with Secondary or Radioactive Beams)

However, in this case one needs to measure the beta decay, see below, with a sensitivity for a branching ratio of the order of 10^{-9} or better. Prediction of the shape of the spectra of delayed alpha-particles from ^{16}N were first published by Baye and Descouvemont [91], see Fig. 30. Note the anomalous interference structure predicted to occur around 1.1 MeV, at a branching ratio at the level of 10^{-9}. The previously measured beta-delayed alpha-particle emission of ^{16}N [92] was analyzed using R-matrix theory by Barker [93] and lately by Ji, Filippone, Humblet and Koonin [94]. They conclude, as shown in Fig., 29a that the data measured at higher energies is dominated by the quasi bound state in ^{16}O at 9.63 MeV, see Fig. 14, and shows little sensitivity to the interference with the bound

Fig. 25. A mythical blind man attempting to describe the position of the elephant's trunk by holding its tail (artwork by Eric T. Harman)

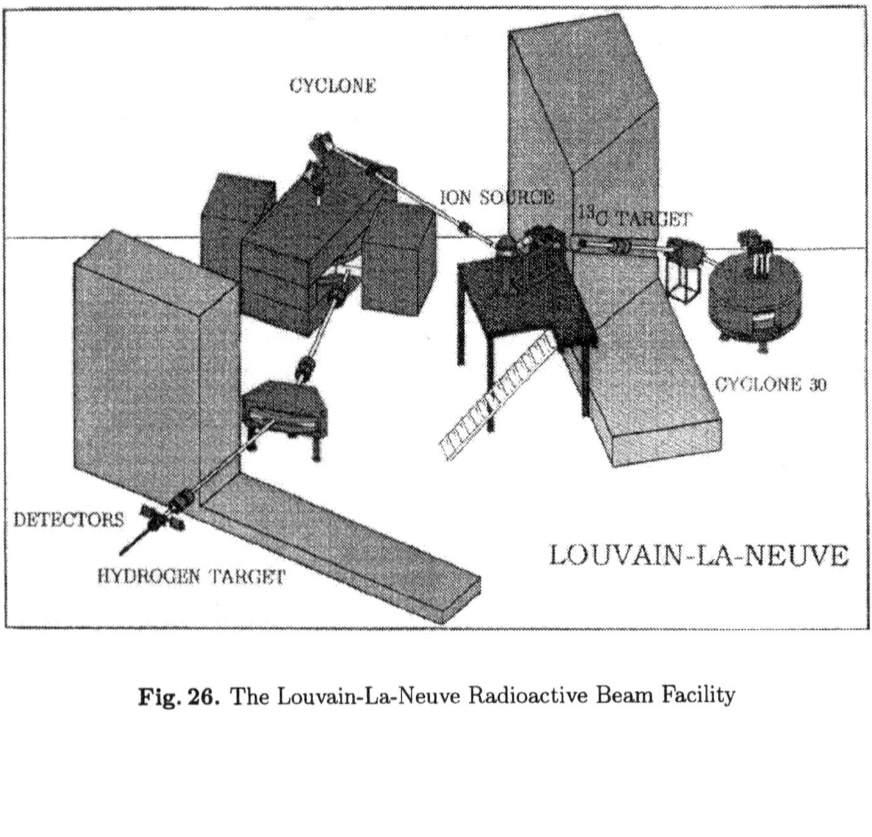

Fig. 26. The Louvain-La-Neuve Radioactive Beam Facility

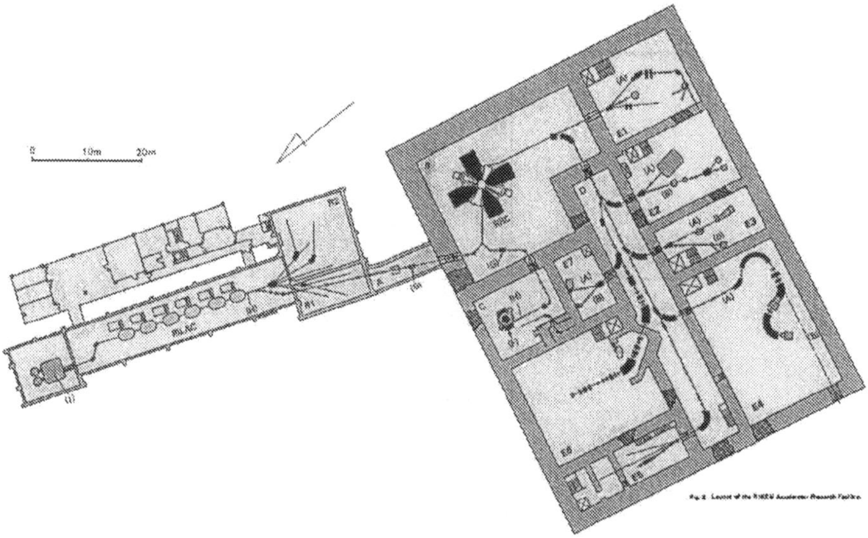

Fig. 27. The Riken-RIPS facility and the setup used for the Coulomb Dissociation of ^{8}B, the Rikkyo-Riken-Yale-Tokyo-Tsukuba-LLN collaboration [90]

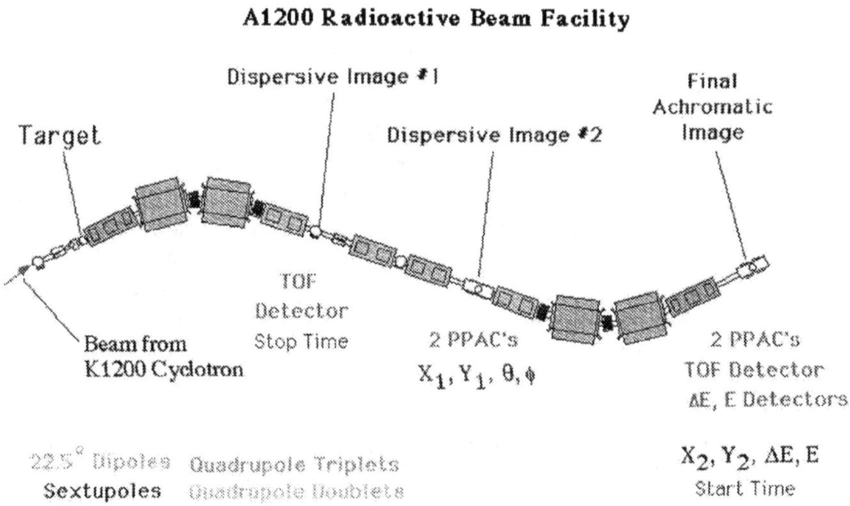

Fig. 28. The Michigan State University A1200 RNB facility

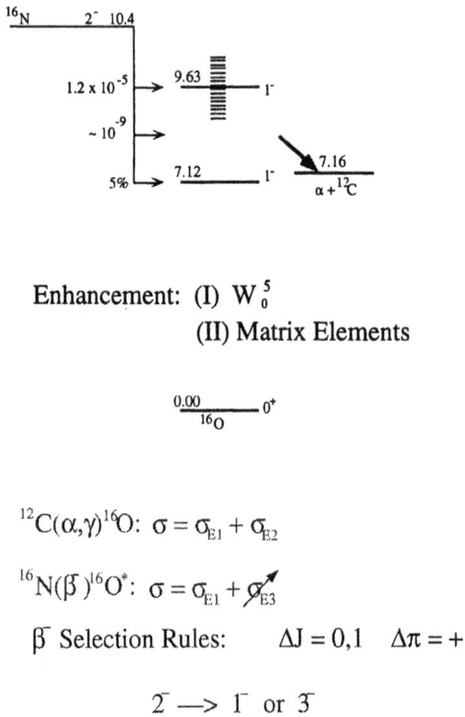

Fig. 29. Nuclear States involved in the beta-delayed alpha-particle emission of ^{16}N

1^- state. The data measured at low energies is predicted to have large sensitivity to the anomalous interference with the bound 1-state. Similar prediction were also given by a K-matrix analysis of Humblet, Filippone, and Koonin [95] of the same early data on ^{16}N [92]. However, it is clear that the interference phase measured in the beta-delayed alpha-particle emission of ^{16}N is not necessarily related to the one measured in $^{12}C(\alpha,\gamma)^{16}O$. Hence, a-priori we might already conclude that while the data on ^{16}N may prove useful for extracting the reduced alpha-width of bound 1^- state, it may be more difficult to exract from it the E1 astrophysical cross section factor.

As shown in Fig. 29, the beta decay can only measure the p-wave S-factor of the $^{12}C(\alpha,\gamma)^{16}O$ reaction, and it also includes (small) contribution from an f-wave. The contribution of the f-wave have to be determined empirically and appears to be very small and leads to additional uncertainty in the quoted S-factor [55,56]. The extraction of the total S- factor of the $^{12}C(\alpha,\gamma)^{16}O$ reaction could then be performed from the knowledge of the E2/E1 ratio which is better known then the individual quantities. An experimental program to study the beta-delayed alpha-particle emissionof ^{16}N (and other nuclei) was carried out at Yale [55,56] and at TRIUMF [57]. From an R-matrix analysis the TRIUMF collaboration quoted a value for the p-wave astrophysical cross section factor of 79 $\pm$ 21 [96]. The Yale study was continued [58,59] and it was found to

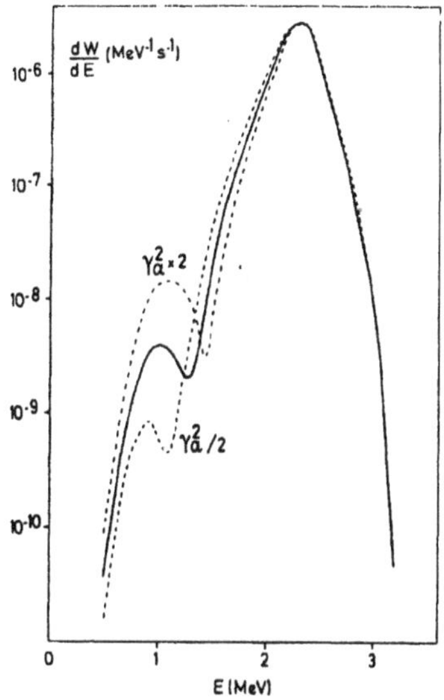

Fig. 30. Spectrum of the beta-delayed alpha-particle emission of ^{16}N, predicted by Baye and Descouvemont [91], some five years before the observation of the interference anomaly [55–57]

be inconsistent with the TRIUMF result [57,96], see Fig. 31. In contrast to the rather small error bar quoted by the TRIUMF collaboration (±20%) an R-matrix analyses of the data by Gerry Hale [60] showed that the ^{16}N data does not rule out a small S-factor. We conclude that the p-wave S-factor for the $^{12}C(\alpha,\gamma)^{16}O$ reaction is in fact not known with the accuracy claimed by Buchmann *et al.* [57] and Azuma *et al.* [96]. In order to determine both the p- and d- wave S-factors of the $^{12}C(\alpha,\gamma)^{16}O$ one can not resort to indirect measurements such the beta-delayed alpha-particle emission of ^{16}N and one must measure the cross section of the $^{12}C(\alpha,\gamma)^{16}O$ reaction at energies as low as possible. In the next section we discuss such a possibility using a new High Intensity Gamma Source (HIGS) at TUNL/Duke.

5.1 The Duke/TUNL Experiment: $^{16}O(\gamma,\alpha)^{12}C$

For determination of the cross section of the $^{12}C(\alpha,\gamma)^{16}O$ at very low energies, as low as E_{cm} = 700 KeV, considerably lower than measured till now, it is very useful to have an experimental setup with three conditions: an amplified cross section, high luminosity and low background. It turns out that the use of the inverse process, the $^{16}O(\gamma,\alpha)^{12}C$ reaction may indeed satisfy all three conditions. The cross section of $^{16}O(\gamma,\alpha)^{12}C$ reaction (with polarized photons) at the kinematical region of interest (photons approx 8-8.5 MeV) is larger by a factor of 50 than the cross section of the direct $^{12}C(\alpha,\gamma)^{16}O$ reaction that occurs in for example Red Giants. Note that the polarization yield an extra factor of two

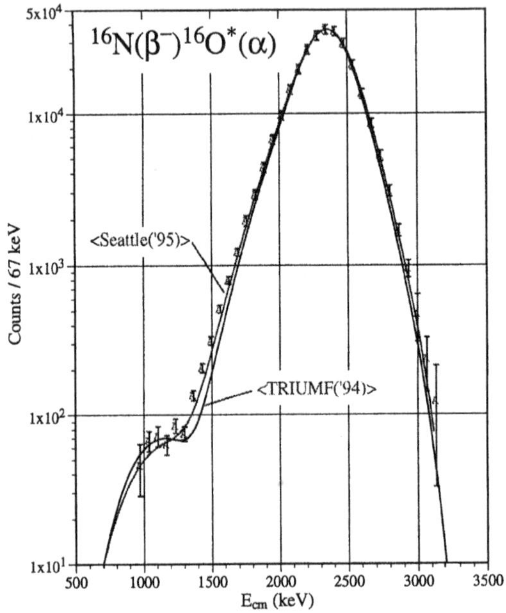

Fig. 31. The newly measured Spectrum of the beta-delayed alpha-particle emission of ^{16}N [58,59] that appears consistent with the unpublished data of the Seattle group, but disagrees with the TRIUMF data [96]

in the enhancement. Thus for the lowest data point measured at 0.9 MeV with the direct cross secion of approx. 60 pb, the photodissociation cross section is 3 nb. It is evident that with similar luminosities, see below, and similar or lower background, the photodissociation cross section can be measured yet to even lower equivalent energies, as low as 0.7 MeV, where the direct $^{12}C(\alpha,\gamma)^{16}O$ cross section is predicted to be of the order of 1 pb. It is clear that detailed balance aids a great deal in this case for measuring the $^{12}C(\alpha,\gamma)^{16}O$ at yet lower energies. However, with (secondary photons from HIGS, see Fig. 32) one can not observe cascade gamma decay, which are considered to be small at low energies.

The luminosity using for example a 100 cm long target of the gas CO_2 at a pressure of 76 torrs (100 mbar), and with a photon beams of 2×10^9 /sec, we obtain a luminosity of 10^{30} $sec^{-1}cm^{-2}$, or a day long integrated luminosity of $0.1pb^{-1}$. Hence a measurement of the photodissociation of ^{16}O with cross section of 10 pb, with a high efficiency detector would yield one count per day. We conclude that it is conceivable that a facility with such luminosity and low background together with a high efficiency detector may allow us to measure the

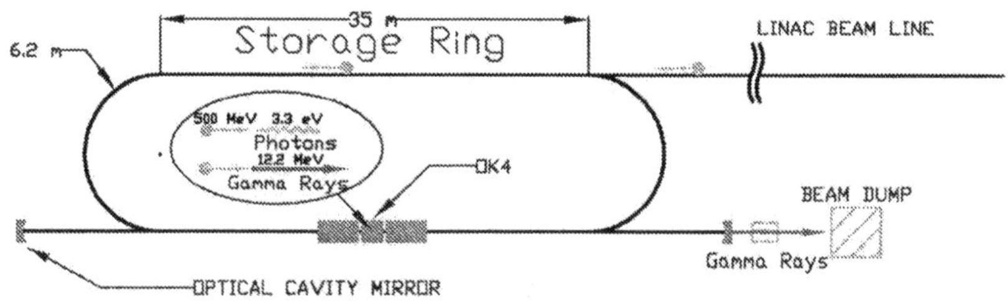

Fig. 32. The electron ring of the Duke High Intensity Gamma Source (HIGS) [97]

photodissociation cross section to a few tens of pb and thus as low as several hundreds of fb for the direct $^{12}C(\alpha,\gamma)^{16}O$ reaction.

The High Intensity Gamma Source (HIGS) [97], in the process of being funded by the USDOE at TUNL/Duke, has already achieved many of its mile stones and it is rapidly approaching its design goal of 2-200 MeV gammas, with 9 MeV gammas at a resolution of 0.1% and intensity in the 10^9 /sec range. The schematical layout of the HIGS facility is shown in Fig. 32. With a 500 MeV pulsed electron beam circulating in the ring, it passes an undulator (OK4) that produces Free Electron Laser photons of 3.3 eV. These photons are reflected back in an optical cavity and arrive in phase for the next pulse in the ring, due to the lasing action. The backscattered photons (of 12.2 MeV) are collimated and used for nuclear physics research at a designated Hall, where we plan to set our experiment. With a Q value of -7.162, our experiment will utilize gammas of energies ranging from 8 to 10 MeV. Note that the emitted photons are linearly polarized [98] and the emitted particles are in a horizontal plane. This simplifies the tracking of particles in this experiment. In addition as the beam is a pulsed, one may use the time information in the trigger of the experiment as well as for using time of flight techniques to further reduce the background.

The main background in such a photodissociation experiment appears to be the large flux of Compton electrons. A promissing detection system would involve the construction of a Time Projection Chamber (TPC). Since the range of available alphas is approximately 8 cm the TPC will be 20 cm wide and one meter long. The TPC could be constructed to be largely insensitive to single Compton electrons, but allow to track both alphas and carbons emitted almost back to back in time correlation. The very different range of alphas and carbons (approx. a factor of 4) aids in the particle identification. Such a TPC detector also allows to measure angular distributions with respect to the polarization vector of the photon, and thus seperate the E1 and E2 components of the $^{12}C(\alpha,\gamma)^{16}O$ reaction.

5.2 The Coulomb Dissociation of ^{14}O (hot CNO) and ^{8}B (Solar Neutrino's):

The Coulomb Dissociation [99] Primakoff [100] process, is the time reverse process of the radiative capture. In this case instead of studying for example the fusion of a proton plus a nucleus (A-1), one studies the disintegration of the final nucleus (A) in the Coulomb field to a proton plus the (A-1) nucleus. The reaction is made possible by the absorption of a virtual photon from the field of a high Z nucleus such as ^{208}Pb. In this case since $\frac{\pi}{k^2}$ for a photon is approximately 1000 times larger than that of a particle beam, the small cross section is enhanced. The large virtual photon flux (typically 100-1000 photons per collision) also gives rise to enhancement of the cross section. Our understanding of the Coulomb Excitation and the virtual photon flux allow us (as in the case of electron scattering) to deduce the inverse nuclear process. However in Coulomb Dissociation since αZ approaches unity (unlike the case in electron scattering), higher order Coulomb effects (Coulomb Post Acceleration) may be non-negligible and they need to be understood [101]. The success of the experiment is in fact hinging on understanding such effects and designing the kinematical conditions so as to minimize such effects.

Hence the Coulomb Dissociation process has to be measured with great care with kinematical conditions carefully adjusted so as to minimize nuclear interactions (i.e. distance of closest approach considerably larger then 20 fm, hence very small forward angles scattering), and measurements must be carried out at high enough energies (many tens of MeV/u) so as to maximize the virtual photon flux [102]. Indeed when such conditions are not carefully selected [103] the measured cross section was shown to be dominated by nuclear effects [104,105], which can not be reliably calculated to allow the extraction of the inverse radiative capture cross section.

Good agreement between measured cross section of radiative capture through a nuclear state, or in the continuum, where achieved for the Coulomb Dissociation of ^{6}Li and the $d(\alpha,\gamma)^{6}Li$ capture reaction [106], and the Coulomb Dissociation of ^{14}O and the $p(^{13}N,\gamma)^{14}O$ capture reaction [87–89]. In addition we note that test experiment on the Coulomb Dissociation of ^{13}N [88] was also found to be in agreement with the $^{12}C(p,\gamma)^{13}N$ capture reaction.

The Coulomb Dissociation of ^{8}B may provide a good opportunity for resolving the issue of the absolute value of the cross section of the $^{7}Be(p,\gamma)^{8}B$ reaction, see chapter 4. The Coulomb Dissociation yield arise from the convolution of the inverse nuclear cross section times the virtual photon flux. While the first one is decreasing as one approaches low energies, the second one is increasing (due to the small threshold of 137 keV). Hence as can be seen in Fig. 33, over the energy region of 400 to 800 keV the predicted measured yield is roughly constant. This is in great contrast to the case of the nuclear cross section that is dropping very fast at low energies, see Fig. 33. Hence measurements at these energies could be used to evaluate the absolute value of the cross section.

An experiment to study the Coulomb Dissociation of ^{8}B was performed during March-April, 1992, at the Riken radioactive beam facility, using the setup

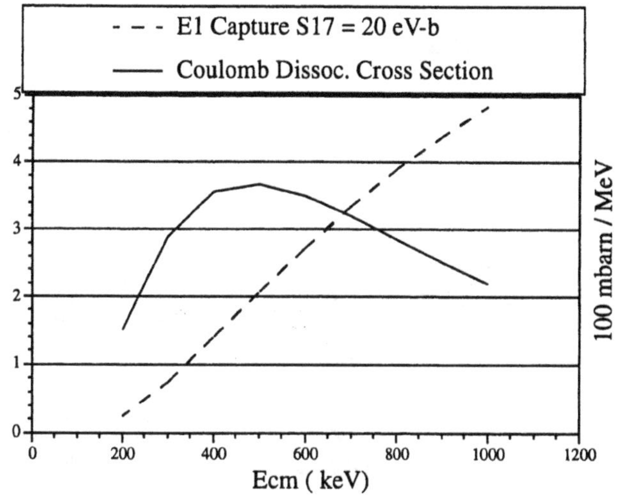

Fig. 33. The cross section for Coulomb Dissociation and E1 capture

shown in Fig. 34. The radioactive beams extracted from the RIPS separator, see Fig. 27, are shown in Fig. 35. Indeed the results of the experiment allow us to measure the radiative capture $^7Be(p,\gamma)^8B$ cross section and the results of the RIKEN I [90] and the RIKEN II [107,108] are consistent with the absolute value of the cross section measured by Filippone et al. [68] and by Vaughn et al. [71], as shown in Fig. 36. This experiment was continued at GSI [109] with similar results at low energy. The results of the RIKEN I [90], RIKEN II [107,108], GSI [109] as well as the MSU result on the E2/E1 [111] are shown in Table II. Note the MSU data suggest an E2 larger than expected from RIKEN I data [110], RIKEN II [107], and GSI data [109].

Table 2. Measured S-factors in Coulomb dissociation experiments

Experiment	$S_{17}(0)$ eV-b	$S_{E2}/S_{E1}(0.6\text{ MeV})$
RIKEN1 [90]	16.9 ± 3.2	$< 7 \times 10^{-4}$ [110]
RIKEN2 [107]	18.9 ± 1.8	$< 4 \times 10^{-5}$ [108]
GSI1 [109]	$20.6 \pm 1.2 \pm 1.0$	$< 3 \times 10^{-5}$
MSU [111]		$6.7\ +2.8\ -1.9 \times 10^{-4}$
ADOPTED	19.4 ± 1.3	$< 3 \times 10^{-5}$

^{8}B Breakup 実験覚え書き

20 Mar. 1992 本林

setup

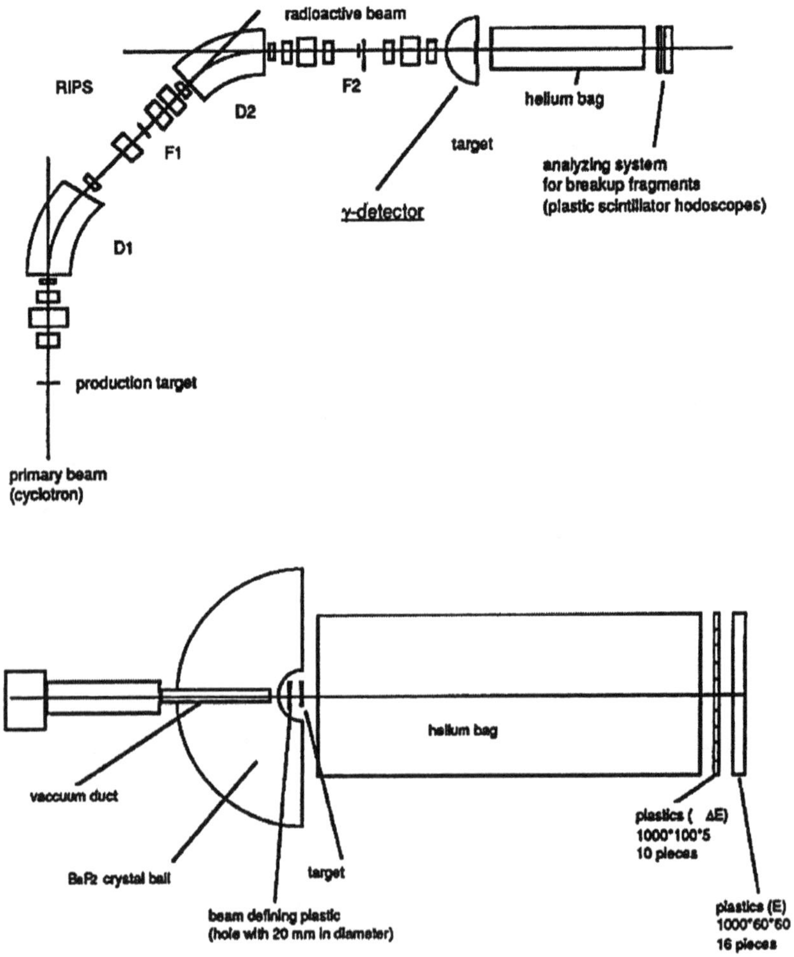

Fig. 34. The experimental setup of the RIKEN Experiments.[90,107,108]

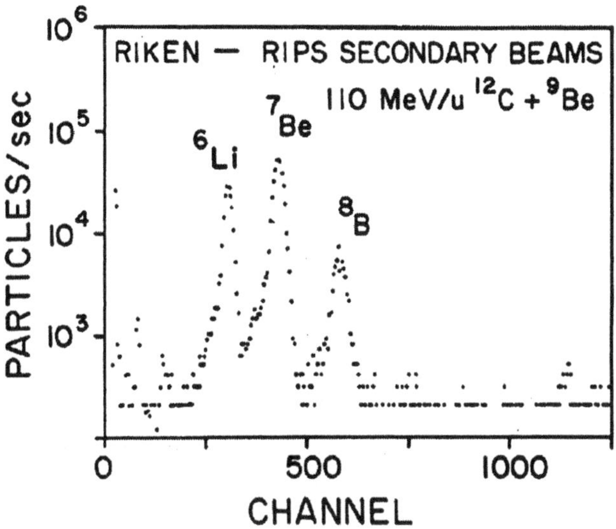

Fig. 35. Radioactive beams extracted from the Riken-RIPS facility and used in the study of the Coulomb Dissociation of ^8B, a Rikkyo-Riken-Yale-Tokyo-Tsukuba-LLN collaboration [90]

5.3 The $^7Be(p,\gamma)^8B$ Reaction Studies with ^7Be Radioactive Beams at LLN:

An experiment to study the $^7Be(p,\gamma)^8B$ reaction with ^7Be radioactive beam is in progress, a UConn-LLN collaboration at LLN [112,113] The experimental detector setup for the UConn-LLN experiment is shown in Fig. 37. The recoil ^8B emerge with a (step) distribution of energies with widths approximately 0.7 MeV, and a stopping spread in aluminum of approximately 0.5 μm. Thus the stopped ^8B are designed to be equally spread over the two aluminum catcher foils (0.5 μm each). The beta-delayed alpha-particle emission of ^8B is measured by measuring coincidence between the two back to back equal energy alpha-particles detected in a pair of detectors, see fig. 37.

In the target region, two monitors measure beam intensity by measuring the elastic scattering off a thin Au foil (evaporated onto a very thin carbon backing) and the recoil protons off the target. The cross section of the $^7Be(p,\gamma)^8B$ reaction will be measured relative to the elastic scattering, thereby removing several systematic uncertainties related to beam-target composition. The hydrogen component of the target is continuously monitored by measuring the recoil protons from the target.

Since two alpha-particles are associated with every decay we calculated a verylarge detection efficiency, approximately 50% of 2π. Our extensive Monte carlo simulations yield a large (98%) coincidence efficiency, and thus approximately 50% total coincidence efficiency for two equal energy correlated back to back alpha-particles. For a ^8B transfer time of 0.07 sec, every 0.5 sec, we obtain

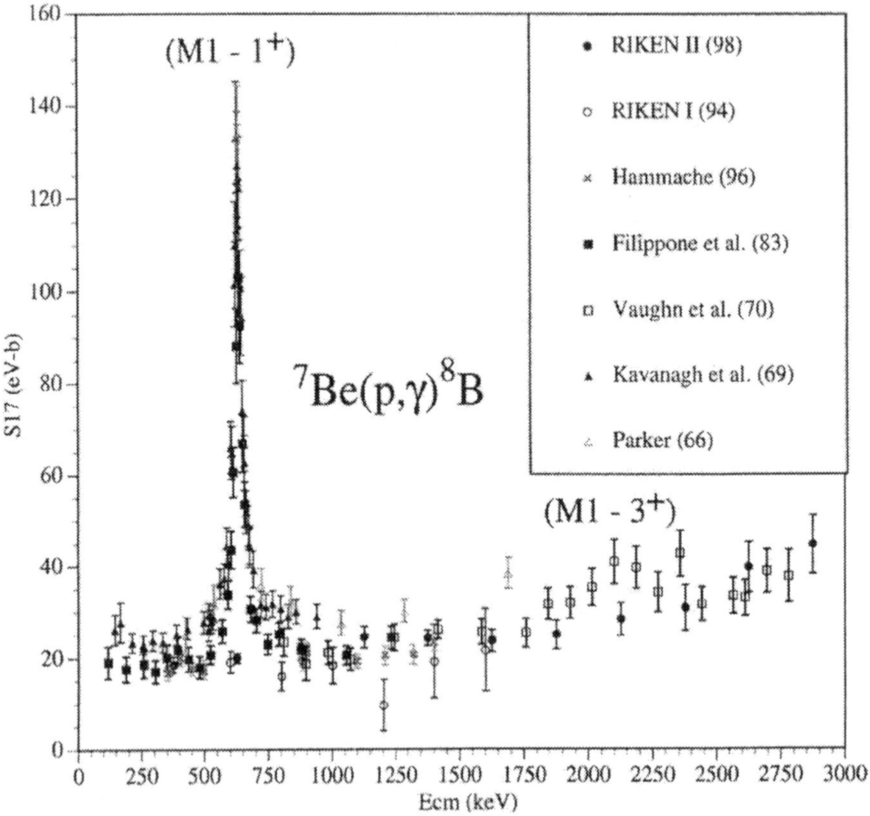

Fig. 36. Extracted $S_17(E)$ cross section factors by the RIKEN experiments as compared to direct measurements

a total alpha-particle detection efficiency of approximately 25%. The closed detection geometry (50% of 4π) with a front and back detectors (a-la calorimetry style) also ensures that the total alpha detection efficiency is nearly independent of the exact location of the collection foils, as long as the two foils remain parallel and at constant distance and the recoil ^8B nuclei are spread equally on both catcher foils.

A beam intensity of 5×10^8 $/sec$ and a 250 $\mu g/cm^2$ CH_2 target ($\Delta E_{cm} = 100\ keV$) containing $2{\times}10^{19}$ $hydrogens/cm^2$ yield a luminosity of 10^{28} $/sec/cm^2$. With expected cross sections of σ = 0.5, 0.4 and 0.2 μb, at E_{cm} = 1.0, 0.8 and 0.5 MeV, respectively, and alpha-particle detection efficiency of 25%, we obtain count rates of approximately 5, 4, and 2 counts per hour. Thus experiments lasting two to three days at E_{cm} = 1.0, 0.8 and 0.5 MeV, respectively, will yield a total count of 240, 192 and 144 counts and statistical uncertainties of 6.4%, 7.2% and 8.3%, respectively. With approved 9 days of experiment we plan to adjust the length of runs to achieve 5% precision at each data point.

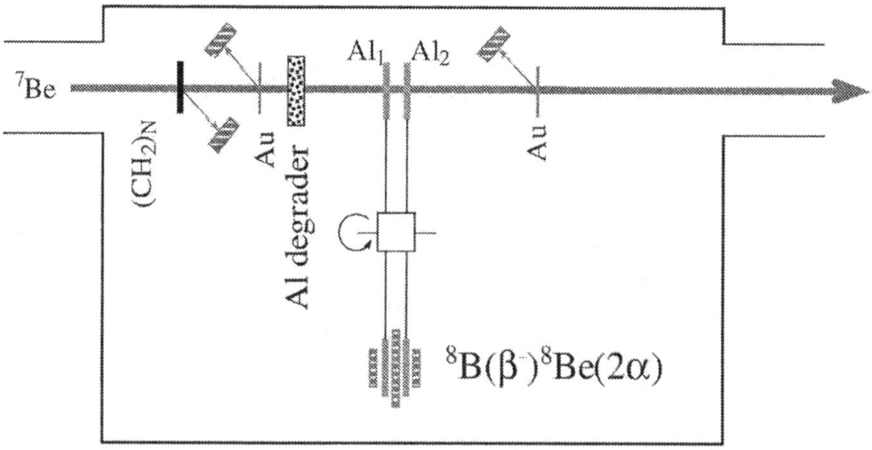

Fig. 37. The Setup of the LLN experiment [112,113]

6 Conclusions and Acknowledgements

We conclude that radioactive beams could be used for carefully planned experiments to solve some of the outstanding and most important problems of nuclear astrophysics today, and hence promise a rich future for low energy nuclear astrophysics studies.

I would like to acknowledge the help of Ralph France III in preparing this manuscript and the work of N. Iwasa, T. Kikuchi, K. Suemmerer, F. Boue and P. Senger on the data analyses of the CD data and Ralph H. France III and James E. McDonald on the data analysis of the $^7Be(p,\gamma)^8B$ reaction. I also acknowledge discussions and encouragements from Professors J.N. Bahcall, C.A. Bertulani, G. Baur, and Th. Delbar.

This work was supported in Part by USDOE Grant No. DE-FG02-94ER40870.

References

1. Dorrit Hoffleit: Yale Alumni Magazine, Nov.(1985) 28.
2. D.C. Wang, and S.M. Gong: 'The historical records of Halley's comet'. In *China, Earth, Moon, and Planets*, David Reidel Publ. Comp. **34**, 55 (1986)
3. Donald D. Clayton: *Principles of Stellar Evolution and Nucleosynthesis* (University of Chicago Press, 1968)
4. C.E. Rolfs, and W.S. Rodney: *Cauldrons in the Cosmos* (University of Chicago Press, 1988)
5. Gordon Baym: *Lectures on Quantum Mechanics* (W.A. Benjamin, Inc., 1969) pp. 431
6. John N. Bahcall: *Neutrino Astrophysics* (Cambridge Univ. Press, 1989)
7. Hans A. Bethe: Physics Today **21:9**, 36 (1968)
8. P.R. Demarque, J.R. Percy: Ap.J. **140**, 541 (1964)

9. S. Chandrasekhar: Rev. Mod. Phys. **56**, 137 (1984)
10. P. Demardue, C.P. Deliyannis, and E. Sarajendini: *Observation Tests of Cosmological Inflation.* ed. by T. Shank et al. (Cluwer Acad. Press, Netherlands, 1991) pp. 111
11. David N. Schramm and Robert V. Wagoner: Ann. Rev. Nucl. Sci. **27**, 37 (1977)
12. Ann Merchant Boesgaard and Gary Steigman: Ann. Rev. Astron. Astrophy. **23**, 319 (1985)
13. R. Rebolo, P. Molaro, and J.F. Beckman: Astron. Astrop. **192**, 192 (1988)
14. David N. Schramm, private communication (1992)
15. A.G. Riess et al.: Astr. Jour. **116,3**, 1009 (1998)
16. E.P. Hubble: *Silliman Lectures* (Yale University, 1936)
17. A.A. Penzias and R.W. Wilson: Ap. J. **142**, 419 (1965)
18. R.H. Dicke, P.J.E. Peebles, P.G. Roll, and D.T. Wilkinson: Ap. J. **142**, 414 (1965)
19. J.H. Applegate, and C.J. Hogan: Phys. Rev. **D31**, 3037 (1985)
20. J.H. Applegate, C.J. Hogan, and R.J. Scherrer: Phys. Rev. **D35**, 1151 (1987)
21. J.H. Applegate, C.J. Hogan, and R.J. Scherrer: Ap. J. **329**, 572 (1988)
22. C.R. Alcock, G.M. Fuller, and G.J. Mathews: Ap. J. **320**, 439 (1987)
23. R.A. Malaney, and W.A. Fowler: Ap. J. **333**, 14 (1988)
24. R.N. Boyd, and T. Kajino: Ap. J. **336**, L 55 (1989)
25. C.R. Alcock, D.S. Dearborn, G.M. Fuller, G.J. Mathews, and B.S. Meyer: Phys. Rev. Lett. **64**, 2607 (1990)
26. Moshe Gai: Phys. Rev. **C45**, 2548 (1992)
27. F.R. Brown, F.P. Butler, H. Chen, N.H. Christ, Z. Dong, W. Schaffer, L.I. Unger, and A. Vaccarino; Phys. Rev. Lett. **65**, 2491 (1990)
28. S.G. Ryan, J.E. Norris, M.S. Bessell, and C.P. Deliyannis: Ap. J. **388** (1992).
29. T.P. Walker, G.J. Mathews, and V.E. Viola: Ap. J. **299**, 745 (1985)
30. W.A. Fowler: Rev. Mod. Phys. **56**. 149 (1984)
31. G.R. Caughlan, and W.A. Fowler: At. Data Nucl. Data Tables **40**, 283 (1988)
32. F.C. Barker, and T. Kajino: *Proc. Int. Workshop on Unstable Beams in Astrophysics, Tokyo, 7-8 June, 1991* (World Scientific, Singapore, 1992) pp. 63
33. W.A. Fowler, G.R. Caughlan, and B.A. Zimmerman: Ann. Rev. Astron. and Ap. **5**, 525 (1967)
34. W.A. Fowler, G.R. Caughlan, and B.A. Zimmerman: Ann. Rev. Astron and Ap. **13**, 69 (1975)
35. P.B. Fernandez, E.G. Adelberger, and A. Garcia: Phys. Rev. **C89**, 1887 (1989)
36. Yoram Alhassid, Moshe Gai, and George F. Bertsch: Phys. Rev. Lett. **49**, 1482 (1982)
37. J.N. Bahcall, and R.K. Ulrich: Rev. Mod. Phys. **60**, 297 (1988)
38. J.N. Bahcall, and M.H. Pinsonneault: Rev. Mod. Phys. **64**, 885 (1992)
39. S. Turck-Chieze, S. Cahen, M. Casse, and C. Doom: Ap. J. **335**, 415 (1988)
40. Sylvaine Turck-Chieze, and Ilidio Lopes: Ap. J. **408**, 347 (1993)
41. S. Turck-Chieze, W. Dappen, E. Fossat, J. Provost, E. Schatzman, and D. Vignaud: Phys. Rep. **230**, 57 (1993)
42. K.S. Hirata et al.: Phys. Rev. Lett. **65**, 1297 (1990)
43. K. Inoue for the Kamiokande III collaboration: *XXVIIIth Recontre de Moriond, Les Arcs, Savoi, France, March 13-20, 1993*
44. Y. Fukuda at al.: Phys. Rev. Lett. **81**, 1158 (1998)
45. A.I. Abazov et al.: Phys. Rev. Lett. **67**, 3332 (1991)
46. P. Anselmann et al.: Phys. Lett. **B314**, 445 (1993); ibid GX 44-1994, February, 1994, submitted to Phys. Lett. B.

47. G.T. Ewan, W.F. Davidson, C.G. Hargrove: Phys. in Canada **48**, 112 (1992)
48. A.B. McDonald: Phys. in Canada **48**, 120 (1992)
49. J. Boger et al.: nucl-ex/9910016, to be published (1999)
50. L. Wolfenstein: Phys. Rev. **D17**, 2369 (1978); ibid **D20**, 2634 (1979)
51. S.P. Mikheyev, and A.Yu. Smirnov: Yad. Fiz. **44**, 847 (1985) [Sov. J. Nucl. Phys. **42**, 913 (1985)]
52. Peter D. Parker: *Proc. fifth Int. Conf. Clust. Nucl.* (Kyoto, 1988) J. Phys. Soc. Jpn. **58** (1989)Suppl. 196.
53. H.M.J. Boffin, G. Paulus, M. Arnould, and N. Mowlavi: Astron. Astrophys. **279**, 173 (1993)
54. B.G. Levi: *Search and Discovery*, Physics Today, July, 1993, pp. 23.
55. Z. Zhao, R.H. France III, K.S. Lai, S.L. Rugari, M. Gai, and E.L. Wilds: Phys. Rev. Lett. **70**, 2066 (1993); ibid **70**, 3524 (1993)
56. Z. Zhao: Ph.D. thesis, Yale University, 1993
57. L. Buchmann, R.E Azuma, C.A. Barnes, J.M. D'Auria, M. Dombsky, U. Giesen, K.P. Jackson, J.D. King, R.G. Korteling, J. Powell, G. Roy, J. Vincent, T.R. Wang, S.S.M. Wong, and P.R. Wrean: Phys. Rev. Lett. **70**, 726 (1993)
58. R.H. France III, E.L. Wilds, N.B. Jevtic, J.E. McDonald, and M. Gai: Nucl. Phys. **A621**, 165c (1997)
59. R.H. France III: Ph.D. thesis, Yale University, 1996
60. G.M. Hale: Nucl. Phys. bf A621, 177c (1997)
61. I. Iben: Ap. J. **196**, 525 (1975); ibid **196**, 549 (1975)
62. T.A. Weaver and S.E. Woosley: Ann. NY Acad. Sci. **336**, 335 (1980)
63. Xi Ze-zong, Po Shu-jen [translated by K.S. Yang]: Science **154**, 597 (1966)
64. Jewan Kim: *Barley existing things* (Mineum Publishing company, Seoul, Korea, 1933)
65. T.A. Weaver, and S.E. Woosley: Phys. Rep. **227**, 65 (1993)
66. F.C. Barker, and H.R. Spear: Ap. J. **307**, 847 (1986)
67. C.W. Johnson, E. Kolbe, S.E. Koonin, and K. Langanke: Ap. J. **392**, 320 (1992)
68. B.W. Filippone, A.J. Elwyn, C.N. Davis, and D.D. Koetke: Phys. Rev. Lett. **50**, 412 (1983); ibid Phys. Rev. **C28**, 2222 (1983)
69. R.W. Kavanagh, T.A. Tombrello, J.M. Mosher, and D.R. Goosman: Bull. Amer. Phys. Soc. **14**, 1209 (1969)
70. P.D. Parker: Phys. Rev. **150**, 851 (1966); ibid Ap. J. **153**, L85 (1968)
71. F.J. Vaughn, R.A. Chalmers, D. Kohler, and L.F. Chase Jr: Phys. Rev. **C2**, 1657 (1970)
72. E.G. Adelberger, S.A. Austin, J.N. Bahcall, A.B. Balantekin, G. Bertsch, G. Bogaert, L. Buchmann, F.E. Cecil, A.E. Champagne, L. de Braeckeleer, C.A. Duba, S.R. Elliott, S.J. Freedman, M. Gai, G. Goldring, C.R. Gould, A. Gruzinov, W.C. Haxton, K.M. Heeger, E. Henley, M. Kamionkowski, R.W. Kavanagh, S.E. Koonin, K. Kubodera, K. Langanke, T. Motobayashi, V. Pandharipande, P. Parker, R.G.H. Robertson, C. Rolfs, R. Sawyer, N. Shaviv, T.D. Shoppa, K. Snover, E. Swanson, R.E. Tribble, S. Turck-Chiez, J.F. Wilkerson.: Rev. of Modern Phys. **70**, 1265 (1998)
73. F. Strieder et al.: Eur. Phys. J. **A3**, 1 (1998)
74. L. Weissman et al.: Nuc. Phys. **A630**, 678 (1998)
75. F. Hammache et al.: Phys. Rev. Lett. **80**, 928 (1998)
76. M. Hass et al.: Phys. Lett. **B462**, 237 (1999)
77. A.Y. Zyuzin et al.: Nucl. Inst. Meth. **A438**, 109 (1999)
78. B.K. Jennings, S. Karataglidis, and T.D. Shoppa, Phys. Rev. **C58**, 3711 (1998)

79. C.A. Barnes et al.: Phys. Lett. **197**, 315 (1987)
80. P. Decrock et al.: Phys. Rev. **C48**, 2057 (1993)
81. T.E. Chupp, R.T. Kozus, A.B. McDonald, P.D. Parker, T.F. Wang, and A.J. Howard: Phys. Rev. **C31**, 1023 (1985)
82. T.F. Wang: Ph.D. thesis, Yale University, unpublished.
83. P. Aguer et al.: *Proc. Int. Symp. Heavy Ions Phys. and Nucl. Astro.*. Ed. by S. Kubono, M. Ishihara, and T. Nomura, (World Scie. 1989) pp. 107.
84. M.S. Smith et al.: Phys. Rev. **C47**, 2740 (1993)
85. M. Gell-Mann, and V.L. Telegdi: Phys. Rev. **91**, 169 (1953)
86. L.A. Radicati: Phys. Rev. **87**, 521 (1952)
87. P. Decrock, Th. Delbar, P. Duhamel, W. Galster, M. Huyse, P. Leleux, I. Licot, E. Lienard, P. Lipnik, M. Loiselet. C. Michotte, G. Ryckewaert, P. Van Duppen, J. Vanhorenbeeck, J. Vervier: Phys. Rev. Lett. **67**, 808 (1991)
88. T. Motobayashi, T. Takei, S. Kox, C. Perrin, F. Merchez, D. Rebreyend, K. Ieki, H. Murakami, Y. Ando, N. Iwasa, M. Murokawa, S. Shirato, J. Ruan (Gen), T. Ichihara, T. Kubo, N. Inabe, A. Goto, S. Kubono, S. Shimoura, and M. Ishihara; Phys. Lett. **B264**, 259 (1991)
89. J. Kiener, A. Lefebvre, P. Aguer, C.O. Bacri, R. Bimbot, G. Bogaert, B. Borderie, F. Calpier, A. Coe, D. Disidier, S. Fortier, C. Grunberg, L. Kraus, I. Linck, G. Pasquier, M.F. Rivet, F. St. Laurent, C. Stephan, L. Tassan-Got, and J.P. Thibaud: Nucl. Phys. **A552**, 66 (1993)
90. T. Motobayashi, N. Iwasa, Y. Ando, M. Kurokawa, H. Murakami, J. Ruan (Gen), S. Shimoura, S. Shirato, N. Inabe, M. Ishihara, T. Kubo, Y. Watanabe, M. Gai, R.H. France III, K.I. Hahn, Z. Zhao, T. Nakamura, T. Teranishi, Y. Futami, K. Furataka, and T. Delbar: Phys. Rev. Lett. **73**, 2680 (1994)
91. D. Baye, and P. Descouvemont: Nucl. Phys. **A481**, 445 (1988)
92. K. Neubeck, H. Schober, and H. Waffler: Phys. Rev. **C10**, 320 (1974)
93. F.C. Barker: Aust. Jour. Phys. **24**, 777 (1971)
94. X. Ji, B.W. Filippone, J. Humblet, and S.E. Koonin: Phys. Rev. **C41**, 1736 (1990)
95. J. Humblet, B.W. Filippone, and S.E. Koonin: Phys. Rev. **C44**, 2530 (1991)
96. R.E. Azuma, et al.: Phys. Rev. **C50**, 1194 (1994)
97. W. Tornow, R. Walter, H.R. Weller, V. Litvinenko, B. Mueller, P. Kibrough: *A Free-electron Laser Generated Gamma-ray Beam for Nuclear Physics* (Duke/TUNL, 1997)
98. V.N. Litvinenko et al.: Phys. Rev. Letts. **78**, 4569 (1997)
99. G. Baur, C.A. Bertulani, and H. Rebel; Nucl. Phys. **A458**, 188 (1986)
100. H. Primakoff: Phys. Rev. **81**. 899 (1951)
101. H. Esbensen and G.F. Bertsch: Phys. Lett. **B359**, 13 (1995); ibid Nucl. Phys. **A600**, 37 (1996)
102. J.D. Jackson: *Classical Electromagnetism* (John Wiley, New York, 1962) Ch. 14
103. J. von Schwarzenberg *et al.*: Phys. Rev. **C53**, R2598 (1996)
104. F.M. Nunes, and I.J. Thompson; Phys. Rev. **C57**, R2818 (1998)
105. C.H. Dasso, S.M. Lenzi, and A. Vitturi: nucl-th/9806002, to be published.
106. J. Kiener, H.J. Gils, H. Rebel, S. Zagromski, G. Gsottschneider, N. Heide, H. Jelitto, and J. Wentz: Phys. Rev. **C44**, 2195 (1991)
107. T. Kikuchi, T. Motobayashi, N. Iwasa, Y. Ando, M. Kurokawa, S. Moriya, H. Murakami, T. Nishio, J. Ruan (Gen), S. Shirato, S. Shimoura, T. Uchibori, Y. Yanagisawa, M. Ishihara, T. Kubo, Y. Watanabe, M. Hirai, T. Nakamura, H. Sakurai, T. Teranishi, S. Kubono, M. Gai, R.H. France III, K.I. Hahn, Th. Delbar, C. Michotte, and P. Lipnik: Phys. Lett. **B391**, 261 (1997)

108. T. Kikuchi, T. Motobayashi, N. Iwasa, Y. Ando, M. Kurokawa,, S. Moriya, H. Murakami, T. Nishio, J. Ruan (Gen), S. Shirato, S. Shimoura, T. Uchibori, Y. Yanagisawa, H. Sakurai, T. Teranishi, Y. Watanabe, M. Ishihara, M. Hirai, T. Nakamura, S. Kubono, M. Gai, R.H. France III, K.I. Hahn, Th. Delbar,P. Lipnik, and C. Michotte: Eur. Phys. J. **A3**, 213 (1998)
109. N. Iwasa, F. Boue, G. Surowka, K. Summerer, T. Baumann, B. Blank, S. Czajkowski, A. Forster, M. Gai, H. Geissel, E. Grosse, M. Hellstrom, P. Koczon, B. Kohlmeyer, R. Kulessa, F. Laue, C. Marchand, T. Motobayashi, H. Oeschler, A. Ozawa, M.S. Pravikoff, E. Schwab, W. Schwab, P. Senger, J. Speer, C. Sturm, A. Surowiec, T. Teranishi, F. Uhlig, A. Wagner, W. Walus, and C.A. Bertulani: Phys. Rev. Lett. **83**, 2910 (1999)
110. Moshe Gai, and Carlos A. Bertulani: Phys. Rev. **C52**, 1706 (1995)
111. B. Davids et al.: Phys. Rev. Lett. **81**, 2209 (1998)
112. M. Gai, J.E. McDonald, R.H. France III, J.S. Schweitzer, C. Angulo, Ch. Barue, S. Cherubini, M. Cogneau, Th. Delbar, M. Gaelens, P. Leleux, M. Loiselet, A. Ninane, G. Ryckewaert, K.B. Swartz, D. Visser: Bull. Amer. Phys. Soc. **44,II**, 1529 (1999)
113. J.E. McDonald; Ph.D. thesis, University of Connecticut, 2000

The Generation of Cosmic Magnetic Fields

Karl-Heinz Rädler

Astrophysikalisches Institut Potsdam
An der Sternwarte 16, D-14482 Potsdam, Germany

Abstract. Most of the magnetic fields of cosmic objects are generated and maintained by dynamo action of the motions of electrically conducting fluids. A brief survey on observational facts concerning cosmic magnetic fields is given. Some basic principles of magnetofluiddynamics are explained. On this basis essential features of the dynamo theory of cosmic objects are developed, first on the kinematic level and later taking into account the full interaction between magnetic field and motion. Particular attention is paid on mean-field electrodynamics and mean-field magnetofluiddynamics and their application to mean-field dynamo models for objects showing irregular or turbulent motions and magnetic fields. A few explanations are given on dynamos in the Earth and the planets, in the Sun and stellar objects and in galaxies.

Preliminary remark

The lectures whose main content is reproduced in this article were planned to give an introduction to the dynamo theory of cosmic magnetic fields. It was not the intention of the lectures, and it is not that of this article to deliver a more or less complete survey on all findings or activities. Other representations of the subject and more results can be found in several monographs [1–7], proceedings of conferences [8–11] and review articles [12–17].

1 Some Observational Facts

At the beginning of the 20th century no other magnetic field of a cosmic object was known than that of the Earth. In 1908 G. E. Hale proposed to interpret particular line splittings in the spectrum of the light coming from sunspots, thinking of the Zeeman-effect, as evidence of magnetic fields at the Sun. In the meantime magnetic fields have been discovered at a large number of very different cosmic objects. We know about magnetic fields of the planets, of several types of main-sequence stars, of white dwarfs and neutron stars, etc. Moreover, in a number of nearby galaxies large-scale magnetic fields have been discovered that penetrate the whole disc and continue into the halo.

Magnetic fields seem to be quite natural attributes of cosmic objects. Together with the gravitation they determine a great part of the structures and processes in the universe. The magnetic fields of cosmic objects show a great variety not only with respect to their magnitudes and spatial extents but also to their geometrical structures and time behaviors. A very rough survey on observed magnetic fields and their features are given in Table 1.

Table 1. Magnetic fields of various cosmic objects and their spatial extents. All values of the magnetic flux densities and the linear dimensions of the objects have to be understood as orders of magnitude only

Object	Magnetic flux density [T]	Linear dimension of the object [m]	Symmetry and time behavior of the magnetic field
Earth	10^{-4}	10^{7} (10^{4} km)	slight deviations from symmetry about rotation axis and equatorial plane, non-oscillatory, reversals
Planets	$10^{-8} \cdots 10^{-3}$	$10^{6} \cdots 10^{8}$ ($10^{3} \cdots 10^{5}$ km)	various degrees of symmetry
Sun	some 10^{-1} (in spots)	10^{9} (10^{6} km)	slight deviations from symmetry about rotation axis and equatorial plane oscillatory, magnetic cycle, grand minima
Cool stars (F, G)		10^{9} (10^{6} km)	sun-like magnetic cycles
Hot stars (A, B)	1	10^{9} (10^{6} km)	oblique rotators
White dwarfs	10^{4}	10^{5} (100 km)	
Neutron stars	10^{8}	10^{4} (10 km)	oblique rotators
Galaxies	10^{-9}	10^{21} (30 kpc)	"axisymmetric" and "bisymmetric" structures

In a crude picture the magnetic field of the Earth is the field of a dipole with the magnetic south pole in the northern hemisphere and the north pole in the southern one and with the dipole axis slightly inclined to the rotation axis. The magnetic flux density at the poles is about 0.6 G, or $0.6 \cdot 10^{-4}$ T [1]. As far as the time variations of the magnetic field of the Earth are concerned we mention only such on large scales. One example are secular variations connected with drifts of the field structures. ¿From paleomagnetic studies we know about the existence of a magnetic field with a dominating dipole part and the present-days order of magnitude since about $3.5 \cdot 10^9$ years. It was, however, occasionally subject to reversals of its polarity, that is, to transitions from phases with the magnetic south pole in the northern hemisphere to such with the north pole in this hemisphere and vice versa. The length of the intervals between reversals lie between 10^5 and 10^7 years, but a reversal lasts only about 10^4 years.

During the last three decades of this century there were spacecraft missions to all planets of the solar system except Pluto, and with them also in-situ measurements of magnetic fields have been carried out. The field of Mercury proved to be much weaker than that of the Earth. Extrapolated to its surface it differs by a factor of about 10^{-3} from the corresponding values for the Earth. No intrinsic magnetic field could be found at Venus. At Mars only a weak magnetic field with strengths comparable to those at Mercury has been measured but the question whether it originates from the interior of the planet is still under debate. The magnetic field of Jupiter shows a geometrical structure very similar to that of the Earth, in particular with almost the same inclination of the dipole axis to the rotation axis, but it is, taken at the surface, stronger by more than a factor 10. Saturn, Uranus and Neptune possess magnetic fields whose strengths at the surfaces are very close to that of the Earth. However, the Saturnian field has a very high degree of axisymmetry about the rotation axis, and the fields of the two other planets mentioned deviate from this symmetry much more than that of the Earth does.

As far as the Sun is concerned not only the sunspots but all phenomena of solar activity such as flares, protuberances, coronal mass ejections etc. are connected with magnetic fields, which are measured with the help of the Zeeman-effect. ¿From the study of sunspots and related phenomena of the solar activity cycle we may conclude that the Sun possesses a general, that is, large-scale magnetic field which consists mainly of two field belts beneath the visible surface with flux densities exceeding at least 10^{-1} T, one in the northern hemisphere and the other, oppositely oriented, in the southern hemisphere. In addition, there is a much weaker poloidal field with only a few 10^{-4} T intersecting the visible surface. This general magnetic field varies periodically in time, more precisely, it changes its polarity with a period which is just two times that of the activity cycle, that is 2×11 years. It is this magnetic cycle which causes and controls all the activity phenomena. Sunspots, for example, occur then as a consequence of instabilities of magnetic flux bundles beneath the visible surface which let

[1] In this article we prefer the international system of units and so the unit Tesla (T) of the magnetic flux density rather than Gauss (G); 1 T = 10^4 G.

these bundles rise and break through the surface. The magnetic cycle affects also the solar corona very strongly and is, for example, responsible for drastic variations of the coronal X-ray emission. When considered over many cycles the solar activity is not strictly periodic. There were several so-called grand minima during the last centuries.

The Sun offers an excellent possibility to study the magnetic phenomena with high resolution. If we could observe the Sun only like a star, that is, as a point-like source of light, it would be impossible, or at least very hard, to detect magnetic fields via Zeeman-effect. It would be then the average of the magnetic flux over the emitting disc which determined the splittings of the magnetically sensible spectral lines, and its smallness makes that the splittings are very small compared to the widths of these lines. This is one of the reasons why there is no direct evidence of magnetic fields at other cool stars comparable to the Sun. However, quite a few features have been observed at a large number of F and G stars which are, according to our knowledge gained in particular by studying the Sun, closely connected with magnetic cycles, for example a cyclic variation of the X-ray emission. There are many good reasons to believe that these stars possess indeed sun-like magnetic cycles.

In the late forties the Zeeman-technique was elaborated for the investigation of stars. On this basis at a number of peculiar A stars magnetic fields with flux densities up to a few T were found. These stars were named "magnetic stars". The flux densities as well as the abundances of particular chemical elements concluded from the spectra show periodic variations with periods of days or weeks. This is interpreted by the model of the "oblique rotator". It assumes structures of the magnetic field and distributions of the chemical elements on its surface which are non-symmetric about the rotation axis and, for an observer moving with the surface, steady. The periodic variations are then simply a consequence of the rotation of the star. Magnetic fields like those of A stars have been observed with some B stars, too.

Much stronger magnetic fields occur at objects corresponding to late stages of the stellar evolution. It was again Zeeman-measurements which revealed that a small fraction of the observed white dwarfs possesses magnetic fields with flux densities up to 10^4 T.

After the discovery of the pulsar phenomenon in the late sixties it turned out that the only acceptable explanation of it can be given by assuming a rapidly rotating neutron star with a very strong magnetic field being non-symmetric about the rotation axis, that is, an oblique rotator. ¿From the observational data flux densities of the order of 10^8 T were derived. In between in a few cases the existence of such strong fields have been confirmed in an independent way by the interpretation of X-ray spectral features as due to electron cyclotron resonance scattering. Recently the observation of anomalous X-ray pulsars suggested that there are even neutron stars with flux densities as large as about 10^{12} T.

Let us now turn from the small objects with extremely strong magnetic fields to extremely large ones with very weak fields. In the last two decades polarization measurements in the radio-range and their interpretation considering the Faraday-effect have shown that many nearby spiral galaxies are penetrated by magnetic fields with flux densities of the order of 10^{-9} T which exhibit simple large-scale spiral patterns covering all the galactic disc. Interestingly enough, two quite different structures of such patterns have been observed, called "axisymmetric" and "bisymmetric" structures. In the first case the structure is roughly symmetric with respect to the rotation axis of the galaxy, and all radial components of the field vectors in the galactic plane point either inward or outward. This implies, of course, that there is magnetic flux out of or into this plane. In the second case the field vectors change its orientation if the pattern is rotated by 180^0 about the axis of the galaxy.

2 The Question of the Origin of Cosmic Magnetic Fields and the Idea of the Cosmic Dynamo

The classical theory of electromagnetism offers two causes for magnetic fields: permanent magnetization of condensed matter and electric currents. Conditions allowing permanent magnetization can be excluded for almost all cosmic objects by several reasons. In particular ferromagnetism is only possible in a range of low temperatures, and even the comparatively cool Earth's core is clearly too hot for that. As a rule, however, the matter in cosmic objects is in a plasma state as, for example, at the Sun, or in some metallic state, as in the Earth's core, and so electric currents are quite possible.

Electric currents in conducting matter are, of course, subject to Ohmic dissipation, which converts the energy stored in the magnetic field into heat. If there is no electromotive force that is able to maintain the currents and so to compensate this energy loss, the currents and the magnetic field are bound to decay. The decay time is proportional to the electric conductivity of the body and to the square of its linear dimensions. As we will see later this time is about 10^4 years for the Earth, and of the order of 10^{11} years for the Sun. Clearly, if there were no electromotive force supporting the electric currents in the Earth's core, the magnetic field would disappear in a time which is extremely short compared to that for which we know about its existence from paleomagnetic studies. The interpretation of steady magnetic fields of objects having solar dimensions and conductivities as "fossil fields", created at the birth of the object and persisting without any electromotive force, cannot generally be excluded but it encounters several difficulties. In any case the Sun's alternating magnetic field can never be explained in this way.

Many candidates for electromotive forces which might be responsible for electric currents and magnetic fields in cosmic bodies have been discussed in the past, for example electromotive forces due to inhomogeneities in the chemical composition or in the temperature of the plasma, those due to different behaviors of electrons and protons under acceleration, etc. Roughly speaking, all possibilities considered but one can be excluded in the explanation of the observed fields, since they lead to much weaker fields only or raise other problems. The only remaining possibility is the generation or maintenance of electric currents by the motion of conducting matter in a magnetic field on the principle of the self-exciting dynamo invented by W. v. Siemens 1866. The idea that the magnetic fields of cosmic objects could result in this way from motions in their conducting interiors was first proposed by J. Larmor [18] in view of the Sun. He also discussed this possibility for the Earth.

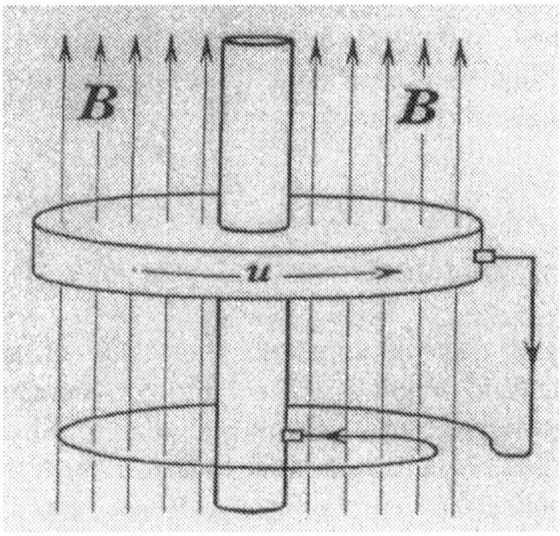

Fig. 1. A disc dynamo. The disc including its axis as well as the wire, with sliding contacts at the rim and the axis of the disc, are electrically conducting, whereas all surroundings are insulating

In order to explain this idea in some more detail consider first a disc dynamo as depicted in Figure 1. If the disc rotates in a given magnetic field, an electromotive force occurs in the disc, which builds up a potential difference between the rim of the disc and the axis. As soon as rim and axis are connected by the conducting wire, this potential difference drives an electric current through the wire. If the latter is properly wound, this current may amplify the original magnetic field. In this way, starting from an arbitrarily weak magnetic field, strong currents and magnetic fields can be produced. Their growth will be limited only by the influence of the forces resulting from currents and magnetic fields on the disc's rotation.

There is a crucial difference between the realization of the dynamo principle in the experimental device considered above and in a cosmic body. For the dynamo action in this device a proper current path is essential, which can easily be fixed by the shape of the conducting wire in its insulating surroundings. A cosmic body, however, is conducting everywhere. So we have to look for a dynamo operating in a medium without insulating regions, which is often called a "homogeneous dynamo". The current paths are then determined by the distribution of the electromotive force given by the fluid motion and the magnetic field, and by the boundary conditions. It was not clear at the beginning whether it was at all possible for currents resulting from this electromotive force to support the magnetic field responsible for it, and it took a long time to learn how the dynamo principle works in cosmic bodies.

3 Magnetofluiddynamics I: Electrodynamic Aspects

In this section we briefly explain some basic principles governing the behavior of the electromagnetic fields in an electrically conducting moving fluid, and add a few remarks of mathematical nature. We consider the motion of the fluid at first as given. The principles governing the motion and the effects of the electromagnetic fields on the motion will be discussed in Section 7.

3.1 Maxwell Equations and Constitutive Equations

We restrict all our considerations to cases with flat space-time. In addition we accept the usual assumptions of magnetofluiddynamics, which we characterize provisionally by high electrical conductivity and non-relativistic velocities of the fluid.

So we require that the electromagnetic fields obey Maxwell's equations in the form

$$\nabla \times \boldsymbol{E} = -\partial_t \boldsymbol{B}\,, \quad \nabla \cdot \boldsymbol{B} = 0\,, \quad \nabla \times \boldsymbol{H} = \boldsymbol{j} \tag{1}$$

and the corresponding constitutive equations in the form

$$\boldsymbol{B} = \mu \boldsymbol{H}\,, \quad \boldsymbol{j} = \sigma(\boldsymbol{E} + \boldsymbol{u} \times \boldsymbol{B} + \boldsymbol{E}^{(e)})\,. \tag{2}$$

We have adopted the international system of units. As usual $\boldsymbol{E}$ means the electric field strength, $\boldsymbol{B}$ the magnetic flux density, $\boldsymbol{H}$ the magnetic field strength, $\boldsymbol{j}$ the electric current density and $\boldsymbol{u}$ the velocity of the fluid. Furthermore, μ is the magnetic permeability of the fluid, always assumed to coincide with that of free space, and σ its electric conductivity. Finally $\boldsymbol{E}^{(e)}$ indicates the place where external or other additional electromotive forces can be included, for example such due to batteries or such describing the effects of the gradients of electron and ion pressure in a plasma, the Hall-effect etc. For the sake of simplicity, however, we ignore $\boldsymbol{E}^{(e)}$, if not indicated otherwise, in the following considerations; the changes which would occur with its inclusion can easily be followed up.

The mentioned assumptions of magnetofluiddynamics can be formulated more precisely by saying that the time ε/σ, where ε means the dielectric constant of free space, is small compared to the characteristic times of the processes considered, and that terms of the order $(u/c)^2$, with c being the speed of light in free space, are negligible in comparison with unity.

Faraday's law (1a)[2] as well as equation (1b), are Maxwell equations in their original forms, that is, are not touched by the assumptions of magnetofluiddynamics. Ampere's law in the form (1c) corresponds to the quasi-steady approximation of electrodynamics, in which the displacement current is ignored, and

[2] If there are several equations in a numbered line (N) we refer to the first one by (Na), to the second one by (Nb) etc.

is just a consequence of these assumptions. Likewise the constitutive equation (2a) can, unless the dielectric constant like the magnetic permeability of the fluid takes its free-space value, only be justified with these assumptions. Finally Ohm's law in the form (2b), with $\boldsymbol{E}^{(e)}$ ignored, can not simply be concluded from the validity of $\boldsymbol{j} = \sigma \boldsymbol{E}$ for an observer moving with fluid. The assumptions of magnetofluiddynamics have to be used in order to justify, for example, the neglect of the convection current $\varrho_e \boldsymbol{u}$, with ϱ_e being the electric charge density, which would otherwise occur.

We note that the equations (1) and (2) together with proper initial or boundary conditions determine the evolution of the electromagnetic fields $\boldsymbol{E}, \boldsymbol{B}, \boldsymbol{H}$ and $\boldsymbol{j}$ if the fluid velocity $\boldsymbol{u}$ is given. In this context (1b) plays only the part of an initial condition, for (1a) implies already $(\partial/\partial t)\nabla \cdot \boldsymbol{B} = 0$.

We have not considered so far the remaining Maxwell equation $\nabla \cdot \boldsymbol{D} = \varrho_e$, where $\boldsymbol{D}$ is the dielectric displacement and ϱ_e again the electric charge density. This equation is not necessary for the calculation of $\boldsymbol{E}, \boldsymbol{B}, \boldsymbol{H}$ and $\boldsymbol{j}$ in the quasi-steady approximation but it allows us, if completed by a constitutive equation connecting $\boldsymbol{D}$ with $\boldsymbol{E}$ and possibly also with $\boldsymbol{u}$ and $\boldsymbol{B}$, to calculate afterwards ϱ_e. By the way, inside a fluid at rest we may put $\varrho_e = 0$ whereas in a moving fluid ϱ_e in general does not vanish.

The assumptions of magnetofluiddynamics imply also simple transformation properties of the electromagnetic fields. Let be $\boldsymbol{B}, \boldsymbol{H}, \boldsymbol{j}$ and $\boldsymbol{E}$ the fields measured in a frame of reference in which the fluid moves with a velocity $\boldsymbol{u}$, and $\boldsymbol{B}', \boldsymbol{H}', \boldsymbol{j}'$ and $\boldsymbol{E}'$ those measured by an observer moving with the fluid. Then we have

$$\boldsymbol{B}' = \boldsymbol{B}\,, \quad \boldsymbol{H}' = \boldsymbol{H}\,, \quad \boldsymbol{j}' = \boldsymbol{j}\,, \quad \boldsymbol{E}' = \boldsymbol{E} + \boldsymbol{u} \times \boldsymbol{B}\,. \tag{3}$$

That is, $\boldsymbol{B}, \boldsymbol{H}$ and $\boldsymbol{j}$ follow simply the Galilean transformation law, and only $\boldsymbol{E}$ the Lorentzian law, specified to small velocities.

In the following we will deal also with fluid bodies surrounded by non-conducting, for instance free space. Then we require the validity of the equations (1) and (2) for all space with the exception that (2b) is replaced by $\boldsymbol{j} = \boldsymbol{0}$ for the non-conducting space. That is, the quasi-steady approximation is used for the non-conducting space too. In particular, electromagnetic waves are generally excluded.

3.2 The Induction Equation

The equations (1) and (2) governing the electromagnetic fields in an electrically conducting fluid can be easily reduced to equations for $\boldsymbol{B}$ alone. Starting from (1a), replacing then $\boldsymbol{E}$ according to (2b) by $\boldsymbol{j}/\sigma - \boldsymbol{u} \times \boldsymbol{B}$ and $\boldsymbol{j}$ in turn according to (1c) and (2a) by $(1/\mu)\nabla \times \boldsymbol{B}$ we arrive at

$$\nabla \times (\eta \nabla \times \boldsymbol{B}) - \nabla \times (\boldsymbol{u} \times \boldsymbol{B}) + \partial_t \boldsymbol{B} = \boldsymbol{0}\,, \quad \nabla \cdot \boldsymbol{B} = 0 \tag{4}$$

with

$$\eta = 1/\mu\sigma\,. \tag{5}$$

We call (4a) the induction equation and η the magnetic diffusivity, or magnetic viscosity. If η is independent of position we have simply

$$\eta \nabla^2 \boldsymbol{B} + \nabla \times (\boldsymbol{u} \times \boldsymbol{B}) - \partial_t \boldsymbol{B} = \boldsymbol{0}, \quad \nabla \cdot \boldsymbol{B} = 0. \tag{6}$$

Equations (4) or (6) poses an initial-boundary value problem for $\boldsymbol{B}$. As soon as a solution $\boldsymbol{B}$ is known, $\boldsymbol{H}$, $\boldsymbol{j}$ and $\boldsymbol{E}$ can be calculated from (1) and (2) without any further integration.

The time variation of the magnetic field, $\partial_t \boldsymbol{B}$, is determined by two physical effects: some kind of diffusion of the field, coupled with dissipation, described by the term $\nabla \times (\eta \nabla \times \boldsymbol{B})$, or $\eta \nabla^2 \boldsymbol{B}$, and a transport of the field, or advection, described by $\nabla \times (\boldsymbol{u} \times \boldsymbol{B})$. The relative importance of advection and dissipation effects can be characterized by the magnetic Reynolds number R_{m}, defined by

$$R_{\mathrm{m}} = UL/\eta_c, \tag{7}$$

where U is a characteristic fluid velocity, L a characteristic length of the process considered and η_c a characteristic value of the magnetic diffusivity. If $R_{\mathrm{m}} \ll 1$ the behavior of the magnetic field is dominated by dissipation, if $R_{\mathrm{m}} \gg 1$ by advection. Under laboratory conditions values of R_{m} exceeding unity can only be reached with enormous efforts, whereas in cosmic objects the values of R_{m} are in general, already as a consequence of the large dimensions, extremely high. Examples are given in Table 2.

Let us consider the time scales on which a magnetic field evolves. The quantities entering the induction equation, $\boldsymbol{u}$ and η, with the characteristic values U and η_c introduced above, together with a characteristic length L allow us to define two times,

$$T_\eta = L^2/\eta_c, \quad T_u = L/U, \tag{8}$$

the first of which we call "diffusion time" or "dissipation time" and the second one "kinematic time", in special context also "turn-over time". By the way, they satisfy $T_\eta/T_u = R_{\mathrm{m}}$. Examples of numerical values, both for laboratory devices and for cosmic objects, are also given in Table 2, too.

We may write the induction equation with dimensionless space and time coordinates. Let us measure the space and time coordinates in units of L and T, respectively, and replace $\boldsymbol{u}$ by $\boldsymbol{u}U$ where $\boldsymbol{u}$ is now dimensionless, and η by $\tilde{\eta}\eta_c$ with $\tilde{\eta}$ being dimensionless too. When identifying T with T_η we then have

$$\nabla \times (\tilde{\eta} \nabla \times \boldsymbol{B}) + R_{\mathrm{m}} \nabla \times (\boldsymbol{u} \times \boldsymbol{B}) - \partial_t \boldsymbol{B} = \boldsymbol{0}, \tag{9}$$

or, identifying T with T_u,

$$R_{\mathrm{m}}{}^{-1} \nabla \times (\tilde{\eta} \nabla \times \boldsymbol{B}) + \nabla \times (\boldsymbol{u} \times \boldsymbol{B}) - \partial_t \boldsymbol{B} = \boldsymbol{0}. \tag{10}$$

3.3 The Magnetic Energy

Before discussing more consequences of the induction equation we deal briefly with the energy stored in the magnetic field. Under the assumptions introduced

Table 2. Values of the magnetic Reynolds number R_m and the diffusion and kinematic times T_η and T_u for laboratory devices as well as the Earth and the Sun. As a comparison value for the electric conductivities σ we note that for copper: $6 \cdot 10^7$ S/m. For the laboratory devices U and L are arbitrarily chosen. For the Earth's core U gives a plausible magnitude of the internal motion, and L corresponds to about one third of the radius. As far as the the convection zone of the Sun is concerned, for granules U and L give their typical scales at the surface, and for sunspots L reflects their typical horizontal extension at the surface. For the consideration concerning the interior of the Sun, L is taken as roughly one third of the solar radius. More comments concerning the values for the Earth's core and the Sun's interior are given in Section 3.4, and concerning the values for the Sun's convection zone in Section 5.7

	σ [S/m] η [m²/s]	U [m/s]	L [m]	R_m	T_η [s]	T_u [s]
Mercury 18°C	$1.04 \cdot 10^6$ $7.65 \cdot 10^{-1}$	1	1	1.3	1.3	1
Sodium 100°C	$1.03 \cdot 10^7$ $7.73 \cdot 10^{-2}$	1	1	12.9	12.9	1
Earth's core	$3 \cdot 10^5$ 2.65	10^{-3}	10^6	$3.8 \cdot 10^2$	$3.8 \cdot 10^{11}$ ($1.2 \cdot 10^4$ yrs)	10^9 (32 yrs)
Sun's convection zone	$3 \cdot 10^3$ $2.65 \cdot 10^2$					
granules		$2 \cdot 10^2$	$2 \cdot 10^6$	$1.5 \cdot 10^6$	$1.5 \cdot 10^{10}$ ($4.8 \cdot 10^2$ yrs)	10^4 (2.8 h)
sunspots			10^7		$3.8 \cdot 10^{11}$ ($1.2 \cdot 10^4$ yrs)	
Sun's interior	10^8 $8.0 \cdot 10^{-3}$		$2 \cdot 10^8$		$5.0 \cdot 10^{18}$ ($1.6 \cdot 10^{11}$ yrs)	

the magnetic energy density is given by $\boldsymbol{B}^2/2\mu$ and the total magnetic energy by the integral of this quantity over all space. We may conclude from the basic equations (1) and (2a) by standard manipulations that

$$\frac{\partial}{\partial t}\left(\frac{\boldsymbol{B}^2}{2\mu}\right) = -\boldsymbol{j} \cdot \boldsymbol{E} - \nabla \cdot \boldsymbol{S}\,, \tag{11}$$

where $\boldsymbol{S}$ is the Poynting vector,

$$\boldsymbol{S} = \boldsymbol{E} \times \boldsymbol{H}\,. \tag{12}$$

The quantity $\boldsymbol{j} \cdot \boldsymbol{E}$ can be interpreted as the work on the charged particles constituting the electric current done by the electric field, and $\nabla \cdot \boldsymbol{S}$ as flow of magnetic energy out of or into a volume element. For an electric conductor, where Ohm's law (2b) applies, relation (11) can be specified to take the form

$$\frac{\partial}{\partial t}\Big(\frac{\boldsymbol{B}^2}{2\mu}\Big) = -\frac{\boldsymbol{j}^2}{\sigma} - \boldsymbol{u} \cdot (\boldsymbol{j} \times \boldsymbol{B}) - \nabla \cdot \boldsymbol{S}\,. \tag{13}$$

Here $\boldsymbol{j}^2/\sigma$ describes the Joule heat production and $\boldsymbol{u} \cdot (\boldsymbol{j} \times \boldsymbol{B})$, if positive, the work on the fluid done by the Lorentz force or, if negative, the work done by the fluid against the Lorentz force.

Considering the variation of the total magnetic energy in time we admit that the conducting body occupies only a part of the space and the remaining part is non-conducting. We integrate both sides of (11) over all space. The integral with $\boldsymbol{j} \cdot \boldsymbol{E}$ reduces itself to one over the conducting body only, where we can use Ohm's law (2b) as we have done in (13). We further accept the reasonable assumption that $\boldsymbol{S}$ vanishes at infinity stronger than $\boldsymbol{O}(r^{-2})$ where r means the distance from a given point. This applies in any case if $\boldsymbol{E}$ and $\boldsymbol{H}$ vanish at least like the fields of an electric charge and of a magnetic dipole. Then the integral over $\nabla \cdot \boldsymbol{S}$ proves to be zero. Thus we obtain

$$\frac{\mathrm{d}}{\mathrm{d}t}\int\limits_{\infty} \frac{\boldsymbol{B}^2}{2\mu}\,dv = -\int\limits_{\mathcal{V}} \frac{\boldsymbol{j}^2}{\sigma}\,\mathrm{d}v - \int\limits_{\mathcal{V}} \boldsymbol{u} \cdot (\boldsymbol{j} \times \boldsymbol{B})\,\mathrm{d}v\,, \tag{14}$$

where $\mathcal{V}$ denotes the region occupied by the fluid body. This result implies that in the absence of a fluid motion any magnetic field is bound to decay. For the maintenance of a magnetic field sufficiently powerful fluid motions are needed.

3.4 The Special Case of a Conductor at Rest

In the absence of motions the magnetic flux density $\boldsymbol{B}$ in a fluid has to obey the equations (4) with $\boldsymbol{u} = \boldsymbol{0}$, that is,

$$\nabla \times (\eta \nabla \times \boldsymbol{B}) + \partial_t \boldsymbol{B} = \boldsymbol{0}\,, \quad \nabla \cdot \boldsymbol{B} = 0\,. \tag{15}$$

Let us restrict our attention to magnetic fields $\boldsymbol{B}$ vanishing at infinity at least like a dipole field. Then the condition concerning $\boldsymbol{S}$ used in the derivation of the magnetic energy balance (14) is fulfilled, and we may conclude that any magnetic field $\boldsymbol{B}$ must decay in the course of time. We speak here, in the absence of fluid motions, of free decay.

We consider first the case in which the conducting fluid is homogeneous and occupies all space, for which the solution $\boldsymbol{B}$ of (15) can readily be given for an arbitrary initial condition.

In view of a later application we deal first with the more general problem which occurs by the inclusion of an arbitrary electromotive force $\boldsymbol{E}^{(e)}$ as mentioned in the context of (2). So we start here from

$$\eta \nabla^2 \boldsymbol{B} - \partial_t \boldsymbol{B} = -\nabla \times \boldsymbol{E}^{(e)}\,, \quad \nabla \cdot \boldsymbol{B} = 0\,. \tag{16}$$

Each Cartesian component of the first of these equations is analogous to a heat conduction equation of the form

$$\eta \Delta T - \partial_t T = -q \,, \tag{17}$$

where T means a temperature field, η now a temperature conduction coefficient independent of position and time, and q stands for heat sources. We consider (17) as valid in all space and assume that q vanishes at infinity, and we look for solutions $T = T(\boldsymbol{x}, t)$ vanishing at infinity too. As it is well known the solution of the initial value problem defined by a given $T = T(\boldsymbol{x}, t_0)$ for some initial time t_0 can be written in in the form

$$\begin{aligned} T(\boldsymbol{x}, t) = & \int_\infty G(\boldsymbol{x} - \boldsymbol{x}', t - t_0) \, T(\boldsymbol{x}', t_0) \, \mathrm{d}^3x' \\ & + \int_{t_0}^{t} \int_\infty G(\boldsymbol{x} - \boldsymbol{x}', t - t') \, q(\boldsymbol{x}', t') \, \mathrm{d}^3x' \, \mathrm{d}t' \,. \end{aligned} \tag{18}$$

Here $G(\boldsymbol{x}, t)$ means a Green's function defined by

$$\Delta G - \partial_t G = 0 \quad \text{for} \quad t > 0 \quad \text{and} \quad G \to \delta^3(\boldsymbol{x}) \quad \text{as} \quad t \to 0 \,, \tag{19}$$

that is,

$$G(\boldsymbol{x}, t) = (4\pi\eta t)^{-3/2} \exp(-\boldsymbol{x}^2/4\eta t) \,. \tag{20}$$

We conclude from this that the solution of equation (16a) for $\boldsymbol{B}$ can be given in the form

$$\begin{aligned} \boldsymbol{B}(\boldsymbol{x}, t) = & \int_\infty G(\boldsymbol{x} - \boldsymbol{x}', t - t_0) \, \boldsymbol{B}(\boldsymbol{x}', t_0) \, \mathrm{d}^3x' \\ & + \int_{t_0}^{t} \int_\infty G(\boldsymbol{x} - \boldsymbol{x}', t - t') \, (\nabla' \times \boldsymbol{E}^{(e)}(\boldsymbol{x}', t')) \, \mathrm{d}^3x' \, \mathrm{d}t' \,. \end{aligned} \tag{21}$$

It can be easily shown that the condition (16b) is indeed satisfied for all $t > t_0$ if it holds true for $t = t_0$.

If we now put again $\boldsymbol{E}^{(e)} = \boldsymbol{0}$, equation (21) delivers us the mentioned general solution of the initial value problem posed by (16).

Likewise for the cases with a finite fluid body surrounded by free space solutions of the free-decay problem are known. As an example we consider a spherical body with constant electric conductivity. In this case the equations governing $\boldsymbol{B}$ can be solved analytically. The solution for an arbitrary initial distribution of $\boldsymbol{B}$ can be represented as a superposition of independent modes, each of which has the form $\boldsymbol{B}_n(\boldsymbol{x}) \exp(-\lambda_n t)$ with a constant λ_n being its decay rate. The slowest-decaying mode is a dipole field. Its decay-rate, say λ_1, is given by

$$\lambda_1 = \pi^2 \eta / R^2 \,, \tag{22}$$

where R is the radius of the body. The corresponding decay time T_{decay} defined by $\lambda_1 T_{\mathrm{decay}} = 1$ reads

$$T_{\mathrm{decay}} = R^2 / \pi^2 \eta \,. \tag{23}$$

We note that T_{decay} coincides with T_η defined in (8) if we put $L = R/\pi$. This may justify some of our choices of L in Table 2.

3.5 The Magnetic Flux

We return now to the case of moving fluids. It is often useful to consider the magnetic flux Φ_{m} through a given surface $\mathcal{S}$, defined by

$$\Phi_{\mathrm{m}} = \int_{\mathcal{S}} \boldsymbol{B} \cdot \mathrm{d}\boldsymbol{s} \,. \tag{24}$$

Due to the solenoidality of $\boldsymbol{B}$ this quantity must coincide for all surfaces $\mathcal{S}$ with the same contour $\partial\mathcal{S}$.

A quantity of particular interest is the magnetic flux Φ_{m} through a surface $\mathcal{S}$ which moves with the fluid, called "co-moving" or "material" surface in the following. The variation of Φ_{m} in time depends then on the variations of both $\boldsymbol{B}$ and $\mathcal{S}$. Simple geometrical considerations, using the solenoidality of $\boldsymbol{B}$, show that

$$\frac{\mathrm{d}\Phi_{\mathrm{m}}}{\mathrm{d}t} = \int_{\mathcal{S}} (\partial_t \boldsymbol{B} - \nabla \times (\boldsymbol{u} \times \boldsymbol{B})) \cdot \mathrm{d}\boldsymbol{s} \,. \tag{25}$$

The second term under the integral is due to the motion of the surface $\mathcal{S}$.

Replacing now $\partial_t \boldsymbol{B}$ under the integral in (25) by $-\nabla \times \boldsymbol{E}$, employing Stokes' theorem and using Ohm's law (2b) we find

$$\frac{\mathrm{d}\Phi_{\mathrm{m}}}{\mathrm{d}t} = -\int_{\partial\mathcal{S}} \frac{\boldsymbol{j}}{\sigma} \cdot \mathrm{d}\boldsymbol{l} \,, \tag{26}$$

where the orientation of the contour $\partial\mathcal{S}$ defined by $d\boldsymbol{l}$ is assigned to the surface element $d\boldsymbol{s}$ introduced with (24) in the sense of a right-handed screw. Equation (26) is very useful for studying induction processes in moving fluids.

3.6 The High-Conductivity Limit

As explained above, in many studies of processes in cosmic objects we are faced with very high values of the magnetic Reynolds number R_{m}. In the limit $R_{\mathrm{m}} \to \infty$, which we call the high-conductivity limit, equations (4) turn into

$$\partial_t \boldsymbol{B} - \nabla \times (\boldsymbol{u} \times \boldsymbol{B}) = \boldsymbol{0} \,, \quad \nabla \cdot \boldsymbol{B} = 0 \,. \tag{27}$$

This can be most easily concluded from (10). We note that (4) and (27) differ in the order of the highest derivatives and so in the boundary conditions needed. The solutions of (27) can be readily given as soon as the paths $\boldsymbol{x} = \boldsymbol{x}(t)$ of the fluid elements, that is, the solutions of $\mathrm{d}\boldsymbol{x}/\mathrm{d}t = \boldsymbol{u}(\boldsymbol{x}, t)$ are known.

Remembering (25) we conclude from (27), or we can derive directly from (26), that in this limit

$$\frac{\mathrm{d}\Phi_{\mathrm{m}}}{\mathrm{d}t} = 0 \tag{28}$$

for any material surface $\mathcal{S}$. That is, the magnetic flux through such surfaces is conserved.

The equations (27) and(28) are equivalent to each other in the sense that, if (25) is given, (27) implies (28), and the validity of (28) for any surface $\mathcal{S}$ allows us to conclude (27a).

Let us consider a magnetic flux tube defined such that its boundary is not intersected by magnetic field lines. As a consequence of the solenoidality of the magnetic flux density, $\nabla \cdot \boldsymbol{B} = 0$, the magnetic flux through each cross-section of the tube is the same. Let us mark the fluid which is at a given time enclosed in a given flux tube and consider the regions in which it occurs due to its motion at a later time. Since in the limit considered the magnetic flux through material surfaces is conserved, this region must be again a flux tube. In other words, the fluid flow transforms flux tubes into flux tubes. An analogous conclusion is that two fluid elements, if they are at a given time connected by a magnetic field line, are always connected by a field line. In that sense we speak of "frozen magnetic fields", in particular of "frozen magnetic field lines".

An direct consequence of this is that the topology of field lines in an ideal conductor can never change.

Another interesting consequence of the magnetic flux conservation in an ideal conductor was pointed out by Bondi and Gold [19]. Consider a fluid body which occupies a finite simply connected region surrounded by free space, and a magnetic field penetrating this body and continuing in outer space. Imagine a sphere so that the body lies completely in it. In the space outside this sphere the magnetic field can be represented by a multipole expansion, that is, in the form $\boldsymbol{B} = -\nabla \Phi$ with Φ being a sum of terms $c_l^m r^{-(l+1)} P_l^m(\cos\theta)\exp(im\phi)$ where the c_l^m are complex coefficients, the P_l^m associated Legendre polynomials, and r, θ and ϕ spherical coordinates; $l = 1$ corresponds to a dipole, $l = 2$ to a quadrupole, etc. Due to the fluid motion inside the body the magnetic field may well change in time. As a consequence of the magnetic flux conservation at the boundary, however, the c_l^m are bounded, that is, the $|c_l^m|$ do not exceed certain values determined by the initial magnetic field. So the magnetic field in outer space can not grow arbitrarily.

3.7 Magnetic Field and Differential Rotation

The concept of frozen magnetic flux is also useful in order to form pictures on how magnetic fields in a conducting fluid evolve under the influence of its motion, even for cases with a finite magnetic Reynolds number. We demonstrate this for magnetic fields which penetrate a conducting spherical body showing differential rotation, that is rotation with an angular velocity varying with radius or latitude. We will rely on this example in our explanations on dynamos later.

There is a crucial difference in the behavior of fields being symmetric or non-symmetric about the rotation axis.

With axisymmetric fields the effect of differential rotation can easily be followed up. In the example depicted in the left half of Figure 2 we start from a magnetic field of dipole-type whose symmetry axis coincides with the rotation axis of the body. As a consequence of the rotational shear the magnetic field lines are stretched and wound up. The resulting field configuration inside the body

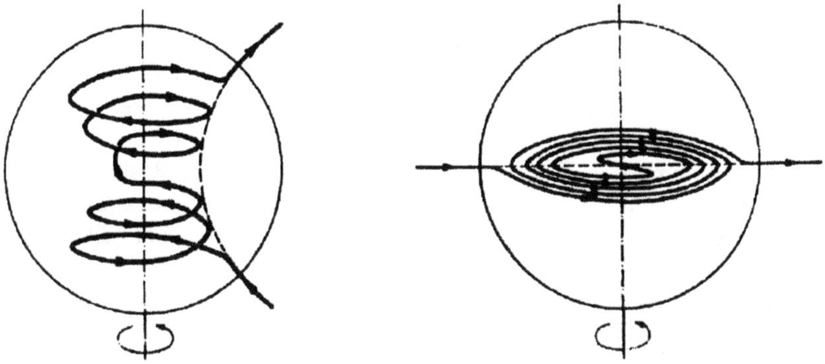

Fig. 2. The influence of a differential rotation on axisymmetric and non-axisymmetric magnetic fields. It is assumed that the inner parts of the body rotate in the indicated way whereas the surface is at rest. Left: a field line of an axisymmetric, initially purely poloidal field. Right: a field line of a non-axisymmetric, initially purely poloidal field. The dotted lines show the initial field lines

can be understood as a superposition of the original field and two oppositely oriented field belts in the two hemispheres created by the differential rotation. The field in outer space, as a continuation of the original field, remains unaffected by the differential rotation. If the original field is maintained, for example by a proper electromotive force, the field in the belts evolves in competition with the Ohmic dissipation and reaches a steady state. Its magnitude in this state is determined by the magnetic Reynolds number which we have to ascribe to the differential rotation. Arbitrarily strong fields can be obtained if only this Reynolds number is sufficiently high.

With non-axisymmetric magnetic fields the effect of differential rotation is more complex. In the example shown in the right half of Figure 2 we start again from a dipole field but suppose its axis to lie in the equatorial plane of the rotating body. Again the field lines are stretched and wound up by the differential rotation. In contrast to the axisymmetric case, however, this leads to a configuration in which oppositely oriented field lines lie very close together. As a consequence of the small-scale structures generated an enhancement of the dissipation occurs. The amplification of the magnetic field by stretching of the field lines then competes with the enhanced dissipation, and it is impossible to reach high field strengths. The field continuing in outer space is weakened too.

The outlined difference in the behavior of axisymmetric and non-axisymmetric magnetic fields has many interesting consequences [20–23].

3.8 Symmetry Properties of the Basic Equations

As it is well known Maxwell's equations together with constitutive equations with constant coefficients show certain symmetry properties, which allow us to derive from a given solution other ones by subjecting all field quantities to changes like translations, time shifts, rotations or reflections. For later use we formulate here such properties for our basic equations (1) and (2).

First we define such changes for an arbitrary vector field $\boldsymbol{F}$. We denote the fields that occur with translations, time shifts, rotations or reflections by $\boldsymbol{F}^{\mathrm{tr}}$, $\boldsymbol{F}^{\mathrm{ts}}$, $\boldsymbol{F}^{\mathrm{rot}}$ or $\boldsymbol{F}^{\mathrm{ref}}$. Then we have $\boldsymbol{F}^{\mathrm{tr}}(\boldsymbol{x},t) = \boldsymbol{F}(\boldsymbol{x}+\Delta\boldsymbol{x},t)$ with a constant vector $\Delta\boldsymbol{x}$, and $\boldsymbol{F}^{\mathrm{ts}}(\boldsymbol{x},t) = \boldsymbol{F}(\boldsymbol{x},t+\Delta t)$ with a constant Δt. Restricting ourselves on rotations about an axis running trough the point $\boldsymbol{x} = \boldsymbol{0}$ and on reflections about planes containing this point we have $\boldsymbol{F}^{\mathrm{rot}}(\boldsymbol{x},t) = \boldsymbol{D}^{-1}\boldsymbol{F}(\boldsymbol{D}\boldsymbol{x},t)$ where $\boldsymbol{D}$ is a matrix with $\det(\boldsymbol{D}) = 1$, and $\boldsymbol{F}^{\mathrm{ref}}(\boldsymbol{x},t) = \boldsymbol{D}^{-1}\boldsymbol{F}(\boldsymbol{D}\boldsymbol{x},t)$ with another $\boldsymbol{D}$ with $\det(\boldsymbol{D}) = -1$. The last relation applies also for the reflection at the point $\boldsymbol{x} = \boldsymbol{0}$ and takes then the particular form $\boldsymbol{F}^{\mathrm{ref}}(\boldsymbol{x},t) = -\boldsymbol{F}(-\boldsymbol{x},t)$. We note that a reflection about a plane can always be composed of an reflection about a point in this plane and a 180^0 rotation about an axis intersecting this plane perpendicularly in this point.

Returning now to Maxwell's equations and the constitutive equations in the form (1) and (2) we recall that μ was introduced as a constant, and we assume here in addition σ to be independent on position and time too. Let us suppose that these equations are satisfied with the fields $\boldsymbol{B}, \boldsymbol{H}, \boldsymbol{E}, \boldsymbol{j}$ and $\boldsymbol{u}$. Then the same holds true after replacing these fields with $\boldsymbol{B}^{\mathrm{tr}}, \boldsymbol{H}^{\mathrm{tr}}, \boldsymbol{E}^{\mathrm{tr}}$, $\boldsymbol{j}^{\mathrm{tr}}$ and $\boldsymbol{u}^{\mathrm{tr}}$, with $\boldsymbol{B}^{\mathrm{ts}}, \boldsymbol{H}^{\mathrm{ts}}, \boldsymbol{E}^{\mathrm{ts}}$, $\boldsymbol{j}^{\mathrm{ts}}$ and $\boldsymbol{u}^{\mathrm{ts}}$, with $\boldsymbol{B}^{\mathrm{rot}}, \boldsymbol{H}^{\mathrm{rot}}, \boldsymbol{E}^{\mathrm{rot}}$, $\boldsymbol{j}^{\mathrm{rot}}$ and $\boldsymbol{u}^{\mathrm{rot}}$, or, in formal contrast to this, with $-\boldsymbol{B}^{\mathrm{ref}}, -\boldsymbol{H}^{\mathrm{ref}}, \boldsymbol{E}^{\mathrm{ref}}$, $\boldsymbol{j}^{\mathrm{ref}}$ and $\boldsymbol{u}^{\mathrm{ref}}$. The peculiarity with the signs in the last case does not indicate a physically relevant symmetry breaking but is a consequence of the definition of the curl operation. Note that it is defined either with reference to an right-handed coordinate system or in a coordinate-independent way via Stokes' theorem using then a connection between the direction of the normal vector of a surface and the orientation of its contour in the right-hand sense.

For the induction equation, which can be derived from (1) and (2) the situation is simpler. Since the equations (4) are linear and homogeneous in $\boldsymbol{B}$ its validity remains untouched by changing the sign of $\boldsymbol{B}$. Consequently, if these equations apply with the fields $\boldsymbol{B}$ and $\boldsymbol{u}$ they do so also after replacing them with $\boldsymbol{B}^{\mathrm{tr}}$ and $\boldsymbol{u}^{\mathrm{tr}}$, with $\boldsymbol{B}^{\mathrm{ts}}$ and $\boldsymbol{u}^{\mathrm{ts}}$, with $\boldsymbol{B}^{\mathrm{rot}}$ and $\boldsymbol{u}^{\mathrm{rot}}$, and also with $\boldsymbol{B}^{\mathrm{ref}}$ and $\boldsymbol{u}^{\mathrm{ref}}$.

The above-mentioned peculiarity with reflected fields is often taken as a reason to introduce the concept of polar and axial vectors, in which $\boldsymbol{E}$, $\boldsymbol{j}$ and $\boldsymbol{u}$ occur as polar and $\boldsymbol{B}$ and $\boldsymbol{H}$ as axial vectors. So far we have considered changes of given vector fields but never any coordinate transformations. The definition of polar and axial vectors is based on the behaviors of their component representations under coordinate transformations. The statements made above have counterparts on the level of the behavior of the component representations of the equations considered. We prefer, however, to draw our conclusions primarily by considering changes of the fields rather than changes of coordinate systems, and will only occasionally comment them in terms of polar and axial vectors.

3.9 Poloidal and Toroidal Vector Fields

In the discussion of special problems with vector fields like $\boldsymbol{B}$ or $\boldsymbol{u}$ it proves to be advantageous to consider them as sums of poloidal and toroidal parts. If the fields are axisymmetric the definitions of these parts are very simple. In the absence of this symmetry the situation is more complex, and the generalizations to this case are in a sense restricted to spherical problems. We explain here the definitions of poloidal and toroidal fields and mention their most important properties. For more details and proofs we refer to other representations, e.g. [24,2,25].

Let us start with an axisymmetric vector field, $\boldsymbol{F}$, and adopt a cylindrical coordinate system s, ϕ, z, or a spherical one r, θ, ϕ, such that the components of $\boldsymbol{F}$ with respect to these systems do not depend on ϕ. We put then

$$\boldsymbol{F} = \boldsymbol{F}^{\mathrm{P}} + \boldsymbol{F}^{\mathrm{T}}, \tag{29}$$

call $\boldsymbol{F}^{\mathrm{P}}$ and $\boldsymbol{F}^{\mathrm{T}}$ poloidal and toroidal fields and define them by

$$\boldsymbol{F}^{\mathrm{P}} = \boldsymbol{F} - (\boldsymbol{F}\cdot\boldsymbol{e}_\phi)\boldsymbol{e}_\phi\,, \quad \boldsymbol{F}^{\mathrm{T}} = (\boldsymbol{F}\cdot\boldsymbol{e}_\phi)\boldsymbol{e}_\phi\,, \tag{30}$$

where $\boldsymbol{e}_\phi$ means the unit vector in ϕ-direction. This definition implies several interesting properties of poloidal and toroidal fields. For example, we have $\nabla\cdot\boldsymbol{F}^{\mathrm{T}} = 0$, and $\nabla\times\boldsymbol{F}^{\mathrm{P}}$ and $\nabla\times\boldsymbol{F}^{\mathrm{T}}$ are toroidal and poloidal, respectively. As a consequence $\nabla^2\boldsymbol{F}^{\mathrm{P}}$ is poloidal and $\nabla^2\boldsymbol{F}^{\mathrm{T}}$ toroidal.

In the special case where $\boldsymbol{F}$ is solenoidal, $\nabla\cdot\boldsymbol{F} = 0$, in addition to $\nabla\cdot\boldsymbol{F}^{\mathrm{T}} = 0$ we have also $\nabla\cdot\boldsymbol{F}^{\mathrm{P}} = 0$. Then $\boldsymbol{F}^{\mathrm{P}}$ can be expressed with the help of a vector potential, which has to be toroidal, that is,

$$\boldsymbol{F}^{\mathrm{P}} = \nabla\times(G\boldsymbol{e}_\phi) = \nabla(sG)\times\frac{\boldsymbol{e}_\phi}{s} \tag{31}$$

with some scalar quantity G. We note that $s = r\cos\theta$.

When identifying $\boldsymbol{F}$ with the magnetic flux density $\boldsymbol{B}$, which has to be solenoidal, and adopting the usual notation we have

$$\boldsymbol{B} = \boldsymbol{B}^{\mathrm{P}} + \boldsymbol{B}^{\mathrm{T}}, \quad \boldsymbol{B}^{\mathrm{P}} = \nabla\times(A\boldsymbol{e}_\phi) = \nabla(sA)\times\frac{\boldsymbol{e}_\phi}{s}, \quad \boldsymbol{B}^{\mathrm{T}} = B\boldsymbol{e}_\phi\,, \tag{32}$$

with two scalars A and B. As can easily be shown $2\pi s_0 A(s_0, z_0)$ is just the magnetic flux through a surface whose contour is the circle defined by $s = s_0$ and $z = z_0$. The field lines of $\boldsymbol{B}^{\mathrm{P}}$ are given by $sA = \text{const}$ together with $\phi = \text{const}$, those of $\boldsymbol{B}^{\mathrm{T}}$ are concentric circles around the axis of the coordinate system.

Let us now switch to a general, not necessarily axisymmetric vector field $\boldsymbol{F}$. We first remark that any such field can be represented in the form

$$\boldsymbol{F} = \boldsymbol{r}\times\nabla U + \boldsymbol{r}V + \nabla W\,, \tag{33}$$

where $\boldsymbol{r}$ means the radius vector with $\boldsymbol{r} = \boldsymbol{0}$ at the origin of the coordinate system, and U, V and W are scalar functions depending on the three coordinates r, θ and ϕ. The determination of U, V and W for a given $\boldsymbol{F}$ requires in general the integration of a system of partial differential equations with respect to θ and ϕ on surfaces $r = \text{const}$. Clearly, $\boldsymbol{F}$ is invariant under certain gauge transformations of these functions. The only possibilities for such transformations are $U \to U + u$ and $V \to V - \mathrm{d}w/\mathrm{d}r$ in combination with $W \to W + w$, with u and w depending only on r but not on θ or ϕ. They leave not only $\boldsymbol{F}$ unchanged but also $\boldsymbol{r} \times \nabla U$ and $\boldsymbol{r}V + \nabla W$.

When working with representations like (33) it is useful to recall the vector relations

$$\begin{aligned} \nabla \times (\boldsymbol{r}F) &= -\boldsymbol{r} \times \nabla F \\ \nabla \times (\nabla \times (\boldsymbol{r}F)) &= -\nabla \times (\boldsymbol{r} \times \nabla F) = -\boldsymbol{r}\,\Delta F + \nabla \frac{\partial}{\partial r}(rF) \\ \nabla \times (\nabla \times (\nabla \times (\boldsymbol{r}F)) &= \nabla^2 (\boldsymbol{r} \times \nabla F) = \boldsymbol{r} \times \nabla \Delta F \\ \boldsymbol{r} \times (\boldsymbol{r} \times \nabla F) &= \frac{\boldsymbol{r}}{r} \frac{\partial}{\partial r} (r^2 F) - \nabla (r^2 F) , \end{aligned} \tag{34}$$

where F is any scalar.

We split now again $\boldsymbol{F}$ according to (29) into poloidal and toroidal parts, $\boldsymbol{F}^{\mathrm{P}}$ and $\boldsymbol{F}^{\mathrm{T}}$. These are uniquely defined by requiring that they can be represented in the form

$$\boldsymbol{F}^{\mathrm{P}} = \boldsymbol{r}V + \nabla W , \quad \boldsymbol{F}^{\mathrm{T}} = \boldsymbol{r} \times \nabla U , \tag{35}$$

or, in components with respect to the spherical coordinate system,

$$\boldsymbol{F}^{\mathrm{P}} = (rV + \frac{\partial W}{\partial r}, \frac{1}{r}\frac{\partial W}{\partial \theta}, \frac{1}{r\sin\theta}\frac{\partial W}{\partial \phi}), \quad \boldsymbol{F}^{\mathrm{T}} = (0, -\frac{1}{\sin\theta}\frac{\partial U}{\partial \phi}, \frac{\partial U}{\partial \theta}), \tag{36}$$

by three scalars U, V and W. In contrast to the definition of $\boldsymbol{F}^{\mathrm{P}}$ and $\boldsymbol{F}^{\mathrm{T}}$ given for the axisymmetric case our generalized one is no longer local but considers $\boldsymbol{F}$ on a whole surface $r = \text{const}$. It implies again remarkable properties of poloidal and toroidal fields:

(i) If, on a surface $r = \text{const}$, $\boldsymbol{F} = \boldsymbol{0}$ then also $\boldsymbol{F}^{\mathrm{P}} = \boldsymbol{F}^{\mathrm{T}} = \boldsymbol{0}$ and vice versa.
(ii) If f is a scalar depending only on r but not on θ or ϕ, then $f\,\boldsymbol{F}^{\mathrm{P}}$ is poloidal and $f\,\boldsymbol{F}^{\mathrm{T}}$ is toroidal.
(iii) $\boldsymbol{r} \times \boldsymbol{F}^{\mathrm{P}}$ is toroidal and $\boldsymbol{r} \times \boldsymbol{F}^{\mathrm{T}}$ poloidal.
(iv) $\boldsymbol{F}^{\mathrm{T}}$ is solenoidal, that is $\nabla \cdot \boldsymbol{F}^{\mathrm{T}} = 0$.
(v) $\nabla \times \boldsymbol{F}^{\mathrm{P}}$ is toroidal and $\nabla \times \boldsymbol{F}^{\mathrm{T}}$ poloidal.
(vi) If, on a surface $r = \text{const}$, $\boldsymbol{r} \cdot (\nabla \times \boldsymbol{F}^{\mathrm{T}}) = 0$ then $\boldsymbol{F}^{\mathrm{T}} = \boldsymbol{0}$.
(vii) $\boldsymbol{F}^{\mathrm{P}}$ and $\boldsymbol{F}^{\mathrm{T}}$ are orthogonal to each other in the sense of $\langle \boldsymbol{F}^{\mathrm{P}} \cdot \boldsymbol{F}^{\mathrm{T}} \rangle = 0$ where $\langle \cdots \rangle$ means averaging over the full solid angle.

Again we may conclude that $\nabla^2 \boldsymbol{F}^{\mathrm{P}}$ is poloidal and $\nabla^2 \boldsymbol{F}^{\mathrm{T}}$ toroidal.

Let us again consider a solenoidal field $\boldsymbol{F}$. With conclusions analogous to those used in the axisymmetric case we find

$$\boldsymbol{F}^{\mathrm{P}} = \nabla \times (\boldsymbol{r} \times \nabla G) = -\nabla \times (\nabla \times \ (\boldsymbol{r}G)) , \tag{37}$$

with some scalar G.

Finally we identify again $\boldsymbol{F}$ with the magnetic flux density $\boldsymbol{B}$. Adopting the usual notation to arrive at

$$\boldsymbol{B} = \boldsymbol{B}^{\mathrm{P}} + \boldsymbol{B}^{\mathrm{T}} \tag{38}$$
$$\boldsymbol{B}^{\mathrm{P}} = -\nabla \times (\boldsymbol{r} \times \nabla S) = \nabla \times (\nabla \times \ (\boldsymbol{r}S)), \quad \boldsymbol{B}^{\mathrm{T}} = -\boldsymbol{r} \times \nabla T = \nabla \times \ (\boldsymbol{r}T),$$

with two scalars S and T, called "defining scalars". Whereas the field lines of $\boldsymbol{B}^{\mathrm{P}}$ have complex three-dimensional patterns, those of $\boldsymbol{B}^{\mathrm{T}}$ are simply defined by $T = \text{const}$ together with $r = \text{const}$. The magnetic energy in a spherical shell or the total magnetic energy in all space, given by integrals over $\boldsymbol{B}^2/2\mu$, can always be split into two parts, one depending on $\boldsymbol{B}^{\mathrm{P}}$ and the other on $\boldsymbol{B}^{\mathrm{T}}$ only.

4 The Kinematic Dynamo Problem

In this section we give first a mathematical formulation of the kinematic dynamo problem. For the sake of simplicity we restrict ourselves to the case of a finite fluid body surrounded by free space. Our formulation can easily be modified to cover cases with other surroundings of the fluid or with an infinitely extended fluid. We further mention theorems excluding dynamo action with simple geometries, or symmetries, of the magnetic field or the motion, and report on successful attempts to construct dynamo models.

4.1 The Mathematical Formulation of a Typical Problem

Let us consider the dynamo problem for a finite electrically conducting body surrounded by free space. We denote the region occupied by the fluid by $\mathcal{V}$, its boundary by $\partial\mathcal{V}$, all outer space by $\mathcal{V}'$, and the distance of a any point from a given one of the fluid region by a.

We start with a mathematical formulation of the problem on the level of the Maxwell equations (1) and and the constitutive equations (2). We require that

$$\nabla \times \boldsymbol{E} = -\partial_t \boldsymbol{B}\,, \quad \nabla \cdot \boldsymbol{B} = 0\,, \quad \nabla \times \boldsymbol{B} = \mu \boldsymbol{j} \quad \text{everywhere} \tag{39}$$
$$\boldsymbol{j} = \sigma(\boldsymbol{E} + \boldsymbol{u} \times \boldsymbol{B}) \quad \text{in} \quad \mathcal{V}, \quad \boldsymbol{j} = \boldsymbol{0} \quad \text{in} \quad \mathcal{V}' \tag{40}$$
$$\boldsymbol{B} = \boldsymbol{O}(a^{-3}) \quad \text{as} \quad a \to \infty\,. \tag{41}$$

¿From this we may derive a second formulation, which considers no other electromagnetic fields than $\boldsymbol{B}$. It reads

$$\nabla \times (\eta \nabla \times \boldsymbol{B}) - \nabla \times (\boldsymbol{u} \times \boldsymbol{B}) + \partial_t \boldsymbol{B} = \boldsymbol{0}\,, \quad \nabla \cdot \boldsymbol{B} = 0 \quad \text{in} \quad \mathcal{V} \tag{42}$$
$$\nabla \times \boldsymbol{B} = \boldsymbol{0}\,, \quad \nabla \cdot \boldsymbol{B} = 0 \quad \text{in} \quad \mathcal{V}' \tag{43}$$
$$[\boldsymbol{B}] = \boldsymbol{0} \quad \text{across} \quad \partial\mathcal{V} \tag{44}$$
$$\boldsymbol{B} = \boldsymbol{O}(a^{-3}) \quad \text{as} \quad a \to \infty\,, \tag{45}$$

where $[\cdots]$ denotes the jump of a quantity across a surface. The conditions (41) and (45) exclude electric currents at infinity and thus specify a self-exciting dynamo in contrast to a externally excited one. In contrast to the first formulation we exclude in the second one explicitly electric surface currents on the boundary of the fluid body. By the way, if the outer space is simply connected (43) can be replaced by

$$\boldsymbol{B} = -\nabla\Phi , \quad \Delta\Phi = 0 \quad \text{in} \quad \mathcal{V}' . \tag{46}$$

The equations (42)–(45) pose an initial value problem for $\boldsymbol{B}$. We speak of a dynamo if there is a solution of these equations which does not decay in the course of time, that is,

$$\boldsymbol{B} \not\longrightarrow 0 \quad \text{as} \quad t \to \infty . \tag{47}$$

Let us add a remark concerning equations (39)–(41). They are, if surface currents are excluded, sufficient for the determination of $\boldsymbol{B}$. For the determination of $\boldsymbol{E}$, however, we have to add, for example, equations like $\nabla \cdot \boldsymbol{E} = 0$ in $\mathcal{V}'$ and $\boldsymbol{E} = \boldsymbol{O}(a^{-2})$ as $a \to \infty$ and also a condition that fixes the total electric charge on the conducting body.

4.2 Some Comments

4.2.1

As explained already in Section 3.3 in the context of magnetic energy, in the absence of fluid motions any magnetic field whose behavior is described by (42)–(45) is bound to decay. A dynamo requires that the magnetic Reynolds number R_m exceeds some critical value, and it seems plausible that this value is in the order of unity. So a necessary condition for a dynamo reads

$$R_\mathrm{m} \geq R_\mathrm{m\,crit} = O(1) . \tag{48}$$

The exact value of $R_\mathrm{m\,crit}$ depends of course also on the definition of R_m.

4.2.2

We want to stress that our definition of a dynamo refers to situations without any external electromotive force. If we included such an electromotive force corresponding to a non-zero $\boldsymbol{E}^{(e)}$ in Ohm's law (2b), equation (42a) would be no longer homogeneous but had a term $\nabla \times \boldsymbol{E}^{(e)}$ on the right-hand side. Then we may have a non-decaying magnetic field $\boldsymbol{B}$ already in the absence of any fluid motion, that is for $\boldsymbol{u} = \boldsymbol{0}$, and it is well possible that this is markedly amplified by the motion, that is for $\boldsymbol{u} \neq \boldsymbol{0}$. However, we do not include this amplification of a magnetic field in our definition of a dynamo.

4.2.3

A dynamo corresponds to an instability of the non-magnetic state of a physical system in the sense that magnetic perturbations can grow. Consider, as a simple example, a steady fluid flow. Then the magnetic flux density $\boldsymbol{B}$ has to obey the equations (42)–(45) with a velocity $\boldsymbol{u}$ independent of time. We may then look for solutions of the form

$$\boldsymbol{B} = \Re(\hat{\boldsymbol{B}}(\boldsymbol{x}) \exp(pt)) \tag{49}$$

with $\hat{\boldsymbol{B}}$ being a complex steady vector field and p a complex constant. Clearly $\hat{\boldsymbol{B}}$ has to obey the equations (42)–(45) with $\boldsymbol{B}$ replaced by $\hat{\boldsymbol{B}}$, and $\partial_t \boldsymbol{B}$ by $p\,\hat{\boldsymbol{B}}$. These equations pose an eigenvalue problem with the eigenvalue parameter p. We may parameterize the magnitude of the fluid flow by the magnetic Reynolds number R_{m}. Then the eigensolutions $\hat{\boldsymbol{B}}$ and the eigenvalues p depend, of course, on R_{m}. Let us put

$$p = \lambda + \mathrm{i}\omega\,, \tag{50}$$

with real λ and ω, where λ, if positive, is the growth rate of the magnetic field given by the respective eigensolution. We have a dynamo if there is at least one non-negative eigenvalue, that is, one with

$$\lambda \geq 0\,. \tag{51}$$

We call the value of R_{m} for which $\lambda = 0$ for one eigensolution and $\lambda < 0$ for all others the "marginal value" of R_{m}, and correspondingly we speak of "marginally stable" magnetic fields etc.

At the first glance the ansatz (49) seems to be a very special one. In general, however, the eigenvalue problem described here has an infinite set of solutions, $\hat{\boldsymbol{B}}_i$ and p_i. In a wide range of assumptions the $\hat{\boldsymbol{B}}_i$ constitute a complete set of vector functions. Then the general solution of the initial value problem for $\boldsymbol{B}$ posed by (42)–(45) is just given by

$$\boldsymbol{B}(\boldsymbol{x}\,,t) = \Re\Big(\sum_i b_i\,\hat{\boldsymbol{B}}_i\,(\boldsymbol{x})\,\exp(p_i\,t)\Big) \tag{52}$$

where the b_i are constants determined by $\boldsymbol{B}(\boldsymbol{x}\,,0)$.

4.2.4

Let us have a look on the energy balance of a dynamo. We recall relation (14) which describes the time variation of the total magnetic energy. In the case of a dynamo the time derivative of this energy has to be non-negative, that is,

$$\int_{\mathcal{V}} \frac{\boldsymbol{j}^2}{\sigma}\mathrm{d}v \leq -\int_{\mathcal{V}} \boldsymbol{u}\cdot(\boldsymbol{j}\times\boldsymbol{B})\,\mathrm{d}v\,. \tag{53}$$

Estimating the two integrals in the usual way we return to the condition (48).

Relation (14) clearly demonstrates that a dynamo requires a permanent input of kinetic energy, which maintains the flow against the Lorentz forces. The work done against the Lorentz force enhances the magnetic field. This in turn is subject to Ohmic dissipation. So in the course of the dynamo process kinetic energy is permanently converted into heat.

4.2.5

An important question in dynamo theory concerns the time scales on which a magnetic field evolves. According to the considerations in Section 3.2 we may expect that this time scale is given by T_η or T_u or something in between. As it was also shown there, T_η is very large for many objects. Dynamos with time scales of that order then hardly provide us with a satisfying explanation of the magnetic fields of these objects. We have to look for dynamos operating on shorter time scales, for instance of the order of T_u.

With this in mind we distinguish between "slow" and "fast" dynamos. For a definition we consider the dependence of the growth of the magnetic field, defined by the growth rate λ, within a time T_u in the limit of large magnetic Reynolds numbers R_m. If

$$\lambda T_u \to \text{positive value} \quad \text{as} \quad R_\mathrm{m} \to \infty \tag{54}$$

we speak of a fast dynamo, otherwise of a slow dynamo.

4.2.6

We have formulated the dynamo problem by the equations (42)–(45) which pose a problem for all space. Under the assumptions allowing to derive (46) it can be reduced to an "inner problem", that is, to one for the fluid body only given by (42) and proper boundary conditions. The latter, however, are different from the conditions usually considered in mathematical textbooks.

To explain this in more detail we first consider the equations (46). As it is known from potential theory, the function Φ satisfying the Laplace equation $\Delta\Phi = 0$ in $\mathcal{V}'$ is uniquely determined if its normal derivative or, what is the same, the normal component of $\boldsymbol{B}$ on $\partial\mathcal{V}$ is given, which we denote by B_norm in the following. The problem posed in this way is known as the outer Neumann problem. Its solution can be represented in the form

$$\Phi(\boldsymbol{x}) = \int_{\partial\mathcal{V}} \Gamma(\boldsymbol{x}, \boldsymbol{x}')\, B_\mathrm{norm}(\boldsymbol{x}')\, \mathrm{d}s' , \tag{55}$$

where Γ means a proper Green's function.

Suppose now that $\boldsymbol{B}$ in $\mathcal{V}$ is given and recall that $\boldsymbol{B}$ has to be continuous across $\partial\mathcal{V}$. Thus B_norm in (55) may be interpreted as limit obtained by approaching $\partial\mathcal{V}$ from inside, that is out of $\mathcal{V}$. Then (46) with Φ given by this integral defines a continuation of $\boldsymbol{B}$ into $\mathcal{V}'$ such that its normal component is indeed continuous across $\partial\mathcal{V}$. The continuity of the tangential components, however, is not yet guaranteed in this way. Denoting these components, again understood as limit from inside, by $\boldsymbol{B}_\mathrm{tang}$ we have to require that

$$\boldsymbol{B}_\mathrm{tang}(\boldsymbol{x}) = -\nabla_\mathrm{tang}\left(\int_{\partial\mathcal{V}} \Gamma(\boldsymbol{x}, \boldsymbol{x}')\, B_\mathrm{norm}(\boldsymbol{x}')\, \mathrm{d}s'\right) \quad \text{at} \quad \partial\mathcal{V}. \tag{56}$$

This relation plays the part of the boundary condition for the inner problem. It is non-local in the sense that it connects $\boldsymbol{B}_\mathrm{tang}$ in a given point with B_norm in all other points of $\partial\mathcal{V}$.

4.3 Dynamo Theorems

Several types of theorems concerning dynamos have been proved. Some of them provide us with more precise formulations of the necessary condition for a dynamo given with (48) saying that the magnetic Reynolds number has to exceed some critical value. We will not deal with this type of theorems here. Instead we will focus our attention on a few "anti-dynamo theorems" which exclude magnetic fields or flow patterns with special geometries, or symmetries, from dynamo action.

Let us start with Cowling's theorem concerning the magnetic field geometry. As a result of unsuccessful attempts to elaborate simple dynamo models, Cowling [26] proved a theorem, which since has been generalized in several respects [47,28,29,5]; see also [21]. This theorem states that a magnetic field which is symmetric about any axis can never be maintained by dynamo action. That is, a dynamo requires a more complex, three-dimensional magnetic field structure.

Another theorem, which can be considered as a modification of the mentioned one, states the impossibility of a dynamo if both magnetic field and fluid velocity depend on two Cartesian coordinates only; see e.g. [30].

The most interesting theorem concerning the geometry of the fluid motion traces back to Elsasser [31] and Bullard and Gellman [32]; see also e.g. [3]. It applies to spherical bodies in which the magnetic diffusivity is constant or shows a spherically symmetric distribution, that is, η depends only on the radial coordinate r. The theorem states that then a magnetic field can never be maintained by a toroidal motion, that is, with a solenoidal velocity field which lies completely in concentric spherical surfaces r = const, in other words, has no radial components. As long as the assumption concerning the diffusivity applies, in particular dynamo action due to any kind of differential rotation alone has to be excluded.

Here the question arises about the minimal intensity of a radial flow necessary for a dynamo. In this connection an interesting statement was made by Busse [33]. For the case of constant magnetic diffusivity he has shown that a dynamo is only possible if $|\boldsymbol{u} \cdot \boldsymbol{r}|_{\max} / \eta \geq E^{\mathrm{P}} / (E^{\mathrm{P}} + E^{\mathrm{T}})$, where $|\boldsymbol{u} \cdot \boldsymbol{r}|_{\max}$ means the maximal value that $|\boldsymbol{u} \cdot \boldsymbol{r}|$ takes inside the fluid, and E^{P} and E^{T} are the energies stored in the poloidal and toroidal parts of the magnetic field.

We note that in all situations covered by the anti-dynamo theorems mentioned the poloidal part of the magnetic field evolves independently of the toroidal one. It seems that a dynamo requires the full interaction of poloidal and toroidal fields.

4.4 Examples of Working Dynamos

4.4.1

There have been numerous attempts to construct kinematic dynamo models, that is, to find non-decaying solutions of equations like (42)–(45). Many of them failed by several reasons, in particular by such which are now clear from the anti-dynamo theorems proved in the meantime.

The first working kinematic dynamo model was proposed by Herzenberg [34]. In his model the conducting medium occupies a sphere. Apart from two smaller spherical regions inside this sphere the medium is at rest. In each of the two regions it rotates like a rigid body. For a certain range of relative positions of the rotation axes and sufficiently high rotation rates self-excitation occurs. Of course, this model hardly reflects a situation in the interior of a cosmic object. It was, however, in so far very important as it played the role of an existence proof of homogeneous dynamos. Many investigations of models of that kind have been carried out [35–38].

Proceeding to other examples we mention first group a of models which are also rather far from direct applications to cosmic objects, presuppose in particular an infinitely extended conductor and infinitely extended flows, but show certain basic patterns of dynamos.

One example of this kind is a dynamo model proposed by Ponomarenko [39]. It is assumed that the infinitely extended conducting medium is at rest everywhere except in an infinitely long cylinder, and it moves there like a rigid body in full electric contact with the surroundings. The motion consists in a rotation about the cylinder axis and a translation along this axis, that is, it is screw-like. If both components of the motion are sufficiently strong, non-decaying wave-like magnetic fields traveling in axial direction prove to be possible. We will give the condition for that below.

Another interesting example was given by Roberts [40,41]. He considered fluid flows which are spatially periodic in two directions, say the x and y directions in a Cartesian coordinate system, but do not vary in the third one, the z direction. As indicated in Figure 3 in each cell of the flow pattern there is a circulation in the (x, y) plane and a motion along the z axis. These two components of the flow result again in screw-like motions, either right-handed or left-handed in all cells. If both components are sufficiently strong, a non-decaying magnetic field is possible. It does not vanish under averaging over x and y and the averaged field lies in the (x, y) plane. We will return also to this case below.

Many investigations concerning dynamos have been done with a particular class of flows spatially periodic in all three directions, x, y and z, the so-called ABC-flows, named after Arnold, Beltrami and Childress; see e.g. [42]. We do not go into details and note only that the flow patterns investigated by Roberts are closely related to special cases of ABC-flows.

We also mention an interesting model by Gailitis [43]. Again an infinitely extended conducting medium is considered which is at rest everywhere except on the surface of two tori of the same size symmetric about a common axis. The motion consists in a circulation in the meridional planes defined by this axis symmetric about the middle plane between these tori. For the case in which the small radius of these tori is much smaller than their large radius and their distance it was shown that self-excitation of a magnetic field is possible with sufficiently strong circulation. Of course, according to Cowling's theorem the field has to be non-symmetric about any axis and in particular the axis mentioned.

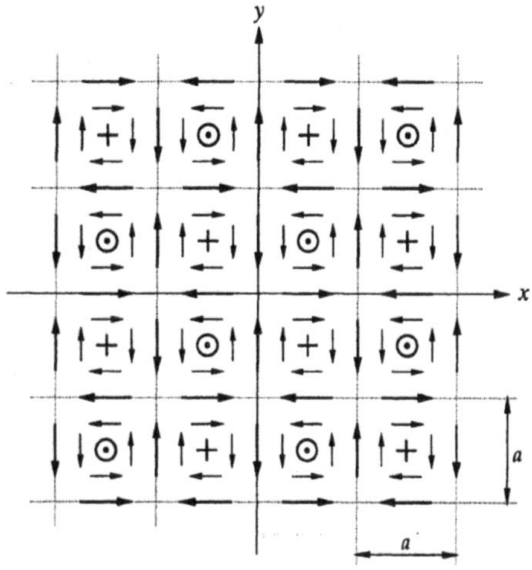

Fig. 3. A flow pattern as used by Roberts

4.4.2

Let us now mention a few dynamo models elaborated with a view to applications, for example, to the Earth or other objects, where the conducting fluid occupies a spherical region and is surrounded by non-conducting space. Pioneering work with respect to such models has already been done by Bullard and Gellman [32]. They in particular developed a proper formalism, known as Bullard-Gellman formalism, for the treatment of the equations governing such models. It uses the representation of the magnetic field and the motion by poloidal and toroidal parts as explained in Section 3.9 and the expansion of the defining scalars in series of spherical harmonics, and it allows the reduction of the governing partial differential equations to an infinite system of ordinary differential equations for functions depending on the radial coordinate only, a truncated version of which has then to be integrated numerically.

Various dynamo models of this kind with many different flow patterns have been investigated. Without going into details we mention here those by Pekeris, Accad and Shkoller [44], by Gubbins [45] and by Kumar and Roberts [46], the results of which has often been discussed in the context of the geodynamo and confirmed repeatedly by independent computations.

4.4.3

Another approach to kinematic dynamo models, which is of high interest in view of cosmical bodies with complex flow patterns, for example of convective or turbulent nature, is based on the concept of mean fields. A particular version of this concept has already been used in the theory of the "nearly symmetric dynamo" developed by Braginsky [27,47,48] with a view to the Earth and widely elaborated later on; see e.g. [49,50]. In a much wider sense it was used in "mean-field electrodynamics", initiated by Steenbeck, Krause and Rädler [51] and likewise elaborated in between in a very general sense [1,2]. It proved to be a useful basis for studying dynamo models which reflect essential features of the magnetic fields observed at the Earth, the Sun and other cosmic objects. It also provided us with a rigorous mathematical formulation of the idea of "cyclonic convection" whose importance for dynamo action was already recognized by Parker [52]. We will explain the essential ideas of mean-field electrodynamics and of the mean-field dynamo theory based on it in Sections 5 and 6.

4.4.4

It would be very desirable to realize and study a homogeneous dynamo in the laboratory. Several experiments designed to approach this goal have been carried out [53]. For true simulations of a homogeneous dynamo mainly flows of liquid sodium are envisaged. As can be seen from the data given in Table 2 huge devices and enormous technical efforts are necessary to reach the values magnetic Reynolds numbers satisfying the self-excitation condition of a dynamo. Two such experiments are under preparation, one in Riga in Latvia [54] and another one in Karlsruhe in Germany [55–59], and few more are planned at other places. The Riga experiment is based on the pattern of the Ponomarenko dynamo, the Karlsruhe experiment on the that of the Roberts dynamo explained above.

By these and other reasons we give some more explanations on these two basic dynamo patterns.

As it was explained above in the model by Ponomarenko [39] the conducting medium is at rest except in an infinite cylinder. Using a proper cylindrical coordinate system (s, ϕ, z) in which this cylinder is given by $s < a$, with a being its radius, we describe the velocity $\boldsymbol{u}$ in its interior by

$$u_s = 0 \,, \quad u_\phi = \omega s \,, \quad u_z = v \,. \tag{57}$$

Here ω is a constant angular velocity and v a constant velocity. We define two dimensionless parameters $R_{\mathrm{m}\perp}$ and $R_{\mathrm{m}\parallel}$ of the type of a magnetic Reynolds number by

$$R_{\mathrm{m}\perp} = |\omega|\, a^2/\eta \,, \quad R_{\mathrm{m}\parallel} = |v|\, a/\eta \,, \tag{58}$$

and put

$$R_\mathrm{m} = \sqrt{R_{\mathrm{m}\perp}^2 + R_{\mathrm{m}\parallel}^2} \,. \tag{59}$$

There are solutions of the relevant equations of the form

$$\boldsymbol{B} = \Re\big(\hat{\boldsymbol{B}}(s) \exp(\mathrm{i}(m\phi + kz) + pt)\big) \tag{60}$$

with $\hat{\boldsymbol{B}}(s)$ being a complex vector field depending on s only, m an integer, and k and p real constants. For a range of sufficiently large $R_{\mathrm{m}\perp}$ and $R_{\mathrm{m}\parallel}$ they do not decay, that is, p is non-negative. The marginal case, $p = 0$, with a minimum value of R_m is given by

$$R_\mathrm{m} = 17.722 \,, \quad R_{\mathrm{m}\perp}/R_{\mathrm{m}\parallel} = 0.7625 \,, \tag{61}$$

and the corresponding solution has a shape determined by

$$m = 1\,, \quad k/a = -0.3875\,; \tag{62}$$

see e.g. [15]. This solution is a helical wave traveling in the direction of the axial flow.

Proceeding now to a special version of the model by Roberts [40,41] we use again a Cartesian coordinate system (x, y, z) and describe the fluid velocity $\boldsymbol{u}$ by

$$u_x = u_\perp \frac{\partial \psi}{\partial y}\,, \quad u_y = -u_\perp \frac{\partial \psi}{\partial x}\,, \quad u_z = u_\| \, \frac{\pi^2}{2a}\, \psi\,,$$
$$\psi = \frac{a}{2}\, \sin(\frac{\pi}{a} x)\, \sin(\frac{\pi}{a} y)\,, \tag{63}$$

where a is the half period length in x or y direction. We further define the dimensionless parameters $R_{\mathrm{m}\,\perp}$ and $R_{\mathrm{m}\,\|}$ of the type of magnetic Reynolds numbers by

$$R_{\mathrm{m}\,\perp} = |u_\perp|\, a/\eta\,, \quad R_{\mathrm{m}\,\|} = |u_\||\, a/\eta\,. \tag{64}$$

There are solutions of the relevant equations of the form

$$\boldsymbol{B} = \Re\big(\hat{\boldsymbol{B}}(x, y)\exp(\mathrm{i}kz + pt)\big) \tag{65}$$

with $\hat{\boldsymbol{B}}$ being a complex vector field and k and p real constants. They do not decay if

$$R_{\mathrm{m}\,\perp}\, R_{\mathrm{m}\,\|}\, \Phi(R_{\mathrm{m}\,\perp}) \geq \frac{32}{\pi} \frac{a}{l}\,, \tag{66}$$

where Φ is a function as depicted in Figure 4 satisfying $\Phi(0) = 1$ and decreasing to zero with growing argument, and l the period length in z direction, that is, $l = 2\pi/k$; see [56]. We point out that the dynamo may work with arbitrarily small non-zero $R_{\mathrm{m}\,\perp}$ or $R_{\mathrm{m}\,\|}$ if only l is sufficiently large.

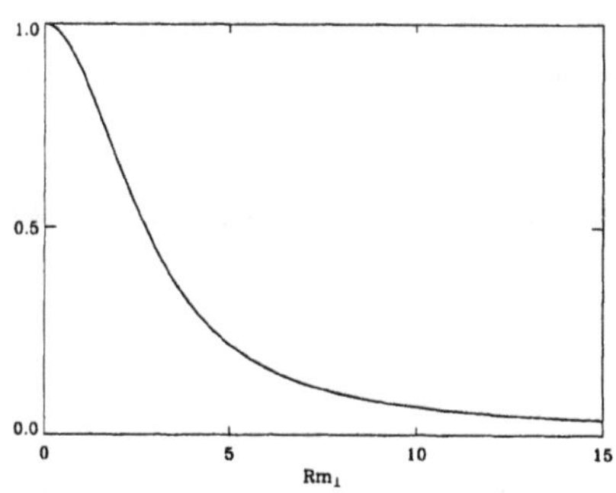

Fig. 4. The dependence of Φ on $R_{\mathrm{m}\,\perp}$

5 Mean-Field Electrodynamics

Let us now focus our attention on electromagnetic processes in an electrically conducting fluid showing an irregular, for instance turbulent motion. Then the electromagnetic fields must show irregular features, too. We may consider both the electromagnetic fields and the motion as superpositions of mean parts with more or less weak variations in space and time and other parts, called "fluctuations", which vary on small scales. A particular question of very high interest for the dynamo processes in cosmic objects concerns the behavior of the mean electromagnetic fields in the presence of a given irregular or turbulent fluid motion. This question is the subject of mean-field electrodynamics. In this section we explain the basic ideas of mean-field electrodynamics and illustrate them by simple examples. For more results we refer also to other representations, e.g. [2,3,12]. Generalizations to cases in which the fluid motions are no longer considered as given are explained in Section 9.

5.1 Definition of Mean Fields and the Reynolds Averaging Rules

Let us start our explanations on mean fields by considering a scalar field F showing some irregular variations in space and time. We write

$$F = \overline{F} + F' . \tag{67}$$

Here $\overline{F}$, which we call "mean field", is understood as an average of F defined by a proper averaging procedure which smoothes the space and time variations or, what means the same, suppresses the contributions with small length and time scales. F', called "fluctuation", contains then all these small-scale contributions to F. Details concerning such averaging procedures will be discussed later.

Analogously we split vector and tensor fields into mean and fluctuating parts. Their mean parts are defined by averaging their components with respect to a given coordinate system using the procedure adopted for scalars. Consider, as an example, a vector field $\boldsymbol{F}$ and a coordinate system with the basic unit vectors $\boldsymbol{e}_i$ so that, with the summation convention adopted, $\boldsymbol{F} = \boldsymbol{e}_i F_i$. Then we have $\overline{\boldsymbol{F}} = \boldsymbol{e}_i \overline{F}_i$. We note that the definition of mean vector or tensor fields depends in that sense on the choice of the coordinate system.

We do not use a specific definition of the averaging procedure in the following but restrict the possibilities by requiring that it ensures the exact or approximative validity of the following Reynolds averaging rules. Let F and G be two arbitrary scalar functions. Firstly we require that averaging is a distributive operation, that is,

$$\overline{F + G} = \overline{F} + \overline{G}. \tag{68}$$

Secondly it has to commute with space and time derivatives,

$$\overline{\partial F / \partial x} = \partial \overline{F} / \partial x \,, \quad \overline{\partial F / \partial t} = \partial \overline{F} / \partial t \,, \tag{69}$$

where x stands for any space coordinate. Thirdly we require that an averaged quantity is invariant under repeated averaging,

$$\overline{\overline{F}} = \overline{F}\,. \tag{70}$$

If (68) applies this is equivalent to $\overline{F'} = 0$. Fourthly we require that an averaged quantity behaves like a constant under further averaging in the sense that

$$\overline{\overline{F}G} = \overline{F}\,\overline{G}\,. \tag{71}$$

For later use we note that (68) and (71) imply

$$\overline{FG} = \overline{F}\,\overline{G} + \overline{F'G'}\,. \tag{72}$$

We give now a few examples of averaging procedures and explain to which extend they satisfy these rules.

(i) Statistical or ensemble averages

In this case we suppose that there is an infinitude of copies of the object considered. The individual copies are labelled by a value of a parameter p, for convenience taken as a continuous variable. In that sense the quantity F to be averaged depends, in addition to the space and time variables, on this parameter p. Then we define

$$\overline{F}(\boldsymbol{x},t) = \int F(\boldsymbol{x},t;p)\,g(p)\,\mathrm{d}p\,, \quad \int g(p)\,\mathrm{d}p = 1\,, \tag{73}$$

where $g(p)$ is some normalized distribution function, and both integrations are over all values of p. Averages of this kind clearly ensure the validity of all four rules (68)–(71). There is, however, a serious difficulty to relate these averages to observable quantities.

(ii) Space averages

A general form of a space average is given by

$$\overline{F}(\boldsymbol{x},t) = \int_\infty F(\boldsymbol{x}+\boldsymbol{\xi},t)\,g(\boldsymbol{\xi})\,\mathrm{d}^3\xi\,, \quad \int_\infty g(\boldsymbol{\xi})\,\mathrm{d}^3\xi = 1\,. \tag{74}$$

Here $g(\boldsymbol{\xi})$ is a normalized weight function which is different from zero only in some region around $\boldsymbol{\xi} = \mathbf{0}$. The integrations, formally over all $\boldsymbol{\xi}$-space, are in fact over this region only. With such averages the two rules (68) and (69) apply exactly but in general (70) and (71) are violated. The two latter can be justified as an approximation if there is a gap in the spectrum of the length scales of F, and all large scales are much larger and all small ones much smaller than the characteristic length of the averaging region. A situation of that kind is sometimes named "two-scale" situation.

There are, however, particular space averages to which all averaging rules apply. Consider, for example, a case in which the variation of F in space is properly described by spherical coordinates r, θ, ϕ, and put

$$\overline{F}(r,\theta,t) = \frac{1}{2\pi} \int_0^{2\pi} F(r,\theta,\phi,t)\,\mathrm{d}\phi\,. \tag{75}$$

When using this average, of course, all mean fields are by definition axisymmetric. As far as this is acceptable for the problem under consideration the average is very useful. Its big advantage is that indeed all four rules (68)–(71) apply exactly.

(iii) Time averages

Similar to space averages we may define time averages by

$$\overline{F}(\boldsymbol{x},t) = \int_\infty F(\boldsymbol{x},t-\tau)\,g(\tau)\,\mathrm{d}\tau\,, \quad \int_\infty g(\tau)\,\mathrm{d}\tau = 1\,, \tag{76}$$

with some normalized weight function $g(\tau)$ different from zero in some neighborhood of $\tau = 0$ so that the integrations are in fact over these τ only. The comments made with the general form of the space average apply analogously.

(iv) Averages based on filtering of spectra

We may, for example, represent the dependency of F on space coordinates by an Fourier integral,

$$F(\boldsymbol{x},t) = \int_\infty \hat{F}(\boldsymbol{k},t)\exp(\mathrm{i}\,\boldsymbol{k}\cdot\boldsymbol{x})\mathrm{d}^3k\,, \tag{77}$$

with the integration over all $\boldsymbol{k}$-space, and then put

$$\overline{F}(\boldsymbol{x},t) = \int_{|\boldsymbol{k}|<K} \hat{F}(\boldsymbol{k},t)\exp(\mathrm{i}\,\boldsymbol{k}\cdot\boldsymbol{x})\mathrm{d}^3k\,, \tag{78}$$

where K means some constant. For averages defined in this way the three rules (68)–(70) apply exactly, and with a sufficiently large gap in the $\boldsymbol{k}$-spectrum and a proper choice of K the remaining rule (71) can again be justified as an approximation. By the way, (78) can be rewritten so that it takes the form of (74) but with a rather complex function g.

The special space average defined by (75) can also be interpreted as one based on filtering a Fourier spectrum with respect to ϕ. Another interesting possibility consists, for example, in filtering the multipole spectrum of vector fields so that the mean fields are just dipole fields, or dipole and quadrupole fields, etc.

5.2 Basic Equations For Mean Fields

Let us now return to electromagnetic processes in an electrically conducting fluid showing an irregular motion and consider both the electromagnetic fields and the velocity of the motion as superpositions of mean and fluctuating parts, for example $\boldsymbol{B} = \overline{\boldsymbol{B}} + \boldsymbol{B}'$ and $\boldsymbol{u} = \overline{\boldsymbol{u}} + \boldsymbol{u}'$. We rely on Maxwell's and the constitutive equations in the form (1) and (2), again with $\boldsymbol{E}^{(e)}$ ignored, and subject them to averaging. Using the rules (68)–(71) we obtain

$$\nabla \times \overline{\boldsymbol{E}} = -\partial_t \overline{\boldsymbol{B}}\,, \quad \nabla \cdot \overline{\boldsymbol{B}} = 0\,, \quad \nabla \times \overline{\boldsymbol{H}} = \overline{\boldsymbol{j}} \tag{79}$$

and

$$\overline{\boldsymbol{B}} = \mu \overline{\boldsymbol{H}}\,, \quad \overline{\boldsymbol{j}} = \sigma(\overline{\boldsymbol{E}} + \overline{\boldsymbol{u}} \times \overline{\boldsymbol{B}} + \boldsymbol{\mathcal{E}}) \tag{80}$$

where

$$\boldsymbol{\mathcal{E}} = \overline{\boldsymbol{u}' \times \boldsymbol{B}'}\,. \tag{81}$$

In the same way we may conclude from (4), or derive from (79)–(80), that

$$\nabla \times (\eta \nabla \times \overline{\boldsymbol{B}}) - \nabla \times (\overline{\boldsymbol{u}} \times \overline{\boldsymbol{B}} + \boldsymbol{\mathcal{E}}) + \partial_t \overline{\boldsymbol{B}} = \boldsymbol{0}\,, \quad \nabla \cdot \overline{\boldsymbol{B}} = 0\,. \tag{82}$$

Obviously the mean electromagnetic fields together with the mean motion satisfy essentially the same equations as the original fields with the original motion. The only deviation is the additional mean electromagnetic force $\boldsymbol{\mathcal{E}}$ due to the fluctuations of motion and magnetic field, $\boldsymbol{u}'$ and $\boldsymbol{B}'$, just at the place where $\boldsymbol{E}^{(e)}$ occurred in the original equations.

So the crucial point in the elaboration of mean-field electrodynamics is the determination of the mean electromagnetic force $\boldsymbol{\mathcal{E}}$. Since $\boldsymbol{u}'$ is considered as given, we have to look for the determination of $\boldsymbol{B}'$. Starting with the original induction equation (4), replacing there $\boldsymbol{B}$ and $\boldsymbol{u}$ by $\overline{\boldsymbol{B}} + \boldsymbol{B}'$ and $\overline{\boldsymbol{u}} + \boldsymbol{u}'$, and using the averaged induction equation (82) together with (81) we find

$$\begin{aligned} \nabla \times (\eta \nabla \times \boldsymbol{B}' - \overline{\boldsymbol{u}} \times \boldsymbol{B}' - \boldsymbol{u}' \times \boldsymbol{B}' + \overline{\boldsymbol{u}' \times \boldsymbol{B}'}) + \partial_t \boldsymbol{B}' &= \nabla \times (\boldsymbol{u}' \times \overline{\boldsymbol{B}})\,, \\ \nabla \cdot \boldsymbol{B}' &= 0\,. \end{aligned} \tag{83}$$

These equations together with proper initial and boundary conditions determine $\boldsymbol{B}'$ if $\overline{\boldsymbol{u}}$, $\boldsymbol{u}'$ and $\overline{\boldsymbol{B}}$ are given. Considered in this way, the first line is an inhomogeneous equation with the inhomogeneity depending on $\overline{\boldsymbol{B}}$. So we can write the solution in the form

$$\boldsymbol{B}' = \boldsymbol{B}'^{(0)} + \boldsymbol{B}'^{(\overline{B})}\,, \tag{84}$$

where $\boldsymbol{B}'^{(0)}$ stands for a solution of the homogeneous version of this equation and $\boldsymbol{B}'^{(\overline{B})}$ for a particular solution of the full equation. $\boldsymbol{B}'^{(0)}$ depends on $\overline{\boldsymbol{u}}$ and $\boldsymbol{u}'$ but not on $\overline{\boldsymbol{B}}$. More precisely, it is a functional of these quantities in the sense that $\boldsymbol{B}'^{(0)}$ in a given point in space and time depends on $\overline{\boldsymbol{u}}$ and $\boldsymbol{u}'$ in other points, too. $\boldsymbol{B}'^{(\overline{B})}$ is a functional of $\overline{\boldsymbol{u}}, \boldsymbol{u}'$ and $\overline{\boldsymbol{B}}$, which has obviously a linear dependence on $\overline{\boldsymbol{B}}$. We may specify $\boldsymbol{B}'^{(\overline{B})}$ without any loss of generality so that it is not only linear but also homogeneous in $\overline{\boldsymbol{B}}$, that is, it is equal to zero if $\overline{\boldsymbol{B}}$ vanishes everywhere in space and time.

With this in mind we write now

$$\boldsymbol{\mathcal{E}} = \boldsymbol{\mathcal{E}}^{(0)} + \boldsymbol{\mathcal{E}}^{(\overline{B})}\,. \tag{85}$$

Here $\boldsymbol{\mathcal{E}}^{(0)}$ is, again in the sense explained above, a functional of $\overline{\boldsymbol{u}}$ and $\boldsymbol{u}'$, which depends on $\boldsymbol{u}'$, of course, via averaged quantities only. $\boldsymbol{\mathcal{E}}^{(\overline{B})}$ is a functional of $\overline{\boldsymbol{u}}, \boldsymbol{u}'$ and $\overline{\boldsymbol{B}}$, which is linear and homogeneous in $\overline{\boldsymbol{B}}$.

In view of $\boldsymbol{\mathcal{E}}^{(0)}$ it is of interest whether the homogeneous version of the equations (83) for $\boldsymbol{B}'$, that is the version with $\overline{\boldsymbol{B}} = \boldsymbol{0}$, have only decaying solutions or also non-decaying ones. In the first case $\boldsymbol{\mathcal{E}}^{(0)}$, if initially non-zero, decays to zero, too. The second case corresponds to a dynamo working on the scales of the turbulence, which requires, of course, a sufficiently high Reynolds number for these scales. Then $\boldsymbol{\mathcal{E}}^{(0)}$ needs no longer to decay to zero. But as it will become clear in Section 5.4 there are well conditions under which $\boldsymbol{\mathcal{E}}^{(0)}$ has then to be equal to zero by other reasons. If $\boldsymbol{\mathcal{E}}^{(0)}$ does not vanish it is surely of some interest as long as $\overline{\boldsymbol{B}}$ is small but it will loose its importance as soon as $\overline{\boldsymbol{B}}$ has grown up to a magnitude for which $\boldsymbol{\mathcal{E}}^{(\overline{B})}$ is much larger than $\boldsymbol{\mathcal{E}}^{(0)}$. With this in mind, for the sake of simplicity we ignore $\boldsymbol{\mathcal{E}}^{(0)}$ in all what follows, that is, we put $\boldsymbol{\mathcal{E}} = \boldsymbol{\mathcal{E}}^{(\overline{B})}$.

With this simplification the mean electromotive force $\boldsymbol{\mathcal{E}}$ due to fluctuations of motion and magnetic field has to be considered as a functional of $\overline{\boldsymbol{u}}, \boldsymbol{u}'$ and $\overline{\boldsymbol{B}}$, which is linear and homogeneous in $\overline{\boldsymbol{B}}$. We can express this by writing

$$\mathcal{E}_i(\boldsymbol{x}, t) = \int_0^\infty \int_\infty K_{ij}(\boldsymbol{x}, t; \boldsymbol{\xi}, \tau) \, \overline{B}_j(\boldsymbol{x} - \boldsymbol{\xi}, t - \tau) \, \mathrm{d}^3\xi \, \mathrm{d}\tau \,. \tag{86}$$

Here we think of Cartesian coordinates and adopt again the summation convention. K_{ij} is a kernel determined by $\overline{\boldsymbol{u}}$ and $\boldsymbol{u}'$, where the dependence on $\boldsymbol{u}'$ is again via averaged quantities only. On the basis of solutions of the equations (83) derived under special assumptions explicit expressions for the kernel K_{ij} can indeed be constructed; an example will be given in Section 5.6.

Let us now consider situations in which the fluctuations of the fluid velocity and thus those of the magnetic field are of turbulent nature. A typical feature of turbulence is that the correlations of two fluctuating quantities in different points in space and time deviate markedly from zero only if their distances in space and time are not too large, more precisely not much larger than a properly defined correlation length and time. Accepting this we may conclude that the kernel K_{ij} in (86) is markedly non-zero only if $|\boldsymbol{\xi}|$ and $|\tau|$ do not exceed the order of the correlation length and time.

We introduce now in addition the assumption that the mean magnetic flux density $\overline{\boldsymbol{B}}$ varies only weakly in space and time so that $\overline{B}_j(\boldsymbol{x} - \boldsymbol{\xi}, t - \tau)$ in (86) can be replaced by some of the first terms of its Taylor series with respect to $\boldsymbol{\xi}$ and τ,

$$\overline{B}_j(\boldsymbol{x} - \boldsymbol{\xi}, t - \tau) = \overline{B}_j(\boldsymbol{x}, t) - \frac{\partial \overline{B}_j(\boldsymbol{x}, t)}{\partial x_k} \, \xi_k - \frac{\partial \overline{B}_j(\boldsymbol{x}, t)}{\partial t} \, \tau \cdots \,. \tag{87}$$

For the sake of simplicity we consider here only the first two terms, that is, we make the simplest assumption concerning the spatial variation and ignore any time variation of $\overline{\boldsymbol{B}}$ in the relevant regions determined by correlation length and time. In this way we arrive at

$$\mathcal{E}_i = a_{ij}\,\overline{B}_j + b_{ijk}\,\frac{\partial \overline{B}_j}{\partial x_k}\,, \tag{88}$$

where we have dropped the arguments $\boldsymbol{x}$ and t everywhere. The tensors a_{ij} and b_{ijk} are again determined by $\overline{\boldsymbol{u}}$ and $\boldsymbol{u}'$ only. ¿From (86)–(88) we can easily conclude that

$$a_{ij} = \int_0^\infty \int_\infty K_{ij}(\boldsymbol{x},t;\boldsymbol{\xi},\tau)\,\mathrm{d}^3\xi\,\mathrm{d}\tau$$

$$b_{ijk} = -\int_0^\infty \int_\infty K_{ij}(\boldsymbol{x},t;\boldsymbol{\xi},\tau)\,\xi_k\,\mathrm{d}^3\xi\,\mathrm{d}\tau\,. \tag{89}$$

When applying (88) to a specific situation we have, of course, to check whether the neglect of further terms is justified.

5.3 Definitions Concerning Symmetry Properties of Turbulent Fields

In the following we want to discuss the mean electromotive force $\boldsymbol{\mathcal{E}}$ under the assumption that the fluctuating velocity field $\boldsymbol{u}'$ corresponds to a turbulence. Let us first give some definitions concerning properties of turbulence.

For this purpose we consider the behavior of mean quantities depending on the $\boldsymbol{u}'$-field under changes of this field. Simple examples of such mean quantities are the scalar $\overline{\boldsymbol{u}'^2(\boldsymbol{x},t)}$ or the two-point correlation tensor $\overline{u_i'(\boldsymbol{x},t)u_j'(\boldsymbol{x}+\boldsymbol{\xi},t+\tau)}$, other examples are the tensors a_{ij} or b_{ijk} introduced above. By changes of the $\boldsymbol{u}'$-field we mean translations, time shifts, rotations about an axis, or reflections about a plane or a point as explained in Section 3.8.

We call a turbulence "homogeneous" if all averaged quantities depending on the $\boldsymbol{u}'$-field are invariant under arbitrary translations of this field, and "steady" if the same applies with arbitrary time shifts. We call a turbulence "axisymmetric" about a given axis if all averaged quantities are invariant under arbitrary rotations of the field about this axis, and "isotropic" with respect to a given point if this applies to all axes running through this point. Finally we call a turbulence "reflectionally symmetric", or "mirror-symmetric", about a given plane or point if all averaged quantities are invariant under reflection of the field about this plane or point. We note that these definitions depend on the way in which the averages are defined.

Of course, a homogeneous isotropic turbulence is isotropic in all points. Likewise a homogeneous isotropic reflectionally symmetric turbulence, which is sometimes called "gyrotropic" turbulence, is reflectionally symmetric about all planes and all points.

5.4 The Mean Electromotive Force for Homogeneous Isotropic Turbulence

Let us now consider the mean electromotive force $\boldsymbol{\mathcal{E}}$ as given by (88) for the case in which there is no mean motion, $\overline{\boldsymbol{u}} = \boldsymbol{0}$, but an irregular one, described by the velocity field $\boldsymbol{u}'$, which corresponds to a homogeneous isotropic turbulence. For the sake of simplicity we assume that the magnetic diffusivity η is independent of position.

As a consequence of the homogeneity and isotropy of the turbulence the components of the tensors a_{ij} and b_{ijk}, as averaged quantities, must be invariant under arbitrary translations of $\boldsymbol{u}'$ and under arbitrary rotations about arbitrary axes. So far we did not speak about changes of the coordinate system. Let us now subject the coordinate system always to the same transformations, that is translation and rotation, as the $\boldsymbol{u}'$-field. Then the representation of the original field in the original system coincides with that of the transformed field in the transformed coordinate system. Consequently the components of the tensors a_{ij} and b_{ijk} in both systems have to coincide, too. Taking this together with the invariance of these components under transformations of the $\boldsymbol{u}'$-field alone we arrive at the conclusion that the same invariance must exist with respect to transformations of the coordinate system alone. So the homogeneity of the turbulence implies that a_{ij} and b_{ijk} are independent of position, and its isotropy that they are isotropic tensors, whose defining property is just the invariance of their components under arbitrary rotations of the coordinate system. Isotropic tensors of the second and the third rank can differ only by scalar factors from the Kronecker tensor δ_{ij} and the Levi-Civita tensor ϵ_{ijk}. So we have

$$a_{ij} = \alpha\,\delta_{ij}\,, \quad b_{ijk} = \beta\,\epsilon_{ijk}\,, \tag{90}$$

with α and β independent of position and determined by $\boldsymbol{u}'$ only.

Returning with this result to (88) we find

$$\boldsymbol{\mathcal{E}} = \alpha\overline{\boldsymbol{B}} - \beta\nabla\times\overline{\boldsymbol{B}}\,. \tag{91}$$

By the way, if we had not already ignored the contribution $\boldsymbol{\mathcal{E}}^{(0)}$ to $\boldsymbol{\mathcal{E}}$ we would have to conclude here that it is an isotropic quantity in the above sense and, since there is no isotropic vector, is equal to zero.

Using the result (91) Ohm's law (80b) can be written in the form

$$\overline{\boldsymbol{j}} = \sigma_{\mathrm{m}}(\overline{\boldsymbol{E}} + \alpha\overline{\boldsymbol{B}}) \tag{92}$$

with

$$\sigma_{\mathrm{m}} = \frac{\sigma}{1+\mu\sigma\beta}\,. \tag{93}$$

Note that $\mu\sigma\beta = \beta/\eta$. Analogously, the induction equation (82a) can be rewritten so that we have

$$\eta_{\mathrm{m}}\,\nabla^2\overline{\boldsymbol{B}} + \alpha\nabla\times\overline{\boldsymbol{B}} - \partial_t\overline{\boldsymbol{B}} = \boldsymbol{0}\,, \quad \nabla\cdot\overline{\boldsymbol{B}} = 0 \tag{94}$$

where $\eta_{\mathrm{m}} = 1/\mu\sigma_{\mathrm{m}}$, or

$$\eta_{\mathrm{m}} = \eta + \beta\,. \tag{95}$$

The occurrence of a contribution to the mean electromotive force $\boldsymbol{\mathcal{E}}$ of the form $\alpha\overline{\boldsymbol{B}}$, that is, parallel or antiparallel to the mean magnetic field, is called

"α-effect". We will see soon that it is the central element of mean-field dynamo theory. The other contribution, $-\beta \nabla \times \overline{\boldsymbol{B}}$, can be interpreted by introducing a mean-field conductivity $\sigma_{\rm m}$ different from the conductivity σ of the fluid in the usual sense, or a mean-field diffusivity $\eta_{\rm m}$ different from the diffusivity η. We will discuss these issues in more detail later.

It is, of course, important to know whether or under which conditions the coefficients α and β are indeed non-zero, and in which way they depend on the velocity field $\boldsymbol{u}'$. With this in mind we study first the behavior of α and β under reflections of the $\boldsymbol{u}'$-field. We start from relation (91) with $\boldsymbol{\mathcal{E}}$ expressed by its definition (81),

$$\overline{\boldsymbol{u}' \times \boldsymbol{B}'} = \alpha(\boldsymbol{u}')\, \overline{\boldsymbol{B}} - \beta(\boldsymbol{u}')\, \nabla \times \overline{\boldsymbol{B}}\,. \tag{96}$$

The notation should stress the dependence of α and β on $\boldsymbol{u}'$. This relation can be understood as a consequence of the connections between $\boldsymbol{u}'$, $\boldsymbol{B}'$ and $\overline{\boldsymbol{B}}$ given by equations (83) with $\overline{\boldsymbol{u}} = \boldsymbol{0}$. If, however, $\boldsymbol{u}'$, $\boldsymbol{B}'$ and $\overline{\boldsymbol{B}}$ satisfy these equations then, as explained in Section 3.8, $\boldsymbol{u}'^{\rm ref}$, $\boldsymbol{B}'^{\rm ref}$ and $\overline{\boldsymbol{B}}^{\rm ref}$ defined by any reflection of them have to do so, too. Consequently, (96) must also apply if we replace $\boldsymbol{u}'$, $\boldsymbol{B}'$ and $\overline{\boldsymbol{B}}$ by $\boldsymbol{u}'^{\rm ref}$, $\boldsymbol{B}'^{\rm ref}$ and $\overline{\boldsymbol{B}}^{\rm ref}$. We restrict the discussion of (96) now to the origin $\boldsymbol{x} = \boldsymbol{0}$ of the coordinate system, what does not imply any loss of generality, and consider reflections just at this point, that is, $\boldsymbol{u}'^{\rm ref}(\boldsymbol{x}) = -\boldsymbol{u}'(-\boldsymbol{x})$, $\boldsymbol{B}'^{\rm ref}(\boldsymbol{x}) = -\boldsymbol{B}'(-\boldsymbol{x})$ and $\overline{\boldsymbol{B}}^{\rm ref}(\boldsymbol{x}) = -\overline{\boldsymbol{B}}(-\boldsymbol{x})$; the argument t is dropped here. Specifying (96) to $\boldsymbol{x} = \boldsymbol{0}$ we have

$$\overline{\boldsymbol{u}'(\boldsymbol{0}) \times \boldsymbol{B}'(\boldsymbol{0})} = \alpha(\boldsymbol{u}')\, \overline{\boldsymbol{B}}(\boldsymbol{0}) - \beta(\boldsymbol{u}')\, (\nabla \times \overline{\boldsymbol{B}})(\boldsymbol{0})\,. \tag{97}$$

Doing the same with the version of (96) for reflected fields we obtain

$$\overline{\boldsymbol{u}'^{\rm ref}(\boldsymbol{0}) \times \boldsymbol{B}'^{\rm ref}(\boldsymbol{0})} = \alpha(\boldsymbol{u}'^{\rm ref})\, \overline{\boldsymbol{B}}^{\rm ref}(\boldsymbol{0}) - \beta(\boldsymbol{u}'^{\rm ref})\, (\nabla \times \overline{\boldsymbol{B}}^{\rm ref})(\boldsymbol{0})\,. \tag{98}$$

Expressing on the left-hand side $\boldsymbol{u}'^{\rm ref}$ and $\boldsymbol{B}'^{\rm ref}$ by $\boldsymbol{u}'$ and $\boldsymbol{B}'$, on the right-hand side $\overline{\boldsymbol{B}}^{\rm ref}$ by $\overline{\boldsymbol{B}}$, and taking into account that $(\nabla \times \overline{\boldsymbol{B}}^{\rm ref})(\boldsymbol{0}) = (\nabla \times \overline{\boldsymbol{B}})(\boldsymbol{0})$, we find

$$\overline{\boldsymbol{u}'(\boldsymbol{0}) \times \boldsymbol{B}'(\boldsymbol{0})} = -\alpha(\boldsymbol{u}'^{\rm ref})\, \overline{\boldsymbol{B}}(\boldsymbol{0}) - \beta(\boldsymbol{u}'^{\rm ref})\, (\nabla \times \overline{\boldsymbol{B}})(\boldsymbol{0})\,. \tag{99}$$

Comparing this with (97) we conclude

$$\alpha(\boldsymbol{u}'^{\rm ref}) = -\alpha(\boldsymbol{u}')\,, \quad \beta(\boldsymbol{u}'^{\rm ref}) = \beta(\boldsymbol{u}')\,. \tag{100}$$

That is, α changes its sign but β remains untouched under reflections of the $\boldsymbol{u}'$-field. If the turbulence is not only homogeneous and isotropic but also reflectionally symmetric then α, as an averaged quantity, has to be equal to zero. A necessary condition for the occurrence of the α-effect is therefore a violation of the reflectional symmetry of the turbulence.

In a rough picture we may understand a turbulent motion as a superposition of eddies with simple flow patterns. We consider in particular eddies with helical, that is screw-like motions, which are roughly characterized by a flow along an axis and a circulation around it, and we distinguish between right-handed and left-handed motions. We note that under reflection a right-handed structure turns into a left-handed one and vice versa. In a homogeneous isotropic turbulence the distribution of the eddies is such that no point in space is preferred over another point and no direction of their axes over another direction. In a reflectionally symmetric turbulence we have in addition an equipartition of right-handed and left-handed motions, and for a turbulence lacking reflectional symmetry this equipartition is violated. The α-effect just requires the violation of this equipartition.

Of course, the case of a homogeneous isotropic but not reflectionally symmetric turbulence is in a sense unrealistic, for under conditions compatible with homogeneity and isotropy there are hardly reasons for a preferred generation of either right-handed or left-handed motions. Turbulent motions on rotating bodies in general violate reflectional symmetry because the Coriolis force generates, depending on the special conditions, preferably either right-handed or left-handed motions. However, apart from homogeneity, these motions lack also isotropy, for already the angular velocity that defines the Coriolis force introduces a preferred direction. Nevertheless the study of the case of homogeneous isotropic but not reflectionally symmetric turbulence is very instructive. It reveals aspects of turbulent motions lacking reflectional symmetry which occur also in the absence of homogeneity or isotropy.

5.5 Dynamo Action of Homogeneous Isotropic Turbulence

Let us now demonstrate that the α-effect as it may occur with a homogeneous isotropic turbulence lacking reflectional symmetry is indeed capable of dynamo action. We consider an infinitely extended fluid and assume that equations (94) for $\overline{\boldsymbol{B}}$ with constant α and η_{m} apply in all space. Anticipating later results, we suppose η_{m} to be positive.

Let us look for solutions of (94) of the form

$$\overline{\boldsymbol{B}} = \Re\big(\hat{\boldsymbol{B}} \exp(\mathrm{i}\,\boldsymbol{k} \cdot \boldsymbol{x} + pt)\big)\,, \tag{101}$$

with $\hat{\boldsymbol{B}}$ being a complex constant vector, $\boldsymbol{k}$ a real wave vector and p a real parameter describing, if positive, a growth rate. With (94) we find

$$(\eta_{\mathrm{m}} \boldsymbol{k}^2 + p)\hat{\boldsymbol{B}} + \mathrm{i}\alpha \boldsymbol{k} \times \hat{\boldsymbol{B}} = \boldsymbol{0}\,, \quad \boldsymbol{k} \cdot \hat{\boldsymbol{B}} = 0\,, \tag{102}$$

or, using a Cartesian coordinate system (x, y, z) in which $\boldsymbol{k} = (0, 0, k)$,

$$(\eta_{\mathrm{m}} k^2 + p)\hat{B}_x - \mathrm{i}\alpha k \hat{B}_y = 0\,, \quad \mathrm{i}\alpha k \hat{B}_x - (\eta_{\mathrm{m}} k^2 + p)\hat{B}_y = 0\,, \quad \hat{B}_z = 0\,. \tag{103}$$

There are non-trivial solutions $\hat{\boldsymbol{B}}$ only if the determinant $(\eta_{\mathrm{m}} k^2 + p)^2 - \alpha^2 k^2$ is equal to zero, that is, if

$$p = -\eta_{\mathrm{m}} k^2 \pm |\alpha k|\,. \tag{104}$$

For convenience we may restrict our discussion to non-negative k. The solution $\overline{\boldsymbol{B}}$ of (94) corresponding to the lower sign in (104) decays for all k. The one corresponding to the upper sign grows for $k < |\alpha|/\eta_{\rm m}$, is steady for $k = 0$ and $k = |\alpha|/\eta_{\rm m}$, and decays for $k > |\alpha|/\eta_{\rm m}$. Note that $\overline{\boldsymbol{B}}$ is a homogeneous field if $k = 0$, and its variability with z increases with k.

Let us introduce a dimensionless parameter R_α built after the pattern of the magnetic Reynolds number,

$$R_\alpha = |\alpha|\, l/\eta_{\rm m}\,, \tag{105}$$

where l is a wave-length defined by $l = 2\pi/k$. Then our result says that a dynamo is possible as soon as

$$R_\alpha \geq 2\pi\,. \tag{106}$$

Note that this condition can be fulfilled with arbitrarily small $|\alpha|$ if only l is sufficiently large.

A simple mean-field dynamo model with homogeneous isotropic not reflectionally symmetric turbulence in a spherical fluid body surrounded by free space has been proposed by Krause and Steenbeck [60]; see also [2]. Although in a sense unrealistic, it helps to understand how an α-effect dynamo works. In addition it provides us with a useful introduction into the mathematical treatment of spherical mean-field dynamo models, which can be done analytically in this particular case.

In the model under consideration $\overline{\boldsymbol{B}}$ has to satisfy (94) inside the fluid body and to continue in outer space as a solenoidal potential field vanishing at infinity. The general solution $\overline{\boldsymbol{B}}$ is a superposition of independent modes of the form $\boldsymbol{B}_n(\boldsymbol{x}) \exp(p_n t)$ where the $\boldsymbol{B}_n$ are fields consisting of both poloidal and toroidal parts and the p_n are their growth rates. We introduce here a dimensionless parameter R_α by

$$R_\alpha = |\alpha|\, R/\eta_{\rm m}\,, \tag{107}$$

where R is the radius of the fluid sphere. The model works as dynamo if

$$R_\alpha \geq 4.49\,. \tag{108}$$

The most easily excitable mode, which is steady for $R_\alpha = 4.49$ and grows if $R_\alpha > 4.49$, has a poloidal part of dipolar structure.

5.6 Approximative Calculation of the Mean Electromotive Force

We present now a method for an approximate calculation of the electromotive force $\boldsymbol{\mathcal{E}}$ for turbulent fluid motions. For the sake of simplicity we restrict ourselves again to an infinitely extended fluid without mean motion, $\overline{\boldsymbol{u}} = \boldsymbol{0}$, and assume that the magnetic diffusivity η is independent of position and time. As far as $\boldsymbol{u}'$ is concerned, however, we admit now an arbitrary turbulence. Only in the next section we will specify the results to a homogeneous isotropic one.

Under the assumptions adopted equations (83) can be written in the form

$$\eta\, \nabla^2 \boldsymbol{B}' - \partial_t \boldsymbol{B}' = -\nabla \times (\boldsymbol{u}' \times \overline{\boldsymbol{B}} + (\boldsymbol{u}' \times \boldsymbol{B}')')\,, \quad \nabla \cdot \boldsymbol{B}' = 0\,, \tag{109}$$

where $(\boldsymbol{u}' \times \boldsymbol{B}')'$, of course, means $\boldsymbol{u}' \times \boldsymbol{B}' - \overline{(\boldsymbol{u}' \times \boldsymbol{B}')}$. For a first step of approximation we cancel the term $(\boldsymbol{u}' \times \boldsymbol{B}')'$ in (109). For a second step we could take it into account with $\boldsymbol{B}'$ as resulting from the first step, and analogously we could carry out further steps. Calculations of this kind are, however, very tedious, and therefore we restrict ourselves here to the first step. The approximation defined in this way is often called "first-order smoothing" or, by reasons which will become visible soon, "second-order correlation approximation". A sufficient condition for its validity is obviously $|\boldsymbol{B}'|/|\overline{\boldsymbol{B}}| \ll 1$, which we will express in another form later. There are reasons to assume that the approximation applies also in some region beyond this condition, but this is a rather complex issue, which we do not want to discuss here.

Equations (109), if simplified as mentioned, agree formally with (16), whose general solution has been given with (21). Following this pattern we write the solution of (109) for $\boldsymbol{B}'$ in the form

$$B'_k(\boldsymbol{x},t) = \int_\infty G(\boldsymbol{x}-\boldsymbol{x}', t-t_0)\, B'_k(\boldsymbol{x}', t_0)\, \mathrm{d}^3x' \qquad (110)$$
$$+\epsilon_{klm}\,\epsilon_{mpq} \int_{t_0}^{t} \int_\infty G(\boldsymbol{x}-\boldsymbol{x}', t-t')\, (\partial/\partial x'_l)\, (u'_p(\boldsymbol{x}',t')\overline{B}_q(\boldsymbol{x}',t'))\, \mathrm{d}^3x'\, \mathrm{d}t' ,$$

where $\boldsymbol{B}'(\boldsymbol{x}, t_0)$ is assumed to be solenoidal. With a change of the integration variables and an integration by parts this turns into

$$B'_k(\boldsymbol{x},t) = \int_\infty G(\boldsymbol{\xi},\tau)\, B'_k(\boldsymbol{x}-\boldsymbol{\xi}, t_0)\, \mathrm{d}^3\xi \qquad (111)$$
$$+\epsilon_{klm}\,\epsilon_{mpq} \int_0^{t-t_0} \int_\infty \frac{1}{\xi}\frac{\partial G(\boldsymbol{\xi},\tau)}{\partial \xi}\, \xi_l\, u'_p(\boldsymbol{x}-\boldsymbol{\xi}, t-\tau)\, \overline{B}_q(\boldsymbol{x}-\boldsymbol{\xi}, t-\tau)\, \mathrm{d}^3\xi\, \mathrm{d}\tau .$$

Note that G depends on $\boldsymbol{\xi}$ via ξ only.

For the calculation of $\boldsymbol{\mathcal{E}}$ we start from (81), write it in the form

$$\mathcal{E}_i(\boldsymbol{x},t) = \epsilon_{ijk}\, \overline{u'_j(\boldsymbol{x},t)\, B'_k(\boldsymbol{x},t)} \qquad (112)$$

and insert B'_k as given by (111). Restricting our attention to times t far away from the initial time t_0 so that there is no longer any correlation between quantities measured at these different times, we omit the contribution to $\mathcal{E}_i(\boldsymbol{x},t)$ which contains $B'_k(\boldsymbol{x}', t_0)$ and let $t_0 \to -\infty$. Then a straightforward calculation leads just to a representation of $\mathcal{E}_i(\boldsymbol{x},t)$ in the form of (86) with

$$K_{ij}(\boldsymbol{x},t;\boldsymbol{\xi},\tau) = (\epsilon_{ilm}\delta_{nj} - \epsilon_{ilj}\delta_{mn})\, \frac{1}{\xi}\frac{\partial G(\boldsymbol{\xi},\tau)}{\partial \xi}\, Q_{lm}(\boldsymbol{x},t;-\boldsymbol{\xi},-\tau)\, \xi_n . \qquad (113)$$

Here Q_{lm} means the correlation tensor of second rank for the $\boldsymbol{u}'$-field,

$$Q_{lm}(\boldsymbol{x},t;\boldsymbol{\xi},\tau) = \overline{u'_l(\boldsymbol{x},t)\, u'_m(\boldsymbol{x}+\boldsymbol{\xi}, t+\tau)} . \qquad (114)$$

By the way, omitting the term with $B'_k(\boldsymbol{x}', t_0)$ in (110) or (111) corresponds just to the neglect of the contribution of $\boldsymbol{\mathcal{E}}^{(0)}$ to $\boldsymbol{\mathcal{E}}$ introduced above.

Specifying now the relations (89) for a_{ij} and b_{ijk} by the result (113) we obtain

$$\begin{aligned} a_{ij} &= (\epsilon_{ilm}\delta_{nj} - \epsilon_{ilj}\delta_{mn}) \int_0^\infty \int_\infty \frac{1}{\xi}\frac{\partial G(\boldsymbol{\xi},\tau)}{\partial \xi} Q_{lm}(\boldsymbol{x},t;-\boldsymbol{\xi},-\tau)\,\xi_n\,\mathrm{d}^3\xi\,\mathrm{d}\tau \\ &= -(\epsilon_{ilm}\delta_{nj} - \epsilon_{ilj}\delta_{mn}) \int_0^\infty \int_\infty G(\boldsymbol{\xi},\tau)\,\frac{\partial Q_{lm}(\boldsymbol{x},t;-\boldsymbol{\xi},-\tau)}{\partial \xi_n}\,\mathrm{d}^3\xi\,\mathrm{d}\tau \qquad (115) \\ b_{ijk} &= -(\epsilon_{ilm}\delta_{nj} - \epsilon_{ilj}\delta_{mn}) \int_0^\infty \int_\infty \frac{1}{\xi}\frac{\partial G(\boldsymbol{\xi},\tau)}{\partial \xi} Q_{lm}(\boldsymbol{x},t;-\boldsymbol{\xi},-\tau)\,\xi_n\,\xi_k\,\mathrm{d}^3\xi\,\mathrm{d}\tau\,. \end{aligned}$$

As can easily be seen from (109) the above-mentioned condition $|\boldsymbol{B}'|/|\overline{\boldsymbol{B}}| \ll 1$ for the validity of the second-order correlation approximation used here can be expressed by

$$\min\left(\sqrt{\overline{\boldsymbol{u}'^2}}\,\tau_\mathrm{c}/\lambda_\mathrm{c}\,,\,\sqrt{\overline{\boldsymbol{u}'^2}}\,\lambda_\mathrm{c}/\eta\right) \ll 1\,, \qquad (116)$$

where λ_c and τ_c mean correlation length and time. In higher approximations in addition to second-order correlations higher-order ones occur in (113) and (115).

For the further evaluation of the relations (115) it is useful to replace the integration variables $\boldsymbol{\xi}$ and τ by dimensionless variables defined on the basis of λ_c and τ_c, and to express any dependence on η as one on the dimensionless parameter

$$q = \lambda_\mathrm{c}^2/\eta\tau_\mathrm{c}\,. \qquad (117)$$

If we equate $\sqrt{\overline{\boldsymbol{u}'^2}}$ to $\lambda_\mathrm{c}/\tau_\mathrm{c}$ the parameter q can be interpreted as the magnetic Reynolds number $\sqrt{\overline{\boldsymbol{u}'^2}}\,\lambda_\mathrm{c}/\eta$ for the turbulent motion.

Two limiting cases defined via q are of particular interest, which allow much simpler representations of a_{ij} and b_{ijk} . In the high-conductivity limit, $q \to \infty$, the integrals in (115) reduce to such over τ only, which contain Q_{lm} and its derivatives only with $\boldsymbol{\xi} = \boldsymbol{0}$, and the condition (116) to $\sqrt{\overline{\boldsymbol{u}'^2}}\,\tau_\mathrm{c}/\lambda_\mathrm{c} \ll 1$. In the low-conductivity limit, $q \to 0$, we have integrals over $\boldsymbol{\xi}$ only, which contain Q_{lm} only with $\tau = 0$, and $\sqrt{\overline{\boldsymbol{u}'^2}}\,\lambda_\mathrm{c}/\eta \ll 1$. We will give particular results for such limiting cases in the following section.

5.7 α-Effect and Mean-Field Conductivity in the Case of Homogeneous Isotropic Turbulence

Returning to the case of homogeneous isotropic turbulence we first conclude from (90) that

$$\alpha = \frac{1}{3}\,a_{ii}\,, \quad \beta = \frac{1}{6}\,\epsilon_{ijk}\,b_{ijk}\,. \qquad (118)$$

Using then (115) we find

$$\begin{aligned} \alpha &= \frac{1}{3}\int_0^\infty \int_\infty \frac{1}{\xi}\frac{\partial G(\boldsymbol{\xi},\tau)}{\partial \xi}\,\overline{(\boldsymbol{u}'(\boldsymbol{x},t)\times \boldsymbol{u}'(\boldsymbol{x}+\boldsymbol{\xi},t-\tau))}\cdot\boldsymbol{\xi}\,\mathrm{d}^3\xi\,\mathrm{d}\tau \\ &= -\frac{1}{3}\int_0^\infty \int_\infty G(\boldsymbol{\xi},\tau)\,\overline{\boldsymbol{u}'(\boldsymbol{x},t)\cdot(\nabla\times\boldsymbol{u}'(\boldsymbol{x}+\boldsymbol{\xi},t-\tau))}\,\mathrm{d}^3\xi\,\mathrm{d}\tau \qquad (119) \\ \beta &= -\frac{1}{3}\int_0^\infty \int_\infty \xi\,\frac{\partial G(\boldsymbol{\xi},\tau)}{\partial \xi}\,\overline{u'_\xi(\boldsymbol{x},t))\,u'_\xi(\boldsymbol{x}+\boldsymbol{\xi},t-\tau)}\,\mathrm{d}^3\xi\,\mathrm{d}\tau\,, \end{aligned}$$

where u'_ξ means $\boldsymbol{u}' \cdot \boldsymbol{\xi}/\xi$. We note that due to the isotropy of the turbulence the quantities $\overline{(\boldsymbol{u}'(\boldsymbol{x},t) \times \boldsymbol{u}'(\boldsymbol{x}+\boldsymbol{\xi},t+\tau)) \cdot \boldsymbol{\xi}}$, $\overline{(\boldsymbol{u}'(\boldsymbol{x},t) \cdot (\nabla \times \boldsymbol{u}'(\boldsymbol{x}+\boldsymbol{\xi},t+\tau))}$ and $\overline{u'_\xi(\boldsymbol{x},t)\, u'_\xi(\boldsymbol{x}+\boldsymbol{\xi},t+\tau)}$ do not depend on the direction of $\boldsymbol{\xi}$ and that the last one is equal to $\frac{1}{3}\,\overline{\boldsymbol{u}'(\boldsymbol{x},t) \cdot \boldsymbol{u}'(\boldsymbol{x}+\boldsymbol{\xi},t+\tau)}$.

Evaluating this for the high-conductivity limit, $q \to \infty$, we obtain

$$\begin{aligned} \alpha &= \frac{1}{3} \int_0^\infty \frac{1}{\xi^2} \overline{(\boldsymbol{u}'(\boldsymbol{x},t) \times \boldsymbol{u}'(\boldsymbol{x},t-\tau)) \cdot \boldsymbol{\xi}}\, \mathrm{d}\tau \\ &= -\frac{1}{3} \int_0^\infty \overline{\boldsymbol{u}'(\boldsymbol{x},t) \cdot (\nabla \times \boldsymbol{u}'(\boldsymbol{x},t-\tau))}\, \mathrm{d}\tau \\ \beta &= \frac{1}{3} \int_0^\infty \overline{u'_\xi(\boldsymbol{x},t)\, u'_\xi(\boldsymbol{x},t-\tau)}\, \mathrm{d}\tau\,. \end{aligned} \tag{120}$$

Remarkably enough, both α and β remain non-zero values in this limit. We write (120) in the simpler form

$$\alpha = -\frac{1}{3} \overline{\boldsymbol{u}' \cdot (\nabla \times \boldsymbol{u}')}\, \tau_{\mathrm{c}}^{(\alpha)}\,, \quad \beta = \frac{1}{9} \overline{\boldsymbol{u}'^2}\, \tau_{\mathrm{c}}^{(\beta)}\,, \tag{121}$$

with correlation times $\tau_{\mathrm{c}}^{(\alpha)}$ and $\tau_{\mathrm{c}}^{(\beta)}$ defined just by equating the integrals in (120) to $\overline{\boldsymbol{u}' \cdot (\nabla \times \boldsymbol{u}')}\, \tau_{\mathrm{c}}^{(\alpha)}$ or $\frac{1}{3}\overline{\boldsymbol{u}'^2}\, \tau_{\mathrm{c}}^{(\beta)}$; the exceptional case in which the first integral in (120) is unequal but $\overline{\boldsymbol{u}' \cdot (\nabla \times \boldsymbol{u}')}$ equal to zero is excluded here. The quantity $\overline{\boldsymbol{u}' \cdot (\nabla \times \boldsymbol{u}')}$ is called "helicity" of the turbulent motion.

For the low-conductivity limit $q \to 0$ we find

$$\begin{aligned} \alpha &= -\frac{1}{4\pi\eta} \int_\infty \overline{\boldsymbol{u}'(\boldsymbol{x},t) \cdot (\nabla \times \boldsymbol{u}'(\boldsymbol{x}+\boldsymbol{\xi},t))}\, \frac{\mathrm{d}^3\xi}{\xi}\,, \\ \beta &= \frac{1}{4\pi\eta} \int_\infty \overline{u'_\xi(\boldsymbol{x},t) u'_\xi(\boldsymbol{x}+\boldsymbol{\xi},t)}\, \frac{\mathrm{d}^3\xi}{\xi}\,, \end{aligned} \tag{122}$$

With a view to interesting alternative relations for α and β we note that $\boldsymbol{u}'$ can be represented by $\boldsymbol{u}' = \nabla \times \boldsymbol{a} + \nabla\varphi$ with a vector potential $\boldsymbol{a}$ satisfying $\nabla \cdot \boldsymbol{a} = 0$ and a scalar potential φ, which are normalized such that $\overline{\boldsymbol{a}} = 0$ and $\overline{\varphi} = 0$. Using this we can rewrite (122) in

$$\alpha = -\frac{1}{3\eta} \overline{\boldsymbol{a} \cdot (\nabla \times \boldsymbol{a})}\,, \quad \beta = \frac{1}{3\eta} (\overline{\boldsymbol{a}^2} - \overline{\varphi^2})\,. \tag{123}$$

It is often said that dynamo action of turbulent motions is in a simple way connected with their helicity, and that the coefficient α in the mean electromotive force is, apart from a factor, just the helicity. We want to stress that this applies only under rather special conditions. Apart from homogeneity and isotropy of the turbulence the second-order correlation approximation and the restriction to the high-conductivity limit are necessary to justify a statement of that kind. Our results show that the situation changes already in a remarkable way if we replace the high-conductivity limit by the low-conductivity limit.

Let us add a few remarks concerning the mean-field conductivity $\sigma_{\rm m}$ defined in (93), which depends on β. We restrict ourselves to the high-conductivity limit $q \to \infty$. Remarkably enough, since β does not vanish in this limit, $\sigma_{\rm m}$ remains finite even if $\sigma \to \infty$. If we use β in the form (121) and accept that $\mu\sigma\overline{\boldsymbol{u}'^2}\tau_{\rm c}^{(\beta)} \gg 1$ we find

$$\sigma_{\rm m} = 9/\mu\overline{\boldsymbol{u}'^2}\tau_{\rm c}^{(\beta)} . \tag{124}$$

The condition $\mu\sigma\overline{\boldsymbol{u}'^2}\tau_{\rm c}^{(\beta)} \gg 1$ coincides with $q \gg 1$ if $\sqrt{\overline{\boldsymbol{u}'^2}}$ is of the order of $\lambda_{\rm c}/\tau_{\rm c}$ and $\tau_{\rm c}^{(\beta)}$ of that of $\tau_{\rm c}$. Of course, the relation (121) must be considered with some care since the applicability of (120) is possibly not well justified if $\sqrt{\overline{\boldsymbol{u}'^2}}$ is not much smaller than $\lambda_{\rm c}/\tau_{\rm c}$.

As an example we consider the situation in the convection zone of the Sun. as characterized in Table 2. Using the value of σ given there and choosing for $\sqrt{\overline{\boldsymbol{u}'^2}}$ and $\tau_{\rm c}^{(\beta)}$ a typical velocity and a typical life time of a granule, that is 200 m/s and 600 s, we conclude from (121) that $\sigma_{\rm m}/\sigma \simeq 10^{-4}$. Even if this has to be considered as a rough estimate only, it clearly demonstrates that the mean-field conductivity, which is relevant to large-scale phenomena, is much smaller than the conductivity in the usual sense, which determines small-scale processes. This finding points a way to resolve the conflict between the value of T_η given in Table 2 for sunspots, which is about 10^4 years, and their real life time of at least 2 months. When calculating T_η with $\sigma_{\rm m}$ instead of σ we find about one year, which is at least much closer to the real life time of sunspots.

We add a remark concerning the simple spherical mean-field dynamo model mentioned in Section 5.5. We have seen here that α and β need not to vanish in the high-conductivity limit, and it is well possible that $|\alpha|R/\eta_{\rm m} > 4.49$ even in this limit. For this case the model allows magnetic fields that grow exponentially with time everywhere, also outside the fluid body. This, however, is in conflict with the statement by Bondi and Gold explained in Section 3.6. Of course, the assumption of a mean electromotive force corresponding to a homogeneous and isotropic turbulence also in the close neighborhood of the boundary of the conducting body, which was used in this model, is obviously incorrect. Indeed a consequent treatment of a modified model taking into account deviations from homogeneous isotropic turbulence near the boundary has resolved the conflict [61]. In the modified model a dynamo proves to be possible even in the high-conductivity limit but its magnetic field is then completely confined inside the fluid body [62].

5.8 The Mean Electromotive Force for Axisymmetric Turbulence

Let us briefly deal with the case in which the turbulence is no longer necessarily homogeneous and isotropic but axisymmetric. The preferred axis may be defined, for example, by an gradient of the intensity or of any other property of the turbulence given by an averaged quantity, or by the angular velocity responsible for the Coriolis force. The unit vector parallel to this axis is denoted by $\boldsymbol{\kappa}$.

Starting again with the representation (88) for $\boldsymbol{\mathcal{E}}$ and modifying properly the arguments used in the case of homogeneous isotropic turbulence in Section 5.4 we conclude that the tensors a_{ij} and b_{ijk} have to be axisymmetric in the sense that their components are invariant under rotations of the coordinate system about an axis parallel to $\boldsymbol{\kappa}$. The general form of such tensors is a linear combination of all tensors which can be built up from the isotropic tensors δ_{ij} and ϵ_{ijk} and the vector κ_i,

$$\begin{aligned} a_{ij} &= a_1\,\delta_{ij} + a_2\,\epsilon_{ijl}\,\kappa_l + a_3\,\kappa_i\,\kappa_j \\ b_{ijk} &= b_1\,\epsilon_{ijk} + b_2\,\delta_{ij}\,\kappa_k + b_3\,\delta_{ik}\,\kappa_j + b_4\,\delta_{jk}\,\kappa_i \\ &\quad + b_5\,\epsilon_{ijl}\,\kappa_k\,\kappa_l + b_6\,\epsilon_{ikl}\,\kappa_j\,\kappa_l + b_7\,\epsilon_{jkl}\,\kappa_i\,\kappa_l + b_8\,\kappa_i\,\kappa_j\,\kappa_k\,. \end{aligned} \tag{125}$$

The coefficients $a_1, a_2, \ldots b_8$ are determined by $\boldsymbol{u}'$ and may vary with the space coordinate along $\boldsymbol{\kappa}$. Since $(\epsilon_{ijl}\,\kappa_k + \epsilon_{jkl}\,\kappa_i + \epsilon_{kil}\,\kappa_j)\kappa_l = \epsilon_{ijk}$ we may put for example $b_5 = b_6$ without any loss of generality. ¿From (88) and (125) we then obtain

$$\begin{aligned} \boldsymbol{\mathcal{E}} = {} & a_1\,\overline{\boldsymbol{B}} - a_2\,\boldsymbol{\kappa}\times\overline{\boldsymbol{B}} + a_3\,(\boldsymbol{\kappa}\cdot\overline{\boldsymbol{B}})\,\boldsymbol{\kappa} \\ & - b_1\,\nabla\times\overline{\boldsymbol{B}} + b_2\,(\boldsymbol{\kappa}\cdot\nabla)\,\overline{\boldsymbol{B}} + b_3\,\nabla(\boldsymbol{\kappa}\cdot\overline{\boldsymbol{B}}) \\ & - b_5\,\boldsymbol{\kappa}\times((\boldsymbol{\kappa}\cdot\nabla)\,\overline{\boldsymbol{B}} + \nabla(\boldsymbol{\kappa}\cdot\overline{\boldsymbol{B}})) - b_7\,(\boldsymbol{\kappa}\cdot(\nabla\times\overline{\boldsymbol{B}}))\boldsymbol{\kappa} + b_8\,(\boldsymbol{\kappa}\cdot\nabla(\boldsymbol{\kappa}\cdot\overline{\boldsymbol{B}}))\boldsymbol{\kappa}\,. \end{aligned} \tag{126}$$

Because of $\nabla\cdot\overline{\boldsymbol{B}} = 0$ there is no contribution with b_4. Using the identity $\boldsymbol{\kappa}\times(\nabla\times\overline{\boldsymbol{B}}) = \nabla(\boldsymbol{\kappa}\cdot\overline{\boldsymbol{B}}) - (\boldsymbol{\kappa}\cdot\nabla)\,\overline{\boldsymbol{B}}$ we turn (126) into the form

$$\begin{aligned} \boldsymbol{\mathcal{E}} = {} & -\alpha_1\,\overline{\boldsymbol{B}} - \alpha_2\,(\boldsymbol{\kappa}\cdot\overline{\boldsymbol{B}})\,\boldsymbol{\kappa} - \gamma\,\boldsymbol{\kappa}\times\overline{\boldsymbol{B}} \\ & - \beta_1\,\nabla\times\overline{\boldsymbol{B}} - \beta_2\,(\boldsymbol{\kappa}\cdot(\nabla\times\overline{\boldsymbol{B}}))\,\boldsymbol{\kappa} - \delta\,\boldsymbol{\kappa}\times(\nabla\times\overline{\boldsymbol{B}}) \\ & - \beta_1^\kappa\,\nabla(\boldsymbol{\kappa}\cdot\overline{\boldsymbol{B}}) - \beta_2^\kappa\,(\boldsymbol{\kappa}\cdot\nabla(\boldsymbol{\kappa}\cdot\overline{\boldsymbol{B}}))\boldsymbol{\kappa} - \delta^\kappa\,\boldsymbol{\kappa}\times\nabla(\boldsymbol{\kappa}\cdot\overline{\boldsymbol{B}}) \end{aligned} \tag{127}$$

with new coefficients $\alpha_1, \alpha_2, \ldots \delta^\kappa$ being linear combinations of $a_1, a_2, \ldots b_8$, chosen with a view to later generalizations.

We now rely on Ohm's law (80b) with $\overline{\boldsymbol{u}} = \boldsymbol{0}$ and insert there $\boldsymbol{\mathcal{E}}$ as given by (127). We further split $\overline{\boldsymbol{j}}$ and analogously $\overline{\boldsymbol{E}}$, $\overline{\boldsymbol{B}}$ and ∇ in the two parts $\overline{\boldsymbol{j}}_\parallel = (\boldsymbol{\kappa}\cdot\overline{\boldsymbol{j}})\,\boldsymbol{\kappa}$ and $\overline{\boldsymbol{j}}_\perp = \overline{\boldsymbol{j}} - \overline{\boldsymbol{j}}_\parallel$. In this way we obtain

$$\begin{aligned} \overline{\boldsymbol{j}}_\parallel &= \sigma_{\mathrm{m}\,\parallel}\,(\overline{\boldsymbol{E}}_\parallel - (\alpha_1+\alpha_2)\,\overline{\boldsymbol{B}}_\parallel - (\beta_1^\kappa + \beta_2^\kappa)\,\nabla_\parallel \overline{B}_\parallel) \\ \overline{\boldsymbol{j}}_\perp + c\boldsymbol{\kappa}\times\overline{\boldsymbol{j}}_\perp &= \sigma_{\mathrm{m}\,\perp}\,(\overline{\boldsymbol{E}}_\perp - \alpha_1\,\overline{\boldsymbol{B}}_\perp - \gamma\boldsymbol{\kappa}\times\overline{\boldsymbol{B}}_\perp - \beta_1^\kappa\,\nabla_\perp \overline{B}_\parallel - \delta^\kappa\,\boldsymbol{\kappa}\times\nabla_\perp\overline{B}_\parallel) \end{aligned} \tag{128}$$

with $\overline{B}_\parallel$ standing for $\boldsymbol{\kappa}\cdot\overline{\boldsymbol{B}}$ and

$$\sigma_{\mathrm{m}\,\parallel} = \frac{\sigma}{1+\mu\sigma(\beta_1+\beta_2)}\,, \quad \sigma_{\mathrm{m}\,\perp} = \frac{\sigma}{1+\mu\sigma\beta_1}\,, \quad c = \mu\sigma_{\mathrm{m}\,\perp}\,\delta\,. \tag{129}$$

Compared to the corresponding results (91) and (92) for homogeneous isotropic turbulence the situation here is more complex. One remarkable aspect is that there is no longer an isotropic mean-field conductivity. In general $\sigma_{\mathrm{m}\,\parallel}$ and $\sigma_{\mathrm{m}\,\perp}$ are different so that even in the simplest case in which $\alpha_1, \alpha_2, \delta, \beta_1^\kappa, \beta_2^\kappa$ and δ^κ vanish, only $\overline{\boldsymbol{j}}_\parallel$ is parallel to $\overline{\boldsymbol{E}}_\parallel$ and $\overline{\boldsymbol{j}}_\perp$ to $\overline{\boldsymbol{E}}_\perp$, but no longer $\overline{\boldsymbol{j}}$ to $\overline{\boldsymbol{E}}$. If δ is non-zero in addition $\overline{\boldsymbol{j}}_\perp$ and $\overline{\boldsymbol{E}}_\perp$ are inclined to each other. Likewise the α-effect, now described by the two coefficients α_1 and α_2, is in general no longer isotropic. Further new aspects consists in the occurrence of other induction effects described by the γ term, the β_1^κ term, ... in (127) and (128), simply called "γ-effect", "β_1^κ-effect", ... in the following. The γ-effect corresponds to transport of mean magnetic flux as it would occur with a mean motion, which is, however, not taken into account here. The $\beta_1^\kappa, \beta_2^\kappa$, and δ^κ terms depend on derivatives of $\overline{\boldsymbol{B}}$ which cannot be expressed by $\nabla \times \overline{\boldsymbol{B}}$. If β_1^κ is constant the corresponding term is a gradient and can always be compensated by a part of the mean electric field. In contrast to that the β_2^κ and δ^κ-effects can well be sources of mean electric currents. We will come back to the induction effects mentioned here in the more general framework of the next section.

Again to the behavior of the coefficients $\alpha_1, \alpha_2, \ldots \delta^\kappa$ under reflections of the $\boldsymbol{u}'$ field deserves particular interest. We distinguish between reflections at planes perpendicular to $\boldsymbol{\kappa}$ and such at planes containing $\boldsymbol{\kappa}$. Clearly the reflectional symmetry with respect to the first type of planes is broken if there is a gradient of the intensity or of another property of the turbulence, and that with respect to the second type if Coriolis forces act. Modifying properly the arguments used in Section 5.4 we find that α_1, α_2 and γ inverse their signs under reflection of $\boldsymbol{u}'$ at planes perpendicular to $\boldsymbol{\kappa}$ but all other coefficients remain unchanged. Furthermore, $\alpha_1, \alpha_2, \delta, \beta_1^\kappa$ and β_2^κ inverse their signs under reflection at planes containing $\boldsymbol{\kappa}$ and all others remain unchanged.

Thus an α-effect, that is non-zero α_1 or α_2, is only possible if the reflectional symmetry of $\boldsymbol{u}'$ with respect to both types of planes is broken. A turbulence with a gradient of its intensity or of another property under the influence of Coriolis forces opens the possibility of an α-effect but never such a gradient alone or Coriolis forces alone.

As it was demonstrated in Section 5.5 the α-effect is capable of dynamo action. We note that the δ-effect together with a shear in the mean motion may also work as a dynamo; see also Section 6.3. In contrast to the α-effect the δ-effect can be non-zero even in the case of symmetry with respect to the planes perpendicular to $\boldsymbol{\kappa}$ if only that with respect to planes containing $\boldsymbol{\kappa}$ is broken. This is possible in a homogeneous turbulence subject to Coriolis forces. That is, under conditions which do not allow for an α-effect dynamo another kind of mean-field dynamo is well possible.

Using the results of Section 5.6 we may easily find relations comparable with (119) which connect the coefficients $\alpha_1, \alpha_2, \ldots \delta^\kappa$ with averaged quantities depending on $\boldsymbol{u}'$ [63]. We refrain from giving them here.

5.9 The Mean Electromotive Force for More Complex Forms of the Turbulence

We leave now the special cases of homogeneous isotropic and of axisymmetric turbulence and admit again a mean motion as well as arbitrary kinds of turbulent motions. For the discussion of the electromotive $\boldsymbol{\mathcal{E}}$ in such more general cases it is useful to express its connection with $\overline{\boldsymbol{B}}$ and its spatial derivative as given so far by (88) in another form.

Considering first the last term on the right-hand side of (88) we note that the tensor $\partial B_j/\partial x_k$ can be split in a symmetric part, denoted by $(\nabla\overline{\boldsymbol{B}})^{\mathrm{s}}_{jk}$ in the following, and an antisymmetric part, which can be represented in the form $\epsilon_{jkl}V_l$ with a vector $\boldsymbol{V}$. The latter is given by $V_l = -\frac{1}{2}\epsilon_{lmn}\partial\overline{B}_m/\partial x_n$, that is, $\boldsymbol{V} = -\frac{1}{2}\nabla\times\overline{\boldsymbol{B}}$. Thus that last term in (88) can be replaced by the sum of two terms, one of the form $b_{ij}(\nabla\times\overline{\boldsymbol{B}})_j$ and the other of the form $c_{ijk}(\nabla\overline{\boldsymbol{B}})^{\mathrm{s}}_{jk}$. We note that $b_{ij} = -\frac{1}{4}\,\epsilon_{jkl}\,b_{ikl}$ and $c_{ijk} = \frac{1}{2}\,(b_{ijk}+b_{ikj})$. Let us now modify the last term in (88) in this way. We may in addition split the tensors a_{ij} and b_{ij}, too, in symmetric and antisymmetric parts and express the latter by vectors. In this way it becomes clear that the representation of $\boldsymbol{\mathcal{E}}$ given by (88) is equivalent to

$$\boldsymbol{\mathcal{E}} = -\boldsymbol{\alpha}\cdot\overline{\boldsymbol{B}} - \boldsymbol{\gamma}\times\overline{\boldsymbol{B}} - \boldsymbol{\beta}\cdot(\nabla\times\overline{\boldsymbol{B}}) - \boldsymbol{\delta}\times(\nabla\times\overline{\boldsymbol{B}}) - \boldsymbol{\kappa}\cdot(\nabla\overline{\boldsymbol{B}})^{\mathrm{s}}\,, \tag{130}$$

where $\boldsymbol{\alpha}$ and $\boldsymbol{\beta}$ are symmetric tensors of second rank, $\boldsymbol{\gamma}$ and $\boldsymbol{\delta}$ vectors, and $\boldsymbol{\kappa}$ is a tensor of third rank. The latter may be assumed to be symmetric in the indices connecting it with $(\nabla\overline{\boldsymbol{B}})^{\mathrm{s}}$, and contributions producing terms with $\nabla\cdot\overline{\boldsymbol{B}}$ can be omitted. Of course, $\boldsymbol{\alpha},\boldsymbol{\beta},\boldsymbol{\gamma},\boldsymbol{\delta}$ and $\boldsymbol{\kappa}$ are again determined by the fluid motion, that is, by $\overline{\boldsymbol{u}}$ and $\boldsymbol{u}'$. The choice of the signs in (130) is not compelling but follows certain conventions. When combining Ohm's law (80b) for mean fields with (130) we find

$$\overline{\boldsymbol{j}} = \boldsymbol{\sigma}_{\mathrm{m}}\cdot(\overline{\boldsymbol{E}} + (\overline{\boldsymbol{u}}-\boldsymbol{\gamma})\times\overline{\boldsymbol{B}} - \boldsymbol{\alpha}\cdot\overline{\boldsymbol{B}} - \boldsymbol{\delta}\times(\nabla\times\overline{\boldsymbol{B}}) - \boldsymbol{\kappa}\cdot(\nabla\overline{\boldsymbol{B}})^{\mathrm{s}})\,, \tag{131}$$

where $\boldsymbol{\sigma}_{\mathrm{m}}$ is now a conductivity tensor defined by

$$\sigma_{\mathrm{m}\,ij} = \sigma\,(\delta_{ij} + \mu\sigma\beta_{ij})^{-1}\,. \tag{132}$$

We may also include the effect of the term $\boldsymbol{\delta}\times(\nabla\times\overline{\boldsymbol{B}})$ in the conductivity tensor and write

$$\overline{\boldsymbol{j}} = \tilde{\boldsymbol{\sigma}}_{\mathrm{m}}\cdot(\overline{\boldsymbol{E}} + (\overline{\boldsymbol{u}}-\boldsymbol{\gamma})\times\overline{\boldsymbol{B}} - \boldsymbol{\alpha}\cdot\overline{\boldsymbol{B}} - \boldsymbol{\kappa}\cdot(\nabla\overline{\boldsymbol{B}})^{\mathrm{s}}) \tag{133}$$

with

$$\tilde{\sigma}_{\mathrm{m}\,ij} = \sigma\,(\delta_{ij} + \mu\sigma(\beta_{ij} + \epsilon_{ijk}\delta_k))^{-1}\,. \tag{134}$$

In contrast to $\boldsymbol{\sigma}_{\mathrm{m}}$ the tensor $\tilde{\boldsymbol{\sigma}}_{\mathrm{m}}$ is no longer symmetric.

We speak here again of "α-effect" if there is a contribution to the electromotive force $\boldsymbol{\mathcal{E}}$ having the form $-\boldsymbol{\alpha}\cdot\overline{\boldsymbol{B}}$. Of course, this contribution is in general neither parallel nor antiparallel to the magnetic field. If we want to stress this we use also the notation "anisotropic α-effect" in contrast to "isotropic" or "ideal α-effect" as it occurs, for instance, with homogeneous isotropic turbulence.

Like a mean motion also an inhomogeneous turbulence is able to transport mean magnetic flux. Such effects of turbulent motions are described here by the velocity $\boldsymbol{\gamma}$. They may consist in an expulsion of magnetic flux from regions of enhanced turbulence, discussed as "turbulent diamagnetism" [64,2,65–67], or in the transport of flux through a layer with convective motions, discussed as "pumping of magnetic flux" [68,69].

Anisotropies in the turbulence give rise to an anisotropic mean-field conductivity described by the conductivity tensors $\boldsymbol{\sigma}_{\mathrm{m}}$, determined by $\boldsymbol{\beta}$, or $\tilde{\boldsymbol{\sigma}}_{\mathrm{m}}$, determined by $\boldsymbol{\beta}$ and $\boldsymbol{\delta}$.

The contribution to $\boldsymbol{\mathcal{E}}$ given by $-\boldsymbol{\kappa} \cdot (\nabla \overline{\boldsymbol{B}})^{s}$ is difficult to interpret but can well be a source of mean electric currents.

Many calculations of components of $\boldsymbol{\alpha}, \boldsymbol{\beta}, \boldsymbol{\gamma}, \ldots$ have been carried out under assumptions which more or less reflect the situations in cosmic objects; see e.g. [2,63,21].

6 Kinematic Mean-Field Dynamo Models

Let us now use the findings of mean-field electrodynamics for the elaboration of kinematic dynamo models that reflect essential features of the Earth and the planets, the Sun or other stellar objects. After a few general explanations we will focus our attention to "conventional" models with simple symmetries in their structures and in the motion of the fluid, and roughly summarize results of the numerous numerical investigations of such models. We refer also to more detailed representations, e.g. [2,21,22].

6.1 Basic Equations

We consider again, as a typical example, a magnetic field penetrating a conducting fluid body surrounded by free space and assume that the electromagnetic fields satisfy the equations (39)–(41) or (42)–(45). We assume in addition that the fluid motion and therefore the electromagnetic fields, too, show irregular or even turbulent features, and rely on the mean-field concept. Subjecting the equations mentioned to averaging, and adopting the Reynolds rules (68)–(71), we arrive at

$$\nabla \times \overline{\boldsymbol{E}} = -\partial_t \overline{\boldsymbol{B}}\,, \quad \nabla \cdot \overline{\boldsymbol{B}} = 0\,, \quad \nabla \times \overline{\boldsymbol{B}} = \mu \overline{\boldsymbol{j}} \quad \text{everywhere} \tag{135}$$

$$\overline{\boldsymbol{j}} = \sigma(\overline{\boldsymbol{E}} + \overline{\boldsymbol{u}} \times \overline{\boldsymbol{B}} + \boldsymbol{\mathcal{E}}) \quad \text{in} \quad \mathcal{V}\,, \quad \overline{\boldsymbol{j}} = \boldsymbol{0} \quad \text{in} \quad \mathcal{V}' \tag{136}$$

$$\overline{\boldsymbol{B}} = \boldsymbol{O}(a^{-3}) \quad \text{as} \quad a \to \infty\,, \tag{137}$$

or alternatively,

$$\nabla \times (\eta \nabla \times \overline{\boldsymbol{B}}) - \nabla \times (\overline{\boldsymbol{u}} \times \overline{\boldsymbol{B}} + \boldsymbol{\mathcal{E}}) + \partial_t \overline{\boldsymbol{B}} = \boldsymbol{0}\,, \quad \nabla \cdot \overline{\boldsymbol{B}} = 0 \quad \text{in} \quad \mathcal{V} \tag{138}$$

$$\nabla \times \overline{\boldsymbol{B}} = \boldsymbol{0}\,, \quad \nabla \cdot \overline{\boldsymbol{B}} = 0 \quad \text{in} \quad \mathcal{V}' \tag{139}$$

$$[\boldsymbol{B}] = \boldsymbol{0} \quad \text{across} \quad \partial\mathcal{V} \tag{140}$$

$$\boldsymbol{B} = \boldsymbol{O}(a^{-3}) \quad \text{as} \quad a \to \infty\,. \tag{141}$$

$\boldsymbol{\mathcal{E}}$ is again the mean electromotive force due to fluctuations defined by (81). Difficulties which might arise with space averages if the averaging region contains the boundary have been ignored; they have to be discussed with the applications.

These equations define the dynamo problem on the mean-field level. We speak of a "mean-field dynamo" if the mean magnetic flux density does not decay to zero in the course of time, that is,

$$\overline{\boldsymbol{B}} \not\to 0 \quad \text{as} \quad t \to \infty\,. \tag{142}$$

We stress, however, that the notation "mean-field dynamo" has to be used with care. It does not refer to a real physical object but to a particular model of such an object only. The existence of mean-field dynamo in the sense of (142) always implies the existence of a dynamo in the original sense of (47).

It is very important to note that mean fields are not subject to Cowling's theorem as explained in Section 4.3. The proofs of this theorem cannot be repeated if the original equations are replaced with the mean-field equations; a possible exception are cases with $\boldsymbol{\mathcal{E}} \cdot \overline{\boldsymbol{B}} = 0$. That is why mean-field dynamos may well be axisymmetric. The deviation of $\boldsymbol{B}$ from axisymmetry, which is necessary for a dynamo, need not to occur in $\overline{\boldsymbol{B}}$. It is sufficient to have it in $\boldsymbol{B}'$.

Let us finally have a look on the magnetic energy. As a consequence of the Reynolds rules, the energy density $\overline{\boldsymbol{B}^2}/2\mu$ can be splitted uniquely into the two parts $\overline{\boldsymbol{B}}^2/2\mu$ and $\overline{\boldsymbol{B}'^2}/2\mu$, which can be attributed to the mean and the fluctuating parts of the magnetic field. For the total energy stored in the mean magnetic field we find, starting from (135)–(137) and repeating manipulations as done in Section 3.8,

$$\frac{\mathrm{d}}{\mathrm{d}t} \int_{\infty} \frac{\overline{\boldsymbol{B}}^2}{2\mu} \, \mathrm{d}v = - \int_{\mathcal{V}} \frac{\overline{\boldsymbol{j}}^2}{\sigma} \, \mathrm{d}v - \int_{\mathcal{V}} \overline{\boldsymbol{u}} \cdot (\overline{\boldsymbol{j}} \times \overline{\boldsymbol{B}}) \, \mathrm{d}v + \int_{\mathcal{V}} \overline{\boldsymbol{j}} \cdot \boldsymbol{\mathcal{E}} \, \mathrm{d}v \, . \tag{143}$$

Note that the integrals over $\overline{\boldsymbol{j}}^2/\sigma$ and $\overline{\boldsymbol{u}} \cdot (\overline{\boldsymbol{j}} \times \overline{\boldsymbol{B}})$ describe only parts of the total Joule heat production and of the work done by or against the Lorentz force. There are other parts resulting from fluctuating fields, which do not occur here.

6.2 Conventional Mean-Field Dynamo Models

Many mean-field dynamo models have been developed for various objects like the Earth and the planets, the Sun and several types of stars, or for galaxies. In almost all cases simple symmetries were assumed with respect to the shape of the conducting bodies, to the distributions of the electric conductivity and to the fluid motions.

We will formulate here rather general assumptions of this kind from which we will then draw conclusions concerning the possible structures of the magnetic fields. When doing so we suppose that an axis and a plane perpendicular to it are given, which we call rotation axis and equatorial plane in the following. We assume that the shape of the fluid body and the distribution of the electric conductivity, or of the magnetic diffusivity η, are

- symmetric about the rotation axis,
- symmetric about the equatorial plane,
- steady.

In addition we assume that all averaged quantities depending on the velocity field $\boldsymbol{u}$, that is $\overline{\boldsymbol{u}} + \boldsymbol{u}'$, are invariant under

- rotations of $\boldsymbol{u}$ about the rotation axis,
- reflections of $\boldsymbol{u}$ about the equatorial plane,
- time shifts in $\boldsymbol{u}$.

As the simplest consequence of these last assumptions we note that the mean velocity $\overline{\boldsymbol{u}}$ is symmetric about both the rotation axis and the equatorial plane and steady. Another simple consequence is, for example, that the helicity $\overline{\boldsymbol{u}' \cdot (\nabla \times \boldsymbol{u}')}$ of the fluctuating motions is symmetric about the rotation axis but antisymmetric about the equatorial plane and steady.

The assumptions introduced allow us, however, also far-reaching conclusions concerning the mean magnetic field.

According to our explanations in Section 3.8 the equations (42)–(45), if satisfied with fields $\boldsymbol{B}$ and $\boldsymbol{u}$, are also satisfied with fields generated from them by rotation about the rotation axis, by reflection about the equatorial plane, or by time shift. Then the mean-field equations (138)–(141), if valid with a mean magnetic field $\overline{\boldsymbol{B}}$, must apply for all such fields generated from that by rotation, reflection or time shift in the above sense, with the given velocities $\overline{\boldsymbol{u}}$ and $\boldsymbol{u}'$. For $\overline{\boldsymbol{B}}$ as an averaged quantity cannot be influenced by corresponding changes of $\overline{\boldsymbol{u}}$ and $\boldsymbol{u}'$.

If a given field $\overline{\boldsymbol{B}}$ as well as the reflected one satisfies the equations (138)–(141) then their sum and their difference do so, too. The sum is symmetric, the difference antisymmetric about the equatorial plane. Thus it implies no loss of generality to look from the very beginning for symmetric and antisymmetric fields only, for all others can then be gained by superposition.

We may decompose any field $\overline{\boldsymbol{B}}$ into its Fourier modes with respect to the azimuthal coordinate ϕ so that $\overline{\boldsymbol{B}} = \sum_{m \geq 0} \Re(\hat{\boldsymbol{B}}^m \exp(\mathrm{i}m\phi))$, with complex axisymmetric vectors $\hat{\boldsymbol{B}}^m$. The fact that together with a given field $\overline{\boldsymbol{B}}$ satisfying equations (138)–(141) all fields generated by rotation must do so, too, allows us to conclude that any individual Fourier mode $\Re(\hat{\boldsymbol{B}}^m \exp(\mathrm{i}m\phi))$ is a solution of these equations. So it means no loss of generality to restrict the attention on these modes only, for again all other fields can be gained by superposition.

Finally, the fact that together with a field $\overline{\boldsymbol{B}}$ which satisfies (138)–(141) also the corresponding ones gained by time shift do so leads to the conclusion that $\overline{\boldsymbol{B}}$ has to vary like $\Re(\tilde{\boldsymbol{B}} \exp(pt))$ with time, where $\tilde{\boldsymbol{B}}$ is a complex vector field and p a complex constant.

Taking all these findings together we see that it is sufficient to look for solutions of the equations (138)–(141) having the form

$$\overline{\boldsymbol{B}} = \Re\big(\hat{\boldsymbol{B}} \exp(\mathrm{i}m\phi + (\lambda + \mathrm{i}\omega)t)\big) . \tag{144}$$

All other solutions can be gained by superposition. Here $\hat{\boldsymbol{B}}$ means a complex vector field being antisymmetric or symmetric about the equatorial plane, symmetric about the rotation axis and steady, m is a non-negative integer, and λ and ω are real constants. We denote the solutions of the form (144) by A or S according to their antisymmetry or symmetry about the equatorial plane, and add the parameter m to characterize the symmetry with respect to the rotation axis. Examples of field pictures of Am and Sm modes are given in Figure 5. Clearly λ is the growth rate of the solution considered. A mean-field dynamo requires

$$\lambda \geq 0 . \tag{145}$$

If $\omega = 0$ the solution varies monotonously with time, if $\omega \neq 0$ oscillatory. Axisymmetric modes, $m = 0$, with $\omega \neq 0$ are intrinsically oscillatory. A non-axisymmetric mode, $m \neq 0$, with $\omega \neq 0$ has the form of a wave traveling in azimuthal direction. Its field configuration rotates like a rigid body with the angular velocity $-\omega/m$ and is, of course, steady in a co-rotating frame of reference.

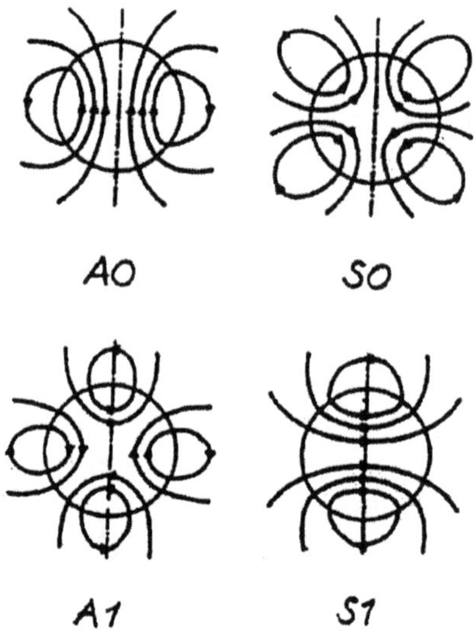

Fig. 5. Schematic representation of poloidal magnetic field lines of A0, S0, A1 and S1 modes in meridian planes of spherical models. In the case of A0 and S0 modes the patterns agree for all such planes. In the case of the A1 and S1 modes the special planes have been chosen which are not crossed by field lines

We have drawn our conclusions on the solutions $\overline{\boldsymbol{B}}$ of the mean-field equations from the general assumptions formulated above concerning the shape of the fluid body, the distribution of the magnetic diffusivity η and the properties of the fluid velocity $\boldsymbol{u}$, without any specification of the form of the mean electromotive force $\boldsymbol{\mathcal{E}}$. Let us now add again the assumption used in Sections 5.2–5.9 according to which $\boldsymbol{\mathcal{E}}$ in a given point is determined by $\overline{\boldsymbol{B}}$ and its first spatial derivatives in the same point only. Then $\boldsymbol{\mathcal{E}}$ can be represented in the form (130). The general assumptions introduced here, however, imply special properties of the quantities $\boldsymbol{\alpha}, \boldsymbol{\beta}, \boldsymbol{\gamma}, \boldsymbol{\delta}$ and $\boldsymbol{\kappa}$.

In order to formulate these properties we introduce two vectors $\hat{\boldsymbol{\omega}}$ and $\hat{\boldsymbol{g}}$ describing preferred directions in the fluctuating velocity field. We identify the first one, $\hat{\boldsymbol{\omega}}$, with the unit vector in the direction of the rotation axis of the fluid body and the second one, $\hat{\boldsymbol{g}}$, for example with the unit vector in the direction opposite to the gravitational force but put it equal to zero where such a direction cannot be defined. Whereas $\hat{\boldsymbol{\omega}}$ is independent of position, $\hat{\boldsymbol{g}}$ varies in space but is symmetric about the rotation axis and the equatorial plane. We write then

$$
\begin{aligned}
\boldsymbol{\alpha}\cdot\overline{\boldsymbol{B}} = {}& \alpha_1\,(\hat{\boldsymbol{\omega}}\cdot\hat{\boldsymbol{g}})\,\overline{\boldsymbol{B}} + \alpha_2\,(\hat{\boldsymbol{\omega}}\cdot\hat{\boldsymbol{g}})\,(\hat{\boldsymbol{g}}\cdot\overline{\boldsymbol{B}})\,\hat{\boldsymbol{g}} + \alpha_3\,(\hat{\boldsymbol{\omega}}\cdot\hat{\boldsymbol{g}})\,(\hat{\boldsymbol{\omega}}\cdot\overline{\boldsymbol{B}})\,\hat{\boldsymbol{\omega}} \\
&+\alpha_4\,((\hat{\boldsymbol{\omega}}\cdot\overline{\boldsymbol{B}})\,\hat{\boldsymbol{g}} + (\hat{\boldsymbol{g}}\cdot\overline{\boldsymbol{B}})\,\hat{\boldsymbol{\omega}}) \\
&+\alpha_5\,(\hat{\boldsymbol{\omega}}\cdot\hat{\boldsymbol{g}})((\hat{\boldsymbol{\lambda}}\cdot\overline{\boldsymbol{B}})\,\hat{\boldsymbol{g}} + (\hat{\boldsymbol{g}}\cdot\overline{\boldsymbol{B}})\,\hat{\boldsymbol{\lambda}}) \\
&+\alpha_6\,((\hat{\boldsymbol{\lambda}}\cdot\overline{\boldsymbol{B}})\,\hat{\boldsymbol{\omega}} + (\hat{\boldsymbol{\omega}}\cdot\overline{\boldsymbol{B}})\,\hat{\boldsymbol{\lambda}}) \\
\boldsymbol{\beta}\cdot(\nabla\times\overline{\boldsymbol{B}}) = {}& \beta_1\,\nabla\times\overline{\boldsymbol{B}} + \beta_2\,(\hat{\boldsymbol{g}}\cdot(\nabla\times\overline{\boldsymbol{B}}))\,\hat{\boldsymbol{g}} + \beta_3\,(\hat{\boldsymbol{\omega}}\cdot(\nabla\times\overline{\boldsymbol{B}}))\,\hat{\boldsymbol{\omega}} \\
&+\beta_4\,(\hat{\boldsymbol{\omega}}\cdot\hat{\boldsymbol{g}})((\hat{\boldsymbol{\omega}}\cdot(\nabla\times\overline{\boldsymbol{B}}))\,\hat{\boldsymbol{g}} + (\hat{\boldsymbol{g}}\cdot(\nabla\times\overline{\boldsymbol{B}}))\,\hat{\boldsymbol{\omega}}) \qquad (146)\\
&+\beta_5\,((\hat{\boldsymbol{\lambda}}\cdot(\nabla\times\overline{\boldsymbol{B}}))\,\hat{\boldsymbol{g}} + (\hat{\boldsymbol{g}}\cdot(\nabla\times\overline{\boldsymbol{B}}))\,\hat{\boldsymbol{\lambda}}) \\
&+\beta_6\,(\hat{\boldsymbol{\omega}}\cdot\hat{\boldsymbol{g}})((\hat{\boldsymbol{\lambda}}\cdot(\nabla\times\overline{\boldsymbol{B}}))\,\hat{\boldsymbol{\omega}} + (\hat{\boldsymbol{\omega}}\cdot(\nabla\times\overline{\boldsymbol{B}}))\,\hat{\boldsymbol{\lambda}}) \\
\boldsymbol{\gamma}\times\overline{\boldsymbol{B}} = {}& \gamma_1\,\hat{\boldsymbol{g}}\times\overline{\boldsymbol{B}} + \gamma_2\,(\hat{\boldsymbol{\omega}}\cdot\hat{\boldsymbol{g}})\,\hat{\boldsymbol{\omega}}\times\overline{\boldsymbol{B}} + \gamma_3\,\hat{\boldsymbol{\lambda}}\times\overline{\boldsymbol{B}} \\
\boldsymbol{\delta}\times(\nabla\times\overline{\boldsymbol{B}}) = {}& \delta_1\,(\hat{\boldsymbol{\omega}}\cdot\hat{\boldsymbol{g}})\,\hat{\boldsymbol{g}}\times(\nabla\times\overline{\boldsymbol{B}}) \\
&+\delta_2\,\hat{\boldsymbol{\omega}}\times(\nabla\times\overline{\boldsymbol{B}}) + \delta_3\,(\hat{\boldsymbol{\omega}}\cdot\hat{\boldsymbol{g}})\,\hat{\boldsymbol{\lambda}}\times(\nabla\times\overline{\boldsymbol{B}})\,,
\end{aligned}
$$

where $\hat{\boldsymbol{\lambda}}$ means $\hat{\boldsymbol{\omega}}\times\hat{\boldsymbol{g}}$. As for the term $\boldsymbol{\kappa}\cdot(\nabla\overline{\boldsymbol{B}})^{\mathrm{s}}$ we note that it can be represented as a sum of four contributions $\boldsymbol{\beta}^g\cdot\boldsymbol{V}^g$, $\boldsymbol{\beta}^\omega\cdot\boldsymbol{V}^\omega$, $\boldsymbol{\delta}^g\times\boldsymbol{V}^g$ and $\boldsymbol{\delta}^\omega\times\boldsymbol{V}^\omega$ with tensors $\boldsymbol{\beta}^g$ and $\boldsymbol{\beta}^\omega$ and vectors $\boldsymbol{\delta}^g$ and $\boldsymbol{\delta}^\omega$ analogous to $\boldsymbol{\alpha}$ and $\boldsymbol{\beta}$ and to $\boldsymbol{\gamma}$ and $\boldsymbol{\delta}$, respectively, and $\boldsymbol{V}^g$ and $\boldsymbol{V}^\omega$ standing for $(\nabla\overline{\boldsymbol{B}})^{\mathrm{s}}\cdot\hat{\boldsymbol{g}}$ and $(\nabla\overline{\boldsymbol{B}})^{\mathrm{s}}\cdot\hat{\boldsymbol{\omega}}$.

As a consequence of the general assumptions formulated above the coefficients $\alpha_1, \alpha_2, \ldots \delta_3$ as well as $\beta_1^g, \beta_2^g, \ldots \delta_3^\omega$ are symmetric about the rotation axis and the equatorial plane and steady.

Comparing $\boldsymbol{\mathcal{E}}$ as obtained for homogeneous isotropic turbulence, that is (91), with our result (130) and (146) we see that the contribution $\alpha\,\overline{\boldsymbol{B}}$ there, describing the isotropic α-effect, corresponds to $-\alpha_1\,(\hat{\boldsymbol{\omega}}\cdot\hat{\boldsymbol{g}})\,\overline{\boldsymbol{B}}$ here, which is, however, accompanied by other contributions causing an anisotropy of the α-effect. We will use the notation α in the following also in the sense of $\alpha = -\alpha_1\,(\hat{\boldsymbol{\omega}}\cdot\hat{\boldsymbol{g}})$. Clearly α is then, in contrast to α_1, antisymmetric about the equatorial plane.

6.3 The Basic Dynamo Mechanisms

In all dynamo models investigated so far in which poloidal and toroidal parts of the magnetic field can be defined an interplay between these parts proved to be crucial. This applies both to dynamos in the original sense and to mean-field dynamos. So we may characterize the various mean-field dynamo mechanisms by the induction processes which are dominant in the generation of the poloidal field from the toroidal one and of the toroidal field from the poloidal one.

The α-effect is capable to generate both a poloidal field from a toroidal one and vice versa. This leads to a dynamo mechanism, which we call "α^2-mechanism". Figure 6 demonstrates it for a spherical body and axisymmetric magnetic fields of dipole and quadrupole type, that is, A0 and S0 modes. For the sake of simplicity no other contribution to the electromotive force $\boldsymbol{\mathcal{E}}$ is considered than $\alpha\,\overline{\boldsymbol{B}}$ with $\alpha > 0$ in the northern and $\alpha < 0$ in the southern hemisphere. As it can be readily followed up in the figure the α-effect with the toroidal field leads to toroidal currents which just support the poloidal field. Likewise the α-effect with the poloidal field results in poloidal currents, which in turn support the toroidal field. In this way, a sufficiently strong α-effect is able to maintain magnetic fields with the configurations envisaged or make them grow. If the signs of α are inverted the orientation of either the poloidal or the toroidal fields have to be inverted too. For dynamos of that kind the poloidal and the toroidal fields are of the same order of magnitude. A view on the energy balance (143) shows that in each hemisphere the signs of α and $\overline{\boldsymbol{B}} \cdot (\nabla \times \overline{\boldsymbol{B}})$ have to coincide to an extent which ensures that the last integral is positive.

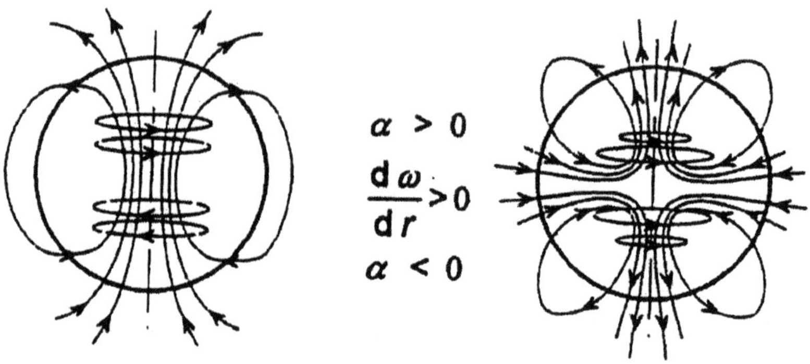

Fig. 6. Axisymmetric poloidal and toroidal magnetic field configurations of dipole and quadrupole type as can be maintained by α^2 or $\alpha\omega$-mechanisms

We now admit a differential rotation of the fluid body, that is, assume a mean velocity $\overline{\boldsymbol{u}}$ corresponding to a rotation with an angular velocity ω varying, for example, with the radial coordinate r. As explained in Section 3.7 by this kind of motion magnetic field lines are wound up so that, if a poloidal field exists, a toroidal one is generated. If poloidal field configurations as in Figure 6 are given and $\mathrm{d}\omega/\mathrm{d}r > 0$, toroidal fields as shown there occur even in the absence of the α-effect. Of course, a differential rotation can well be more efficient in the generation of the toroidal field than the α-effect. This opens the possibility of another dynamo mechanism, in which as before the poloidal field is generated by the α-effect from the toroidal one, but the toroidal field predominantly by differential rotation from the poloidal one. If the α-effect is indeed negligible in this last generation process we speak of an "$\alpha\omega$-mechanism". In this case the toroidal field is much stronger than the poloidal one, and the energy input is mainly due to the differential rotation, described by the second integral rather than the third on the right-hand side of (143).

In general, of course, both α-effect and differential rotation take part in the generation of the toroidal field. With regard to this the extreme cases without any differential rotation or with a very strong one considered so far are sometimes labelled as "pure α^2-mechanism" or "pure $\alpha\omega$-mechanism", and the more general case as "$\alpha^2\omega$-mechanism".

The first dynamo models working with α^2-mechanism and the $\alpha\omega$-mechanism have been proposed and elaborated with a view to the Earth and the Sun by Steenbeck and Krause [70,71]. A large number of spherical and other dynamo models working with these mechanisms have been studied later on, taking into account various contributions to the electromotive force $\boldsymbol{\mathcal{E}}$ as indicated in (130) and (146) and various forms of the mean velocity $\overline{\boldsymbol{u}}$, and considering both axisymmetric and non-axisymmetric magnetic fields. The results have been summarized at several places [2,21,22,12]. We will mention a few general features of them below.

Before doing so we want to point out that, in addition to the α-effect mechanisms discussed so far, other mechanisms due to mean-field induction effects proved to be possible. For example, contributions to $\boldsymbol{\mathcal{E}}$ described by particular components of the tensor $\boldsymbol{\beta}$ or by the vector $\boldsymbol{\delta}$ imply couplings between poloidal and toroidal magnetic fields, too. From the energy balance (143) we may conclude, however, that a dynamo without other induction effects than those described by $\boldsymbol{\beta}$ can be excluded as long as the conductivity tensor $\boldsymbol{\sigma}$ is positive definite, which has to be assumed in all realistic cases, and that a dynamo due to effects described by $\boldsymbol{\delta}$ only is in any case impossible. In combination with a differential rotation, however, these effects are capable of dynamo action. This has been demonstrated by investigations of a number of models [72–74,21,22,75]. The relevance of these other mechanisms for cosmic objects, however, is still an open question.

Returning now to dynamo models with α-effect and differential rotation we introduce the two dimensionless parameters R_α and R_ω measuring the magnitudes of these induction effects,

$$R_\alpha = \alpha_c \, L/\eta_{\mathrm{m\,c}} \, , \quad R_\omega = \Delta_c\omega \, L^2/\eta_{\mathrm{m\,c}} \, , \tag{147}$$

where α_c means a characteristic value of α in the northern hemisphere, $\Delta_c\omega$ a characteristic difference of the angular velocities between outer and inner layers, $\eta_{\mathrm{m\,c}}$ a characteristic value of η_{m}, and L a characteristic linear dimension of the conducting body.

Let us first consider spherical dynamo models as elaborated in view of the Earth and the planets as well as the Sun and stellar objects, with the outer space being non-conducting.

We start with the pure α^2-mechanism, that is $R_\omega = 0$. In a number of simple models no other contribution to $\boldsymbol{\mathcal{E}}$ has been included than that corresponding to the idealized α-effect, that is, $\alpha\overline{\boldsymbol{B}}$ with a scalar α depending on radius and latitude. In these models the excitation conditions for the A0, S0, A1 and S1 modes, that is, the marginal values of R_α, proved to be very close together, often with a slight preference for the A0 mode; the A2, S2, A3, S3, ... modes are less easily excitable. The axisymmetric modes, A0 and S0, are non-oscillatory, the non-axisymmetric ones show, depending on the specific form of α, either eastward or westward migrations. In models involving anisotropies of the α-effect or the γ-effect, however, a clear preference for A1 or S1 modes over all other modes has been observed in a wide range of reasonable assumptions [76,77,22,78]. In particular the anisotropies of the α-effect due to rapid rotation of the body act in that sense [78,79]. Results of that kind suggest that a fairly realistic α^2-mechanism always favors non-axisymmetric field structures. Incidentally, the idealized α-effect together with a weak differential rotation, that is small $|R_\omega/R_\alpha|$, may also lead to a preference of A1 or S1 modes [76,80,22,81].

Proceeding now to models in which differential rotation plays an essential part we first recall the explanations of Section 3.7 according to which it acts in very different ways on axisymmetric and non-axisymmetric magnetic fields. We repeat the essential points here in terms of poloidal and toroidal fields. In the axisymmetric case the differential rotation generates a toroidal field if a poloidal one exists, where the latter remains unaffected. In this way an arbitrarily strong toroidal field can be produced if only the differential rotation is sufficiently strong, that is, $|R_\omega|$ is sufficiently high. In the non-axisymmetric case, again a toroidal field is generated from a poloidal one. In addition, however, both the poloidal and the toroidal fields are deformed so that fields with opposite directions come close together, and thus both fields are subject to an enhanced dissipation. Even with a very strong differential rotation, that is, very large $|R_\omega|$, the ratio of the magnitudes of toroidal and poloidal fields can never exceed the order of unity.

For dynamo models with α-effect and differential rotation both parameters R_α and R_ω are important. It is, however, often useful to consider instead of them their combinations $R_\alpha R_\omega$ and R_α/R_ω.

The pure $\alpha\omega$-mechanism corresponds to the limit $R_\alpha/R_\omega \to 0$. For the reason just explained it works only with axisymmetric fields, that is, supports A0 and S0 modes only. Already in simple models involving only the idealized α-effect and differential rotation both types of modes have been observed with both oscillatory and non-oscillatory time dependence [82,83,22,70,84]. The excitation conditions for the A0 and S0 modes depend on the product $R_\alpha R_\omega$ only, but the ratio of the magnitudes of the poloidal and the toroidal field is given by R_α/R_ω. Which mode is preferably excited, and whether or not it is oscillatory, depends in a complex way on the distribution of α and ω, in particular on the sign of $R_\alpha R_\omega$. For the pure $\alpha\omega$-mechanism anisotropies of the α-effect or of the mean-field conductivity play a minor part. In view of the solar dynamo, models favoring oscillatory A0 modes are of particular interest, which have been extensively studied [85,82,83,22,84,86].

In the general case of the $\alpha^2\omega$-mechanism, that is in the transition region between the pure α^2-mechanism and the pure $\alpha\omega$-mechanism, the situation is more complex. Numerous investigations have been carried out considering this region, in particular again in the context of the solar dynamo [76]. One crucial question arising here concerns the conditions under which the preference of non-axisymmetric fields appearing in the α^2-regime turns into a preference of axisymmetric fields to be expected in the $\alpha\omega$-regime. Some results suggest that this transition may occur at rather low values of $|R_\alpha/R_\omega|$, close to those which seem reasonable for the Sun.

With a view to the large scale magnetic fields observed in numerous nearby galaxies a number of dynamo models with non-spherical geometry have been studied. It was, for example, assumed that the region occupied by the conducting fluid is an oblate ellipsoid [87,88], a torus [89] or an infinitely extended slab [90]. In addition to models with non-conducting outer space others were developed in which the dynamo-active region is embedded in an extended conducting medium without sharp boundaries [91,92,6]. By reasons connected with the conditions in galaxies mainly $\alpha\omega$-mechanisms have been investigated. In the most cases a preference of S0 modes has been observed under reasonable assumptions.

7 Magnetofluiddynamics II: Fluiddynamic Aspects

In all considerations so far on the behavior of magnetic fields in a conducting fluid its motion was assumed as given. The dynamical constraints as well as the back-reaction of magnetic fields on the motion were ignored. We will now very briefly give a few explanations concerning these aspects.

7.1 Basic Equations

We rely on the equations (4) for the magnetic flux density $\boldsymbol{B}$,

$$\partial_t \boldsymbol{B} - \nabla \times (\boldsymbol{u} \times \boldsymbol{B}) = -\nabla \times (\eta\,(\nabla \times \boldsymbol{B}))\,, \quad \nabla \cdot \boldsymbol{B} = 0\,, \tag{148}$$

but consider the velocity $\boldsymbol{u}$ no longer as given. Instead we add the momentum balance and the condition of mass conservation in the form

$$\begin{aligned} \varrho(\partial_t \boldsymbol{u} + (\boldsymbol{u} \cdot \nabla)\,\boldsymbol{u}) &= -\nabla p - 2\varrho \boldsymbol{\Omega} \times \boldsymbol{u} + \boldsymbol{F}^{(\mathrm{f})} + \boldsymbol{F}^{(\mathrm{m})} + \boldsymbol{F}^{(\mathrm{e})} \\ \partial_t \varrho + \nabla \cdot (\varrho \boldsymbol{u}) &= 0\,. \end{aligned} \tag{149}$$

Here ϱ means the mass density of the fluid and p the pressure. We refer to a steadily rotating frame. $\boldsymbol{\Omega}$ is the angular velocity responsible for the Coriolis and centrifugal forces. The centrifugal force is included in the pressure term. $\boldsymbol{F}^{(\mathrm{f})}$ stands for the forces per unit volume due to internal friction. It can be be represented as divergence of the stress tensor $\boldsymbol{S}$,

$$F_i^{(\mathrm{f})} = \frac{\partial S_{ij}}{\partial x_j}\,, \quad S_{ij} = \varrho\nu\,(\frac{\partial u_i}{\partial x_j} + \frac{\partial u_j}{\partial x_i}) + \varrho\nu'\,(\nabla \cdot \boldsymbol{u})\,\delta_{ij}\,, \tag{150}$$

where ν is the kinematic viscosity and ν' another viscosity coefficient. $\boldsymbol{F}^{(\mathrm{m})}$ means the force per unit volume exerted by the electromagnetic field. In the magnetohydrodynamic approximation it is simply the Lorentz force,

$$\boldsymbol{F}^{(\mathrm{m})} = \boldsymbol{j} \times \boldsymbol{B} = \frac{1}{\mu}(\nabla \times \boldsymbol{B}) \times \boldsymbol{B} = \frac{1}{\mu}((\boldsymbol{B} \cdot \nabla)\boldsymbol{B} - \frac{1}{2}\nabla \boldsymbol{B}^2)\,, \tag{151}$$

which can also be written as a divergence of the magnetic part $\boldsymbol{M}$ of the Maxwell stress tensor,

$$F_i^{(\mathrm{m})} = \frac{\partial M_{ij}}{\partial x_j}\,, \quad M_{ij} = \frac{1}{\mu}\,(B_i\,B_j - \frac{1}{2}\,B^2 \delta_{ij})\,. \tag{152}$$

The gradient term in (151), which corresponds to the δ_{ij} term in (152), can also be included in the pressure term of equation (149a). Finally, $\boldsymbol{F}^{(e)}$ stands for external forces which we will specify later. If necessary we include here also the gravitational force $\varrho \boldsymbol{g}$, where $\boldsymbol{g}$ means this force per unit mass. Since ϱ and p have now to be considered as unknown quantities, too, we have to complete these equations by an equation of state,

$$\varrho = \varrho\,(p, T)\,, \tag{153}$$

which in general introduces the temperature T as another unknown quantity. This, in turn, requires to add the heat transport equation,

$$\varrho c_{\mathrm{v}}(\partial_t T + \boldsymbol{u} \cdot \nabla T) = -\nabla \cdot (\kappa_* \nabla T) + q\,, \tag{154}$$

where c_{v} is the specific heat capacity of the fluid for constant volume, κ_* its heat-conductivity coefficient, and q stands for any kind of heat production per unit volume including that by Joule dissipation or internal friction.

These equations together with proper initial and boundary conditions determine the evolution of magnetic field, motion and temperature, that is $\boldsymbol{B}$, $\boldsymbol{u}$ and T, if external forces or heat sources, $\boldsymbol{F}^{(e)}$ or q, are given. In addition to the couplings of these quantities explicitly indicated in the above equations there are in general others, for example by the temperature dependence of the material coefficients.

7.2 The Case of Incompressible Flow and the Boussinesq Approximation

Since the set of equations just given is rather complex it suggests itself to consider it under simplifying assumptions. In that sense we assume first the fluid to be incompressible and homogeneous so that ϱ and ν are constant. Then equations (149)–(152) can be replaced by

$$\begin{gathered} \partial_t \boldsymbol{u} + (\boldsymbol{u} \cdot \nabla)\,\boldsymbol{u} = -\frac{1}{\varrho}\,\nabla p - 2\,\boldsymbol{\Omega} \times \boldsymbol{u} + \nu \nabla^2 \boldsymbol{u} + \frac{1}{\mu\varrho}(\nabla \times \boldsymbol{B}) \times \boldsymbol{B} + \frac{1}{\varrho}\,\boldsymbol{F}^{(e)}\,, \\ \nabla \cdot \boldsymbol{u} = 0\,. \end{gathered} \tag{155}$$

Together with (148) they are sufficient to determine the evolution of magnetic field and motion, $\boldsymbol{B}$ and $\boldsymbol{u}$. We note that in contrast to the Maxwell equation $\nabla \cdot \boldsymbol{B} = 0$ the condition $\nabla \cdot \boldsymbol{u} = 0$ plays not only the part of an initial condition but leads together with the first line of (155) to a relation connecting p and $\boldsymbol{u}$. The equations (148) and (155) are no longer coupled with (153) and (154), which are thus of secondary interest only.

Often the so-called Boussinesq-approximation is used which considers compressibility of the fluid only as far as it is important for buoyancy but ignores it otherwise. In the simplest case it assumes a given steady reference state of the physical system considered with $\boldsymbol{u} = \boldsymbol{B} = \boldsymbol{0}$ and $\varrho = \varrho_0$, $p = p_0$ and $T = T_0$ where ϱ_0, p_0 and T_0 are given functions of the space coordinates, which have of course to satisfy the equation of state (153). For the sake of simplicity we assume again ν and now also κ_* to be constant and q to be independent on $\boldsymbol{u}$ and $\boldsymbol{B}$. The Boussinesq-approximation, which we cannot justify here in detail, is defined by the equations (148) and the equations (155) with ϱ specified to be ϱ_0 and $\boldsymbol{F}^{(\mathrm{e})}$ specified by

$$\boldsymbol{F}^{(\mathrm{e})} = \varrho \alpha \theta , \quad \varrho c_{\mathrm{v}} \left(\partial_t \theta + \boldsymbol{u} \cdot \nabla (T_0 + \theta) \right) = \kappa_* \Delta \theta . \tag{156}$$

and ϱ understood as ϱ_0 everywhere. $\boldsymbol{F}^{(\mathrm{e})}$ is now the buoyancy force, θ means the deviation of T from T_0, that is $\theta = T - T_0$, and α is the volume expansion coefficient introduced with $\varrho = \varrho_0(1 + \alpha\theta)$, which has to be understood as a consequence of the equation of state.

When investigating a problem concerning the behavior of magnetic field and fluid velocity, $\boldsymbol{B}$ and $\boldsymbol{u}$, on the basis of the equations (148) and (155) we have first to fix the viscosity parameters η and ν and the angular velocity $\boldsymbol{\Omega}$ which determines the Coriolis force. We may, however, formulate the problem so that instead only two dimensionless parameters occur, the magnetic Prandtl number P_{m} and the Taylor number Ta, or alternatively the Ekman number E,

$$P_{\mathrm{m}} = \nu / \eta , \quad Ta^{1/2} = 2 \Omega L^2 / \nu , \quad E = Ta^{-1/2} , \tag{157}$$

where $\Omega = |\boldsymbol{\Omega}|$ and L means a characteristic length of the processes considered. If we include in the sense of the Boussinesq-approximation the temperature T, or $T_0 + \theta$, and so equations (156) we have also the quantities κ_* and c_{v}, and so other dimensionless parameters, that is, the (original) Prandtl number P, or alternatively the Roberts number Rb, and the Rayleigh number Ra,

$$P = \nu / \kappa , \quad Rb = \kappa / \eta , \quad Ra = \alpha g (\partial T)_{\mathrm{c}} L^4 / \nu \kappa , \tag{158}$$

where g means the gravitational acceleration, $(\partial T)_{\mathrm{c}}$ a characteristic value of the gradient of T_0, both taken as positive, and κ is a characteristic value of $\kappa_* / \varrho c_{\mathrm{v}}$. If Ra exceeds a critical value the physical system considered is no longer stable but show a convective instability.

We may consider P_{m}, Ta, or E, as well as P and Ra as input parameters specifying the problem formulated on the basis of the equations (148) and (155)–(156). The relations between the magnitudes of the individual terms in these equations can be characterized by other dimensionless parameters defined on the basis of typical values B and U of the magnetic flux density $\boldsymbol{B}$ and the fluid velocity $\boldsymbol{u}$ that occur as solutions.

In addition to the magnetic Reynolds number R_{m} introduced with (7) we have the (original) Reynolds number Re defined by

$$Re = UL / \nu , \tag{159}$$

which gives the ratio of the magnitudes of the inertial term $(\boldsymbol{u} \cdot \nabla)\, \boldsymbol{u}$ and the friction term $\nu \nabla^2 \boldsymbol{u}$ in (155). As a rule a laminar flow looses its stability and turns into a turbulent one if Re exceeds a critical value. We note that $R_{\mathrm{m}} / Re = P_{\mathrm{m}}$.

Concerning the Coriolis force we mention the Rossby number Ro, defined by

$$Ro = U/2\Omega L\,, \tag{160}$$

which gives the ratio of the magnitudes of inertial term $(\boldsymbol{u}\cdot\nabla)\,\boldsymbol{u}$ and Coriolis term $2\,\boldsymbol{\Omega}\times\boldsymbol{u}$.

The effect of the magnetic field on the fluid motion can be characterized by the Alfven number A, the Stuart number N, the Hartmann number H or the Elsasser number Λ,

$$A = B^2/\mu\varrho U^2\,,\; N = \sigma B^2 L/\varrho U\,,\; H = \sqrt{\sigma}BL/\sqrt{\varrho\nu}\,,\; \Lambda = \sigma B^2/2\varrho\Omega\,. \tag{161}$$

Clearly A gives the ratio of magnetic to kinetic energy. N is, apart from a factor P_{m}, the ratio of the magnitude of the Lorentz term $(1/\mu\varrho)\,(\nabla\times\boldsymbol{B})\times\boldsymbol{B}$ to that of the friction term $\nu\nabla^2\boldsymbol{u}$. The same applies to H^2 if the order of $|\nabla\times\boldsymbol{B}|$ is, with a view to Ohm's law, estimated by UB/η. Finally Λ is the product of R_{m} and the ratio of the magnitudes of Lorentz and Coriolis terms, $(1/\mu\varrho)\,(\nabla\times\boldsymbol{B})\times\boldsymbol{B}$ and $2\,\boldsymbol{\Omega}\times\boldsymbol{u}$.

7.3 Rotating Fluids

On rotating bodies the fluid dynamics is in general strongly dominated by Coriolis forces. In that sense we consider now the equations (155) in the limit $E\to 0$ and $Ro\to 0$. They reduce then to

$$\frac{1}{\varrho}\,\nabla p + 2\,\boldsymbol{\Omega}\times\boldsymbol{u} - \frac{1}{\mu\varrho}(\nabla\times\boldsymbol{B})\times\boldsymbol{B} - \frac{1}{\varrho}\,\boldsymbol{F}^{(\mathrm{e})} = \boldsymbol{0}\,,\quad \nabla\cdot\boldsymbol{u} = 0\,. \tag{162}$$

If we introduce in addition $\Lambda\to 0$ the term $(1/\mu\varrho)\,(\nabla\times\boldsymbol{B})\times\boldsymbol{B}$ vanishes. In this case we speak of a "geostrophic balance" and of a "geostrophic flow", in the more general case with this term included of "magnetostrophic balance" and "magnetostrophic flow".

Let us consider the geostrophic case, assume that ϱ does not vary in space and $\boldsymbol{F}^{(\mathrm{e})}$ is a conservative force, that is, has the form of a gradient. Taking then the curl of (162a) we find

$$(\boldsymbol{\Omega}\cdot\nabla)\boldsymbol{u} = \boldsymbol{0}\,. \tag{163}$$

That is, the flow must be two-dimensional. There are no variations in the direction of $\boldsymbol{\Omega}$.

8 The General Dynamo Problem

The kinematic dynamo models of Sections 4 and 6 work with prescribed fluid motions. Although such models contributed enormously to our understanding of cosmic magnetic fields they possess some shortcomings. After a brief discussion of these shortcomings we will explain the dynamo problem in its wider sense as the problem of the evolution of both magnetic fields and motion with some given cause of these motions.

8.1 Shortcomings of the Kinematic Approach

So far we have discussed kinematic dynamo models with several kinds of prescribed fluid flow. We have, however, never asked whether these flows are dynamically at all possible. Suppose that a laminar flow of a certain intensity is given so that the magnetic Reynolds number R_{m} exceeds its marginal value. Imagine that this flow is driven by a given force. If the magnetic Prandtl number P_{m} is much smaller than unity, as must be assumed for many realistic cases, the hydrodynamic Reynolds number Re is much higher than R_{m}. Then, however, the stability of the assumed laminar flow is questionable. A more complex or even turbulent flow has to be expected instead.

But even if this difficulty does not occur there is another issue which limits the validity of the kinematic approach. If the fluid motion is given and R_{m} exceeds its marginal value the magnetic field in a kinematic model grows endlessly. In reality, however, the Lorentz force, which grows too, acts on the fluid and changes the motion. This back-reaction of the magnetic field on the motion limits its growth.

8.2 Scenarios of Dynamo Action

These and other reasons force us to investigate the dynamo problem considering the full interaction of magnetic field and motion as described by the equations (148)–(154). Instead of the fluid flow then its causes have to be given, for example in the form of conditions allowing of motions due to thermal or other instabilities.

Let consider in some more detail the possibility of a dynamo driven by thermal instabilities. Consider, with a view to a planet or a star, a rotating body of a compressible conducting fluid with some density stratification. Assume some heat source in the inner part of this body and allow the heat to escape in outer space. In Figure 7 cases with different heat production rates are considered. As long as the heat production rate and thus the temperature gradient inside the body are sufficiently small there is no reason for convective motions or magnetic fields; any motion and any field vanish in the course of time. If the heat production rate and the temperature gradient grow, the stratification becomes unstable and convective motion sets in. As long as this motion is very weak there is still no cause for a magnetic field. If the heat production rate grows further and the convection becomes more vigorous, the non-magnetic state of the body becomes unstable and a magnetic field develops. The further evolution of motion and magnetic field is then essentially influenced by their interaction.

This consideration should underline the fact that the occurrence of magnetic fields in cosmic bodies is not a consequence of very special or exceptional circumstances but is as natural as the development of convective motions. The inset of convection requires that a parameter of the type of the Rayleigh number Ra lies above a critical value, and growing magnetic fields occur if the magnetic Reynolds number R_{m} exceeds another critical value. In many cosmic objects these conditions are satisfied.

Another cause of motions capable of dynamo action is the instability of a shear flow, for example in the case of differential rotation. Interestingly enough,

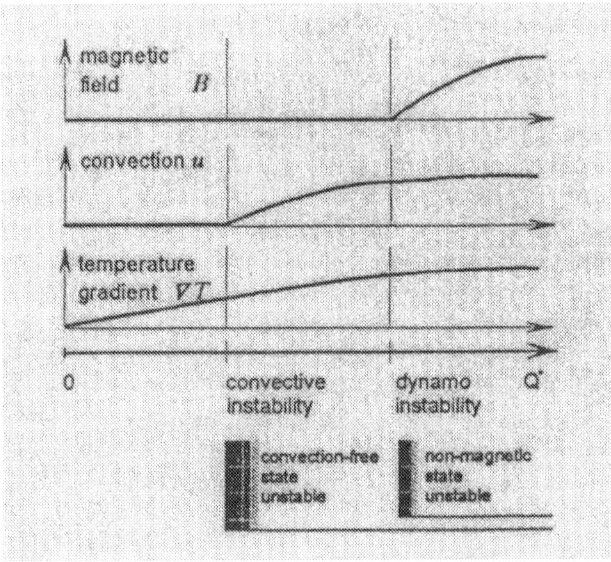

Fig. 7. Schematic representation of the dependence of temperature gradient, convection and magnetic field in a rotating body with density stratification on the heat production rate, $\dot{Q}$, in its inner part

a shear flow which is stable in the absence of a magnetic field can become unstable in the presence of a magnetic field [93].

9 Mean-Field Magnetofluiddynamics and Nonlinear Mean-Field Dynamo Models

Mean-field electrodynamics as explained in Section 5 proved to be a useful tool in the dynamo theory of cosmic objects. It suggests itself to extend the mean-field concept to magnetofluiddynamics in the sense of Section 7, that is, to establish a "mean-field magnetofluiddynamics". We will sketch here a few basic ideas and discuss their implications for mean-field dynamo models.

9.1 Basic Equations for the Mean Fields

For the sake of simplicity we restrict our explanations either to the case of an incompressible fluid or to cases in which the Boussinesq-approximation applies and start therefore from the equations (148) and (155), again with constant ν. We suppose that the force $\boldsymbol{F}^{(\mathrm{e})}$ has a fluctuating part and interpret it as a random force connected with instabilities which drive fluctuating motions and thus the fluctuating magnetic fields, too. Taking the average of (148) we obtain as in Section 5.2, see (81)–(82),

$$\begin{aligned} \partial_t \overline{\boldsymbol{B}} - \nabla \times (\overline{\boldsymbol{u}} \times \overline{\boldsymbol{B}}) &= -\nabla \times (\eta \nabla \times \overline{\boldsymbol{B}} - \boldsymbol{\mathcal{E}}) \\ \nabla \cdot \overline{\boldsymbol{B}} &= 0 \end{aligned} \tag{164}$$

with an electromotive force $\boldsymbol{\mathcal{E}}$ due to fluctuations of the motion and the magnetic field,

$$\boldsymbol{\mathcal{E}} = \overline{\boldsymbol{u}' \times \boldsymbol{B}'} \, . \tag{165}$$

Extending averaging to (155) we find in addition

$$\begin{aligned} \partial_t \overline{\boldsymbol{u}} + (\overline{\boldsymbol{u}} \cdot \nabla) \overline{\boldsymbol{u}} &= -\frac{1}{\varrho} \nabla \overline{p} - 2\varrho \boldsymbol{\Omega} \times \overline{\boldsymbol{u}} + \boldsymbol{g} + \nu \nabla^2 \overline{\boldsymbol{u}} \\ &+ \frac{1}{\mu\varrho} (\nabla \times \overline{\boldsymbol{B}}) \times \overline{\boldsymbol{B}} + \frac{1}{\varrho} (\overline{\boldsymbol{F}^{(\mathrm{e})}} + \boldsymbol{\mathcal{F}}) \\ \nabla \cdot \overline{\boldsymbol{u}} &= 0 \end{aligned} \tag{166}$$

with a ponderomotive force $\boldsymbol{\mathcal{F}}$ due to fluctuations of the motion and the magnetic field,

$$\boldsymbol{\mathcal{F}} = -\varrho \, \overline{(\boldsymbol{u}' \cdot \nabla) \, \boldsymbol{u}'} + \frac{1}{\mu} \overline{(\nabla \times \boldsymbol{B}') \times \boldsymbol{B}'} \, , \tag{167}$$

or

$$\mathcal{F}_i = \frac{\partial T_{ij}}{\partial x_j} \, , \quad T_{ij} = -\varrho \, \overline{u'_i u'_j} + \frac{1}{\mu} (\overline{B'_i B'_j} - \frac{1}{2} \overline{B'^2} \delta_{ij}) \, . \tag{168}$$

When considering equations (166) and (167), or (168), in the special case without any magnetic field, $\boldsymbol{B} = \boldsymbol{0}$, we are just on the level of Reynolds' theory of turbulent flows. The tensor $-\varrho \, \overline{u'_i u'_j}$ describes the Reynolds stresses due to the fluctuating motion. In the general case, that is, in the presence of a magnetic field, the tensor $(1/\mu)(\overline{B'_i B'_j} - (1/2)\overline{B'^2} \delta_{ij})$, has to be added describing the Maxwell stresses of the magnetic field fluctuations.

Clearly the calculation of the mean fields $\overline{\boldsymbol{B}}$ and $\overline{\boldsymbol{u}}$ requires the determination of the quantities $\boldsymbol{\mathcal{E}}$ and $\boldsymbol{\mathcal{F}}$. Following the pattern of Section 5.2 we may derive a system of equations governing the behavior of $\boldsymbol{B}'$ and $\boldsymbol{u}'$,

$$\begin{aligned} \partial_t \boldsymbol{B}' - \nabla \times (\overline{\boldsymbol{u}} \times \boldsymbol{B}' + \boldsymbol{u}' \times \overline{\boldsymbol{B}} + (\boldsymbol{u}' \times \boldsymbol{B}')') &= -\nabla \times (\eta \nabla \times \boldsymbol{B}') \\ \nabla \cdot \boldsymbol{B}' &= 0 \end{aligned} \tag{169}$$

$$\begin{aligned} \partial_t \boldsymbol{u}' + (\overline{\boldsymbol{u}} \cdot \nabla) \boldsymbol{u}' + (\boldsymbol{u}' \cdot \nabla) \overline{\boldsymbol{u}} + ((\boldsymbol{u}' \cdot \nabla) \boldsymbol{u}')' &= -\frac{1}{\varrho} \nabla p' - 2 \boldsymbol{\Omega} \times \boldsymbol{u}' + \nu \nabla^2 \boldsymbol{u}' \\ + \frac{1}{\mu\varrho} ((\nabla \times \overline{\boldsymbol{B}}) \times \boldsymbol{B}' + (\nabla \times \boldsymbol{B}') \times \overline{\boldsymbol{B}} &+ ((\nabla \times \boldsymbol{B}') \times \boldsymbol{B}')') + \frac{1}{\varrho} \boldsymbol{F}^{(\mathrm{e})\prime} \\ \nabla \cdot \boldsymbol{u}' &= 0 \, . \end{aligned}$$

We conclude from these equations that $\boldsymbol{B}'$ and $\boldsymbol{u}'$ and, consequently, $\boldsymbol{\mathcal{E}}$ and $\boldsymbol{\mathcal{F}}$ are functionals of $\overline{\boldsymbol{B}}$, $\overline{\boldsymbol{u}}$ and $\boldsymbol{F}^{(\mathrm{e})\prime}$. Of course, the dependency on $\boldsymbol{F}^{(\mathrm{e})\prime}$ is via averaged quantities only. Dependencies on $\boldsymbol{\Omega}$ and $\boldsymbol{g}$ are considered as obvious and not explicitly mentioned in the following.

We assume now that $\boldsymbol{F}^{(e)\,\prime}$ does not depend on $\overline{\boldsymbol{B}}$ and $\overline{\boldsymbol{u}}$. Let us consider for a moment the special case of (169) in which $\overline{\boldsymbol{B}} = \overline{\boldsymbol{u}} = \boldsymbol{0}$ and denote the corresponding solutions by $\boldsymbol{B}'^{(0)}$ and $\boldsymbol{u}'^{(0)}$. The turbulence that occurs in this special case, with velocity fields $\boldsymbol{u}'^{(0)}$ and magnetic fields $\boldsymbol{B}'^{(0)}$, is called "original turbulence" in the following discussion. Using the third of the equations (169) we can express $\boldsymbol{F}^{(e)\,\prime}$ everywhere by $\boldsymbol{u}'^{(0)}$ and $\boldsymbol{B}'^{(0)}$. Consequently, $\boldsymbol{\mathcal{E}}$ and $\boldsymbol{\mathcal{F}}$ can be considered as functionals of $\overline{\boldsymbol{u}}$ and $\overline{\boldsymbol{B}}$ and of $\boldsymbol{u}'^{(0)}$ and $\boldsymbol{B}'^{(0)}$, where the latter occur only in the form of averaged quantities. In contrast to our considerations in Section 5, however, $\boldsymbol{\mathcal{E}}$ is in general no longer linear in $\overline{\boldsymbol{B}}$.

Here the question arises whether in the special case $\overline{\boldsymbol{B}} = \overline{\boldsymbol{u}} = \boldsymbol{0}$ non-decaying solutions $\boldsymbol{B}'^{(0)}$ of (169) exist. As explained in Section 5.4 this would mean that there are small-scale dynamos. If we exclude this possibility we may, at least for times sufficiently far away from the initial instant, put $\boldsymbol{B}'^{(0)} = 0$, and so $\boldsymbol{\mathcal{E}}$ and $\boldsymbol{\mathcal{F}}$ loose their dependencies on $\boldsymbol{B}'^{(0)}$. Otherwise, of course, these dependencies have to be taken into account.

Following the ideas explained in Section 5 we may draw far-reaching conclusions concerning the structures of $\boldsymbol{\mathcal{E}}$ and $\boldsymbol{\mathcal{F}}$ from assumptions on $\overline{\boldsymbol{u}}$ and $\overline{\boldsymbol{B}}$ and on symmetry properties of the original turbulence fields $\boldsymbol{u}'^{(0)}$ or $\boldsymbol{B}'^{(0)}$, and we can calculate essential parameters that connect $\boldsymbol{\mathcal{E}}$ and $\boldsymbol{\mathcal{F}}$ with $\overline{\boldsymbol{u}}$ and $\overline{\boldsymbol{B}}$. We do not want to go into details but explain only a few aspects in the following.

9.2 The Mean Electromotive Force

Let us consider first the simple special case in which the original fluctuating velocity $\boldsymbol{u}'^{(0)}$ corresponds to a homogeneous isotropic turbulence but there is no original fluctuating magnetic field, that is, $\boldsymbol{B}'^{(0)} = \boldsymbol{0}$. We further assume that there is no mean motion, $\overline{\boldsymbol{u}} = \boldsymbol{0}$, and that the mean magnetic field, whose magnitude may be arbitrarily high, varies only weakly in space and not at all in time so that $\boldsymbol{\mathcal{E}}$ in a given point in space and time depends in an arbitrary way on $\overline{\boldsymbol{B}}$ in this point but only linearly on its first spatial derivatives and not on any higher ones. The problem of determining the structure of $\boldsymbol{\mathcal{E}}$ is analogous to that of the evaluation of (88) done in Section 5.8 for a turbulence possessing one preferred direction. The tensors a_{ij} and b_{ijk} must be again axisymmetric, that is, must have the form given with (125), where the preferred direction is now defined by $\overline{\boldsymbol{B}}$. Like the Lorentz force they must also be invariant under the inversion of the sign of $\overline{\boldsymbol{B}}$. So we arrive at

$$\boldsymbol{\mathcal{E}} = (\alpha - \tilde{\alpha}(\overline{\boldsymbol{B}} \cdot (\nabla \times \overline{\boldsymbol{B}})))\,\overline{\boldsymbol{B}} - (\gamma_1 \nabla \overline{\boldsymbol{B}}^2 + \gamma_2 (\overline{\boldsymbol{B}} \cdot \nabla)\overline{\boldsymbol{B}}\,) \times \overline{\boldsymbol{B}} - \beta \nabla \times \overline{\boldsymbol{B}}\,, \quad (170)$$

with coefficients $\alpha, \tilde{\alpha}, \gamma_1, \gamma_2$ and β determined by $\boldsymbol{u}'^{(0)}$ and $|\overline{\boldsymbol{B}}|$. In the limit of small $|\overline{\boldsymbol{B}}|$ the coefficients α and β turn into those discussed in Sections 5.4 and 5.7, and the terms with $\tilde{\alpha}, \gamma_1$ and γ_2 vanish.

There are several investigations which show that $|\alpha|$ and also β in general decrease with growing $|\overline{\boldsymbol{B}}|$; see e.g. [94]. Such reductions of $|\alpha|$ or β under the influence of the mean magnetic field are called "α-quenching" or "β-quenching".

Let us relax the above assumption on the absence of original magnetic field fluctuations, and assume that both $\boldsymbol{u}'^{(0)}$ and $\boldsymbol{B}'^{(0)}$ correspond to a homogeneous isotropic magnetofluiddynamic turbulence. Then (170) remains its validity but the coefficients $\alpha, \tilde{\alpha}, \gamma_1, \gamma_2$ and β are now determined by $\boldsymbol{u}'^{(0)}, \boldsymbol{B}'^{(0)}$ and $|\overline{\boldsymbol{B}}|$. Each of them possesses contributions depending on $\boldsymbol{B}'^{(0)}$ and $|\overline{\boldsymbol{B}}|$ only or on $\boldsymbol{u}'^{(0)}$ and $|\overline{\boldsymbol{B}}|$ only. To give an example we consider again the limit of small $|\overline{\boldsymbol{B}}|$ and adopt in addition the high conductivity limit as explained in Section 5.6, that is, $\lambda_{\mathrm{c}}^2/\eta\tau_{\mathrm{c}} \to 0$, and the analogously defined high-viscosity limit, $\lambda_{\mathrm{c}}^2/\nu\tau_{\mathrm{c}} \to 0$, with λ_{c} and τ_{c} being as above correlation length and time. For this case it turns out that

$$\alpha = \alpha^{(\mathrm{k})} + \alpha^{(\mathrm{m})}$$

$$\alpha^{(\mathrm{k})} = -\frac{1}{3}\overline{\boldsymbol{u}'^{(0)} \cdot (\nabla \times \boldsymbol{u}'^{(0)})}\,\tau_{\mathrm{c}}^{(\alpha\,\mathrm{k})}, \quad \alpha^{(\mathrm{m})} = \frac{1}{3\mu\varrho}\overline{\boldsymbol{B}'^{(0)} \cdot (\nabla \times \boldsymbol{B}'^{(0)})}\,\tau_{\mathrm{c}}^{(\alpha\,\mathrm{m})}. \tag{171}$$

with properly defined correlation times $\tau_{\mathrm{c}}^{(\alpha\,\mathrm{k})}$ and $\tau_{\mathrm{c}}^{(\alpha\,\mathrm{m})}$. There are many investigations of the α-effect and related effects for cases with original magnetic field fluctuation as considered here [95,7].

9.3 The Mean Ponderomotive Force

Proceeding now to the ponderomotive force $\mathcal{F}$ we assume at first that there is no magnetic field, $\boldsymbol{B} = \boldsymbol{0}$, and that the original turbulence described by $\boldsymbol{u}'^{(0)}$ is homogeneous and isotropic. The correlation tensor $\overline{u_i' u_j'}$ in general depends on $\overline{\boldsymbol{u}}$. There are, however, good reasons to assume that this dependence vanishes if the mean velocity $\overline{\boldsymbol{u}}$ is independent of position and time. We now assume that the mean velocity $\overline{\boldsymbol{u}}$ varies only weakly in space and not at all in time so that the correlation tensor $\overline{u_i' u_j'}$ depends linearly on the first spatial derivatives and not at all on any other ones. So we have

$$\overline{u_i' u_j'} = \frac{1}{3}\overline{\boldsymbol{u}'^{(0)2}}\,\delta_{ij} - \nu_{\mathrm{t}}\,(\frac{\partial \overline{u}_i}{\partial x_j} + \frac{\partial \overline{u}_j}{\partial x_i}) \tag{172}$$

with some constant coefficient ν_{t} determined by $\boldsymbol{u}'^{(0)}$. This leads to

$$\mathcal{F} = \varrho\,\nu_{\mathrm{t}}\,\nabla^2\overline{\boldsymbol{u}}. \tag{173}$$

Under the assumptions adopted here the effect of $\mathcal{F}$ is the same as that of replacing the kinematic viscosity ν by a mean-field viscosity ν_{m} defined by $\nu_{\mathrm{m}} = \nu + \nu_{\mathrm{t}}$. We call ν_{t} the "turbulent viscosity" or "eddy-viscosity". The theory of the mean-field viscosity has been widely elaborated; see e.g. [96].

Let us change our assumption so that the original turbulence is no longer homogeneous and isotropic but, as to be expected on rotating bodies, inhomogeneous and influenced by Coriolis forces. Then the correlation tensor $\overline{u_i' u_j'}$ has not only contributions corresponding to those given in (172) but a number of other ones. We adopt here the notation introduced in Section 6.2, that is, use the unit vectors $\hat{\boldsymbol{\omega}}$ and $\hat{\boldsymbol{g}}$ parallel to $\boldsymbol{\Omega}$ and $\boldsymbol{g}$ and put $\boldsymbol{\lambda} = \hat{\boldsymbol{\omega}} \times \hat{\boldsymbol{g}}$. Restricting our attention to one of these contributions we write

$$\overline{u_i' u_j'} = \cdots + v\,(\lambda_i\,\hat{g}_j + \lambda_j\,\hat{g}_i) \tag{174}$$

where v means a coefficient varying like the turbulence intensity with the space coordinates. As a consequence we have

$$\boldsymbol{\mathcal{F}} = \cdots - \varrho\,((\hat{\boldsymbol{g}} \cdot \nabla v)\boldsymbol{\lambda} + (\boldsymbol{\lambda} \cdot \nabla v)\hat{\boldsymbol{g}})\,. \tag{175}$$

The contribution $-\varrho\,(\hat{\boldsymbol{g}} \cdot \nabla v)\boldsymbol{\lambda}$ describes a force acting in azimuthal direction, which can drive a differential rotation. The other contribution is of minor interest; it vanishes if v has no azimuthal dependence. The occurrence of this azimuthal force, called "Λ-effect", is the crucial point in a widely elaborated theory of stellar rotation initiated by Rüdiger [96].

To give an example of a contribution to $\boldsymbol{\mathcal{F}}$ due to turbulent magnetic field fluctuations $\boldsymbol{B}'$ we start from an original turbulence which is homogeneous and isotropic with respect to $\boldsymbol{u}'^{(0)}$ but has no magnetic part, $\boldsymbol{B}'^{(0)} = \boldsymbol{0}$. We admit, however, a non-zero mean magnetic field that corresponds to a homogeneous isotropic turbulence. Then the correlation tensor $\overline{B_i' B_j'}$ is non-zero, too. Since the Lorentz force is invariant under inversion of the sign of $\boldsymbol{B}$ the tensor $\overline{B_i' B_j'}$ can contain only even powers of $\overline{\boldsymbol{B}}$. Assuming that $\overline{\boldsymbol{B}}$ is sufficiently weak we put

$$\overline{B_i' B_j'} = \varepsilon_1 \overline{\boldsymbol{B}}^2\,\delta_{ij} + \varepsilon_2\,\overline{B}_i\,\overline{B}_j\,, \tag{176}$$

with small dimensionless constants ε_1 and ε_2. This leads to a contribution to $\boldsymbol{\mathcal{F}}$ of the form

$$\boldsymbol{\mathcal{F}} = \cdots + \frac{1}{\mu}\,(\varepsilon_2\,(\nabla \times \overline{\boldsymbol{B}}) \times \overline{\boldsymbol{B}} - \frac{\varepsilon_1}{2}\,\nabla\overline{\boldsymbol{B}}^2)\,. \tag{177}$$

When dropping all terms which have the form of a gradient this contribution turns simply into $(\varepsilon_2/\mu)\,(\overline{\boldsymbol{B}}\cdot\nabla)\overline{\boldsymbol{B}}$. It corresponds, depending on the sign of ε_2, to a slight attenuation or amplification of the Lorentz force resulting immediately from the mean magnetic field as given in (166).

9.4 Implications for Mean-Field Dynamo Models

In most of the kinematic mean-field dynamo models investigated so far independent assumptions are used on α-effect and differential rotation. However, both the electromotive force $\boldsymbol{\mathcal{E}}$ and the ponderomotive force $\boldsymbol{\mathcal{F}}$ and thus both α-effect and differential rotation depend on the small-scale motions. That is, they cannot be completely independent, and their connections should be taken into account. With a view to the Sun indeed mean-field dynamo models have been developed with α-effect and differential rotation derived from the same assumptions on an underlying turbulence [97].

In the kinematic mean-field dynamo models discussed in Section 6 any back-reaction of the magnetic field on the fluid motion has been ignored. Therefore the results apply, strictly speaking, only in the limit of vanishing magnetic fields. Such models have been modified by taking into account this back-reaction in the form of α-quenching, or also β-quenching. Even if symmetries of the models as introduced in Section 6.2 in the limit of vanishing magnetic field exist they are in general perturbed for finite magnetic fields. Then the solutions of the governing equations have no longer the simple form (144). Both the geometrical structure and the time behavior are more complex. We may, of course, understand each solution as a superposition of parts with symmetries as characterized above by Am and Sm, but these parts are no longer solutions. In very simple cases there is an evolution toward stable steady states or comparable states in which the field configuration rotates like a rigid body. We cannot go into more details but refer to a few examples of investigations of that kind [98,99,12,100].

The magnetic field influences not only the small-scale motions, which are taken into account in the form of α-effect and related effects but it modifies or even generates mean motions. With this in mind dynamically more or less consistent mean-field dynamo models have been studied within the framework of the mean-field versions of induction equation and momentum balance; see e.g. [101,98,102].

10 Dynamo Models for Specific Objects

On the basis explained in the preceding sections much research work has been done on dynamos in the Earth and in planets, in the Sun and other stars or in galaxies. We cannot present detailed results here but make only a few remarks on such results and on open questions.

10.1 The Geodynamo and Planetary Dynamos

Let us start our explanations on the geodynamo with a look at the structure of the Earth. We distinguish between an inner and an outer core, both being metallic, and the mantle consisting mainly of silicates. The boundary between the inner and outer core is at a radius of about 2300 km, that between core and mantle at 3500 km, and the mantle reaches almost until the Earth's surface at 6370 km. The inner core is solid but the outer one liquid and allows therefore internal motions. The mantle is highly viscous and admits only very slow internal motions. Compared to the metallic electrical conductivity of the core, see Table 2, the conductivity of the mantle is very small.

The crucial point for the geodynamo are convective motions inside the outer core. The most obvious reason for them consists in the temperature gradient across this layer which causes an unstable stratification and thus drives convection. It is then thermal energy resulting for example from radio-active decay, which is transformed into kinetic energy of these motions. Another reason for convective motions, which has been extensively discussed during the last years, is connected with the so-called "chemical differentiation" of the liquid; see e.g. [103]. As a consequence of the pressure and temperature situation close to the inner core boundary the liquid looses there a part of its heavier components, which solidify and make the inner core slowly grow, and the remaining lighter fluid rises, enriches itself with heavier components at the core-mantle boundary, sinks again to the inner core boundary, etc. In this way the mass is redistributed inside the Earth, and as a result gravitational energy is transformed into kinetic energy. In contrast to the first-mentioned "thermal convection" we speak here of "compositional convection". In the first case only a part of the thermal energy available can be transformed into kinetic one. The upper limit is given by the Carnot efficiency defined by the relevant temperature difference and the maximum absolute temperature in this process. In the second case no comparable limitation exists. Since there are some doubts whether the available thermal energy is sufficient to operate the geodynamo the compositional convection has found particular interest. Another possible cause of motions inside the outer core is the precession of the rotation axis of the Earth [103]. The relevance of this source of energy for the geodynamo is however still under debate. Due to the rotation of the Earth the motions in the outer core are subject to Coriolis forces, that is, they have necessarily helical features.

It suggests itself to consider the geodynamo first in the framework of the mean-field concept and to describe the induction effect of the helical convective motions inside the outer core by an α-effect. On this level simple kinematic models of the geodynamo were first proposed by Steenbeck and Krause [71]. These models, restricted to axisymmetric magnetic fields only, gave at least some idea on how the motions in the core can maintain the Earth's magnetic field. As already mentioned in Section 6.3 many investigation of kinematic mean-field dynamo models, in many respects more sophisticated and taking into account non-axisymmetric magnetic fields too, have been carried out, and the results have been discussed in view of the Earth; see e.g. [2,21,22,12].

Another but in some sense similar approach to kinematic models of the geodynamo, the theory of the "nearly symmetric dynamo" was proposed by Braginsky already in 1964 [27,47,48]. This concept, which we mentioned in Section 4.4, has been widely elaborated; see e.g. [49,50].

Much research work has been done in view of dynamically consistent dynamo models of the Earth which explain the magnitude and reflect essential aspects of the geometrical structure and of the complex spectrum of variations of the geomagnetic field. This implies many studies of the behavior of fluids in a rotating shell and on convection in the presence of magnetic fields; see e.g. [104]. We refer here to review articles on the theory of the geodynamo; see e.g. [105,106].

In the last years considerable progress in numerical simulations of the geodynamo on the basis of the relevant equations for magnetic field, fluid motion, temperature etc. has been achieved [107–109,111,110,112,113]. Many difficulties in numerical computations result from the fact that the requirements for space and time resolution grow enormously when parameters of the model like the magnetic Prandtl number or the Ekman number approach realistic values. Although by such reasons the simulations do not meet the situation in the Earth correctly they reproduce in an impressive way quite a few essential features of the geometrical structure and the time behaviors of the geomagnetic field including its reversals.

Turning now to the planets we first note that the absence of an intrinsic magnetic field at Venus is plausible from the fact that due to its very slow rotation there are probably not sufficiently strong helical features of motions on this planet. As explained in Section 6.3 in spherical mean-field dynamo models there is in general no preference of axisymmetric magnetic fields except for a high degree of isotropy of the small-scale motions or a sufficiently strong differential rotation [114,77,22,78,79]. So it is at least not very surprising that the magnetic fields of Uranus and Neptune deviate drastically from symmetry about the rotation axes.

Many systematic numerical studies of dynamos in spherical shells, which are of interest in view of the planets, have been carried out [115,116].

10.2 Solar and Stellar Dynamos

As explained in Section 1 the observational facts on magnetic phenomena on the Sun give evidence for a large-scale magnetic field consisting essentially of two strong field belts beneath the visible surface of the Sun, one in the north hemisphere and another one with opposite orientation in the southern hemisphere, and a much weaker poloidal field penetrating the surface. Roughly speaking, this large-scale field is symmetric about the rotation axis, antisymmetric about the equatorial plane, and oscillatory with a period of about 2×11 years. There are good reasons to assume that it is due to a dynamo which operates in the convection zone, ranging from a radius of about 500 000 km until the photosphere and chromosphere at 696 000 km, or in the overshoot layer underneath the convection zone. As everywhere in the Sun the matter in these layers is electrically conducting, see Table 2, and it shows convective motions influenced by Coriolis forces as well as a differential rotation.

It suggests itself to discuss this dynamo within the framework of mean-field dynamo theory. Let us start by considering kinematic mean-field models with the simple symmetry properties described in Section 6.2. We clearly have to relate the large-scale field mentioned to an oscillatory A0 mode generated by an $\alpha\omega$-dynamo operating in a layer with both the α-effect and differential rotation. Dynamo models of this kind were first proposed by Steenbeck and Krause [70]. In these and a large number of more sophisticated models developed later the dynamo was assumed to work completely within the convection zone and not in the overshoot layer. They were able to represent many features of the solar magnetic cycle; for reviews see e.g. [85,117–120,97,121,86,122].

In addition to the requirements concerning symmetries and the time behavior of the magnetic field, the solar dynamo models must meet other observational constraints. They should, for example, reproduce the equatorward migration of the toroidal field belts during each half-cycle, which determines the shape of the butterfly diagram, and they should also satisfy some phase relation between the radial and the azimuthal field components derived from observations.

The direction in which the toroidal field belts migrate depends on the sign of α and the radial dependence of the angular velocity ω. They migrate equatorward if the signs of α and $\partial\omega/\partial r$ in the northern hemisphere are opposite. Since there are good reasons to assume that $\alpha > 0$ in the northern hemisphere, it was concluded that $\partial\omega/\partial r < 0$. This is, however, in conflict with recent helioseismological results, which strongly suggest that there is nearly no radial dependence of ω inside the convection zone but rather a strong gradient of ω at its lower boundary, with $\partial\omega/\partial r < 0$ only at higher and $\partial\omega/\partial r > 0$ at lower latitudes [123]. Another discrepancy arises from estimates showing that the magnetic flux produced in the convection zone should leave it so quickly that the dynamo could not work there.

By these and other reasons it was proposed to assume that the solar dynamo operates mainly in the overshoot layer below the convection zone; see e.g. [97,124]. For this layer several different approaches lead to $\alpha < 0$ in the northern hemisphere [97,125] so that at lower latitudes, where $\partial\omega/\partial r > 0$, again an equatorward migration of the toroidal field belts is to be expected. In the overshoot layer a sufficient storage of magnetic flux seems to be possible; see e.g. [126,127]. Several models for dynamos in the overshoot layer have been investigated; e.g. [128,97,129,125]. The question on the site of the solar dynamo is, however, still under debate, and at present even on the kinematic level no completely satisfying solar dynamo model is available.

In many of the solar dynamo models investigated so far independent assumptions were made on the dependence of the α-tensor and related quantities and of the angular velocity on the space coordinates. As explained in Section 9, however, all these quantities are determined by the properties of the small-scale motions. In some recent models all these quantities are indeed derived from the same assumptions on the small-scale motions; see e.g. [97].

The solar cycle exhibits stochastic features, too. There are a few investigations of dynamo models which try to mimic them by assumed stochastic fluctuations of the α-effect; see e.g. [130,131,125].

There is a large number of investigations of more or less simple solar dynamo models in the nonlinear regime. They consider deviations of the solar magnetic field from simple symmetries and from a constant amplitude oscillation; see e.g. [122]. In this way they provide us with some understanding of the deviations of the butterfly diagram from the north-south symmetry and of the grand minima of the solar activity.

As explained already in Section 1 there is some observational material giving evidence of magnetic cycles at other active stars. Many investigations on solar dynamos have been extended to these cases; for reviews see e.g. [132–134,122].

10.3 Galactic Dynamos

As far as the magnetic fields observed in galaxies are concerned the idea of their primordial origin has been extensively discussed. There are, however, quite a few reasons to reject it; see e.g. [135]. As an alternative the idea of generation and maintenance of such fields by dynamo action within the interstellar medium has been elaborated. The fact that the electric conductivity of this medium is extremely small compared to planetary or stellar interiors is in a sense compensated by the huge dimensions of the galaxies.

Again mean-field dynamo models are considered working with an α-effect due to turbulent motions of the interstellar medium under the influence of Coriolis forces and with the differential rotation of the galactic disc; see e.g. [136,6,137]. An essential source of turbulent motions are supernova explosions. The α-effect has been estimated from simple assumptions on such motions [138], and also on the basis of numerical simulations [139]. Estimates of that kind together with the known data on the rotational shear and the dimensions of a galactic disc lead to values of R_α and R_ω which justify the assumption that dynamos of $\alpha\omega$-type may well operate in galaxies [6].

A number of disc-like mean-field dynamo models have been studied. Many of them satisfy the symmetry assumptions used in Section 6.2, which ignore, of course, any effect of structures like spiral arms. There are several models of that kind in which the conducting fluid is surrounded by free space [89,90,141,87,88]. In addition models have been investigated in which the dynamo-active region is embedded in an extended conducting medium so that there are no sharp boundaries [91,140,6]. Some more recent models of galactic dynamos consider also structures like spiral arms [142,143].

Acknowledgement

The author thanks his colleagues Dr. H. Fuchs and Dr. M. Rheinhardt for reading the manuscript and many helpful comments, Dr. M. Schüler for technical support.

References

1. F. Krause and K.-H. Rädler: 'Elektrodynamik der mittleren Felder in turbulenten leitenden Medien und Dynamotheorie', in *Ergebnisse der Plasmaphysik und der Gaselektronik*, volume 2, ed. by R. Rompe and M. Steenbeck,(Akademie–Verlag Berlin, 1971) pp. 1–154
2. F. Krause and K.-H. Rädler: *Mean–Field Magnetohydrodynamics and Dynamo Theory*, (Akademie–Verlag Berlin and Pergamon Press Oxford, 1980)
3. H. K. Moffatt: *Magnetic Field Generation in Electrically Conducting Fluids.* (Cambridge University Press, 1978)
4. M. R. E. Proctor and A. D. Gilbert: *Lectures on Solar and Planetary Dynamos*, (Cambridge University Press, 1994)
5. P. H. Roberts: *An Introduction to Magnetohydrodynamics* (Longmans, Green and Co. Ltd., 1967)
6. A. A. Ruzmaikin, A. M. Shukurov, and D. D. Sokoloff: *Magnetic Fields of Galaxies*, (Astrophysics and Space Science Library Vol. 133. Kluwer Academic Publishers, 1988)
7. Ya. B. Zeldovich, A. A. Ruzmaikin, and D. D. Sokoloff: *Magnetic Fields in Astrophysics*, (The Fluid Mechanics of Astrophysics and Geophysics Vol.3. Gordon and Breach Science Publishers, 1983)

8. *The Cosmic Dynamo*, ed. by F. Krause, K.-H. Rädler, and G. Rüdiger, (Kluwer Academic Publishers Dordrecht Bosten London, 1993)
9. *Stellar Dynamos: Nonlinearity and Chaotic Flows.*, ed. by M. Nunez and A. Ferriz-Mas, (Astronomical Society of the Pacific, 1999)
10. *Solar and Planetary Dynamos*, ed. by M. R. E. Proctor, P. C. Matthews, and A. M. Rucklidge, (Cambridge University Press, 1993)
11. *Stellar and Planetary Magnetism*, ed. by A. M. Soward, (Gordon and Breach Science Publishers, 1983)
12. K.-H. Rädler: Rev. Mod. Astron. **8**, 295 (1995)
13. P. H. Roberts: 'Dynamo theory', in *Lectures on Applied Mathematics*, volume 14, ed. by W. H. Reid, (Amer. Mathematics Soc., Providence, 1971) pp. 129–206
14. P. H. Roberts: 'Dynamo theory', in *Astrophysical Fluid Dynamics*, ed. by J.-P. Zahn and J. Zinn-Justin, (Elsevier Science Publishers B.V., 1993)
15. P. H. Roberts: 'Fundamentals of dynamo theory', in *Lectures on Solar and Planetary Dynamos*, ed. by M. R. E. Proctor and A. D. Gilbert, (Cambridge University Press, 1994), pp. 1–58
16. P. H. Roberts and A. M. Soward: Ann. Rev. Fluid Mech. **24**, 459 (1992)
17. N. O. Weiss: Y. J. Roy. Astron. Soc. **12**, 432 (1971)
18. J. Larmor: *How could a rotating body such as the Sun become a magnet?*, Rep. Brit. Assoc. Adv. Sc., pp. 159–160 (1919)
19. H. Bondi and T. Gold: Mon. Not. R. Astron. Soc. **110**, 607 (1950)
20. D. Moss: Mon. Not. R. Astron. Soc. **257**, 593 (1992)
21. K.-H. Rädler: Astron. Nachr. **301**, 101 (1980)
22. K.-H. Rädler: Astron. Nachr. **307**, 89 (1986)
23. K.-H. Rädler: 'On the effect of differential rotation on axisymmetric and non-axisymmetric magnetic fields of cosmical bodies', in *Plasma-Astrophysics* (Proceedings of the Joint Varenna-Abastumani International School and Workshop, Sukhumi, ESA SP-251, 1986), pp. 569–574
24. H.-J. Bräuer and K.-H. Rädler: Astron. Nachr. **308**, 27 (1987)
25. K.-H. Rädler: Astron. Nachr. **295**, 73 (1974)
26. T. G. Cowling: Mon. Not. R. Astron. Soc. **94**, 39 (1934)
27. S. I. Braginskij: Geomagn. Aeron. **4**, 732 (1964)
28. R. Hide and T. N. Palmer: Geophys. Astrophys. Fluid Dyn. **19**, 301 (1982)
29. R. W. James, P. H. Roberts, and D. E. Winch: Geophys. Astrophys. Fluid Dyn. **15**, 149 (1980)
30. D. Lortz: Phys. Fluids **11**, 913 (1968)
31. W. M. Elsasser: Phys.Rev. **69**, 106 (1946)
32. E. C. Bullard and H. Gellman: Phil. Trans. Roy. Soc. A **247**, 213 (1954)
33. F. H. Busse: J. Geophys. Res. **80**, 278 (1975)
34. A. Herzenberg: Phil. Trans. R. Soc. London, Ser. A **250**, 543 (1958)
35. R. D. Gibson: Q. J. Mech. Appl. Math. **21**, 243 (1968)
36. R. D. Gibson: Q. J. Mech. Appl. Math. **21**, 257 (1968)
37. F. J. Lowes and I. Wilkinson: Nature **198**, 1158 (1963)
38. F. J. Lowes and I. Wilkinson: Nature **219**, 717 (1968)
39. Yu. B. Ponomarenko: PMTF **1973(6)**, 47 (1973) in Russian
40. G. O. Roberts: Phil. Trans. R. Soc. London, Ser. A **266**, 535 (1970)
41. G. O. Roberts: Phil. Trans. Roy. Soc. London A **271**, 411 (1972)
42. S. Childress and A. D. Gilbert: *Stretch, Twist, Fold: The Fast Dynamo*, (Lecture Notes in Physics. Springer-Verlag Berlin Heidelberg New York, 1995)
43. A. Gailitis: Magnetohydrodynamics **6**, 14 (1970)

44. C. L. Pekeris, Y. Accad, and B. Shkoller: Phil. Trans. R. Soc. London, Ser. A **275**, 425 (1973)
45. D. Gubbins: Phil. Trans. R. Soc. London, Ser. A **274**, 493 (1973)
46. S. Kumar and P. H. Roberts: Proc. R. Soc. London, Ser. A **344**, 235 (1975)
47. S. I. Braginskij: Sov. Phys. JETP **20**, 726 (1964)
48. S. I. Braginskij: Sov. Phys. JETP **20**, 1462 (1964)
49. S. I. Braginskij: Phys. Earth Planet. Inter. **11**, 191 (1976)
50. S. I. Braginskij: 'The nonlinear dynamo and model-z'. In *Lectures on Solar and Planetary Dynamos*, ed. by M. R. E. Proctor and A. D. Gilbert, (Cambridge University Press, 1994) pp. 267–304
51. M. Steenbeck, F. Krause, and K.-H. Rädler: Z. Naturforsch. **21a**, 369 (1966)
52. E. N. Parker: Astrophys. J. **122**, 293 (1955)
53. O. Lielausis: Astron. Nachr.**315**, 303 (1994)
54. A. Gailitis: 'Current status of liquid sodium MHD dynamo experiment in Riga', in *Proceedings of the International Conference "Transfer Phenomena in Magnetohydrodynamic and Electroconducting Flows"* (Aussois, France, 1997). pp. 33–38
55. F. H. Busse: Geophys. J. R. Astron. Soc. **42**, 437 (1975)
56. K.-H. Rädler, E. Apstein, M. Rheinhardt, and M. Schüler: Studia geoph. et geod. **42**, 224 (1998)
57. K.-H. Rädler, E. Apstein, and M. Schüler: 'The alpha–effect in the Karlsruhe dynamo experiment' in *Proceedings of the International Conference "Transfer Phenomena in Magnetohydrodynamic and Electroconducting Flows" Held in Aussois, France, 1997*, pp. 9–14
58. A. Tilgner: Phys. Lett. A **226**, 75 (1996)
59. A. Tilgner: Acta Astron. et Geophys. Univ. Comenianae **19**, 51 (1997)
60. F. Krause and M. Steenbeck: Z. Naturforsch. **22a**, 671 (1967)
61. K.-H. Rädler: Geophys. Astrophys. Fluid Dyn. **20**, 191 (1982)
62. K.-H. Rädler and U. Geppert: 'Turbulent dynamo action in the high–conductivity limit: A hidden dynamo.', in *Stellar Dynamos: Nonlinearity and Chaotic Flows*, ed. by M. Nunez and A. Ferriz-Mas, (ASP Conference Series Vol .178, 1999) pp. 151–163. .
63. K.-H. Rädler: 'Mean–field magnetohydrodynamics as a basis of solar dynamo theory', in *Basic Mechanisms of Solar Activity*, ed. by V. Bumba and J. Kleczek, (D. Reidel Publishing Company Dordrecht Holland, 1976) pp. 223–344
64. T. S. Ivanova and A. A. Ruzmaikin: Sov. Astron. **20**, 227 (1976)
65. K.-H. Rädler: Z. Naturforsch. **23a**, 1841 (1968)
66. K.-H. Rädler: Z. Naturforsch. **23a**, 1851 (1968)
67. Ya. B. Zeldovich: J. Exptl. Theoret. Phys.**31**, 154 (1956)
68. E. M. Drobyshevski, E. N. Kolesnikova, and V. S. Yuferev: J. Fluid Mech. **101**, 65 (1980)
69. E. M. Drobyshevski and V. S. Yuferev: J. Fluid Mech. **65**, 33 (1974)
70. M. Steenbeck and F. Krause: Astron. Nachr. **291**, 49 (1969)
71. M. Steenbeck and F. Krause: Astron. Nachr. **291**, 271 (1969)
72. H. K. Moffatt and M. R. E. Proctor: Geophys. Astrophys. Fluid Dyn. **21**, 265 (1982)
73. K.-H. Rädler: Monatsber. Dtsch. Akad. Wiss. Berlin **11**, 272 (1969)
74. K.-H. Rädler: Monatsber. Dtsch. Akad. Wiss. Berlin **12**, 468 (1970)
75. P. H. Roberts: Phil. Trans. R. Soc. London, Ser. A **272**, 663 (1972)
76. A. Brandenburg, I. Tuominen, and K.-H. Rädler: Geophys. Astrophys. Fluid Dyn. **49**, 45 (1989)

77. K.-H. Rädler: Mem. Soc. Roy. Soc. Liege **VIII**, 109 (1975)
78. G. Rüdiger: Astron. Nachr. **301**, 181 (1980)
79. G. Rüdiger and D. Elstner: Astron. Astrophys., 1 (1993)
80. T. T. Ivanova and A. A. Ruzmaikin: Astron. Nachr. **306**, 177 (1985)
81. P. H. Roberts and M. Stix: Astron. Astrophys. **18**, 453 (1972)
82. W. Deinzer and M. Stix: Astron. Astrophys. **12**, 111 (1971)
83. W. Deinzer, H.-U. v. Kusserow, and M. Stix: Astron. Astrophys. **36**, 69 (1974)
84. M. Stix: Astron. Astrophys. **24**, 275 (1973)
85. A. Brandenburg and I. Tuominen: 'The solar dynamo'. In *The Sun and Cool Stars: Activity, Magnetism, Dynamos*, ed. by I. Tuominen, D. Moss, and G. Rüdiger, (Lecture Notes in Physics, Springer Verlag, 1991)
86. M. Stix: Geophys. Astrophys. Fluid Dyn. **62**, 211 (1991)
87. M. Stix: Astron. Astrophys. **47**, 243 (1975)
88. M. P. White: Astron. Nachr. **299**, 209 (1978)
89. W. Deinzer, H. Grosser, and D. Schmitt: Astron. Astrophys. **273**, 405 (1993)
90. K.-H. Rädler and E. Wiedemann: 'Mean-field models of galactic dynamos admitting axisymmetric and non-axisymmetric magnetic field structures' in *Galactic and Intergalactic Magnetic Fields (Proceedings of the IAU Symposium No 140, Heidelberg 1990)*, ed. by R. Beck, P. P. Kronberg, and R. Wielebinski, (Kluwer Academic Publishers, 1990), pp. 107 – 112
91. D. Elstner, R. Meinel, and G. Rüdiger: Geophys. Astrophys. Fluid Dyn. **50**, 85 (1990)
92. G. Rüdiger, D. Elstner, and M. Schultz: 'The galactic dynamo: Modes and models', in *The Cosmic Dynamo*, ed. by F. Krause, G. Rüdiger, and K.-H. Rädler, (Kluwer Academic Publishers, 1993), pp. 321–331
93. S. A. Balbus and J. F. Hawley: Rev. Mod. Phys. **70**, 1 (1998)
94. G. Rüdiger and L.L. Kichatinov: Astron. Astrophys. **269**, 581 (1993)
95. G. B. Field, E. G. Blackman, and H. Chou: Astrophys. J. **513**, 638 (1999)
96. G. Rüdiger: *Differential Rotation and Stellar Convection*, (Akademie Verlag Berlin, 1989)
97. G. Rüdiger: 'Dynamo theory and the period of the solar cycle', in *Solar Magnetic Fields*, ed. by M. Schüssler and W. Schmidt, (Cambridge University Press, 1994), pp. 77–93
98. A. Brandenburg and I. Tuominen: Geophys. Astrophys. Fluid Dyn. **49**, 129 (1989)
99. R. Hollerbach: Geophys. Astrophys. Fluid Dyn. **60**, 245 (1991)
100. K.-H. Rädler, E. Wiedemann, A. Brandenburg, R. Meinel, and I. Tuominen: Astron. Astrophys. **239**, 413 (1990)
101. D. M. Barker and D. Moss: Astron. Astrophys.**283**, 1009 (1994)
102. H. Fuchs, K.-H. Rädler, M. Rheinhardt, and M. Schüler: Chaos, Solitons & Fractals **5**, 2013 (1995)
103. W. V. R. Malkus: 'Energy sources for planetary dynamos', in *Lectures on Solar and Planetary Dynamos*, ed by M. R. E. Proctor and A. D. Gilbert, (Cambridge University Press, 1994), pp. 161–179
104. M. R. E. Proctor: 'Convection and magnetoconvection in a rapidly rotating sphere', in *Lectures on Solar and Planetary Dynamos*, ed. by M. R. E. Proctor and A. D. Gilbert, (Cambridge University Press, 1994) pp. 97–115
105. R. Hollerbach: Phys. Earth Planet. Inter. **98**, 163 (1996)
106. A. M. Soward: Geophys. Astrophys. Fluid Dyn. **62**, 191 (1991)
107. U. Christensen and P. Olson: Geophys. Res. Lett. **25**, 1565 (1998)
108. U. Christensen, P. Olson, and G. A. Glatzmaier: Geophys. J. Int. **138**, 393 (1999)

109. G. A. Glatzmaier and P. H. Roberts: Nature **377**, 203 (1995)
110. G. A. Glatzmaier and P. H. Roberts: Physica D **97**, 81 (1996)
111. G. A. Glatzmaier and P. H. Roberts: Science **274**, 1887 (1996)
112. E. Grote, F. H. Busse, and A. Tilgner: Phys. Earth Planet. Inter. **117**, 259 (2000)
113. W. Kuang and J. Bloxham: Nature **389**, 371 (1997)
114. D. Moss and A. Brandenburg: Geophys. Astrophys. Fluid Dyn. **80**, 229 (1995)
115. F. H. Busse, E. Grote, and A. Tilgner: Studia geoph. et geod. **42**, 211 (1998)
116. D. R. Fearn: 'Nonlinear planetary dynamos', in *Lectures on Solar and Planetary Dynamos*, ed. by M. R. E. Proctor and A. D. Gilbert, editors, (Cambridge University Press, 1994), pp. 219–244
117. E. E. DeLuca and P. A. Gilman: *The Solar Dynamo*, (The University of Arizona Press Tucson, 1992), pp. 275–303.
118. P. A. Gilman: 'What can we learn about solar cycle mechanisms from observed velocity fields?', in *The Solar Cycle*, (ASP Conference Series 27, 1992), pp. 241–254
119. K.-H. Rädler: 'The solar dynamo', in *Inside the Sun*, ed. by G. Berthomieu and M. Cribier, (Proceedings of the IAU Colloquium No 121, Kluwer Academic Publishers, 1990), pp. 385–402
120. R. Rosner and N. O. Weiss: 'The origin of the solar cycle', in *The Solar Cycle*, ed. by K. Harvey, (Astronomical Society of the Pacfic, 1992), pp. 511–531
121. M. Stix: 'Dynamo theory and the solar cycle', in *Basic Mechanisms of Solar Activity*, ed. by V. Bumba and J. Kleczek, (D. Reidel Publishing Company Dordrecht Holland, 1976), pp. 367–388
122. N. O. Weiss: 'Solar and stellar dynamos', in *Lectures on Solar and Planetary Dynamos*, ed. by M. R. E. Proctor and A. D. Gilbert, (Cambridge University Press, England, 1994), pp. 59–95
123. K. G. Libbrecht: 'The excitation and damping of solar oscillations', in *Seismology of the Sun and Sun-Like Stars*, ed. by E.J. Rolfe, (ESA Publications Division, 1988) pp. 3–10
124. D. Schmitt: 'The solar dynamo', in *The Cosmic Dynamo*, ed. by F Krause, K.-H. Rädler, and G. Rüdiger, (Kluwer Academic Publishers, 1993), pp. 1–12
125. D. Schmitt, A. Ferriz-Mas, and M. Schüssler: 'An α-effect due to instability of toroidal magnetic flux tubes', in *Solar Magnetic Fields*, ed. by M. Schüssler and W. Schmidt, (Cambridge University Press, 1994), pp. 101–103
126. F. Moreno-Insertis, M. Schüssler, and A. Ferriz-Mas: 'Storage of magnetic flux in the overshoot region', in *The Cosmic Dynamo*, ed. by F. Krause, K.-H. Rädler, and G. Rüdiger, (Kluwer Academic Publishers, 1993) pp. 41–44
127. M. Schüssler: 'Flux tubes and dynamos', in *The Cosmic Dynamo*, ed, by F. et Al. Krause, (IAU, 1993) pp. 27– 39
128. A. Brandenburg: 'Simulating the solar dynamo', In *The Cosmic Dynamo*, ed. by F. Krause, K.-H. Rädler, and G. Rüdiger, (Kluwer Akademic Publishers, 1993) pp. 111–121
129. G. Rüdiger and A. Brandenburg: Astron. Astrophys. **296**, 557 (1995)
130. P. Hoyng: Astron. Astrophys. **272**, 321 (1993)
131. P. Hoyng, D. Schmitt, and L. J. W. Teuben: Astron. Astrophys. **289**, 265 (1994)
132. G. Belvedere: 'Solar and stellar cycles'. In *Inside the Sun*, ed. by M. Cribier and G. Berthomieu, (Kluwer Academic Publishers, 1990) pp. 371–382
133. E. H. Levy: 'Physical assessment of stellar dynamo theory', in *Cool Stars, Stellar Systems, and the Sun*, ed. by M. S. Giampapa and J. A. Bookbinder, (ASP Conference Series 26, 1992) pp. 223–239,

134. N. O. Weiss: 'Dynamo processes in stars', in *Accretion Discs and Magnetic Fields in Astrophysics*, ed. by G. Belvedere, (Kluwer Academic Publishers, 1989), pp. 11–29
135. D. Elstner: 'The galactic dynamo - success and limitations of current models'. in *Plasma Turbulence and Energetic Particles in Astrophysics (Proceedings of the International Conference Cracow 1999)*, ed. by M. Ostrowski and R. Schlickeiser, (Colonel, Poland, 1999), pp. 74–83
136. R. Beck, A. Brandenburg, D. Moss, A. Shukurov, and D. Sokoloff: Ann. Rev. Astron. Astrophys.**34**, 155 (1996)
137. R. Wielebinski and F. Krause: The Astron. Astrophys. Rev. **4**, 449 (1993)
138. K. Ferrière: Astron. Astrophys. **310**, 438 (1996)
139. U. Ziegler: Astron. Astrophys. **313**, 448 (1996)
140. D. Schultz, M. Elstner and G. Rüdiger: Astron. Astrophys. (1993)
141. A. M. Soward: Astron. Nachr. **299**, 25 (1978)
142. D. Moss: Mon. Not. R. Astron. Soc. **297**, 860 (1998)
143. R. Rhode, R. Beck, and D. Elstner: Astron. Astrophys. **350**, 423 (1999)

Neutron Stars and Strong-Field Effects of General Relativity

Włodek Kluźniak

Copernicus Astronomical Center, ul. Bartycka 18, 00-716 Warszawa,Poland

Abstract. The basic observed properties of neutron stars are reviewed. I suggest that neutron stars in low-mass X-ray binaries are the best of all known sites for testing strong-field effects of general relativity.

1 Validity of General Relativity

General Relativity (GR) is the correct description of gravity and space-time. The phenomena verified with three classic tests of GR are so well established that they are now used as tools in every-day astronomical practice and even in technological applications.

The gravitational bending of light, famously detected in Eddington's solar eclipse expedition is today used to determine the stellar content of our Galaxy and the Magellanic Clouds (from stellar micro-lensing events detected by the OGLE, MACHO and EROS experiments). Lensing of distant galaxies by intervening galaxy clusters is used to determine the (dark) matter distribution in the latter.

Gravitational redshift, first observed in spectra of the white dwarf Sirius B in 1925, has since been detected in the laboratory (Pound-Rebka experiment) and is now of necessity taken into account in surveying practice (the GPS system). The effect is also essential in timing radio pulsars – when compared to some millisecond pulsars, terrestrial clocks clearly run slower at full moon than at new moon.

The magnitude of precession of the perihelion of Mercury is dwarfed by the same effect in the Hulse-Taylor pulsar, where the periastron shifts by $4.2°$ per year. A similar system, Wolszczan's binary pulsar, allows a confirmation of the Shapiro delay.

Of course, GR also provides the framework for understanding the evolution of our expanding Universe. All these successes allow us to confidently use general relativity, even in domains where its validity has not yet been strictly proven.

Observations of certain X-ray binaries (e.g., Cygnus X-1 and the so called X-ray novae), as well as of stellar motions in our Galaxy, and of velocities in the inner cores of other galaxies, strongly suggest the existence of black holes. However, the laws of GR have not yet been truly tested in the strong field regime.

1.1 Why Neutron Stars

The strength of gravity is conveniently parametrized by the mass to size ratio, $(M/R)(G/c^2)$. For black holes, of course, $GM/(Rc^2) \sim 1$, as for the Schwarzschild radius $R_{\rm Sch} = 2MG/c^2$. For the Sun, $GM_\odot/c^2 \approx 1.5\,{\rm km}$, while the solar radius $R_\odot \approx 300000$ km, which yields $M_\odot/R_\odot \sim 10^{-5}$ (in units of c^2/G). A similar value is obtained for mass/distance in the binary Hulse-Taylor pulsar, where relativistic effects in the orbital motion are so clearly detected (because the pulsar period is so short $\approx 0.06\,{\rm s}$, and known to 10 significant figures). For white dwarfs, $M/R \sim 10^{-3}$. But for neutron stars, $M/R \sim 10^{-1}$, and GR effects just outside their surface are about as important as near the black hole surface.

As a testbed for GR, neutron stars have one great advantage over black holes – they have a tangible surface which can support magnetic fields and can emit X-rays and other radiation. A great deal can be learned about neutron stars without assuming the validity of GR. Hence, a great deal can be learned about GR by observing neutron stars. Today, about 1000 radio pulsars are known and about 100 X-ray binaries containing neutron stars, so also in sheer numbers neutron stars have an advantage over black holes.

1.2 Basic References

The narrative presented in Sections 1 and 2, to a large extent relies on well established observations and theories, which have made their way into excellent textbooks, where detailed references can be found to the literature. Among those, particularly useful in the context of these lectures are the ones by Shapiro and Teukolsky [1], Lipunov [2], Mészáros [3], Glendenning [4], and Frank, King and Raine [5].

2 A Brief History of Neutron Stars

Before discussing in detail the properties of rapidly rotating, (at most) weakly magnetized, compact stars – which are ideal astrophysical objects for testing strong-field predictions of General Relativity – let us recount how they were identified.

2.1 Key Dates

The basic chronology of the discovery of neutron stars can be found, together with the references, e.g., in [1]. The following selection reflects my bias of what seems particularly important with the hindsight of today.

1914: Adams discovered that the rather dim, $L \approx 3 \times 10^{-3} L_\odot$, star Sirius B (orbiting Sirius), whose mass had been determined to be $M \approx 0.85 \pm 0.10 M_\odot$, has the spectrum of a "white" star – hence the name white dwarf. The unusual combination of low luminosity and high temperature implied a small radius,

$R \approx 2 \times 10^4$ km. This conclusion was based on an application of the black-body formula

$$L = 4\pi\sigma_B R^2 T^4. \tag{1}$$

1925: Adams measures the redshift, z, of certain lines in Sirius B. Applying general relativity, one can infer the value of M/R from z, and from the known mass a value of the stellar radius, $R \sim 10^4$ km. The agreement with the spectroscopically determined value was a great triumph of GR.

1926: The Fermi-Dirac statistic is discovered.

1926 (December): Fowler identifies the agent holding up white dwarfs against gravity – it is the degeneracy pressure of electrons.

1930: Chandrasekhar discovers theoretical models of white dwarfs, from which the maximum value for white dwarf mass follows, the famous $1.4M_\odot$. Incidentally, $M \sim 1M_\odot$ and $R \sim$ few $\times 10^3$ km imply a density $\rho \sim 10^6\,\mathrm{g/cm^3}$, which in turn implies a minimum period of possible rotation or vibration of a few seconds: $(G\rho)^{-1/2} \sim 3$ s.

1932: Chadwick discovers the neutron.

1932: Landau discusses cold, degenerate stars composed of neutrons.

1934: Baade and Zwicky write: "With all reserve we advance the view that supernovae represent the transition from ordinary star to neutron stars." This remains a remarkable contribution – two years after the discovery of neutrons, Baade and Zwicky correctly explain the mechanism of Supernovae (type II) explosions, find the correct value for the gravitational binding energy released in the creation of a neutron star, $\sim 10^{53}$ erg, and even identify a site where a neutron star is present (and was discovered 35 years later!): the Crab nebula.

1938: Landau discusses the energy released inside ordinary stars with neutron-star cores (a theoretical precursor of what is now known as a Thorne-Żytkow object). At the time, the energy source of the Sun was not known. The great contribution here is the pointing out of the enormous energy released in accretion onto neutron stars.

1939: Oppenheimer and Volkoff solve the relativistic equations of stellar structure for a fermi gas of neutrons, and thus construct the first detailed model of a neutron star. They find a maximum mass ($\approx 0.7M_\odot$, lower than the one for modern equations of state), above which the star is unstable to collapse. Thus the road to the theoretical discovery of black holes is paved.

1940's are lost to the Second World War.

1950's: The basic physics of the interior of neutron stars is worked out by the Soviet school, including a detailed understanding of the superfluid phase.

1962: Giacconi et al. discover the first extrasolar source of X-rays, Sco X-1.

1967: Shklovsky derives a model for Sco X-1, in which the X-ray source is an accreting neutron star in a binary system.

1967: Pacini points out that neutron stars should rotate with periods $P <<$ 1 s, and may have magnetic fields of surface value $B \sim 10^{12}$ G. The ensuing dipole radiation is not directly observable, as its frequency $2\pi/P$ is below the plasma frequency of interstellar space.

1967: Radio pulsars with $P \leq 3$ s discovered by Hewish, Bell et al.

1968: Gold gives the "lighthouse" model of radio pulsars.

1968: Spin-down of radio pulsars is measured, $\dot{P} > 0$. From this moment, it is clear that pulsars are rotating, compact objects, ultimately powered by the kinetic energy of their rotation.

1971: Giacconi et al. discover the first of accreting counterparts of radio pulsars, the X-ray pulsar Cen X-3, of period 4.84 s. Today, many are known, in the period range $0.7\,\mathrm{s} \leq P \leq 10000\,\mathrm{s}$.

1978: Trümper et al. discover the $\sim 40\,\mathrm{keV}$ cyclotron line in the spectrum of the accreting X-ray pulsar Her X-1. From the formula $h\nu = 1\,\mathrm{keV} \times (B/10^8\,\mathrm{G})$, the inferred value of the magnetic field at the stellar surface is $B_p = \mathrm{few} \times 10^{12}\,\mathrm{G}$, in agreement with the estimates of the dipole strength of ordinary radio pulsars.

1982: The discovery of millisecond pulsars by Backer, Kulkarni et al.

1996: The discovery of kHz quasi-periodic oscillations (QPOs) in the X-ray flux of low-mass X-ray binaries (LMXBs).

1998: The discovery of 2.5 ms pulsar in the transient LMXB SAX J 1808.4-3658 by Wijnands and van der Klis.

2.2 The Physics of Identifying Neutron Stars

It should be apparent from the above review, that the basic physics behind identifying neutron stars is fairly simple. Of course, the discovery was possible only after decades of sustained technological development, particularly in the field of radio and X-ray detectors, as well as much observational effort. Also, the existence of neutron stars would not have been so readily accepted without the solid theoretical foundations laid down over a period of many years. But the basic, incontrovertible, observational arguments are really based on two or three simple formulae.

Let us accept the theoretical result, that a neutron star is a body of mass $M \sim 1 M_\odot$ and radius $R \sim 10\,\mathrm{km}$, hence of mean density $\bar{\rho} > 10^{14}\,\mathrm{g/cm^3}$. How can we be certain that such bodies have been discovered?

a) The mass can be determined directly in some binary systems by methods of classical astronomy (as developed for spectroscopic binaries), essentially by an application of Kepler's laws. For the binary X-ray pulsars, the errors are rather large, but it is clear that one or two solar masses is the right value. For the binary radio pulsars (the Hulse-Taylor and Wolszczan pulsars), where the pulse phase can be determined very precisely and relativistic effects give much redundancy, the mass has been measured very accurately (to $0.01 M\odot$) and is close to $1.4 M_\odot$. For binary (millisecond) radio pulsars with white dwarf companions, the mass function is always consistent with these values.

b) In bright, steady, X-ray sources, and especially in X-ray bursters (where the X-ray flux briefly saturates at a certain peak value), one can assume that the radiative flux is limited, at the so called Eddington value, by a balance between radiation pressure on electrons and gravitational pull on protons. Since both forces are proportional to $(\mathrm{distance})^{-2}$, there is a direct relation between flux and mass. Again, $M \sim 1 M_\odot$ is obtained, for $L_X \approx 10^{38}\,\mathrm{erg/s}$.

c) The radius can be determined whenever a thermal spectrum is detected, by a combination of the black-body formula, (1), and of Wien's law giving the characteristic temperature of a body emitting the thermal spectrum. Thus, for X-ray pulsars, such as Her X-1, the spectrum gives a characteristic temperature of $T \sim 10\,\mathrm{keV} \sim 10^8\,\mathrm{K}$, which in combination with the luminosity $L_X \sim 10^{37}\,\mathrm{erg/s}$ gives an area of $\sim 10^{10}\,\mathrm{cm}^2$, consistent with the area of a "polar cap." This is the area through which open magnetic field lines pass for a $R \sim 10\,\mathrm{km}$ star, rotating at $P = 1.24$ s, with a $B \sim 10^{12}\,\mathrm{G}$ field.

For the non-pulsating bright X-ray source Sco X-1, $T \sim 1\,\mathrm{keV}$, $L \sim 10^{38}\,\mathrm{erg/s}$, i.e., $R \sim 10\,\mathrm{km}$ directly, as expected if the accreting material is spread over the whole surface.

d) For pulsars, an upper limit to the stellar radius follows from causality, $\omega R < c$, hence $R < cP/(2\pi)$. For millisecond pulsars, this gives $R < 100\,\mathrm{km}$.

e) The moment of inertia of certain pulsars (if they are powered by rotation) can be measured directly in "cosmic calorimeters." If the luminosity of the Crab nebula ($\approx 5 \times 10^{38}\,\mathrm{erg/s}$) is equated to $I\omega\dot{\omega}$, for the known period ($P = 33\,\mathrm{ms}$) and its derivative of the Crab pulsar (or the known age of the nebula), the value $I \approx 10^{45}\,\mathrm{g{\cdot}cm^2}$ is obtained. A similar, but less secure, argument can be given for the famous eclipsing pulsar PSR 1957+20 ($P = 1.6\,\mathrm{ms}$, $\dot{P} \approx 10^{-19}$). It is thought that the power needed to ablate the $0.02 M_\odot$ companion is $\sim 10^{38}\,\mathrm{erg/s}$ (assuming isotropic emission from the pulsar). Again, $I \sim 10^{45}\,\mathrm{g{\cdot}cm^2}$ is obtained.

f) Finally, a lower limit to the density can at once be derived for rotating objects from Newton's formula for keplerian orbital motion: $\omega_K = \sqrt{GM/R^3} = \sqrt{4\pi\bar{\rho}/3}$. Since $2\pi/P = \omega \leq \omega_K$, for any star rotating at a period P, the mean density satisfies $\bar{\rho} \geq 3\pi G^{-1} P^{-2}$. With the known value of Newton's constant, this gives directly $\bar{\rho} > 2 \times 10^{14}\,\mathrm{g/cm^3}$, for SAX J 1808.4-3658 ($P = 2.5\,\mathrm{ms}$) or the millisecond pulsars, such as PSR 1957+20 ($P = 1.6\,\mathrm{ms}$).

These basic results are subject to many consistency checks, which in all cases support the basic result that objects with a solid or fluid surface (i.e., they are not black holes!) have been identified of dimensions $M \sim 1 M_\odot$ and $R \sim 10\,\mathrm{km}$:

i) The gravitational energy released in accretion $L \sim GM\dot{M}/R$ is consistent (for the discussed values $M \sim M_\odot$ and $R \sim 10\,\mathrm{km}$) with the mass accretion rate inferred from theoretical studies of binary evolution.

ii) In some X-ray bursters, the photosphere clearly expands. Again, spectral fits for the temperature and for the radius of the photosphere (1), assuming Eddington luminosity, constrain the M–R relationship, in a manner consistent with the values discussed above.

iii) The surface magnetic field measured from the cyclotron line in X-ray pulsars agrees, to an order of magnitude ($B_p \sim 10^{12\pm1}\,\mathrm{G}$), with the one inferred for radio pulsars, by applying the notion that the spin down in the latter sources is obtained through balancing the energy loss in the simple dipole formula $\dot{E} = -2|\ddot{m}|^2/(3c^2)$, where $|m| = B_p R^3/2$, with the kinetic energy loss of a body of moment of inertia $I = 10^{45}\,\mathrm{g{\cdot}cm^2}$.

Incidentally, for millisecond pulsars, the value inferred from spin-down, $B_p \sim 10^{9\pm1}\,\mathrm{G}$, is consistent with the absence of polar cap accretion (and of associated pulsations) in X-ray bursters and other LMXBs. Thus, as far as the magnetic field is concerned, two or three classes of neutron stars are known – ordinary radio pulsars and accreting X-ray pulsars ($B \sim 10^{12\pm1}$ G), millisecond radio pulsars ($B \sim 10^{9\pm1}$ G), and low-mass X-ray binaries, where there is no evidence for such strong magnetic fields (i.e., $B < 10^9$ G).

iv) The observed long-term spin-up and spin-down of accreting X-ray pulsars is also consistent with a moment of inertia $I \sim 10^{45}\,\mathrm{g{\cdot}cm^2}$, for torques which are expected at the mass-accretion rates derived from the observed X-ray flux, assumed to be $L_X \sim GM\dot{M}/R \sim 0.1\dot{M}c^2$, and the assumption that the lever arm corresponds to an Alfvenic radius, obtained by balancing the ram pressure with the dipole magnetic pressure, i.e., $B^2/(8\pi) \sim \rho v_r^2$ at $r = r_A$, $\dot{M} = \epsilon 4\pi r^2 \rho v_r$, $B = B_p R^3/r^3$, where $\epsilon \sim 1$ is a geometric factor.

3 The Maximum Mass of Compact Stars

3.1 Neutron Stars or Quark Stars?

It is clear that radio pulsars and some accreting X-ray sources contain compact objects of properties closely resembling those known from theoretical models of neutron stars. Specifically, there can be no doubt that rotating stars of $M \sim M_\odot$ and $R \sim 10\,\mathrm{km}$ exist. However their internal constitution is not yet known. The expected mass and radius of "strange" (quark) stars is similar, the main difference being in that quark stars of small masses would have small radii – unlike neutron stars whose radius generally grows with decreasing mass – [6]. The observed "neutron stars" could be made up mostly of neutrons, but some of them could also be composed partly, or even mostly, of quark matter.

From the point of view of testing GR, the internal constitution of static (non-rotating) stars would matter little, as their external metric, directly accessible to observations, would be independent of their nature – the only parameter in the unique static, spherically symmetric, asymptotically flat solution (the Schwarzschild metric) is the gravitational mass, M, of the central body. However, for rapidly rotating stars, the metric does vary with properties of the body other than its mass, and it would be good to know the precise form of the equation of state (e.o.s.) of matter at supranuclear density.

As we have seen, at least some low-mass X-ray binaries (LMXBs) contain stellar remnants of extremely high density, exceeding $10^{14}\,\mathrm{g\,cm^{-3}}$, and many of them are not black holes because they exhibit X-ray bursts of the type thought to result from a thermonuclear flash on the surface of an ultra-compact star. Further, in these long-lived accreting systems the mass of the compact star is thought to have increased over time by several tenths of a solar mass above its initial value, and in the process the stars should have been spun up to short rotational periods. The compact objects in the persistent LMXBs are expected to be the most massive stellar remnants other than black holes, hence the most stringent limits on the e.o.s. of dense matter is expected to be derived from the mass of the X-ray sources in low-mass X-ray binaries. Before we discuss how this can be done, let us turn to the maximum mass.

3.2 The Maximum Mass of Neutron Stars

One quantity that depends sensitively on the e.o.s. is the maximum mass of a fluid configuration in hydrostatic equilibrium. For neutron stars this maximum mass, and in general the mass-radius relationship, is known from integrating the TOV equations for a wide variety of e.o.s. [7]. The mass of rotating configurations is also known [8]. Here, I will only briefly review the basic physics behind the existence of the maximum mass and then give an example for strange stars, where the e.o.s. is so simple that the variation of mass with the parameter describing the interactions can be determined analytically.

As we know from the work of Chandrasekhar and others, the maximum mass is reached when the adiabatic index reaches a sufficiently low value that the star becomes unstable to collapse. In the Newtonian case, this critical index is 4/3, corresponding to the extreme relativistic limit for fermions supplying the degeneracy pressure, when the formula for kinetic energy of a particle $E = \sqrt{p^2c^2 + m_f^2c^4}$ reduces to $E = pc$.

The very simple argument explaining the instability, due to Landau, goes like this. There is a balance between the increasingly negative gravitational binding energy when a massive sphere of fermions is compressed, and the increasing kinetic energy of each fermion as it is squeezed into an increasingly confined volume – each fermion likes to live in phase space of volume $\sim \hbar$. Of course, as the star is compressed when its mass is increased, the fermion momenta increase and the extreme relativistic regime is approached, with a corresponding softening of the adiabatic index. The total energy of N particles in a star of mass M bound by gravity, is up to factors of order unity, $E_{\text{tot}} = -GM^2/R + N\bar{E}$, where $\bar{E}$ is the mean kinetic energy of the particles. If the particles are fermions, of mass m_f, their momentum following from the uncertainty principle is $p = \hbar(N/V)^{1/3}$, and we can take $V = R^3$ for the volume of the star. In the non-relativistic case, $E = p^2/(2m_f)$ so $N\bar{E} = \hbar^2 N^{5/3}/(2m_f R^2)$, and a stable configuration can be found by minimizing E_{tot} with respect to R. But in the extreme relativistic case, $E = pc$, $N\bar{E} = \hbar c N^{4/3}/R$, and both terms in E_{tot} are now proportional to $1/R$, so no minimum energy configuration is found.

In reality, to find the maximum mass configuration, one has to solve the TOV equations using a plausible e.o.s. The TOV equations have essentially the same scaling properties as the familiar equations of Newtonian hydrostatic equilibrium

$$\frac{\mathrm{d}P}{\mathrm{d}r} = -\frac{Gm\rho}{r^2},$$

$$\frac{\mathrm{d}m}{\mathrm{d}r} = 4\pi r^2 \rho,$$

i.e., if the pressure and density scale with some fiducial density, $P \propto \rho \propto \rho_0$, then $m \propto r \propto \rho_0^{-1/2}$. Such scalings allow some general statements to be made about the maximum mass, such as the Rhoads-Ruffini limit: $M < 3M_\odot$, if $\rho \geq \rho_0 > 2 \times 10^{14}\,\text{g/cm}^3$.

3.3 Quark Stars

Conversion of some up and down quarks into strange quarks is energetically favorable in bulk quark matter (because the Fermi energy is so high) and it has been suggested that at large atomic number, matter in its ground state is in the form of "collapsed nuclei" with strangeness about equal to the baryon number [9]. On this assumption, Witten [10] discussed the possible transformation of neutron stars to stars made up of matter composed of up, down, and strange quarks in equal proportions, and found the maximum mass of such quark stars as a function of the density of (self-bound) quark matter at zero pressure is $\rho_0 \geq 4 \times 10^{14}\,\mathrm{g/cm^3}$. Detailed models of these "strange" stars have been constructed [6,11]. Here, I discuss only the maximum mass of such stars.

Following Alcock [12], take a gas of any relativistic particles – the e.o.s. is $P_g = \rho_g c^2/3$. If these are moving in a background of vacuum with uniform energy density $\rho_v c^2 = B$, i.e., negative pressure $p_v = -B$, then the e.o.s. connecting the total pressure $p = p_g + p_v$, with the total density $\rho = \rho_g + \rho_v$, is

$$p = (\rho - \rho_0)c^2/3, \tag{2}$$

with $\rho_0 c^2 = 4B$. Witten [10] showed that for this simple e.o.s. the maximum mass from the TOV equation is $M = 2M_\odot\sqrt{\rho_1/\rho_0}$, with $\rho_1 \equiv 4.2 \times 10^{14}\,\mathrm{g/cm^3}$. The scaling $\rho_0^{-1/2}$ is discussed in the previous subsection.

The physical interpretation of the result is that the relativistic particles are in fact quarks, and the "bag constant" B, is a device invented at MIT to simulate their confinement. The e.o.s. $p = (\rho - \rho_0)c^2/3$, then, describes interacting quarks in an approximation to quantum chromodynamics (QCD) known as the MIT bag model [13]. Thus, the maximum mass found is the maximum mass of static strange (quark) stars. However, it still depends on the free parameter ρ_0.

3.4 The Maximum Mass of Strange Stars

To illustrate the utility of the scaling law, I will now discuss the maximum mass of a strange star. First, as already noted by Oppenheimer and Volkoff [14], the stellar mass decreases with the fermion mass, so to find the maximum mass of a quark star it is enough to consider massless quarks. In view of the scaling of TOV equations, the question reduces to that of finding ρ_0, the density of strange matter at zero pressure. In short, the maximum mass of a strange star in the model considered is $M_{\max} = 1.98 M_\odot \times (59.8\mathrm{MeV}/B)^{1/2}$, and the least upper bound to the mass of the strange star is given by the same formula, with $B = B_{min}$, the lowest possible value of the bag constant. Realistically, the actual maximum mass of a (non-rotating) strange star will be smaller by about 10% because, because in fact, $m_s > 0$.

Currently, the actual value of B cannot be reliably derived from fits to hadronic masses of the quark-model of nucleons. Its lowest possible value can be found by requiring that neutrons do not combine to form plasma of deconfined up and down quarks, or equivalently, that quark matter composed of up and down quarks in 1:2 ratio is unstable to emission of neutrons through the reaction $u + 2d \to n$. This implies that the baryonic chemical potential at zero pressure of such quark matter satisfies [15]

$$\mu_{u,d}(0) > 939.57\,\mathrm{MeV}. \tag{3}$$

As we neglect the masses of up and down quarks in our considerations, the baryonic chemical potential at pressure P is given by the expression [16]

$$\mu(P) = (P + \rho c^2)/n = 4(A/3)^{3/4}(P + B)^{1/4}, \tag{4}$$

where n is the baryon number density, and $\rho c^2 = An^{4/3} + B$ is the energy density. For matter (not in beta equilibrium) composed of deconfined up and down quarks in 1:2 ratio, $n = n_u = n_d/2$ and hence $A = (1 + 2^{4/3})(3\hbar c/4)\pi^{2/3}C^{-1/3}$, i.e., $\mu(0) \propto (B/C)^{1/4}$, where $C \equiv 1 - 2\alpha_c/\pi$ and α_c is the QCD coupling constant. Inequality (1) then becomes

$$\frac{B}{C} > 58.9\,\mathrm{MeV\,fm}^{-3} \equiv B_1. \tag{5}$$

Thus, $B_{min} = (1 - 2\alpha_c/\pi)B_1$, through lowest order in quark-gluon coupling. So, for massless interacting quarks, the energy density at zero pressure is $\rho_0 c^2 = 4B \geq (1 - 2\alpha_c/\pi)\rho_1 c^2$. For massive quarks the expression for minimum density becomes more complicated, but we will not need it to determine the upper bound to the mass of a static strange star in the MIT bag model – it is enough to consider the e.o.s. of an ultrarelativistic Fermi gas in a volume with vacuum energy density $B > 0$.

For strange matter in beta equilibrium the number densities of the (massless for now) up, down, and strange quarks are equal, $n_u = n_d = n_s$, and the energy density is $\rho c^2 = A_s n^{4/3} + B_1$, with $A_s = 9\hbar c\pi^{2/3}C^{-1/3}/4$, as is appropriate for three colors per flavor. This gives an equation of state identical to that of non-interacting quarks, (2), the only difference being in that the lower bound on the density at zero pressure, following from conditions of neutron stability (3,5), is decreased by the factor C with respect to the value for an ideal Fermi gas in a bag:

$$\rho_0(\alpha_c) = \left(1 - \frac{2\alpha_c}{\pi}\right)\rho_0(0).$$

Thus, through lowest order in the QCD interaction, the fiducial density is changed, but not the e.o.s. Since the stellar mass scales as $\rho_0^{-1/2}$, this implies that the least upper bound on the mass of the star as a function of the QCD coupling constant is given for non-rotating strange stars by

$$M_{\max}(\alpha_c) = \left(1 - \frac{2\alpha_c}{\pi}\right)^{-1/2} M_{\max}(0) \tag{6}$$

through first order in α_c. For $\alpha_c = 0.6$ this gives a maximum strange star mass of $2.54M_\odot$, higher by 27% than the maximum mass which is obtained for $\alpha = 0$.

4 Measuring the Mass of Accreting Neutron (or Strange) Stars

Finally, we have to confront the question how the mass of the compact objects in LMXBs may be determined. Hopefully, a mass will be measured which will eliminate a class o equations of state of dense matter. Unfortunately, application to X-ray bursters of standard methods for determining the mass function of the binary – and hence constraining the mass of the compact X-ray source – is exceedingly difficult, as the optical emission is usually dominated by that of the accretion disk (see, e.g., [17]), However, reliable mass values obtained by this method may soon become available, particularly for transient sources, such as the accreting millisecond pulsar SAX J1808.4-3658.

The mass of the compact object in an X-ray binary may also be determined by studying the time variability of the radiation flux formed in the accretion flow. Specifically, for sufficiently weakly magnetized stars, a maximum frequency is expected corresponding to the presence of the innermost (marginally) stable circular orbit allowed in general relativity [18]. It has been reported that such a maximum frequency may have been observed, at least in one system where quasi periodic oscillations (QPOs) in the X-ray flux saturate at a particular value [19]. In this manner, several e.o.s. were excluded [20] on the understanding that the maximum observed kHz QPO frequency implies a mass in excess of $2M_\odot$ (see also [21]). Similar considerations [22] exclude static (or slowly rotating) quark stars if the minimum density of quark matter is $\rho_0 > 4.2 \times 10^{14}\,\mathrm{g/cm^3}$, and the quark matter is taken to be described by the MIT bag model.

The overall conclusion [20] is that neutron-star matter may be composed simply of neutrons with some protons, electrons and muons, as models of more exotic neutron-star matter (including hyperons or pion and kaon condensates) do not agree with the simplest interpretation of the kHz QPO data, namely that the maximum frequency observed in the low-mass X-ray binary 4U 1820-30, i.e., 1066 Hz [19], is attained in the marginally stable orbit around a neutron star. If the compact stellar remnants in these systems are slowly rotating, the same conclusion would apply to ultra-dense matter in general, at densities greater than $4.2 \times 10^{14}\,\mathrm{g/cm^3}$, as matter composed of massless quarks would also be excluded for such densities [22]. However, as we have seen, minimum densities smaller than $4.2 \times 10^{14}\,\mathrm{g/cm^3}$ seem possible for more realistic models of self-bound quark matter, and this would change the conclusion.

For rapidly rotating strange stars the conclusion may be drastically different, as the metric is greatly modified by a pronounced flattening of the star (this effect is less important for neutron stars). In general, the marginally stable orbit is pushed out by this effect, and a fairly low orbital frequency can be obtained for a low mass star. This is illustrated in Fig. 1 (taken from [23]) which exhibits the frequency in the innermost (marginally) stable circular orbit of general relativity (ISCO) as a function of stellar mass, M, for the Schwarzschild metric [the hyperbola $f_+ = 2.2\,\mathrm{kHz}(M_\odot/M)$], as well as the ISCO frequency for strange stars rotating at Keplerian frequencies (i.e., maximally rotating, at the equatorial mass-shedding limit), for various values of the density at zero pressure, ρ_0 of

(2). It turns out that for these maximally rotating models, the ISCO is always at 1.7 to 1.8 km above the stellar surface, the increase of the ISCO orbital frequency for these models can then be understood in terms of Kepler's law: $2\pi f \sim \sqrt{G\bar{\rho}}$, where $\bar{\rho}$ is the mean density of matter inside the orbit.

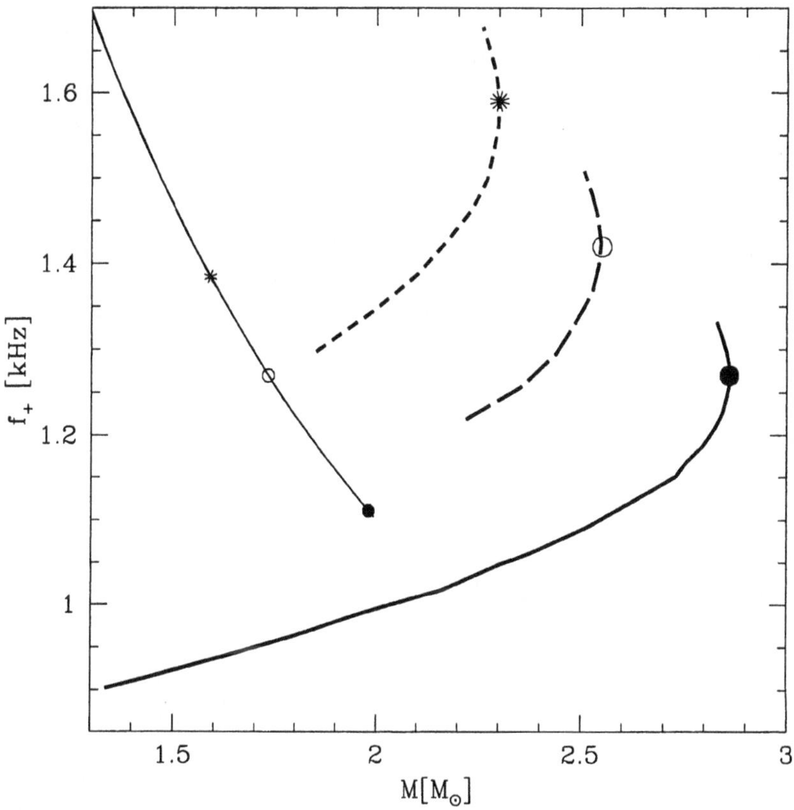

Fig. 1. The frequency of the co-rotating innermost stable circular orbit as a function of mass for static models (thin, continuous line) and for strange stars rotating at the equatorial mass-shedding limit (thick lines, in the style of Fig. 1). For the static models, this frequency is given by the keplerian value at $r = 6GM/c^2$, i.e., by $f_+ = 2198\,\mathrm{Hz}(M_\odot/M)$, and the minimum ISCO frequency corresponds to the maximum mass, denoted by a filled circle, an empty circle, and a star, respectively for $\rho_0/(10^{14}\,\mathrm{g\,cm^{-3}}) = 4.2$, 5.3, and 6.5. Note that the ISCO frequencies for rapidly rotating strange stars can have much lower values, and $f_+ < 1\,\mathrm{kHz}$ can be achieved for strange stars of fairly modest mass, e.g. $1.4M_\odot$, if the star rotates close to the equatorial mass-shedding limit. This figure is from [23]

5 Testing Strong-Field General Relativity with Accreting Neutron Stars

There are really two types of objects where strong-field effects of general relativity are crucial: black holes and accreting neutron (or quark) stars. Black holes are both attractive and difficult in this context – on the one hand, their very existence would be impossible in many other theories of gravity, on the other, their existence is a hypothesis which must experimentally verified. Possibly this will eventually be achieved by careful observations of motions in the inner accretion disk in AGNs and/or black hole binaries.

The existence of neutron stars (or quark stars) would be perfectly possible in Newtonian gravity (although their detailed properties would be different from those expected in a general-relativistic world). But from the point of view of determining the metric, they have the great advantage, that not only their mass can be measured (as for binary black holes), but also, at least in some cases, other basic parameters such as the rotational period and the radius can be determined directly. Hopefully, this would allow relativistic effects in the accretion flow to be unambiguously resolved.

One class of phenomena which may be helpful in pinning down the external metric of accreting sources is the relativistic trapping of vibrational modes in the inner accretion disk. Indeed, it has been suggested that the 67 Hz oscillation seen in the source GRS 1915+105 has this origin, and is a signature of the Kerr metric [24]. This is perhaps the most convincing relativistic effect discovered to date in accreting sources. Unfortunately, the mass of GRS 1915+105 is not known, and there is no independent knowledge of its angular momentum (the source is a black hole candidate [25]).

Another promising avenue is the search for the marginally stable orbit (ISCO), expected to exist in accreting neutron stars [26] and to show up as a maximum frequency in the X-ray spectra of LMXBs [18]. Indeed, the recently discovered kHz QPOs in X-ray bursters and other probable neutron star systems do show some features which are consistent with their observed frequency being the Keplerian frequency in an accretion disk terminating close to the marginally stable orbit [21,19,20]. But with the data gathered to date, it seems easier to constrain the e.o.s. of dense matter, on the assumption that the QPO frequency saturates in the ISCO, than to show that this assumption is indeed correct. One difficulty is that the physics of accretion disks is still very poorly understood.

New data is being gathered daily and new experiments are planned which may lead to a break-through in this field.

Acknowledgements

I thank the organizers of this School for their wonderful hospitality in Guanajuato and the United States-Mexico Foundation for Sciences for financial support.

References

1. S.L. Shapiro, S.A. Teukolsky: *Black Holes, White Dwarfs, and Neutron Stars* (Wiley, New York 1983)
2. W.M. Lipunov: *Astrophysics of neutron stars* (Springer, Berlin 1992)
3. P. Mészáros: *High-Energy Radiation from Magnetized Neutron Stars* (Chicago University Press, Chicago 1992)
4. N. Glendenning: *Compact stars* (Springer, Berlin 1997)
5. J. Frank, A.R. King, D.J. Raine: *Accretion Power in Astrophysics* (Cambridge University Press, Cambridge 1985)
6. C. Alcock, E. Farhi, A. Olinto: Astrophys. J. **310**, 261, (1986)
7. W.D. Arnett, R.L. Bowers: Astrophys. J. Suppl. **33**, 415, (1977)
8. G.B. Cook, S.L. Shapiro, S.A. Teukolsky: Astrophys. J. **424**, 823, (1994)
9. A.R. Bodmer: Phys. Rev. **4**, 1601, (1971)
10. E. Witten: Phys. Rev. **30**, 272, (1984)
11. P. Haensel, J.L. Zdunik, R. Schaefer: Astron. Astrophys. **160**, 121, (1986)
12. C. Alcock: Nucl. Phys. B (Proc. Suppl.) **24B**, 93, (1991)
13. E. Farhi, R.L. Jaffe: Phys. Rev D **30**, 2379, (1984)
14. J.R. Oppenheimer, G.M. Volkoff: Phys. Rev. **55**, 374, (1939)
15. P. Haensel: Acta Phys. Pol. **B18**, 739, (1987)
16. G. Chapline, M. Nauenberg: Nature **264**, 235, (1976)
17. J. van Paradijs, E.P.J. van den Heuvel, E. Kuulkers: in *Compact stars in binaries: IAU Symposium 165* (Kluwer: Dordrecht 1996)
18. W. Kluźniak, P. Michelson, R.V. Wagoner: Astrophys. J. **358**, 538 (1990)
19. W. Zhang et al.: Astrophys. J. Lett. **482**, L167, (1998)
20. W. Kluźniak: Astrophys. J. Lett. **509**, L37 (1998)
21. P. Kaaret et al.: Astrophys. J. Lett. **480**, L27, (1997)
22. T. Bulik, D. Gondek-Rosińska, W. Kluźniak: Astron. Astrophys. **344**, L71, (1999)
23. N. Stergioulas, W. Kluźniak, T. Bulik: Astron. Astropys. **352**, L116, (1999)
24. M. Nowak et al.: Astrophys. J. Lett. **477**, L91, (1997)
25. L. F. Rodríguez: 'Observational Astronomy: the search for black-holes'. In *Nuclear and Particle Astrophysics*, ed. by J. G. Hirsch and D. Page (Cambridge University Press, Cambridge 1998), pp. 1 – 26
26. W. Kluźniak, R.V. Wagoner: Astrophys. J. **297**, 548 (1985)

Gamma Ray Astronomy at High Energies

Trevor C. Weekes

Harvard-Smithsonian Center for Astrophysics, Whipple Observatory, P.O. Box 97, Amado AZ 85645-0097, USA

Abstract. The recently developed field of high energy γ-ray astronomy (above 30 MeV) is reviewed in terms of the techniques used, the observations reported and future prospects for the field. Galactic and extragalactic sources have been detected up to energies of 50 TeV. More than half the sources detected by EGRET on the Compton Gamma Ray Observatory are unidentified. The best studied sources are the blazar class of AGN in which time variations as short as 15 minutes are seen. The next decade will see a new generation of detectors both in space (GLAST) and on the ground (e.g. VERITAS) with the promise of major advances.

1 Why High Energy Gamma Ray Astronomy?

Our universe is dominated by objects emitting radiation via thermal processes. The blackbody spectrum dominates, be it from the microwave background, the sun or the accretion disks around neutron stars. This is the *ordinary* universe, in the sense that anything on an astronomical scale can be considered ordinary. It is tempting to think of the thermal universe as *THE UNIVERSE* and certainly it accounts for much of what we see. However to ignore the largely unseen, non-thermal, *relativistic*, universe is to miss a major component and one that is of particular interest to the physicist, particularly the particle physicist. The relativistic universe is pervasive but largely unnoticed and involves physical processes that are difficult to emulate in terrestrial laboratories.

The most obvious local manifestation of this relativistic universe is the cosmic radiation, whose origin, 88 years after its discovery, is still largely a mystery (although it is generally accepted, *but not proven*, that much of it arises in shock waves from galactic supernova explosions). The existence of a steady rain of particles, whose power law spectrum attests to their non-thermal origin and whose highest energies extend far beyond that achievable in man-made particle accelerators, attests to the strength and reach of the forces that power this, strange, relativistic radiation. If thermal processes dominate the "ordinary" universe, then truly relativistic processes illuminate the "extraordinary" universe and must be studied, not just for their contribution to the universe as a whole but as the indicators of unique cosmic laboratories where physics is demonstrated under conditions to which we can only extrapolate.

Observations of the extraordinary universe are difficult, not least because it is masked by the dominant thermal foreground. In places, we can see it directly such as in the relativistic jets emerging from AGNs but, even there, we must subtract the foreground of thermal radiation from the host elliptical galaxy. The observation of polarization leads us to identify the processes that emit the radio, optical and X-ray radiation as synchrotron emission from relativistic particles, probably electrons, moving in weak electric fields but polarization is not unique to synchrotron radiation and the interpretation is not always unambiguous. The hard, power-law, spectrum of many of the non-thermal emission processes immediately suggests the use of the highest radiation detectors to probe such processes. Hence hard X-ray and γ-ray astronomical techniques must be the observational disciplines of choice for the exploration of the relativistic universe. Because the earth's atmosphere has the equivalent thickness of a meter of lead for this radiation, the exploitation of this form of astronomy had to await the development of space platforms for X-ray and γ-ray telescopes and the development of new techniques in ground-based γ-ray astronomy.

Although the primary purpose of the astronomy of hard photons (here defined as those above 30 MeV) is the search for new sources, be they point-like, extended or diffuse, it opens the door to the investigation of more obscure phenomenon in high energy astrophysics and even in cosmology and particle physics. Astronomy at energies up to 10 GeV has made dramatic progress since the launch of the Compton Gamma Ray Observatory in 1991 and the development of the atmospheric Cherenkov imaging technique.

2 Gamma Ray Detection Techniques

Laboratory γ-ray detectors were far advanced when the concept of "γ-ray astronomy" was first raised in Phillip Morrison's seminal paper in 1958 [70]. Indeed it was the expected ease of detection and the early promise of strong sources that led to the large concentration of effort in the field, even before the development of X-ray astronomy. Today the number of known γ-ray sources is well under a few hundred whereas there are hundreds of thousands of cataloged X-ray sources. What went wrong? The answer is simple: the detection of cosmic γ-rays was not as easy as expected and the early predictions of fluxes were hopelessly optimistic.

The term "γ-ray" is a generic one and is used to describe photons of energy from 100 keV (10^5 eV) to > 100 EeV (10^{20} eV). A range of fifteen decades is more than all the rest of the known electromagnetic spectrum. A wide variety of detection techniques is therefore necessary to cover this huge range. We will concentrate on the telescopes in the somewhat restricted range from 30 MeV to 100 TeV. There are no credible detections of γ-rays at energies much beyond 50 TeV and the "γ-ray telescope" techniques used beyond these energies are really the same as those used to study charged cosmic rays and will not be discussed here. There are some seven decades which are defined, somewhat arbitrarily, as: the High Energy (HE) range from 30 MeV to 100 GeV and the Very High Energy (VHE) range from 100 GeV to 100 TeV. These ranges are not defined by the physics of their production but by the interaction phenomena and techniques employed in their detection. The HE and VHE ranges use the pair production interaction but in very different ways; HE telescopes identify the electron pair in balloon or satellite-borne detectors, whereas VHE detectors detect the electromagnetic cascade that develops in the earth's atmosphere.

Gamma-ray astronomy is still an observation-dominated discipline and the observations have been driven not so much by the astrophysical expectations (which have often been wrong) as the experimental techniques, which have permitted significant advances to be made in particular energy ranges [34]. Hence the most fruitful observations have come at energies of 100 MeV; these were originally inspired by the prediction of the strong bump in the spectra expected from the decay of π^0's that are created in hadron interactions The energy region was exploited primarily because the detection techniques were simpler and more sensitive. In contrast the Medium Energy region (1–30 MeV) has the potential for very interesting astrophysics with the predicted existence of nuclear emission lines but the development of the field has been slow because the techniques are so difficult.

2.1 Peculiarities of Gamma-Ray Telescopes

There are several peculiarities that uniquely pertain to astronomy in the γ-ray energy regime. These factors make γ-ray astronomy particularly difficult and have resulted in the slow development of the discipline.

Above a few MeV there is no efficient way of reflecting γ- rays and hence the dimensions of the γ-ray detector are effectively the dimensions of the γ-ray telescope. This is only the case when the efficiency for γ-ray detection and identification is high; in practice to discriminate against the charged particle background the efficiency is much lower. Hence at any energy the effective aperture of a γ-ray telescope is seldom greater than 1 m^2 and often only a few cm^2, even though the physical size is much larger. For instance the Compton Gamma Ray Observatory was one of the largest and heaviest scientific satellites ever launched; however its HE telescope had an effective aperture of less than 1,600 m^2. Beam concentration is particularly important when the background scales with detector area. This is the case with γ-ray detectors which must operate in an environment dominated by charged cosmic rays.

The problem of a small aperture is compounded by the fact that the flux of cosmic γ-rays is always small. At energies of 100 MeV the strongest source (the Vela pulsar) gives a flux of only one photon per minute. With weak sources long exposures are necessary and one is still dealing with the statistics of small numbers. Small wonder that γ-ray astronomers have frequently been pioneers in the development of statistical methods and that γ-ray conferences are often dominated by arguments over real statistical significance!

As it is to photons in many bands of the electromagnetic spectrum the earth's atmosphere is opaque to all γ-rays. The radiation length is 38 g cm^{-2} and the total thickness is 1030 g cm^{-2}. Even the highest mountain is many radiation lengths below the top of the atmosphere so that it is virtually impossible to consider the direct detection of cosmic γ-rays without the use of a space platform. However the charged cosmic rays constitute a significant background and limit the sensitivity of such measurements. Large balloons can carry the bulky detectors to near the top of the atmosphere and much of the pioneering work in the field was done in this way.

The background can take many forms. In deep space it is the primary cosmic radiation itself, mostly protons, heavier nuclei and electrons. This background can be accentuated by secondary interactions in the spacecraft itself. Careful design and shielding can reduce this effect, as can active anti-coincidence shields. In balloons the secondary cosmic radiation from the cosmic ray interactions above the detector seriously limit the sensitivity and were the initial reason for the slow development of the field. Huge balloons that carry the telescopes to within a few grams of residual atmosphere are a partial solution, but it is still impossible to trust the measurement of absolute diffuse fluxes.

2.2 Pair Production Telescopes

The spark chamber, long obsolete for high energy physics experiments, has been the workhorse detector for γ-ray astronomy in the energy range 30 MeV to 10 GeV from the early sixties through the end of the century. The three experiments, which provided almost all the results during this period, all used the spark chamber as their principal detector. These were the USA's SAS-II (1972–3), Europe's COS-B (1975–1982) and the joint European- USA EGRET on the Compton Gamma Ray Observatory (1991–).

A pair production spark chamber telescope consists of four distinct components as shown schematically in Figure 1:

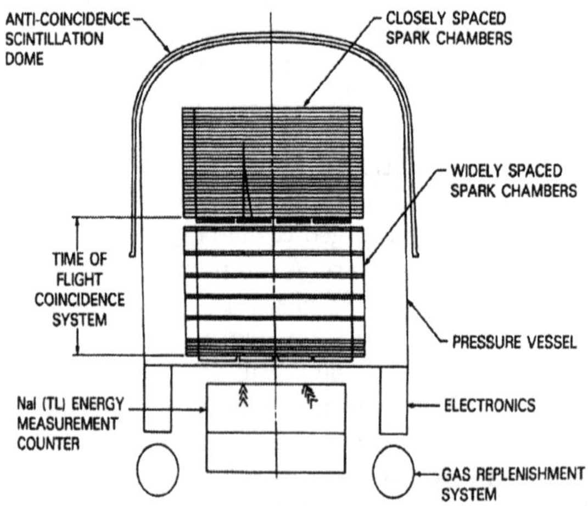

Fig. 1. Example of a spark chamber telescope: EGRET. The telescope is sensitive from 30 MeV to 30 GeV. The field of view is ±20° and the energy resolution is about 20%

(i) The spark chamber consists of a series of parallel metal plates in a closed container; the alternate plates are connected together electrically with one set permanently connected to ground. Upon an indication that a charged particle has passed through the chamber, a high voltage is applied to the second set of plates. The chamber contains a gas at a pressure such that the ionization left behind by the passage of the charged particle permits an electric spark discharge between the plates. The gas is generally a mixture of neon and argon. An electron pair created by a γ-ray interaction in one of the plates is then readily apparent as a pair of sets of sparks that delineate the path of the electron and positron. In practice the tracks are disjointed as the electrons suffer multiple scattering within the plates of the chamber. This limits the thickness of the plates (which should be as thick as possible to ensure that the γ rays interact effectively), but not so thick that the electrons undergo excessive Coulomb scattering in the plate material. Multiple plates ensure that the tracks are effectively mapped. The collection area and angular resolution of the telescope is determined by the spark chamber geometry. In some versions of the spark chamber the plates are replaced by grids of wires, 1 mm apart, which can record the position of the spark to this accuracy; each wire is threaded through a magnetic core memory, which is read out and reset after each event.

(ii) At least one electron must emerge from the spark chamber to ensure that it initiates a trigger that causes the application of the high voltage pulse to the second set of plates to activate the spark chamber. A permanent high voltage difference cannot be maintained between the plates, as the spark discharges will take place spontaneously. This trigger usually consists of an arrangement of scintillation counters and/or a Cherenkov detector so designed as to respond only to downward-going charged particles. It is the need for this trigger which limits the lower energy threshold of the spark chamber telescope. The trigger detection system effectively defines the field of view of the telescope.

(iii) The electrons must be completely absorbed if their energy is to be measured; to achieve this there must be a calorimeter that is some radiation lengths thick. This is generally a NaI(Tl) crystal, whose sole function is to measure the total energy deposited. At the low end of the sensitivity range the energy of the electrons can also be determined by the amount of Coulomb scattering in the plates of the spark chamber.

(iv) Finally the entire assembly must be surrounded by an anti-coincidence detector which signals the arrival of a charged particle, but which has a small interaction cross-section for γ rays. This usually consists of a very thin outer shell of plastic scintillator viewed by photomultipliers.

Although the basic principles of the HE pair production telescope are simple, the detailed design is complex and accounts for the fact that the effective collection area is far smaller than the geometrical cross-section of the telescope. This is illustrated by EGRET, the pair production telescope on the Compton Gamma Ray Observatory (CGRO).

EGRET is the largest and most sensitive high energy γ-ray telescope flown to date; it is the flagship instrument on CGRO. Approximately the size of a compact car and with a total weight of 1,900 kg, the telescope has an effective collection area of 1,600 cm^2 (Figure 1). The basic spark chamber consists of 28 wire grids interleaved with plates of 0.02 radiation length thickness. The wires in the grids each have a magnetic core whose readout indicates the proximity of the spark. The spark chamber is triggered by a coincidence between two thin sheets of plastic scintillator with a 60 cm separation (sufficient to recognize and reject upward going charged particles). The electron energy is measured by a NaI(Tl) calorimeter at the base of the telescope. As usual the entire assembly is surrounded by a thin anti-coincidence shield.

The telescope was designed for a two year lifetime. The neon/ethane gas which fills the chamber gradually gets poisoned and must be replenished. It was anticipated that a filling would last six months. Hence only four gas canisters were attached to the instrument for replenishment at yearly intervals. In practice the unprecedented and unexpected success of the mission has meant that even with extending the replenishment intervals, the EGRET instrument is effectively dead after nine years of useful operation.

2.3 VHE Telescopes

Atmospheric Cherenkov Telescopes: When a high energy γ-ray strikes the upper atmosphere, it produces an electron pair (as it does in a spark chamber). However if the energy of the γ ray, and hence of the electron pair, is large enough, an electromagnetic cascade will result which will continue down through the atmosphere with secondary γ-ray and electron production by bremsstrahlung and pair production [109,77]. The cascade will continue along the axis of the trajectory of the original γ ray and the total energy of the secondary particles will be a good representation of its energy.

For γ rays of energy 100 TeV and above, sufficient particles can reach ground level for the shower to be detected by arrays of particle detectors spread over areas of 0.1 km^2. As the secondary particles all move at nearly the speed of light and retain the original trajectory of the primary γ ray, the shower front arrives as a disk which is only a few meters thick. Differential timing between the detectors can then determine the arrival direction and hence the source of the γ radiation.

At lower energies the cascade will die out as the average energy of the secondary particles drops to the point that ionization losses become the major loss process (Figure 2). For a primary γ ray of energy 1 TeV, few secondary particles will reach even mountain altitude. However, as the relativistic particles traverse the atmosphere, they excite the atmosphere to radiate Cherenkov light with high efficiency. Although the fraction of energy that goes into this mode is small, it provides a very easy way to detect the cascade and thence the γ ray. A simple light detector (mirror, plus phototube, plus fast pulse counting electronics) provides an easy way of detecting the cascade. Early telescopes consisted of ex-World War II searchlight mirrors with phototubes at their foci, coupled to fast pulse counting electronics.

The observations are best made from a dark mountain top observatory. Since the Cherenkov angle in air is about 1° and the amount of light is proportional to

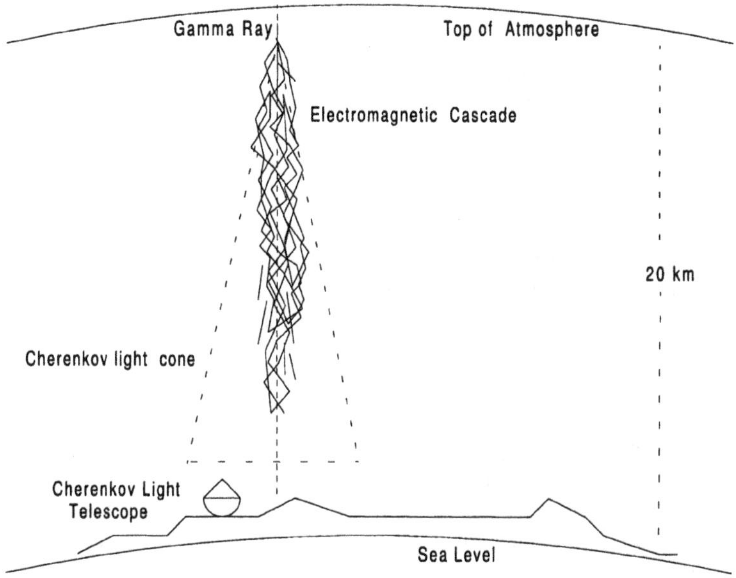

Fig. 2. Schematic of atmospheric air shower detection

the number of particles in the cascade (and hence to the energy of the γ ray), the measurement of the atmospheric Cherenkov component provides a good measure of the energy and arrival direction of the γ ray. Because the light spreads out as it traverses the atmosphere, the collection area for γ-ray detection is as large as the lateral dimensions of the light pool at detector altitude; this can be as much as 50,000 m^2!

This is one of the few astronomical techniques in which the earth's atmosphere plays an essential positive role. However the technique has its drawbacks. Although the atmosphere comes cheap (and the gas does not need to be replenished!), the observer has no control over it; the telescope is wide open to the elements and the detector is susceptible to a troublesome background of light from sun, moon and stars, from the airglow, from lightning and meteors, and from a variety of man-made light sources, from satellites and airplanes to airport beacons and city lights. These limit the sensitivity for γ-ray source detection. However the most troublesome background is that from air showers generated by charged cosmic rays of similar energy to the γ-rays under study. These are thousands of times more numerous and the light flashes are superficially similar to those from γ rays. Because of interstellar magnetic fields, the arrival directions of the charged cosmic rays are isotropic; hence a discrete source of γ rays can stand out only as an anisotropy in an otherwise isotropic distribution of air showers. Unfortunately a γ-ray source would have to be very strong (a few per cent of the cosmic radiation) to be detectable in this way.

2.4 Imaging Detectors

Early attempts to discriminate the electromagnetic showers initiated by γ rays from air showers initiated by charged particles were unsuccessful either using the ground-level arrays of particles detectors or atmospheric Cherenkov detectors [112]. The development of the Cherenkov imaging technique gave the first effective discrimination; an array of photomultipliers in the focal plane of a large optical reflector was used to record a Cherenkov light picture of each air shower. Monte Carlo simulations of the development of air showers from photon and hadron primaries predicted that the images of the former would have somewhat smaller angular dimensions and thus could be identified. The largest optical reflector built for gamma-ray astronomy is the Whipple Observatory 10 m optical reflector (built in 1968) (Figure 3); in 1984 this was equipped with a photomultiplier camera with 37 pixels which was used to detect the Crab Nebula [110]. This first detection led to a rapid development of the imaging technique, with significant improvements in flux sensitivity.

In recent years VHE γ-ray astronomy has seen two major advances: first, the development of high resolution Atmospheric Cherenkov Imaging Telescopes (ACITs) has permitted the efficient rejection of the hadronic background, and second, the construction of arrays of ACITs has improved the measurement of the energy spectra from γ-ray sources. The first is exemplified by the Whipple Observatory 10-m telescope with more modern versions, CAT, a French telescope in Pyrenées [8], and CANGAROO, a Japanese-Australian telescope in Woomera, Australia [42]. The most significant examples of the second are HEGRA, a five telescope array of small imaging telescopes on La Palma in the Canary Islands run by an Armenian-German-Spanish collaboration [29], and the Seven Telescope Array in Utah, which is operated by a group of Japanese institutions [4]. These techniques are relatively mature and the results from contemporaneous observations of the same source with different telescopes are consistent [81]. Vigorous observing programs are now in place at all of these facilities. A vital observing threshold has been achieved whereby both galactic and extragalactic sources have been reliably detected. Many exciting results are anticipated as more of the sky is observed with this present generation of telescopes.

The atmospheric Cherenkov imaging technique has now been adopted at a number of observatories whose properties are summarized in Table 1.

Based on the observations reported from these nine observatories using variants of the Cherenkov imaging technique, the detection of some 13 sources have been claimed, both in the galaxy and beyond [113]. Background rejection of cosmic rays is now in excess of 99.7%, and the technique is effective from energies of 250 GeV to 50 TeV. A signal with significance of 5–10 σ can now be detected from the Crab Nebula in just an hour of observation. Because of the very large collection area associated with the technique, it is particularly powerful for the detection of short transients in TeV γ-ray sources. Cherenkov cameras now often have as many as 600 pixels. In some cases the telescope is an array of small reflectors operated in a stereo mode.

Fig. 3. The Whipple 10m gamma-ray telescope. Note the "10m" refers only to the aperture of the optical reflector; the effective collection area is > 50,000 m^2 so that the γ-ray "aperture" is 120m

There are also two air shower particle detectors which have successfully detected γ rays of a few TeV from the strongest sources. One is a large water Cherenkov detector, MILAGRO near Los Alamos, New Mexico, USA, at an elevation of 2.6 km [92]. The other is a densely packed array of scintillation detectors in Tibet, which operates at an elevation of 4.3 km [5]. Although these telescopes are somewhat less sensitive, they have the advantage over Cherenkov telescopes in that they can operate continuously and hence monitor a large section of the sky.

3 Gamma Ray Sources

Below we present a brief review of γ-ray sources at HE and VHE energies. Division into these energy regions from an observer's perspective is natural since

Table 1. Operating ACIT Observatories c. 1999

Group/Countries	Location	Telescope(s) [Number× Aperture]	Camera [Pixels]	Threshold [TeV]	Epoch [Beginning]
Whipple USA–UK–Irel.	Arizona,USA	10 m	331	250	1984
Crimea Ukraine	Crimea	6×2.4 m	6×37	1	1985
SHALON Russia	Tien Shen, Russia	4 m	244	1.0	1994
CANGAROO Japan – Aust.	Woomera,Aust.	3.8 m	256	0.5	1994
HEGRA Germ.–Arm.–Sp.	La Palma, Sp.	5×3 m	5×271	0.5	1994
CAT France	Pyrenées	3 m	600	0.25	1996
Durham UK	Narrabri,Aust.	3×7 m	1×109	0.25	1996
TACTIC India	Mt.Abu,India	10 m	349	0.3	1997
SevenTA Japan	Utah,USA	7×2 m	7×256	0.5	1998

the observing techniques are quite distinct and since there is currently a gap in coverage in the 10– 100 GeV decade. However the astrophysics obviously spans the complete energy range from 10 MeV to 100 TeV. Since this is primarily an observational review we will divide each source category into HE and VHE sections. There will be more than usual emphasis on VHE observations, representing the bias of the author. In Table 2 the number of sources reported in various categories [43,32,108,18,113] is compared.

4 Galactic Sources: HE

4.1 Unidentified EGRET Sources

Of the 250 sources found in the Third EGRET Catalog [43], almost half of them are galactic as is apparent from their distribution along and centered on the Galactic Plane (Figure 4). Only a small proportion of them have been identified with known galactic objects. The nature of the majority of the objects is completely unknown and is one of the major mysteries and unsolved legacies of the EGRET mission. Although almost half of the sources found by the earlier

Table 2. Status of HE/VHE Sources [113]

Energy Range	10 MeV – 10 GeV	300 GeV – 30 TeV
Platform	Space	Ground
Discrete Sources		
Type	No. of Sources	No. of Sources
AGNs	75	6
Normal Galaxies	1	0
Radiogalaxy	1	0
Pulsars	5	0
SNR Shell	5?	3
SNR Plerion	1	3
Binaries	1	1
Total identified	87	13
Unidentified	165	0
Other Sources		
Galactic Plane	Yes	No
Extragalactic Diffuse	Yes	No
All Sky Survey	Yes	No
Gamma Ray Bursts	5	1?
Other Features		
Flares	hours	minutes
Multiwavelength Correlations	days-weeks	minutes-years
Energy Spectra	moderate	good
Source Location	good	good

COS-B mission were later found to be high points in the galactic diffuse emission, there is little doubt about the reality of these EGRET discrete sources. The angular resolution of the EGRET instrument is such that it gives error boxes of order 1.0° radius. The problem of identification is compounded by the density of objects in the galactic plane, the uneven nature of the diffuse galactic plane distribution and the possibility of source confusion, the time variability of many of the sources and the lack of independent verification of the detections by another γ-ray telescope.

Attempts at identification follow two general lines: statistical association of the distribution of a sub-class of sources with known galactic objects or positional and/or temporal association of an individual source with an object that is well known at other wavelengths. Although the literature contains a number of claims for such identifications they must be regarded as somewhat speculative and only the association with pulsars can be considered definite.

There are 170 unidentified sources in the Third EGRET catalog [43]; their positional information is not good enough to allow unambiguous identification with individual sources. The EGRET exposure is not uniform and there is greater sensitivity to discrete sources away from the galactic plane. There are approximately equal numbers above and below galactic latitude of 10°. Some of them are surely associated with AGNs which have not been identified as conspicuous

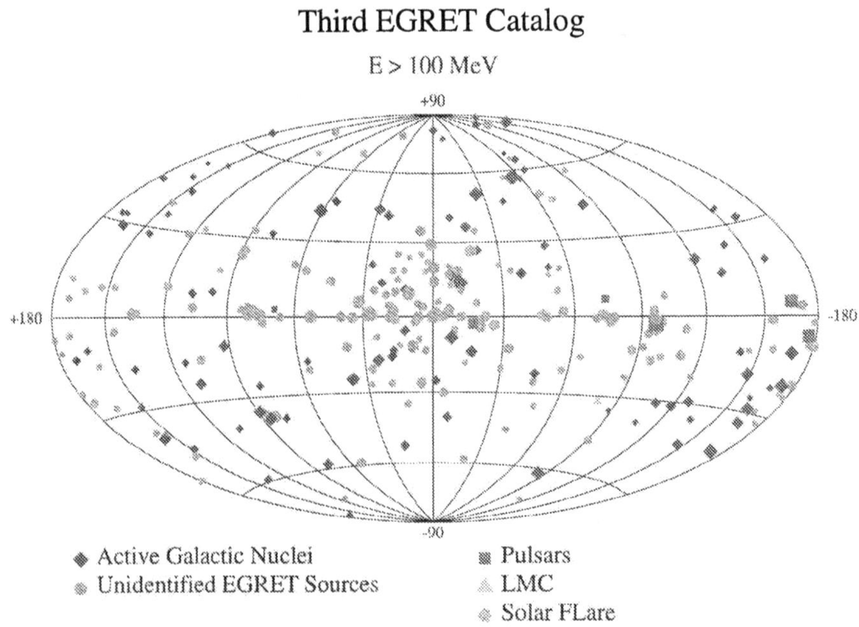

Fig. 4. Distribution of sources seen by EGRET: 3rd Catalog [43]

at other wavelengths. Variability on time-scales of months is a possible clue to their identity. However the distribution of high latitude sources is not consistent with them all being AGNs. The other sources can be sub-divided according to their spatial and spectral characteristics. There may be one or more sub-classes distributed along the galactic plane and another sub-class of weaker (nearer) sources extending out to 30° latitude. These latter roughly correspond to the distribution of stars known as Gould's Belt. These relatively nearby sources have a weak luminosity of 1–5 $\times 10^{32}$ erg s^{-1} for E > 100 MeV. The sources distributed along the galactic plane are more distant (average distance $\approx$ 6 kpc) and hence have a luminosity 7–14 $\times 10^{32}$ erg s^{-1}. Possible associations that have been suggested include SNRs, OB associations, massive stars with stellar winds, accreting black holes, and radio-quiet pulsars.

4.2 Pulsars: HE

Prior to the launch of CGRO, the Crab and Vela pulsars were known sources of pulsed 100 MeV emission. One of the strongest 100 Mev sources was Geminga but its identity as a pulsar was only revealed during the EGRET mission. 2CG342-02 was also known as a 100 MeV source but it took the EGRET experiment to identify it with the pulsar, PSR B1706-44. Two other sources were identified with the pulsars, PSR B1055-52 and PSR B1951+32 on the basis of their positional coincidence and pulsed emission. There are tentative associations with several other pulsars. It is also suggested that some number of the unidentified EGRET sources may be pulsars whose radio beams are not pointing in our direction. The general characteristics of the γ-ray pulsars are that they have flat spectra, that they are steady emitters with long time constants and that their γ-ray luminosity is much less than the rotational energy loss.

Only the Crab pulsar shows a light curve with the γ-ray pulse in phase with the radio pulse. Usually the γ-ray light curve exhibits two peaks that are roughly 180° apart; only for Geminga is the separation exactly 180°. Based on the shape of the light curves it appears that emission from two poles is not the origin of the double peak light curve; it seems more likely that it originates from a hollow cone of emission around a single pole. Where the statistics are good enough it is seen that the spectral shape of the emission changes as a function of phase.

Only the Crab Nebula source has a detectable steady (unpulsed) component at HE energies.

The EGRET pulsar parameters are summarized in Table 3. There is one other γ-ray pulsar, PSR 1509-58 but it is not detected above 1 MeV. The pulsars have several common features. All of them have power spectra that peak at γ-ray energies. All of them turn over or break at some γ-ray energy. Over a large part of their spectrum their emission is characterized by a power law.

The number of pulsars is so small and the range of parameters so large that it is difficult to draw any definite conclusions as to the emission mechanism: in particular it is not possible to differentiate between the favored polar cap or outer gap models. This may be possible with the next generation of γ-ray telescopes.

Table 3. EGRET-detected Gamma-ray Pulsar Parameters

Pulsar	Period (seconds)	Spindown (10^{-15} s/s)	Spectral Index	Luminosity (erg cm^{-2} s^{-1})
Crab	0.033	421	2.15	10×10^{-10}
B1951+32	0.040	5.85	1.74	2.4×10^{-10}
B1706-44	0.102	93	1.72	8.3×10^{-10}
Vela	0.089	125	1.70	71×10^{-10}
B1055-52	0.197	5.83	1.18	4.2×10^{-10}
Geminga	0.237	11.0	1.50	37×10^{-10}

4.3 Pulsars: VHE

There are no confirmed detections from pulsars at VHE energies. Upper limits are found for all the EGRET pulsars that indicate a turnover in the emission spectrum (Figure 5); this turnover is not yet well enough determined to discriminate between models of pulsar emission. The sharpest turnover is seen in PSR B1951+32 [97].

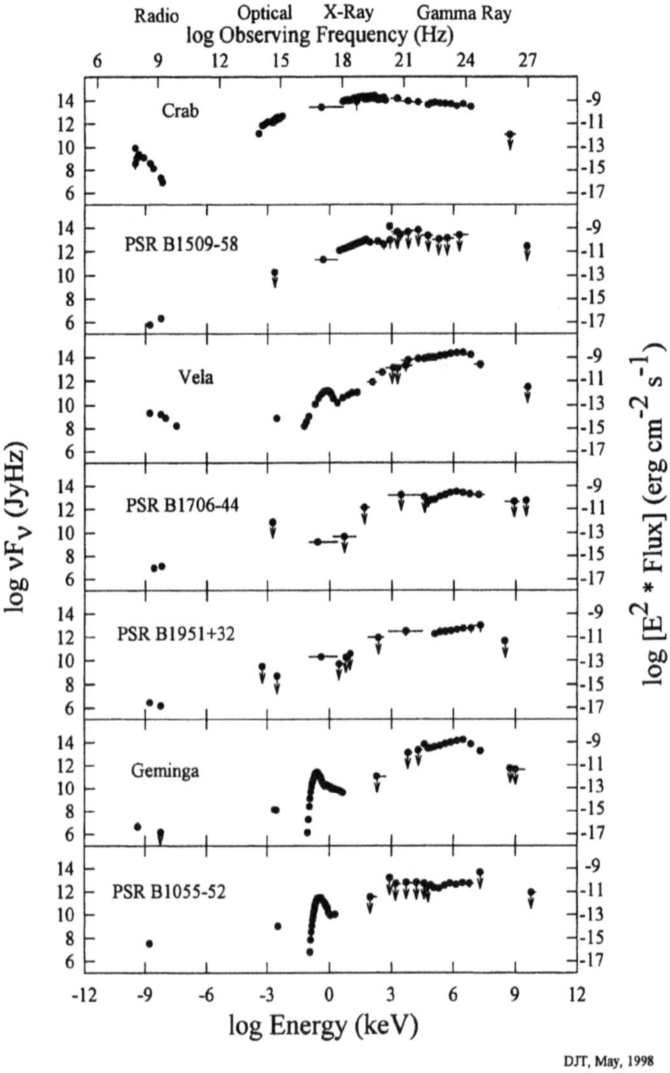

Fig. 5. Power spectra of pulsars detected at γ-ray energies [105]

4.4 Supernova Remnants: HE

Supernova remnants (SNRs) are widely believed to be the sources of hadronic cosmic rays up to energies of approximately $Z \times 10^{14}$ eV, where Z is the nuclear charge of the particle. Supernova blast shocks are among the few galactic mechanisms capable of satisfying the energy required for the production of galactic cosmic rays, although even these must have a high efficiency, ~10% - 30%, for converting the kinetic energy of supernova explosions into high energy particles. The model of diffusive shock acceleration, which provides a plausible mechanism for efficiently converting the explosion energy into accelerated particles, naturally produces a power-law spectrum of $dN/dE \propto E^{-2.1}$. This is consistent with the inferred spectral index at the source for the observed local cosmic-ray spectrum of $dN/dE \propto E^{-2.7}$, after correcting for the effects of propagation in the galaxy.

An indication of shock acceleration of hadronic cosmic rays in SNR shells could come from measurements of γ-ray emission in these objects. Collisions of cosmic-ray nuclei with the interstellar medium result in the production of neutral pions which subsequently decay into γ-rays. The γ-ray spectrum would extend from below 10 MeV up to ~1/10 of the maximum proton energy (> 10 TeV), with a distinctive break in the spectrum near 100 MeV due to the Δ resonance at 1.234 GeV in the cross-section for π^0 production. As γ-ray production requires interaction of the hadronic cosmic rays with target nuclei, this emission will be stronger for those SNR located near, or interacting with, dense targets, such as molecular clouds. The cosmic-ray density, and hence the associated γ-ray luminosity, will increase with time as the SNR passes through its free expansion phase, will peak when the SNR has swept up as much interstellar material as contained in the supernova ejecta (the Sedov phase) and gradually decline thereafter ([30,74]). Thus, γ-ray bright SNRs should be "middle-aged."

Although not nearly as well established as the association with radio pulsars, there is the possible identification of several EGRET sources with known supernova remnants (SNRs). Because such identifications could point to the SNRs as the source of cosmic ray acceleration, these claims have received much attention. The possible identifications [32] are listed in Table 4, together with the source flux (> 100 MeV), and the approximate distance and angular size which are based on measurements at other wavelengths. High densities of gas are required to explain the EGRET emission, of order 100 g cm$^-$3. Subsequent work has shown that other processes must also be considered e.g., bremsstrahlung, inverse Compton [7]. Also if the acceleration of cosmic rays to energies of 100 TeV and above is to occur in these sources, then they should be strong sources of 1 TeV γ rays; as we shall see below, this prediction of the early models is not verified.

4.5 Supernova Remnants: VHE

Plerions: A supernova remnant, with a pulsar at its center which continually fills the remnant with relativistic electrons, is known as a plerion. The distinction

Table 4. Supernova Remnants detected by EGRET

SNR	Flux $\times 10^{-8}$ cm^{-2} s^{-1}	EGRET	Distance (kpc)	Size (arc-min)
W28	56	2EG J1801-2312	1.8-4.0	42
W44	50	2EG J1857+0118	3	30
γ Cygni	126	2EG J2020+4026	1.8	60
IC443	50	2EG J0618+2234	0.7–2.0	45
Monoceros	23	2EG J0635+0521	0.8–1.6	220

between shell-type SNRs and plerions is not sharp but it is useful to make this distinction in the discussion of VHE-emitting SNRs.

The Crab Nebula: The Crab Nebula was the first credible TeV source and it remains the strongest known source in the TeV sky. The observed spectrum is well explained by a Compton-synchrotron model in which the ambient magnetic field is the variable parameter (Figure 6). At least in the 300 GeV to 3 TeV range it is clear that the Crab Nebula, the archetypical plerion, can now be considered a standard VHE candle. There is remarkable agreement between the absolute fluxes and spectral shapes reported from observations of the Crab Nebula by several imaging ACTs; the results from the Whipple, HEGRA, CAT and CANGAROO experiments are shown in Table 5. These are also in agreement with the flux reported in the first detections of the Crab [110,106] but this must be considered fortuitous in view of the large error bars in these early measurements.

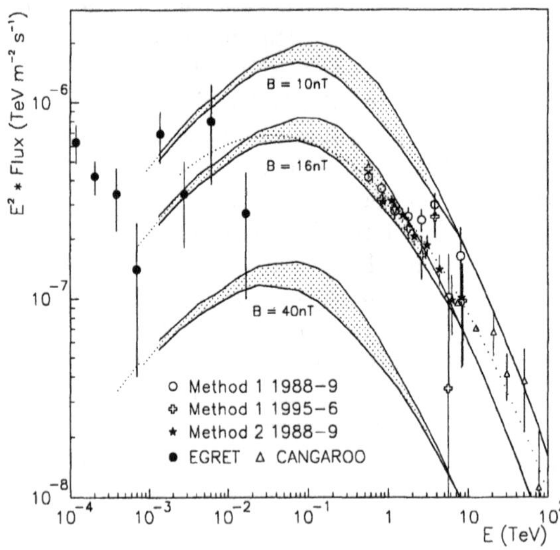

Fig. 6. Gamma-ray spectrum of the Crab Nebula [47]

New observations of the Crab Nebula have been reported at both high and low energies. CELESTE, with a threshold energy of 50 GeV, observed it for just three hours [73] whereas STACEE [78], with an interim threshold of 75 GeV, had a 7 σ detection in 50 hours of observation. Neither experiment could quote a flux value and neither experiment saw any evidence for a pulsed component from the Crab pulsar.

At higher energies the Crab has been seen for the first time by a conventional air shower array (the Tibet High Density Array at 4.5 km) [6]. The energy threshold was 3 TeV and the flux deduced (see Table 5) was a factor of 2–3 higher than that seen in ACT experiments.

Table 5. VHE Flux from the Crab Nebula

Group	VHE Spectrum (10^{-11} photons cm^{-2} s^{-1} TeV^{-1})	E_{th} (TeV)	Reference
Whipple (1991)	$(25(E/0.4TeV))^{-2.4\pm0.3}$	0.4	[106]
Whipple (1998)	$(3.2\pm0.7)(E/TeV)^{(-2.49\pm0.06_{stat}\pm0.04_{syst}}$	0.3	[47]
HEGRA (1999)	$(2.7\pm0.2\pm0.8)(E/TeV)^{-2.60\pm0.05_{stat}\pm0.05_{syst}}$	0.5	[58]
CAT (1999)	$(2.7\pm0.17\pm0.40)(E/TeV)^{-2.57\pm0.14_{stat}\pm0.08_{syst}}$	0.25	[69]
CANGAROO (1998)	$(2.01\pm0.36)\times10^{-2})(E/7TeV)^{-2.53\pm0.18}$	7	[101]
Tibet HD (1999)	$(4.61\pm0.90)\times10^{-1})(E/3TeV)^{-2.62\pm0.17}$	3	[6]

PSR 1706-44: Following the TeV detection of this source by the CANGAROO group [56] and its confirmation by the Durham group [20], there have been no new reports of observations of this source. No periodic emission is seen and it is believed that the VHE emission comes from a weak plerion. Although weaker than the Crab this may be the standard candle for the southern hemisphere.

Vela: The CANGAROO group reported the detection of a 6σ signal from the vicinity of the Vela pulsar [115]. The integral γ-ray flux above 2.5 TeV is 2.5×10^{-12} photons cm^{-2} s^{-1}. There is no evidence for periodicity and the flux limit is about a factor of ten less than the steady flux. The signal is offset (by 0.14°) from the pulsar position which makes it more likely that the source is a synchrotron nebula. Since this offset position is coincident with the birthplace of the pulsar it is suggested that the progenitor electrons are relics of the initial supernova explosion and they have survived because the magnetic field was weak.

Again the source was not confirmed by observations by the Durham group [22]. The upper limit to the γ-ray flux above 300 GeV is 5×10^{-11} photons cm^{-2} s^{-1}. Given the differences in energy and the uncertainties in flux estimates in the two experiments, the Durham group felt the two results were compatible. However it would have been reassuring to see the independent confirmation.

Shell-Type Supernova Remnants. The luminosity of γ-rays from secondary pion production may be detectable with the current generation of ground-based γ-ray detectors, particularly if the objects are located in a region of relatively high density in the interstellar medium [30]. The EGRET detections alone are not sufficient to claim the presence of high energy hadronic cosmic rays. The relatively poor angular resolution of EGRET makes it difficult to definitively identify the detected object with the SNR shell. Background from the diffuse Galactic γ-ray emission complicates spectral measurements. To complicate matters further, with the detection of X-ray synchrotron radiation from SNR shells, the possibility of the production of γ-rays via inverse Compton scattering of ambient soft photons has been realized. Bremsstrahlung radiation may also be a significant source of γ-rays at MeV–GeV energies [35].

Measurements of γ-rays at very high energies may help resolve the puzzle of the γ-ray emission from the EGRET-detected SNRs. VHE γ-ray telescopes have much better angular resolution than EGRET, reducing the source confusion associated with any detection. Also, because the diffuse Galactic γ-ray emission has a relatively steep spectrum, $\propto E^{-2.4}$ to $E^{-2.7}$ ([48]), compared with the expected $\sim E^{-2.1}$ spectrum of γ-rays from secondary pion decay, contamination from background γ-ray emission should be less in the VHE range. Thus, in recent years, searches for emission from shell-type SNRs have been a central part of the observation program of VHE telescopes.

The Whipple Observatory has published the results of observations of six shell-type SNRs (IC 443, γ-Cygni, W 44, W 51, W 63, and Tycho) selected as strong γ-ray candidates based on their radio properties, distance, small angular size, and possible association with a molecular cloud [13]. The small angular size was made a requirement due to the limited field of view (3° diameter) of the Whipple telescope at that time. VHE telescopes can also detect fainter γ-ray sources if they are more compact, because they can reject more of the cosmic-ray background. IC 443, γ-Cygni, and W 44 are also associated with EGRET sources [32]. Despite long observations, no significant excesses were observed, and stringent limits were derived on the VHE flux (see Table 6 and Figure 7).

Table 6. VHE Observations of shell-type supernova remnants

Object Name	Observation Time (min.)	Energy (TeV)	Integral Flux (10^{-11} cm^{-2} s^{-1})	Ref.
Tycho	867.2	>0.3	<0.8	[13]
IC 443	1076.7	>0.3	<2.1	[13]
	678.0	>0.5	<1.9	[46]
W 44	360.1	>0.3	<3.0	[13]
W 51	468.0	<0.3	<3.6	[13]
γ-Cygni	560.0	>0.3	<2.2	[13]
	2820.0	>0.5	<1.1	[46]
W 63	140.0	>0.3	<6.4	[13]

There is another group of shell-type supernova remnants which are observed at TeV energies but in which the progenitors are most likely electrons. These sources have not been detected at MeV–GeV energies.

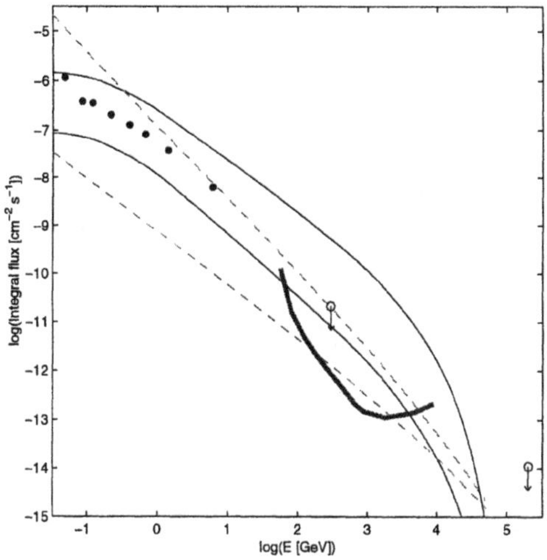

Fig. 7. Gamma-ray spectrum of IC433 [13]

SN1006: In 1997 the CANGAROO Collaboration reported the observation of TeV γ-ray emission from the shell-type SNR, SN 1006 [100]. Observations taken in 1996 and 1997 indicated a statistically significant excess from the northeast rim of the SNR shell. The flux at $> 1.7 \pm 0.5$ TeV was (4.6 $\pm 0.6(sys) \pm 1.4(stat) \times 10^{-12}$ photons cm^{-2} s^{-1}. The observations were motivated by the observation of non-thermal X-rays by the *ASCA* experiment. It represented the first direct evidence of acceleration of particles to TeV energies in the shocks of SNRs.

There is a disturbing report from the Durham group of the failure to detect this source in 40 hours of observation. Their upper limit at 300 GeV was 1.7 $\times 10^{-12}$ photons cm^{-2} s^{-1} and at 1.5 TeV was 1.3 $\times 10^{-12}$ photons cm^{-2} s^{-1}, barely compatible with the CANGAROO observation. They point out that the presence of a bright star near the SNR complicates the measurement.

RXJ1713.7-3946: The detection of TeV gamma-rays from this shell-type SNR was reported by the CANGAROO group [68]. The observations were motivated by the observation of a hard X-ray power-law spectrum by *ASCA*. In this respect, it is very similar to SN1006 but is three times brighter in X-rays. It has a characteristic dimension of 70 arc-min, lies at a distance of 1.1 kpc and has an estimated age of 2,100 years. The γ-ray flux above 2 TeV is 3×10^{-12} photons cm^{-2} s^{-1} with a 5 σ significance. There is evidence that the source is extended in the same direction as the X-ray source.

Cassiopeia A: It is natural that the strongest source in the radio sky should have been one of the first targets of VHE observations [28]. It is appropriate that it should have been eventually detected as a TeV source but only after a very long exposure by the HEGRA group [83]. As with SN1006 and RXJ1713.7-3946, these observations were motivated by observations of a hard X-ray power-law spectrum. The source is a classical shell-SNR of diameter 2.2 arc-min which is effectively point-like to a γ-ray telescope. It is believed to be 300 years old and there is no active pulsar at its center; however there may be an neutron star. The HEGRA observations were made in 1997 and 1998 and comprised some 130 hours on the source. The flux above 1 TeV has not yet been determined but must be of order 3×10^{-12} photons cm^{-2} s^{-1}. The total detection was just less than 5 σ and it is probably the weakest TeV source detected to date.

Upper limits to the TeV emission have been reported by the CAT [40] and Whipple [61] groups. These were at lower energies but, because the exposures were much shorter, the upper limits are compatible with the HEGRA detection. The three results are summarized in Table 7.

Table 7. VHE Observations of Cassiopeia A

Group	E_{th} (TeV)	Exposure (hours)	Flux (10^{-11} photons cm^{-2} s^{-1})
Whipple	500	7.5	< 0.66
CAT	400	24.4	< 0.74
HEGRA	1000	127.9	0.3?

4.6 X-ray Binaries

At one time it appeared that several X-ray binaries (Cygnus X-3, Hercules X-1, etc.) were transient sources of VHE γ rays [19,111]; these observations have not been confirmed nor explained. There is now only one X-ray binary which is still considered a viable candidate source; it is weakly detected at HE and VHE energies.

Centaurus X-3: Cen X-3 contains a 4.8 s pulsar in orbit with a period of 2.1 days. Originally reported as a source of sporadic outbursts of pulsed emission [14,89], it was later found to be a source of steady (unpulsed) weak emission [21]. At this time it was also seen as an unpulsed GeV EGRET source [108]. New observations, taken in 1998 and 1999 by the Durham group [23], do not add to the overall statistical significance of the detections which remain somewhat marginal.

4.7 Diffuse Background

The Galactic Plane is the strongest HE source in the sky and, not surprisingly, it was the first discovered. It was extensively mapped by the SAS-II and COS-B satellite experiments; the EGRET observations have greatly expanded these observations and given a fairly satisfactory match between observations and interpretation. It should be noted that the Galactic Plane is incredibly difficult to study since we are in the middle of it and radiation at many wavelengths is obscured. Radio and γ-ray measurements offer unique windows to the study of the galactic arms and obscured regions. The γ radiation, here as elsewhere, is secondary to the cosmic radiation and hence the study of its distribution is a unique channel of information on the distribution of cosmic rays (hadrons and electrons) throughout the galaxy. The problem is that it is not easy to differentiate between the two classes of progenitor and there are many mechanisms by which the γ radiation can be produced.

To model the γ ray distribution we must know the composition, the distribution and the energy spectrum of the progenitors as well as the density and distribution of the target material in the interstellar medium. The cosmic ray composition and spectrum is initially assumed to be the same as that observed near the solar system. This is perhaps less speculative for the hadron component than for the electron component because the latter is more subject to local source anomalies.

The interstellar gas is mostly (90%) hydrogen which can occur as atomic, molecular or ionized. The 21 cm radio line gives a convenient way of mapping the atomic hydrogen distribution. The molecular hydrogen cannot be seen directly but can be inferred from the 2.6 mm line of Carbon Monoxide. The distribution is uneven and mostly concentrated in large molecular clouds. The ionized component is small and usually ignored. To model Compton scattering of electrons on interstellar photons in the plane it is necessary to know the distribution of visible and infra-red light.

In practice none of these galactic or cosmic ray parameters is known with sufficient accuracy to unambiguously predict the diffuse γ-ray flux. Instead an iterative process is used in which the parameters are roughly estimated and then allowed to vary to get the best fit to the observed distribution [48]. Initially it was assumed that the bulk of the diffuse galactic flux was the result of the interaction of cosmic ray protons with the interstellar gas in the plane. The cosmic ray density was assumed uniform in the galaxy. It was soon apparent that while the basic mechanism might be correct a uniform cosmic ray density did not fit the observations. It is now assumed that the cosmic ray density is uneven and couples locally with the matter density. To model the observed radiation $\pm 2°$ from the plane, a detailed calculation has been made by the Goddard group [48]. The two parameters used as variables are the ratio of density of the CO and H_2 gas and the scale of coupling between the cosmic rays and the matter density. The model fits the observed spatial distribution very well and is used as the basis of determining the background above which point sources are identified as such. The predicted energy spectrum which includes contributions from the proton-proton interactions as well as electron Compton and bremsstrahlung scattering clearly show the π^0 bump near 70 MeV (the only cosmic source that shows this bump) (Figure 8). However at higher energies the observed data points all lie above the predicted spectra from all three mechanisms. This deviation, at energies above 1 GeV, has not been satisfactorily explained.

At higher energies (> 100 GeV) there are no definitive measurements of the Galactic Plane component and the observed upper limits are compatible with a reasonable extrapolation of the EGRET data. VHE telescopes have excellent sensitivity to point sources but are less sensitive to diffuse sources.

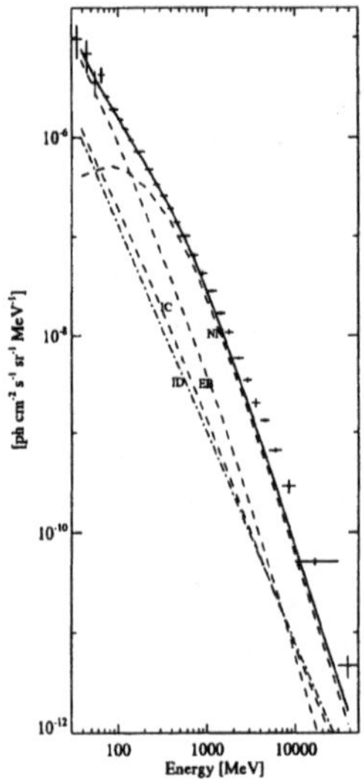

Fig. 8. Differential energy spectrum of diffuse galactic plane emission as measured by EGRET and as predicted for various production processes [48]

5 Extragalactic Sources

5.1 HE Observations

One of the most important results to come from the CGRO mission was the detection of HE γ-ray emission from extragalactic γ-ray sources. In the Third EGRET catalog [43] there are 71 identified AGN sources (and 25 possible identifications); this constitutes the largest sub-class of known HE sources and firmly establishes HE γ-ray astronomy as a true extragalactic discipline. These sources are remarkable for their multitude, their variability, their hard spectra and their great distances – and for the fact that they are mostly associated with one small sub-class of AGNs, the blazars.

Blazars: Although the γ-ray emitting blazars are bright, they were largely unknown until the EGRET mission. COS-B had detected one extragalactic source, the nearby quasar, 3C273. Little attention was paid to this discovery by the AGN community. In fact most of the observing time of the COS-B mission was spent in studying the Galactic Plane where it was felt that the bulk of the interesting sources would lie and the vast off-plane region of the sky was largely unexplored. In fact, 3C273 is not a classic blazar and has a somewhat soft spectrum.

Active galactic nuclei (AGN) are the most energetic on-going phenomena that we see in extragalactic astronomy. The canonical model of these objects is that they contain massive black holes (often at the center of elliptical galaxies) surrounded by accretion disks and that relativistic jets emerge perpendicular to the disks; these jets are often the most prominent observational feature. Blazars are an important sub-class of AGNs because they seem to represent those AGNs which have one of their jets aligned in our direction. Observations of such objects are therefore unique.

The γ-ray AGN astronomer is in the position of the particle physicist who is offered the opportunity to observe the accelerator beam, either head-on or from the side. For the obvious reason that there is more energy transferred in the forward direction the particle physicist usually chooses to put his most important detectors directly in the direction of the beam (or close to it) and its high energy products. While such observations give the best insight into the energetic processes in the jet, they do not give the best pictorial representation. Hence just as it is difficult to visualize the working of a cannon by looking down its barrel, it is difficult to get a picture of the jet by looking at it head-on. Observations at right angles to the jet give us our best low energy view of the jet phenomenon and indeed provide us with the spectacular optical pictures of jets from nearby AGNs (such as M87).

The properties of blazars observed by EGRET have been extensively reviewed (e.g., [71]) and are only briefly summarized here.

- They all have hard spectra with an average differential spectral index of -2.1.
- The redshifts vary from z = 0.03 to 2.4.
- The blazars are mostly radio-selected BL Lacs, indicating a synchrotron peak at soft X-ray energies or ultraviolet.
- Time variations have been observed on time-scales of years to hours.
- The list of detected AGNs includes such prominent objects as 3C273, 3C279, BL Lac, 3C66A, Markarian 421 and W Comae.

Normal Galaxies: The Large Magellanic Cloud is detected as a weak HE source and it is concluded that the cosmic rays are in quasi-equilibrium; the Small Magellanic Cloud is not detected and thereby hangs a tale [93]. If the cosmic radiation observed near the Solar System, and assumed typical of the Galaxy as a whole, is assumed to permeate extragalactic space (as many have assumed), then there is enough target material in the SMC for it to produce detectable amounts of 100 MeV emission. The conclusion drawn is that the extragalactic theory of origin of cosmic rays must be rejected. Andromeda is also not detected. The predicted and observed fluxes are shown in Table 8.

Radiogalaxies: Centaurus A, the closest large radio galaxy at z = 0.0007, has been detected by EGRET [94] as a weak source; no other radio galaxies have been detected. Its detection represents the first evidence for HE emission from a source with a confirmed large-inclination jet. The emission appears steady and the differential spectral index is steeper than most blazars at - 2.40±0.28. The spectrum appears to extend smoothly down to 1 MeV. The intrinsic luminosity is weaker than on-axis AGN sources but since these radio galaxies are more plentiful they may make a significant contribution to the extragalactic background.

Table 8. EGRET Observations of Normal Galaxies

Galaxy	Predicted $F_\gamma(>100\text{MeV})$ $\times 10^{-7}$ cm^{-2} s^{-1}	Observed $F_\gamma(>100\text{MeV})$ $\times 10^{-7}$ cm^{-2} s^{-1}
L.M.C.	2.0 ± 0.4 sr^{-1}	2.3 ± 0.4 sr^{-1}
S.M.C.	< 0.5 sr^{-1}	2.4 sr^{-1}
Andromeda	0.2 sr^{-1}	< 0.5 sr^{-1}

Many more may be detectable with the new generation of instruments such as GLAST and VERITAS.

5.2 VHE Observations

One of the most surprising results to come from VHE γ-ray astronomy was the discovery of TeV-emitting blazars. Unlike the observation of galactic supernovae such as the Crab Nebula, which are essentially standard candles, the VHE light-curves of blazars are highly variable.

MARKARIAN 421 AND 501. Mkn 421 achieved some notoriety largely because it was the first extragalactic source to be identified as a TeV γ-ray emitter [84]. At discovery, its average VHE flux was $\approx$ 30% of the VHE flux from the Crab Nebula. Markarian 421 is the closest example of an AGN which is pointing in our direction. It is a BL Lacertae object, a sub-class of blazars, so-called because they resemble the AGN, BL Lac which is notorious because of the lack of emission lines in its optical spectrum. Because such objects are difficult, and somewhat uninteresting, for the optical astronomer they were largely ignored until they were found to be also strong and variable sources of X-rays and γ-rays.

In Figure 9 the nightly averages of the TeV flux from Markarian 421 (Mkn 421) in 1995 are shown as observed at the Whipple Observatory [12]. Although AGN variability was a feature of the AGNs observed by EGRET at energies from 30 MeV to 10 GeV, the weaker signals (because of the finite collection area) do not allow such detailed monitoring, particularly on short time-scales.

Markarian 501 (Mkn 501), which is similar to Mkn 421 in many ways, was detected as a VHE source by the Whipple group in May 1995 [87]. It was only 8% of the level of the Crab Nebula and was near the limit of detectability of the technique at that time. The discovery was made as part of an organized campaign to observe objects that were similar to Mkn 421 and were at small redshifts.

Variability: Perhaps the most exciting aspect of these detections is the observation of variability on time-scales from minutes to hours. The very large collection areas (> 10,000 m^2) associated with atmospheric Cherenkov telescopes is ideally suited for the investigation of short term variability. The VHE emission from the

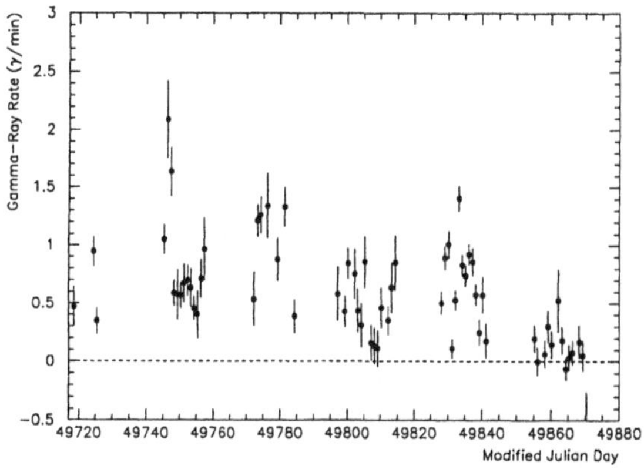

Fig. 9. Daily VHE γ-ray count rates for Mkn 421 during 1995 (from [12])

two best observed sources, Mkn 421 and Mkn 501 (Figure 10), varies by a factor of a hundred. Although many hundreds of hours have now been devoted to their study, the variations are so complex that it is still difficult to characterize their emissions. It has been suggested [12] that for Mkn 421 the emission is consistent with a series of short flares above a baseline that falls below the threshold of the Whipple telescope (Figure 9); the average flare duration is one day or shorter.

The most important observations of Mkn 421 were in May, 1996 when it was found to be unusually active [37]. On May 7, a flare was observed with the largest flux ever recorded from a VHE source. The observations began when the flux was already several times that of the Crab Nebula and it continued to rise over the next two hours before levelling off (Figure 11). Observations were terminated as the moon rose but the following night it was observed at its quiescent level. One week later (May 15) a smaller, but shorter, flare was detected; in this case the complete flare was observed and the doubling time in the rise and fall was $\approx$ 15 minutes. This is the shortest time variation seen in any extragalactic γ-ray source at energies $>$ 10 MeV (apart from that seen in a classical γ-ray burst).

Mkn 501 is also variable, but as at other wavelengths, the characteristic time seems longer. Its baseline emission has varied by a factor of 15 over four years [88] (Figure 10). Hour-scale variability has also been detected but its most important time variation characteristic appears to be the slow variations seen over five months in 1997.

The TeV outburst from Mkn 501 in 1997 merited a Highlight session at the 25th ICRC [81]. Sadly while the conference was taking place the source was already in decline and it has been relatively quiescent ever since. Most of the interest in the source since that time has been in a detailed analysis of the high intensity signal, in particular in the derivation of an accurate energy spectrum.

The 1997 outburst data has been summarized in a number of publications [88,1,85,76]. Variations with doubling times as short as two hours have been

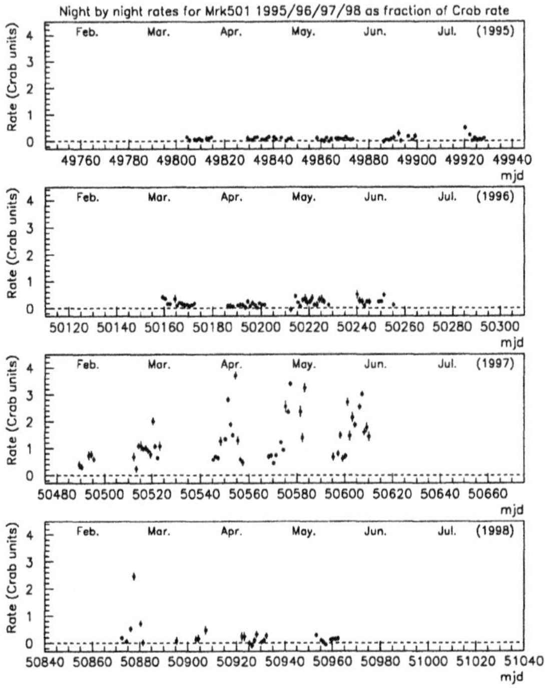

Fig. 10. Average nightly VHE γ-ray flux (in units of VHE Crab flux) for Mkn 501 between 1995 and 1998 (from [88])

reported [88] but in general the variations are not as short as those seen in Markarian 421.

Energy Spectrum: The atmospheric Cherenkov signal is essentially calorimetric and hence it is possible to derive the γ-ray energy spectrum from the observed light pulse spectrum. In practice it is difficult because, unless an array of detectors is used, the distance to the shower core (impact parameter) is unknown. Although the extraction of a spectrum from even a steady and relatively strong source as the Crab Nebula required considerable effort and the development of new techniques, it was relatively easy to measure the spectra of Mkn 421 and Mkn 501 in their high state because the signal was so strong. The general features of the spectra derived from the Whipple observations are in agreement with those derived at the HEGRA telescopes [62].

The May 7, 1996 flare of Mkn 421 provided an excellent data base for the extraction of a spectrum; the data can be fit by a simple power-law ($dN/dE \propto E^{-2.6}$). There is no evidence of a cutoff up to energies of 5 TeV [116] (Figure 12). Because of the possibility of a high energy cutoff due to intergalactic absorption there is considerable interest in the highest energy end of the spectrum. Large

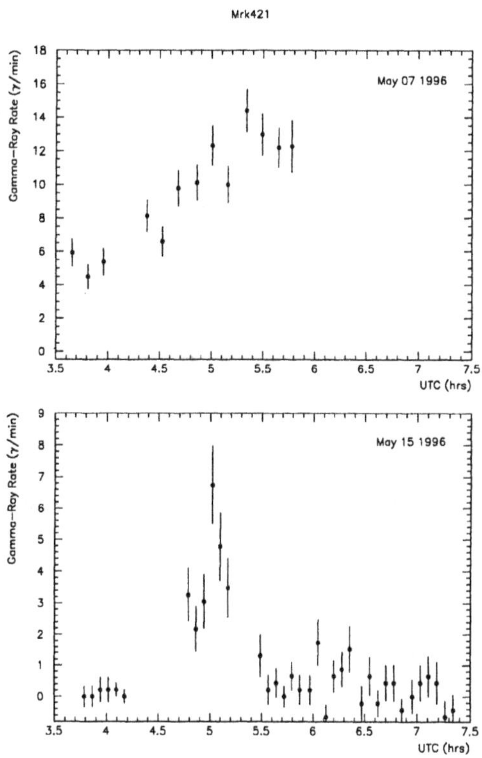

Fig. 11. Mkn 421 flares of 1996 May 7 (left) and May 15 (right) (adapted from [37])

zenith angle observations at Whipple [53] and observations by HEGRA [62] confirm the absence of a cutoff out to 10 TeV.

The energy spectrum of Markarian 421 has been reported by several groups. There is general agreement that it can be fit by a simple power law. While the absolute flux has little meaning since it varies with time, the differential power-law spectral index should be comparable in different experiments unless it is also variable with time. There is good agreement on the indices obtained thus far by CAT ($-2.96 \pm 0.13 \pm 0.05$) [79]; HEGRA ($-3.09 \pm 0.07 \pm 0.10$) [2]; 7TA (-2.81) [114]. However the Whipple group gets consistently harder spectra [54] particularly during flaring e.g (-2.54 ± 0.04) on May 7, 1996. Preliminary analysis of non-flaring data gives a similar result. Obviously further work is required here to ensure that the analysis is free of large systematic errors.

The generally high state of Mkn 501 throughout 1997 give data from the Whipple telescope that can be best fit by a curved spectrum of the form: $\mathrm{d}N/\mathrm{d}E \propto E^{-2.20-0.45\log_{10}E}$ [90] (Figure 12). Here the spectrum extends to at least 10 TeV. The curvature in the spectrum could be caused by the intrinsic emission

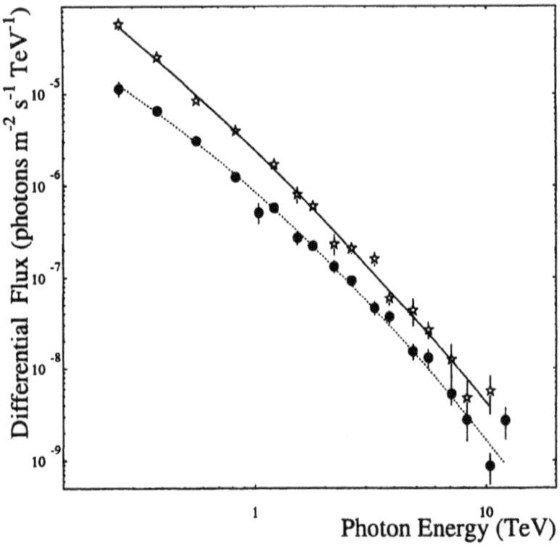

Fig. 12. VHE spectra of Mkn 421 an Mkn 501 as measured with the Whipple Observatory telescope [55]

mechanism or by absorption in the source. Since Mkn 421 and Mkn 501 are virtually at the same redshift it is unlikely that it could be due to intergalactic absorption since Mkn 421 does not show any curvature [55].

Detailed energy spectra come from Whipple observations between 250 GeV and 12 TeV [90,55] and HEGRA data spanning 500 GeV to 20 TeV [60]. The Telescope Array Collaboration has also derived a spectrum over a slightly narrower energy range (600 GeV to 6.5 TeV) [45]. A search for variability in the spectrum revealed no significant changes in spectrum with flux or time [91,60], allowing large data sets to be combined to derive very detailed energy spectra spanning large ranges in energy. The spectra derived by Whipple and HEGRA deviate significantly from a simple power law. For Whipple, the χ^2 probability that a power law is consistent with the measured spectrum is 2.5×10^{-7}. This is the first significant deviation from a power law seen in any VHE γ-ray source and any blazar at energies above 10 MeV. The Whipple spectrum is:

$$\frac{\mathrm{dN}}{\mathrm{dE}} \propto \mathrm{E}^{-2.22\pm0.04_{\mathrm{stat}}\pm0.05_{\mathrm{syst}}-(0.47\pm0.07_{\mathrm{stat}} log_{10}(E))}$$

and the HEGRA spectrum is:

$$\frac{\mathrm{dN}}{\mathrm{dE}} \propto \mathrm{E}^{-1.92\pm0.03_{\mathrm{stat}}\pm0.20_{\mathrm{syst}}} \exp\left[-\frac{\mathrm{E}}{6.2 \pm 0.4_{\mathrm{stat}}(-1.5,+2.9)_{\mathrm{syst}}}\right]$$

where E is in units of TeV. The form of the curvature term in the spectra has no physical significance as the energy resolution of the experiments is not sufficient to resolve particular spectral models. The Whipple spectral form is simply a polynomial expansion in logE v. log(dN/dE) space. The HEGRA form was chosen presumably because attenuation of the VHE γ-rays by pair-production with background IR photons could produce an exponential cut-off. In fact, the Whipple and HEGRA data are completely consistent with each other. The Telescope Array Collaboration derived a spectrum which is well fit by a simple power law (dN/dE $\propto E^{-2.5\pm0.1}$). The data from this spectrum are also consistent with the Whipple and HEGRA spectra.

Multiwavelength Observations: The astrophysics of the γ-ray emission from the jets of AGNs are best explored using multiwavelength observations. These are difficult to organize and execute because of the different observing constraints on radio, optical, X-ray, space-based γ-ray and ground-based γ-ray observatories. Of necessity observations are often incomplete and, when complete coverage is arranged, the source does not always cooperate by behaving in an interesting way!

The first multiwavelength campaign on Mkn 421 coincided with a TeV flare on May 14–15, 1994 and showed some evidence for correlation with the X-ray band; however no enhanced activity was seen in EGRET [63]. A year later, in a longer campaign, there was again correlation between the TeV flare and the soft X-ray and UV data but with an apparent time lag of the latter by one day [12] (Figure 13). The variability amplitude is comparable in the X-ray and TeV emission ($\approx$ 400%) but is smaller in the EUV ($\approx$200%) and optical ($\approx$20%) bands.

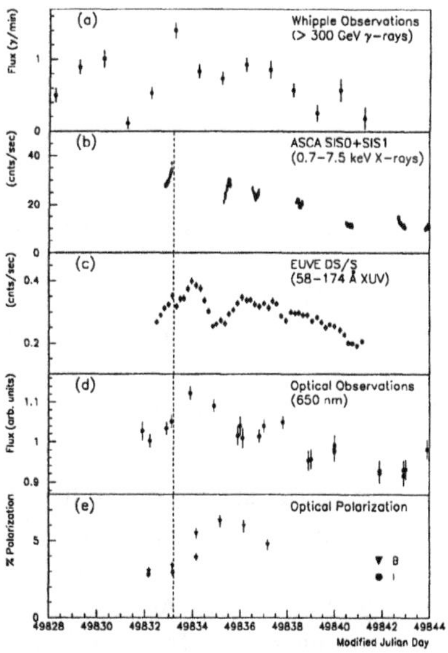

Fig. 13. Multi-wavelength observations of Mrk 421 (from [12]): (a) VHE γ-ray, (b) X-ray, (c) extreme UV, and (d) optical lightcurves taken during the period 1995 April–May

In 1998 there were extensive multiwavelength campaigns on this source between various ground-based gamma-ray observatories and the *ASCA* and *Beppo-SAX* X-ray satellites [99,66]. The most interesting event was the flare seen on April 21, 1998 at the Whipple Observatory [15] and by the *Beppo-SAX* telescopes. Although the flare was observed to rise and peak at the same time in both telescopes, the TeV signal decayed within a few hours whereas the X-ray signal persisted for half a day (Figure 14). It is difficult to model this behavior.

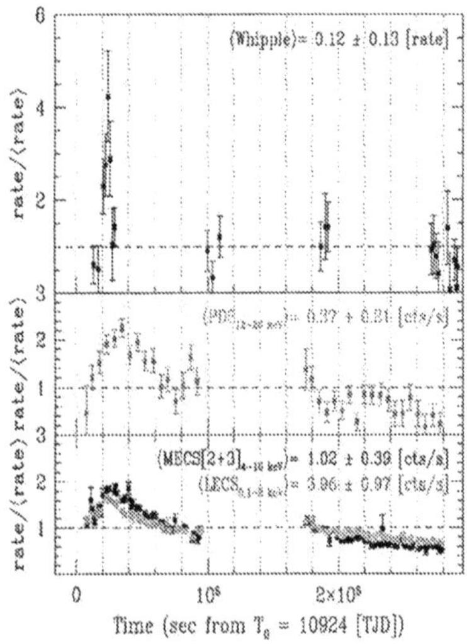

Fig. 14. X-ray and TeV gamma-ray flare as seen by SAX and Whipple, April, 1998

The first multiwavelength campaign on Mkn 501 was undertaken when the TeV signal was seen to be at a high level. The surprising result was that the source was detected by the OSSE experiment on CGRO in the 50–150 kev band (Figure 15). This was the highest flux ever recorded by OSSE from any blazar (it has not detected Mkn 421) but the amplitude of the X-ray variations (≈200%) was less than those of the TeV γ-rays (≈400%) [16].

Power Spectrum: Because of the strong variability in the TeV blazars it is difficult to represent their multiwavelength spectra. In Figure 16 and Figure 17 we show the fluxes plotted as power (ν F_ν) from Mkn 421 and Mkn 501 during flaring as well as the average fluxes. Both sources display the two peak distribution characteristic of Compton-synchrotron models, e.g., the Crab Nebula. Whereas the synchrotron peak in Mkn 421 occurs near 1 keV, that of Mkn 501 occurs beyond 100 keV which is the highest seen from any AGN. In 1998 the synchrotron spectrum peak in Mkn 501 shifted back to 5 keV and the TeV flux fell below the X-ray flux.

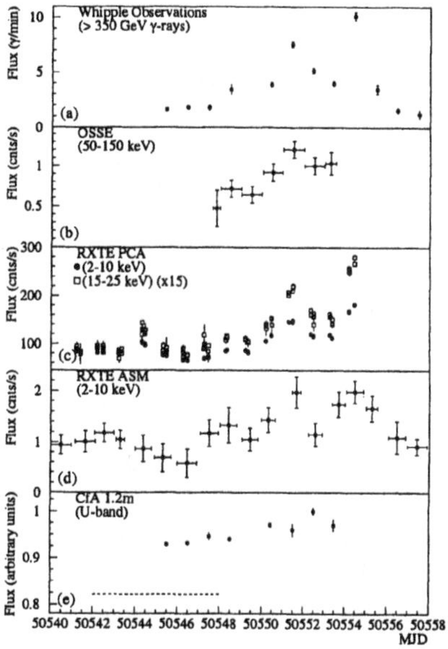

Fig. 15. Multi-wavelength observations of Mkn 501 (adapted from [16]): (a) γ-ray, (b) hard X-ray, (c) soft X-ray, (d) U-band optical taken during the period 1997 April 2–20 (April 2 corresponds to MJD 50540). The dashed line in (d) indicates the optical flux in 1997 March

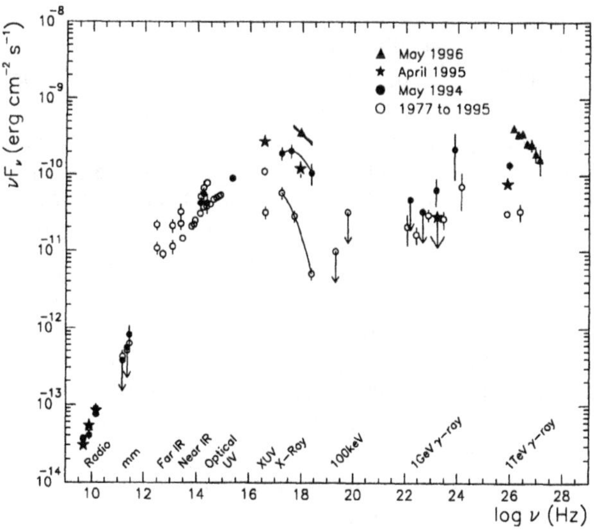

Fig. 16. The multi-wavelength power spectrum of Mkn 421 (adapted from [12]). The dashed line shows an SSC model fit to the data

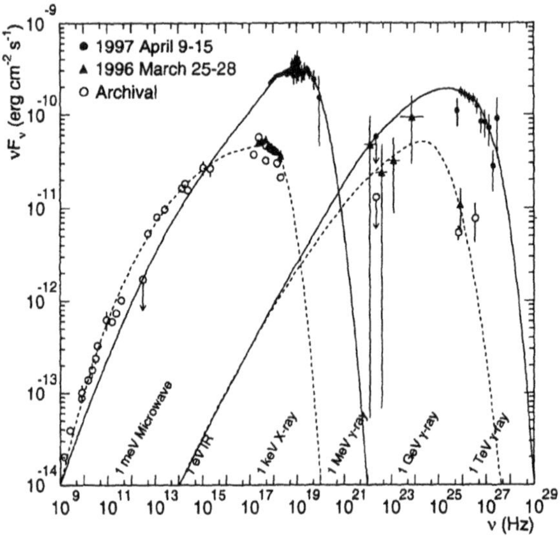

Fig. 17. The multi-wavelength power spectrum of Mkn 501 (adapted from [16])

Periodicity in the 1997 Signal from Mrk 501: Several groups have reported on the apparent periodicity in the TeV γ-ray signal from Mkn 501. The best data base is that of the HEGRA group since they observed during part of the bright period of the moon with one of their telescopes and hence have a database that is less prone to aliases. The reported periodicities occur at 12.7 day [45] and 23–24 day [52,33] and were arrived at using the Lomb method which is recommended for observations made at irregular intervals. The epoch chosen by the HEGRA group for periodicity analysis is a posteriori but coincides with the bulk of the TeV observations and the peak in the γ-ray signal intensity. There is no evidence for periodicity outside this interval, either in 1997 or in other years. A visual inspection shows that the γ-ray signal has a few clearly defined flares with several time constants and the most obvious is at 23 days.

Since all the γ-ray experiments were observing at approximately the same time, they must see the same time variations; hence reports from the separate experiments do not constitute independent confirmations. The important question is whether the observed "periodicity" is really statistically significant given the large number of time variations. It is difficult to arrive at the true statistical significance of the observed effect.

Similar periodicity is seen in the X-ray detector signal from RXTE and it has been suggested that this constitutes independent evidence for the periodicity. However correlation between the X-ray and TeV γ-ray signals from Mkn 421 and 501 on a variety of time-scales now seems to be well-established so that the independent analysis of the RXTE database only confirms this correlation, not the statistical significance of the periodicity.

The conclusion is that while there is apparent periodicity in the TeV/X-ray signals from Mkn 501 for a five month epoch in 1997, it is almost impossible to arrive at a satisfactory statistical significance.

5.3 Observations of Other AGNs

1ES2344+514: Although less well-studied, this X-ray-selected BL Lac at z=0.044 is superficially very similar to the above two sources. Recent X-ray observations by *Beppo-SAX* emphasize this similarity: time variability on times scales of hours has been seen and the putative synchrotron spectrum peaks at energies greater than 10 keV. It was reported as a TeV source [17] primarily on the basis of a flare seen in one night at the 6 σ level; the average flux over that night was $F_\gamma(>350\ \text{GeV}) = (6.6 \pm 1.9) \times 10^{-11}$ photons cm^{-2} s^{-1} which was 60% of the Crab. The averaged flux (including the flare) was at the 5.8 σ level. The source was not detected in the 1996/7 observing season.

Based on the observed behavior of Markarian 421 and Markarian 501 it might have been expected that continued monitoring of 1ES2344+514 would have confirmed this detection and given more information about its properties at high energies. In practice continued monitoring by the Whipple group (M.Catanese, private communication) and HEGRA [59] have not confirmed either the flaring or steady emission.

PKS2155-304: The three sources discussed to date are in the northern hemisphere; it had been predicted that PKS2155-304 would be the best candidate for TeV emission in the southern hemisphere. An X-ray-selected BL Lac, it has been detected by EGRET and has been the object of multiwavelength numerous observing campaigns. The Durham group detected it in 1996 and 1997 [25]; the November 1997 observations were particularly interesting as they coincided with observations by EGRET and RXTE which indicated that the source was active at this time.

More recent observations by the Durham group [24] have not detected the source. Because of its relatively large redshift (z=0.116), the energy spectrum of this source is of particular interest; however none is yet available.

1ES1959+650: The Utah Seven Telescope Array group have reported the detection of the BL Lac, 1ES1959+650 based on 57 hours of observation in 1998 [50]. As with the four AGNs listed above, this is an X-ray-selected BL Lac; its redshift is 0.048. The energy threshold for these observations was 600 GeV. The flux level was not reported but the total signal was at the 3.9 σ level. This is not normally considered high enough to claim the detection of a new source; however within this database there were two epochs which were selected a posteriori which gave signals above the canonical 5 σ level. This source has not yet been confirmed by any other group; it was observed by the Whipple group but no flux was detected.

3C66A: This is potentially the most exciting TeV detection of an AGN as it is quite different from the other AGNs. The source is a radio-selected, EGRET-detected, BL Lac and the redshift is 0.44, i.e., much more distant than the other objects. The Crimean Astrophysical Observatory group using the GT-48 telescope detected this source at the 5 σ level in 1996 [75]. The flux above 900 GeV was $(3 \pm 1) \times 10^{-11}$ photons cm^{-2} s^{-1}. There were previous and later upper limits to the TeV emission from the source, e.g. F_γ (> 350 GeV) < 1.9× 10^{-11} photons cm^{-2} s^{-1} from Whipple in 1993 [51]. Confirmation of this detection is urgently required.

5.4 Implications

The sample of VHE emitting AGNs is still very small but it is possible to draw some conclusions from their properties (summarized in Table 9).

Table 9. Properties of the VHE BL Lac objects

Object	z	EGRET flux (E>100 MeV) (10^{-7} cm^{-2} s^{-1})	Average flux (E>300 GeV) (10^{-12} cm^{-2} s^{-1})	M_v	$\mathcal{F}_X$ (2 keV) μJy	$\mathcal{F}_R$ (5 GHz) (mJy)
Mkn 421	0.031	1.4±0.2	40	14.4	3.9	720
Mkn 501	0.034	3.2±1.3	≥8.1	14.4	3.7	1370
1ES 2344+514	0.044	<0.7	≤8.2	15.5	1.1	220
1ES1959+650	0.048	< 0.5	<13.4	13.7	3.6	252
PKS 2155-304	0.116	3.2±0.8	42	13.5	5.7	310
3C 66A	0.444	2.0±0.3	30	15.5	0.6	806

- The first three objects, all detected by the Whipple group, are the three closest BL Lacs in the northern sky. Some 20 other BL Lacs have been observed with $z < 0.10$ without detectable emission. This could be fortuitous, because they are standard candles and these are closest (but the distance differences are small), or because they suffer the least absorption (but there is no cutoff apparent in their spectra).
- All of the objects are BL Lacs; because such objects do not show emission lines and therefore probably do not have strong optical/infrared absorption close to the source, it is suggested that BL Lacs are preferentially VHE emitters.
- Five of the six sources are classified as XBLs which indicates that they are strong in the X-ray region and that the synchrotron spectrum most likely peaks in that range (and that the Compton spectrum peaks in the VHE γ-ray range). The sixth, 3C 66A, is an RBL, like many of the blazars detected by EGRET; it is believed that these blazars have synchrotron spectra that peak at lower energies and Compton spectra that peak in the HE γ-ray region.

- Only three (Mkn 421, PKS 2155-304 and 3C 66A) are listed in the Third EGRET Catalog; there is a weak detection reported by EGRET for Mkn 501.
- If 3C 66A is confirmed (and to a lesser extent PKS 2155-305), then the intergalactic absorption is significantly less than had been suggested from galactic evolution models.
- There is evidence for variability in all of the sources. The rapid variability seen in Mkn 421 indicates that the emitting region is very small which might suggest it is close to the black hole. In that case the local absorption must be very low (low photon densities). It seems more likely that the region is well outside the dense core.

There are three basic classes of model considered to explain the high energy properties of BL Lac jets: Synchrotron Self Compton (SSC), Synchrotron External Compton (SEC) and Proton Cascade (PC) Models. In the first two the progenitor particles are electrons, in the third they are protons. VHE γ-ray observations have constrained the types of models that are likely to produce the γ-ray emission but still do not allow any of them to be eliminated. For instance, the correlation of the X-ray and the VHE flares is consistent with the first two models where the same population of electrons radiate the X-rays and γ-rays. There is little evidence for the IR component in BL Lac objects which would be necessary in the SEC models as the targets for Compton-scattering, so this particular type of model may not be likely for these objects. The PC models which produce the γ-ray emission through e^+e^- cascades also have great difficulty explaining the rapid cooling observed in the TeV emission from Mkn 421. Also the high densities of unbeamed photons near the nucleus, such as the accretion disk or the broad line region, are required to initiate the cascades and these cause high pair opacities to TeV γ-rays [27].

Significant information comes from the multiwavelength campaigns (although thus far these have been confined to Mkn 421 and Mkn 501). Simultaneous measurements constrain the magnetic field strength (B) and Doppler factor (δ) of the jet when the electron cooling is assumed to be via synchrotron losses. The correlation between the VHE γ-rays and optical/UV photons observed in 1995 from Mkn 421 indicates both sets of photons are produced in the same region of the jet; $\delta > 5$ is required for the VHE photons to escape significant pair-production losses [12]. If the VHE γ-rays are produced in the synchrotron-self-Compton process, $\delta = 15--40$ and $B = 0.03--0.9$G for Mrk 421 [15], [102] and $\delta < 15$ and $B = 0.08--0.2$G for Mkn 501 [90], [102]. On the other hand by assuming protons produce the γ-rays in Mkn 421, Mannheim [65] derives $\delta = 16$ and $B = 90$G. The Mkn 421 values of δ and B are extreme for blazars, but they are still within allowable ranges and are consistent with the extreme variability of Mkn 421.

5.5 Extragalactic Background Light

In traversing intergalactic distances, γ-rays may be absorbed by photon-photon pair production ($\gamma + \gamma \rightarrow e^+ + e^-$) on background photon fields if the center of mass energy of the photon-photon system exceeds twice the rest energy of the electron [41]. The cross-section for this process peaks when

$$E_\gamma \epsilon (1 - \cos\theta) \sim 2(m_e c^2)^2 = 0.52(\text{MeV})^2 \tag{1}$$

where E_γ is the energy of the γ-ray, ϵ is the energy of the low energy photon, θ is the collision angle between the two photons, m_e is the mass of the electron, and c is the speed of light in vacuum. Thus, for photons of energy near 1 TeV, head-on collisions with photons of ~0.5 eV have the highest cross-section, though a broad range of optical-to-IR wavelengths can be important absorbers because the cross-section for pair production is rather broad in energy and spectral features in the extragalactic background density can make certain wavebands more important than the cross-section alone would indicate.

The presence of extragalactic background light (EBL) limits the distance to which VHE γ-ray telescopes can detect sources. This has been put forth as an explanation of the lack of detection of many of the EGRET-detected AGNs (e.g., [96]), as discussed above. The difficulty in understanding the effect of the EBL on the opacity of the universe to VHE γ-rays is that not much is known about the spectrum of the EBL at present, nor how it developed over time. Star formation is expected to be a major contributor to the EBL, with star formation contributing mainly at short wavelengths (1–15 μm) and dust absorption and re-emission contributing at longer wavelengths (15–50 μm). So, measurements of the EBL spectrum can serve as important tracers of the history of the formation of stars and galaxies ([31]). Other, more exotic processes, such as pre-galactic star formation and some dark matter candidates, might also contribute distinctive features to the EBL. Measurements of the EBL have the potential to provide a wealth of information about several important topics in astrophysics.

Experiments that attempt to measure the EBL by directly detecting optical-IR photons, such as the Diffuse Infrared Background Experiment (DIRBE) on the *Cosmic Background Explorer* (*COBE*), are plagued by foreground sources of IR radiation. Emitted and scattered light from interplanetary dust, emission from unresolved stellar components in the Galaxy, and dust emission from the interstellar medium are all significantly more intense than the EBL and must be carefully modelled and subtracted to derive estimates of the EBL. Currently, EBL detections are available only at 140 μm and 240 μm [44]. Tentative detections at 3.5 μm [31] and 400–1000 μm [82] have also been reported.

Because VHE γ-rays are attenuated by optical-IR photons, measurements of the spectra of AGNs provide an indirect means of investigating the EBL that is not affected by local sources of IR radiation [41,96]. The signs of EBL absorption can be cutoffs, but also simple alterations of the spectral index (e.g., [98]), depending on the spectral shape of the EBL and the distance to the source. Like direct measurements of the EBL, this technique has difficulties to overcome. For instance, it requires some knowledge of, or assumptions about, the intrinsic spectrum and flux normalization of the AGNs or the EBL. Also, the AGNs themselves produce dense radiation fields which can absorb VHE γ-rays at the source and thereby mimic the effects of the intergalactic EBL attenuation.

Despite these difficulties, the accurate measurement of VHE spectra with no obvious spectral cut-offs from just the two confirmed VHE-emitting AGNs, Mrk 421 and Mrk 501, has permitted stringent limits to be set on the density of the EBL over a wide range of wavelengths. These limits have been derived from two approaches: (1) assuming a limit to the hardness of the intrinsic spectrum of the AGNs and deriving limits which assume very little about the EBL spectrum (e.g., [10,95]) and (2) assuming some shape for the EBL spectrum, based on theoretical or phenomenological modelling of the EBL, and adjusting the normalization of the EBL density to match the measured VHE spectra (e.g., [49,95]). The latter can be more stringent, but are necessarily more model-dependent. The limits from these indirect methods and from the direct measurements of EBL photons are summarized in Figure 18. At some wavelengths, the TeV limits represent a 50-fold improvement over the limits from DIRBE. These limits are currently well above the predicted density for the EBL from normal galaxy formation [64,80]), but they have provided constraints on a variety of more exotic mechanisms for sources of the EBL (e.g., [10]). They also show that EBL attenuation alone cannot explain the lack of detection of EGRET sources with nearby redshifts at VHE energies, as the optical depth for pair-production does not reach 1 for the stringent limits of [10] until beyond a redshift of $z = 0.1$. With the detection or more AGNs, particularly at higher redshift, and improvements in our understanding of the emission and absorption processes in AGNs, VHE measurements have the potential to set very restrictive limits on the EBL density, and perhaps eventually detect it.

6 Future Prospects

6.1 HE Gamma Ray Astronomy

Although EGRET has still some sensitivity, the mission is essentially over and not much change can be expected in the observational picture until the launch of GLAST in 2005 (hopefully). The intermediate missions, AMS and AGILE, will not significantly improve on the EGRET flux sensitivity and can be considered place-holders for GLAST. The latter will offer an improvement of a factor of 10–20 in most parameters compared to EGRET. The energy coverage anticipated over the next ten years is shown in Table 10.

GLAST (Gamma-ray Large Area Space Telescope)is the next generation pair production telescope that will replace the spark chamber with solid state detectors which will be more compact, more efficient and have better angular and energy resolution. GLAST will operate in the range 20 MeV - 300 GeV, with a scheduled launch date of 2005. GLAST will surpass EGRET by a factor of ten–forty in most parameters (Table 11).

There are two competing technologies for the central pair production detector on GLAST. One (Fiber GLAST) uses crossed planes of scintillation fibers coupled to multi-anode photomultipliers, separated by thin layers of high Z converter plates. The calorimeter uses the same kind of detector but with thicker plates. The fibers are 1.3 m long and the whole detector is 1.8 m high; they have

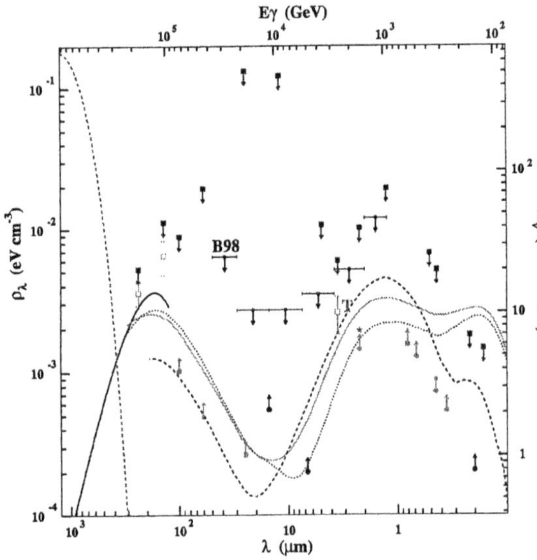

Fig. 18. The diffuse intergalactic infrared background. E_γ is the energy at which the pair-production cross-section peaks for head on collisions with photons of wavelength λ. Upper limits derived from VHE γ-ray spectra are indicated by the horizontal bars with arrows, marked as B98 [10]. Filled squares are upper limits from various experiments measuring the EBL directly. The open squares at 140 μm and 240 μm are detections from DIRBE [44]. (The filled circles are lower limits derived from galaxy counts. The solid curve between 90 μm and 150 μm is a FIRAS detection. The dashed line on the left indicates the 2.7 K cosmic microwave background radiation. The three curves spanning most of the IR wavelengths are different models of [80]. Figure courtesy of V. Vassiliev [107]

a square cross-section of side < 1mm. This technology has already been used in cosmic ray particle experiments and is thus favored by space scientists. The other technology (Silicon GLAST) uses the silicon strip technology that has been used in high energy particle accelerator experiments for a number of years; it has not, so far, been used in space science applications. Again the layers of ionizing particle-sensitive detectors are alternated with thin layers of lead converter. The calorimeter will be made of bars of Caesium Iodide, with individual read outs to give spatial resolution.

Both technologies seem to address equally well the physical demands of GLAST, so it will be a difficult choice to select just one of them. Remarkably both technologies can achieve the dramatic improvement over EGRET, outlined in Table 11, with an instrument that will only be twice as heavy.

6.2 VHE Gamma Ray Astronomy

In contrast to the drought expected in MeV–GeV γ-ray observations in the immediate future, ground-based γ-ray astronomy has never been more active.

Table 10. Future Roadmap for HE/VHE Gamma Ray Astronomy

Energy	MeV 10–100 Space	GeV 0.1–1 Space	GeV 1–10 Space	GeV 10–100 Space/ Ground	TeV 0.1–1 Ground	TeV 1–10+ Ground
Year						
1999	*Comptel*	(EGRET)		**********	**9ACITs*	***+2ASA
2000	****			CEL/STAC	**********	**********
2001	****			**********	**********	**********
2002	Integral			**********	**********	**********
2003	**	AMS/AGILE	**********	*MAGIC**	**********	**********
2003	**	***********	**********	**********	HESS/CAN	**********
2004	**	***********	**********	**********	VERITAS*	**********
2005	*GLAST**	**GLAST**	**GLAST*	*GLAST**	**********	**********
2006	*********	***********	**********	**********	**********	**********
2007	*********	***********	**********	**********	**********	**********
2008	*********	***********	**********	**********	**********	**********

Table 11. Comparison of EGRET and GLAST

Parameter	Units	EGRET (achieved)	GLAST (desired)
Energy Range	MeV	20–30,000	20–300,000
Effective Area	cm^2	1,500	8,000
Field of View	sr	0.5	2
Angular Resolution			
(100 MeV) °	5.8	3.0	
Energy Resolution	%	10	10
Source Sensitivity			
(> 100 MeV)	10^{-7} cm^{-2} s^{-1}	1	4

There are already nine atmospheric Cherenkov imaging telescopes in operation and two air shower arrays; there will be steady improvements in sensitivity in these telescopes over the next decade. One can expect to see a steady increase in the GeV–TeV source catalog (Table 12) from ground-based observations so that even if the GLAST launch were to be delayed there would be a healthy increase in activity in studies of γ-ray astrophysics at these very high energies.

To fully exploit the potential of ground-based γ-ray astronomy the detection techniques must be improved by an order of magnitude. This will happen by extending the energy coverage of the techniques and by increasing their flux sensitivity. Ideally one would like to do both but in practice there must be trade-offs. Reduced energy threshold can be achieved by the use of larger, but cruder, mirrors and this approach is currently being exploited using existing arrays of solar heliostats (STACEE ([26]) and CELESTE ([73]). A German-

Table 12. TeV Source Catalog c.1999 [113]

Source	Type	Discovery	EGRET	Credibility
Galactic				
Crab Nebula	Plerion	1989	yes	A
PSR 1706-44	Plerion?	1995	no	A
Vela	Plerion?	1997	no	B
SN1006	Shell	1997	no	B-
RXJ1713.7-3946	Shell	1999	no	B
Casssiopea A	Shell	1999	no	C
Centuarus X-3	Binary	1998	yes	C
Extragalactic				
Markarian 421	XBL z=0.031	1992	yes	A
Markarian 501	XBL z=0.034	1995	yes	A
1ES2344+514	XBL z=0.044	1997	no	C
PKS2155-304	XBL z=0.116	1999	yes	B
PKS1959+650	XBL z=0.048	1999	no	B-
3C66A	RBL z=0.44	1998	yes	C

Spanish project (MAGIC) ([9]) to build a 17 m aperture telescope has also been proposed. These projects may achieve thresholds as low as 20–30 GeV where they will effectively fill the current gap in the γ-ray spectrum from 20 to 200 GeV. Ultimately this gap will be covered by GLAST with less point source sensitivity at the higher energies. Extension to higher energies (>10 TeV) can be achieved by atmospheric Cherenkov telescopes working at large zenith angles and by particle arrays at very high altitudes.

One of the most ambitious of the Next Generation VHE Telescopes is the Very Energetic Radiation Imaging Telescope Array System (VERITAS) [11]. VERITAS will consist of six telescopes located at the corners of a hexagon of side 80 m with a seventh at the center (Figure 6.2). The telescopes will be similar to the design of the Whipple 10m reflector, which is the most sensitive telescope of its kind.

By employing largely existing technology in the first instance and stereoscopic imaging, VERITAS will achieve the following:

- *Effective area:* >0.1 km^2 at 1 TeV.
- *Effective energy threshold:* <100 GeV with significant sensitivity at 50 GeV.
- *Energy resolution:* 10%–15% for events in the range 0.2 to 10 TeV.
- *Angular Resolution:* <0.05° for individual photons; source location to better than 0.005°.

VERITAS will concentrate on the exciting region between space-based instruments and air shower arrays, with its primary objective being high sensitivity in the 100 GeV to 10 TeV range. The German-French HESS (initially four

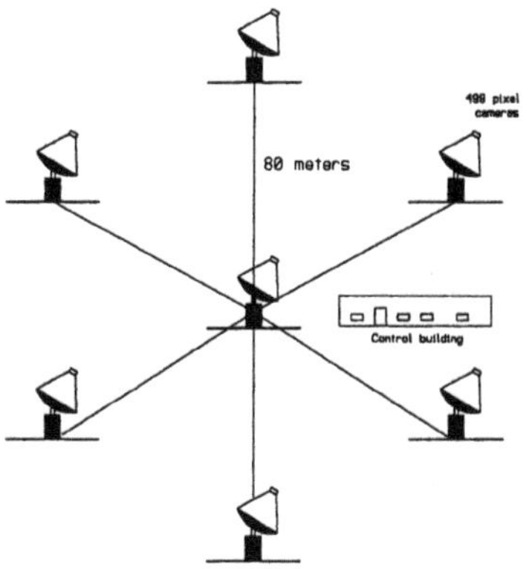

Fig. 19. The seven telescopes of VERITAS, each of 10m aperture, will have the hexagonal distribution shown

and eventually perhaps sixteen 10 m class telescopes) will be built in Namibia [57] and the Japanese NEW CANGAROO array (with three to four telescopes in Australia) [67] will have similar objectives for observations in the southern hemisphere. In each case, the arrays will exploit the high sensitivity of the atmospheric Cherenkov imaging technique and the high selectivity of the array approach. The relative flux sensitivities as a function of energy are shown in Figure 20, where the sensitivities of the wide field detectors are for one year and the ACT are for 50 hours; in all cases a 5σ point source detection is required.

It is apparent from this figure that, on the low energy side, VERITAS will complement the GLAST mission and will overlap with STACEE and CELESTE. At its highest energy it will overlap with the Tibet Air Shower Array [5]. It will cover the same energy range as MILAGRO but with greater flux sensitivity. The wide field coverage of MILAGRO will permit the detection of transient sources which, once detected, can be studied in more detail by VERITAS.

7 Footnote

It is a matter of some disappointment for the many cosmic-ray physicists who entered the field of high energy γ-ray astronomy that none of the sources thus far detected, either at HE or VHE energies, can positively be identified with hadron progenitors. In the early days it was widely believed that γ-ray astronomy would finally solve the mystery of the origin of the cosmic radiation. However with the exception of the Galactic Plane (and perhaps the Large Magellanic Cloud) where we observe, not the source of cosmic radiation but its propagation, every one of the sources detected so far can be attributed to a source in which electrons are the progenitor particles. Only in the case of the Galactic Plane is the much heralded "bump" in the energy spectrum near 70 MeV seen. In some cases there are proponents of plausible models in which hadrons are the progenitors but there are equally vociferous proponents who would advocate electron models

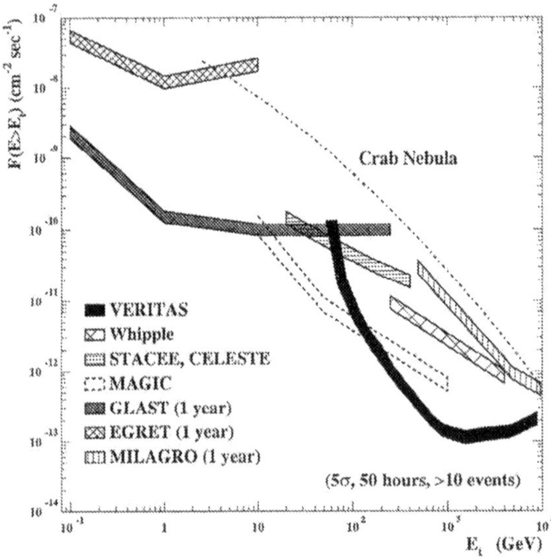

Fig. 20. Comparison of the point source sensitivity of VERITAS to Whipple [110], MAGIC [9], CELESTE/STACEE [86]; [26]; GLAST [38], EGRET [104], and MILAGRO [92]. The sensitivity of MAGIC is based on the availability of new technologies, e.g., high quantum efficiency PMTs, not assumed in the other experiments. EGRET, GLAST and MILAGRO are wide field instruments and therefore ideally suited for all sky surveys

and in many cases these seem the more plausible. Thus in the 40 plus years since the publication of Morrison's seminal paper [70] while we have learned some interesting astrophysics we have come no closer to a definitive model of cosmic-ray origins.

Acknowledgements

Research in VHE γ-ray astronomy at the Smithsonian Astrophysical Observatory is supported by a grant from the U.S. Department of Energy. The United States-Mexico Foundation for Sciences in acknowledged for financial support during my stay in Mexico.

References

1. Aharonian, F.A. et al.: Astron.Astrophys., **342**, 69 (1999)
2. Aharonian, F.A. et al.: Astron.Astrophys., in press (1999)
3. Aharonian, F.A. et al.: in Proc. 26th *Internat. Cosmic Ray Conf.*(Salt Lake City), **3**, 350.
4. Aiso, S., et al.: in Proc. 25th *Internat. Cosmic Ray Conf.*(Durban, South Africa), ed. by M. S. Potgeier, B. C. Raubenheimer, & D. J. van der Walt, **3**, 261 (1997)

5. Amenomori, M. et al.: in Proc. 25th *Internat. Cosmic Ray Conf.*(Durban, South Africa), ed. by M. S. Potgeier, B. C. Raubenheimer, & D. J. van der Walt, **5**, 245 (1997)
6. Amenomori, M. et al.: in Proc. 26th *Internat. Cosmic Ray Conf.*(Salt Lake City), **3**, 456 (1999)
7. Baring, M., et al.: Astrophys.J., **513**, 311 (1999)
8. Barrau, A., et al.: Nucl. Instrum. Methods A, **416**, 278 (1998)
9. Barrio, J.A., et al.: "The MAGIC Telescope", design study, MPI-PhE/98-5 (1998)
10. Biller, S. D., et al.: Phys. Rev. Lett., **80**, 2992 (1998a)
11. Bradbury, S. et al.: in Proc. 26th *Internat. Cosmic Ray Conf.*(Salt Lake City), **5**, 280 (1999)
12. Buckley, J. H., et al.: Astrophys.J.Lettr., **472**, L9 (1996)
13. Buckley, J. H., et al.: Astron.Astrophys., **329**, 639 (1998)
14. Carraminana, A., et al.: in *Timing Neutron Stars*, (Kluwer Acad. Press, Dordrecht 1989) p. 389
15. Catanese, M.: in *BL Lacertae Phenomenon*, ed. by L. O. Takalo & A. Silanpaa (San Francisco: ASP), ASP Conf. Series, **159**, 243 (1999)
16. Catanese, M., et al.: Astrophys.J.Lettr., **487**, L143 (1997)
17. Catanese, M., et al.: 1998, Astrophys.J., **501**, 616 (1998)
18. Catanese, M., Weekes, T.C.: Publ. Ast. Soc. Pac. **111**, 1193 (1999)
19. Chadwick, P.M. et al.: J. Phys. G.: Nucl. Part. Phys. **16**, 1773 (1990)
20. Chadwick, P. M., et al.: in Proc. 25th *Internat. Cosmic Ray Conf.*(Durban), **3**, 189 (1997)
21. Chadwick, P.M. et al.: Astrophys.J., **503**, 391 (1998)
22. Osborne, J.: in Proc. of *Workshop on GeV-TeV Astrophysics*, Snowbird, Colorado (in press) (1999)
23. Chadwick, P.M. et al.: in Proc. 26th *Internat. Cosmic Ray Conf.*(Salt Lake City), **4**, 72 (1999)
24. Chadwick, P.M. et al.: in Proc. 26th *Internat. Cosmic Ray Conf.*(Salt Lake City), **3**, 338 (1999)
25. Chadwick, P.M. et al.: Astrophys.J., **513**, 161 (1999)
26. Chantell, M.C., et al.: Nucl. Instrum. Methods A, **408**, 468 (1998)
27. Coppi, P. S., Kartje, J. F., & Königl, A.: in Proc. of the *Compton Symposium*, ed. by M. Friedlander, N. Gehrels, & D. J. Macomb (New York: AIP), 559 (1993)
28. Chudakov, A.E. et al.: Transl. Consultants Bureau, P.N.Lebedev Phys. Inst. **26**, 99 (1965)
29. Daum, A., et al.: Astropart. Phys., **8**, 1 (1997)
30. Drury, L. O'C., Aharonian, F. A., & Völk, H. J.: Astron.Astrophys., **287**, 959 (1994)
31. Dwek, E., et al.: Astrophys.J., **508**, 106 (1998)
32. Esposito, J. A., et al.: Astrophys.J., **461**, 820 (1996)
33. Fegan, S. et al.: in Proc. of *Workshop on GeV-TeV Astrophysics*, Snowbird, Colorado (in press) (1999)
34. Fichtel, C.E., Trombka, J.I.: "Gamma Ray Astrophysics", NASA Ref. Publ. 1386 (1997)
35. Gaisser, T. K., Protheroe, R. J., & Stanev, T.: Astrophys.J., **492**, 219 (1998)
36. Proceedings of the *Fourth Compton Symposium*, Williamsburg, Virginia, USA, April 1997, ed. by C.D.Dermer, M.S.Strickman, J.D.Kurfess, AIP **410** (1997)
37. Gaidos, J. A., et al.: Nature, **383**, 319 (1996)
38. Gehrels, N., & Michelson, P.: in *TeV Astrophysics of Extragalactic Sources*, ed. by M. Catanese & T. C. Weekes, Astropart. Phys., **11**, 277 (1999)

39. Ghisellini, G., Celotti, A., Fossati, G., Maraschi, L., & Comastri, A.: Mon. Not. Roy.Ast.Soc. **301**, 451 (1998)
40. Goret, P. et al.: in Proc. 26th *Internat. Cosmic Ray Conf.*(Salt Lake City), **3**, 49 (1999)
41. Gould, R. J., & Schréder, G. P.: Phys. Rev., **155**, 1408 (1967)
42. Hara, T., et al.: Nucl. Instrum. Methods A, **332**, 300 (1993)
43. Hartman, R. C., et al.: Astrophys.J.Suppl., **123**, 79 (1999)
44. Hauser, M. G., et al.: Astrophys.J., **508**, 25 (1998)
45. Hayashida, N., et al.: Astrophys.J.Lettr., **504**, L71 (1998)
46. Hess, M. et al.: in Proc. 25th *Internat. Cosmic Ray Conf.*(Durban, South Africa), ed. by M. S. Potgeier, B. C. Raubenheimer, & D. J. van der Walt, **3**, 229 (1997)
47. Hillas, A. M., et al.: Astrophys.J., **503**, 744 (1998)
48. Hunter, S. D., et al.: Astrophys.J., **481**, 205 (1997)
49. de Jager, O. C., Stecker, F. W., & Salamon, M. H.: Nature, **369**, 294 (1994)
50. Kajino, F. et al.: in Proc. 26th *Internat. Cosmic Ray Conf.*(Salt Lake City), **3**, 370 (1999)
51. Kerrick, A. D., et al.: Astrophys.J., **452**, 588 (1995)
52. Kranich, D. et al.: in Proc. 26th *Internat. Cosmic Ray Conf.*(Salt Lake City), **3**, 358 (1999)
53. Krennrich, F. et al.: Astrophys.J., **481**, 758 (1997)
54. Krennrich, F. et al.: in Proc. 26th *Internat. Cosmic Ray Conf.*(Salt Lake City), **3**, 301 (1999)
55. Krennrich, F., et al.: Astrophys.J., **511**, 149 (1999)
56. Kifune, T., et al.: Astrophys.J.Lettr., **438**, L91 (1995)
57. Konopelko, A.K.: in *TeV Astrophysics of Extragalactic Sources*, ed. by M. Catanese & T. C. Weekes, Astropart. Phys. **11**, 263 (1999)
58. Konopelko, A. et al.: in Proc. 26th *Internat. Cosmic Ray Conf.*(Salt Lake City), **3**, 444 (1999)
59. Konopelko, A. et al.: in Proc. 26th *Internat. Cosmic Ray Conf.*(Salt Lake City), **3**, 426 (1999)
60. Konopelko, A.K.: in *TeV Astrophysics of Extragalactic Sources*, ed. by M. Catanese & T. C. Weekes, Astropart. Phys., **11**, 135 (1999)
61. Lessard, R.W. et al.: in Proc. 26th *Internat. Cosmic Ray Conf.*(Salt Lake City), **3**, 488 (1999)
62. Lorenz, E.: in *TeV Astrophysics of Extragalactic Sources*, ed. by M. Catanese & T. C. Weekes, Astropart. Phys., **11**, 131 (1999)
63. Macomb, D. J., et al.: Astrophys.J.Lettr., **449**, L99 (1995)
64. Madau, P., et al.: Mon.Not.Roy.Ast.Soc., **283**, 1388 (1996)
65. Mannheim, K.: Astron.Astrophys., **269**, 67 (1993)
66. Maraschi, L., et al.: in *TeV Astrophysics of Extragalactic Sources*, ed. by M. Catanese & T. C. Weekes, Astropart. Phys., **11**, 189 (1999)
67. Matsubara, Y.: in Proc. of *Towards a Major Atmospheric Cherenkov Detector -V* (Kruger Park), ed. by O.C. deJager, 447 (1997)
68. Muraisi, H. et al.: in Proc. 26th *Internat. Cosmic Ray Conf.*(Salt Lake City), **3**, 500 (1999)
69. Musquere, A. et al.: in Proc. 26th *Internat. Cosmic Ray Conf.*(Salt Lake City), **3**, 460 (1999)
70. Morrison, P.: Il Nuovo Cimento, **7**, 858 (1958)
71. von Montigny, C., et al.: Astrophys.J., **440**, 525 (1995)
72. Mukherjee, R., et al.: Astrophys.J., **490**, 116 (1997)

73. Musquire, A. et al.: in Proc. 26th *Internat. Cosmic Ray Conf.*(Salt Lake City), **3**, 527 (1999)
74. Naito, T., & Takahara, F.: J. Phys. G **20**, 477 (1994)
75. Neshpor, Y. I., et al.: Astron. Letts., **24**, 134 (1998)
76. Nishikawa, D. et al.: 26th *Internat. Cosmic Ray Conf.*(Salt Lake City), **3**, 354 (1999)
77. Ong, R. A.: Physics Reports, **305**, 93 (1998)
78. Oser, S. et al.: 26th *Internat. Cosmic Ray Conf.*(Salt Lake City), **3**, 464 (1999)
79. Piron, F. et al.: 26th *Internat. Cosmic Ray Conf.*(Salt Lake City), **3**, 326 (1999)
80. Primack, J. et al.: Astroparticle Phys. in *TeV Astrophysics of Extragalactic Sources*, ed. by M. Catanese & T. C. Weekes, Astropart. Phys., **11**, 93 (1999)
81. Protheroe, R. J., et al.: in Proc. 25th *Internat. Cosmic Ray Conf.*(Durban), **8**, 317 (1997)
82. Puget, J.-L., et al.: Astron.Astrophys., **308**, L5 (1996)
83. Puelhlhofer, G. et al.: in Proc. 26th *Internat. Cosmic Ray Conf.*(Salt Lake City), **3**, 492 (1999)
84. Punch, M., et al.: Nature, **358**, 477 (1992)
85. Punch, M. et al.: in Proc. 25th *Internat. Cosmic Ray Conf.*(Durban), **3**, 253 (1997)
86. Quebert, J., et al.: in *Towards a Major Atmospheric Cherenkov Detector - IV* (Padova, Italy), ed. by M.Cresti, 248 (1995)
87. Quinn, J., et al.: Astrophys.J.Lettr., **456**, L83 (1996)
88. Quinn, J., et al.: Astrophys.J., **518**, 693 (1998)
89. Raubenheimer, B.C. et al.: Astrophys.J., **336**, 34 (1989)
90. Samuelson, F. W., et al.: Astrophys.J.Lettr., **501**, L17 (1998)
91. Samuelson, F. W.: Ph.D. thesis, Iowa State University (1999)
92. Sinnis, G., et al.: Nucl. Phys. B (Proc. Suppl.), **43**, 141 (1995)
93. Sreekumar, P. et al.: Phys. Rev. Lettr. **70**, 127 (1993)
94. Sreekumar, P. et al.: Astroparticle Phys., G., & Kubo, H. in *TeV Astrophysics of Extragalactic Sources*, ed.by M. Catanese & T. C. Weekes, Astropart. Phys., **11**, 221 (1999)
95. Stanev, T., & Franceschini, A.: Astrophys.J.Lettr., **494**, L59 (1998)
96. Stecker, F. W., de Jager, O. C., & Salamon, M.: Astrophys.J.Lettr., **390**, L49 (1992)
97. Srinivasan, R., et al.: Astrophys.J., **489**, 170 (1997)
98. Stecker, F. W.: in *TeV Astrophysics of Extragalactic Sources*, ed. by M. Catanese & T. C. Weekes, Astropart. Phys., **11**, 83 (1999)
99. Takahashi, T., Madejski, G., & Kubo, H.: in *TeV Astrophysics of Extragalactic Sources*, ed. by M. Catanese & T. C. Weekes, Astropart. Phys., **11**, 177 (1999)
100. Tanimori, T., et al.: Astrophys.J.Lettr., **497**, L25 (1998)
101. Tanimori, T., et al.: Astrophys.J.Lettr., **492**, L33 (1998)
102. Tavecchio, F., Maraschi, L., & Ghisellini, G.: Astrophys.J., **509**, 608 (1998)
103. Tavernet, J.P. et al.: 26th *Internat. Cosmic Ray Conf.*(Salt Lake City), **3**, 322 (1997)
104. Thompson, D. J., et al.: Astrophys.J.Suppl., **86**, 629 (1993)
105. Thompson, D.J.: in *Neutron Stars and Pulsars*, ed. by Shibazaki, N., et al. (Tokyo: Univ. Acad. Press), 273 (1997)
106. Vacanti, G., et al.: Astrophys.J., **377**, 467 (1991)
107. Vassiliev, V.V.: Astroparticle Phys. (in press) (astroph/9908088) (1999)
108. Vestrand, et al.: Astrophys.J.Lettr., **483**, L49 (1997)

109. Weekes, T.C.: Physics Reports, **160**, 1 (1988)
110. Weekes, T.C., et al.: Astrophys.J., **342**, 379 (1989)
111. Weekes, T. C.: Space Sci. Rev., **59**, 315 (1991)
112. Weekes, T.C.: Physica Scripta, **60**, (in press) (1999)
113. Weekes, T.C. : in Proc. of *Workshop on GeV-TeV Astrophysics*, Snowbird, Colorado (in press) (1999)
114. Yamamoto, T. et al.: in Proc. 26th *Internat. Cosmic Ray Conf.*(Salt Lake City), **3**, 297 (1999)
115. Yoshikoshi, T., et al. : Astrophys.J.Lettr., **487**, L65 (1997)
116. Zweerink, J. A., et al.: Astrophys.J.Lettr., **490**, L141 (1997)

Neutrinos in Physics and Astrophysics

Esteban Roulet

Departamento de Física
Universidad Nacional de La Plata
CC67, 1900, La Plata, Argentina

Abstract. An elementary general overview of the neutrino physics and astrophysics is given. We start by a historical account of the development of our understanding of neutrinos and how they helped to unravel the structure of the Standard Model. We discuss why it is so important to establish if neutrinos are massive and we introduce the main scenarios to provide them a mass. The present bounds and the positive indications in favor of non-zero neutrino masses are discussed as well as the major role they play in astrophysics and cosmology.

1 The Neutrino Story

1.1 The Hypothetical Particle

One may trace back the appearance of neutrinos in physics to the discovery of radioactivity by Becquerel one century ago. When the energy of the electrons (beta rays) emitted in a radioactive decay was measured by Chadwick in 1914, it turned out to his surprise to be continuously distributed. This was not to be expected if the underlying process in the beta decay was the transmutation of an element X into another one X' with the emission of an electron, i.e. $X \to X' + e$, since in that case the electron should be monochromatic. The situation was so puzzling that Bohr even suggested that the conservation of energy may not hold in the weak decays. Another serious problem with the 'nuclear models' of the time was the belief that nuclei consisted of protons and electrons, the only known particles by then. To explain the mass and the charge of a nucleus it was then necessary that it had A protons and $A - Z$ electrons in it. For instance, a ^{4}He nucleus would have 4 protons and 2 electrons. Notice that this total of six fermions would make the ^{4}He nucleus to be a boson, which is correct. However, the problem arouse when this theory was applied for instance to ^{14}N, since consisting of 14 protons and 7 electrons would make it a fermion, but the measured angular momentum of the nitrogen nucleus was $I = 1$.

The solution to these two puzzles was suggested by Pauli only in 1930, in a famous letter to the 'Radioactive Ladies and Gentlemen' gathered in a meeting in Tubingen, where he wrote: 'I have hit upon a desperate remedy to save the exchange theorem of statistics and the law of conservation of energy. Namely, the possibility that there could exist in nuclei electrically neutral particles, that I wish to call neutrons, which have spin 1/2 ...'. These had to be not heavier than electrons and interacting not more strongly than gamma rays.

With this new paradigm, the nitrogen nucleus became ^{14}N$= 14p+7e+7$'n', which is a boson, and a beta decay now involved the emission of two particles $X \to X' + e +$ 'n', and hence the electron spectrum was continuous. Notice that no particles were created in a weak decay, both the electron and Pauli's neutron 'n' were already present in the nucleus of the element X, and they just came out in the decay. However, in 1932 Chadwick discovered the real 'neutron', with a mass similar to that of the proton and being the missing building block of the nuclei, so that a nitrogen nucleus finally became just ^{14}N$= 7p + 7n$, which also had the correct bosonic statistics.

In order to account now for the beta spectrum of weak decays, Fermi called Pauli's hypothetised particle the neutrino (small neutron), ν, and furthermore suggested that the fundamental process underlying beta decay was $n \to p+e+\nu$. He wrote [1] the basic interaction by analogy with the interaction known at the time, the QED, i.e. as a vector×vector current interaction:

$$H_{\rm F} = G_{\rm F} \int {\rm d}^3x [\bar{\Psi}_p \gamma_\mu \Psi_n][\bar{\Psi}_e \gamma^\mu \Psi_\nu] + h.c..$$

This interaction accounted for the continuous beta spectrum, and from the measured shape at the endpoint Fermi concluded that m_ν was consistent with zero and had to be small. The Fermi coupling $G_{\rm F}$ was estimated from the observed lifetimes of radioactive elements, and armed with this Hamiltonian Bethe and Peierls [2] decided to compute the cross section for the inverse beta process, i.e. for $\bar{\nu}+p \to n+e^+$, which was the relevant reaction to attempt the direct detection of a neutrino. The result, $\sigma = 4(G_{\rm F}^2/\pi)p_e E_e \simeq 2.3 \times 10^{-44}{\rm cm}^2(p_e E_e/m_e^2)$ was so tiny that they concluded '... This meant that one obviously would never be able to see a neutrino.'. For instance, if one computes the mean free path in water (with density $n \simeq 10^{23}/{\rm cm}^3$) of a neutrino with energy $E_\nu = 2.5$ MeV, typical of a weak decay, the result is $\lambda \equiv 1/n\sigma \simeq 2.5 \times 10^{20}$ cm, which is 10^7AU, i.e. comparable to the thickness of the Galactic disk.

It was only in 1958 that Reines and Cowan were able to prove that Bethe and Peierls had been too pessimistic, when they measured for the first time the interaction of a neutrino through the inverse beta process[3]. Their strategy was essentially that, if one needs 10^{20} cm of water to stop a neutrino, having 10^{20} neutrinos a cm would be enough to stop one neutrino. Since after the second war powerful reactors started to become available, and taking into account that in every fission of an uranium nucleus the neutron rich fragments beta decay producing typically 6 $\bar{\nu}$ and liberating ~ 200 MeV, it is easy to show that the (isotropic) neutrino flux at a reactor is

$$\frac{d\Phi_\nu}{d\Omega} \simeq \frac{2 \times 10^{20}}{4\pi} \left(\frac{\rm Power}{\rm GWatt}\right) \frac{\bar{\nu}}{\rm strad}.$$

Hence, placing a few hundred liters of water near a reactor they were able to see the production of positrons (through the two 511 keV γ produced in their annihilation with electrons) and neutrons (through the delayed γ from the neutron capture in Cd), with a rate consistent with the expectations from the weak interactions of the neutrinos.

1.2 The Vampire

Going back in time again to follow the evolution of the theory of weak interactions of neutrinos, in 1936 Gamow and Teller [4] noticed that the $V \times V$ Hamiltonian of Fermi was probably too restrictive, and they suggested the generalization

$$H_{\rm GT} = \sum_i G_i [\bar{p} O_i n][\bar{e} O^i \nu] + h.c.,$$

involving the operators $O_i = 1,\ \gamma_\mu,\ \gamma_\mu\gamma_5,\ \gamma_5,\ \sigma_{\mu\nu}$, corresponding to scalar (S), vector (V), axial vector (A), pseudoscalar (P) and tensor (T) currents. However, since A and P only appeared here as $A \times A$ or $P \times P$, the interaction was parity conserving. The situation became unpleasant, since now there were five different coupling constants G_i to fit with experiments, but however this step was required since some observed nuclear transitions which were forbidden for the Fermi interaction became now allowed with its generalization (GT transitions).

The story became more involved when in 1956 Lee and Yang suggested that parity could be violated in weak interactions[5]. This could explain why the particles theta and tau had exactly the same mass and charge and only differed in that the first one was decaying to two pions while the second to three pions (e.g. to states with different parity). The explanation to the puzzle was that the Θ and τ were just the same particle, now known as the charged kaon, but the (weak) interaction leading to its decays violated parity.

Parity violation was confirmed the same year by Wu [6], studying the direction of emission of the electrons emitted in the beta decay of polarized ^{60}Co. The decay rate is proportional to $1 + \alpha \boldsymbol{P} \cdot \hat{p}_e$. Since the Co polarization vector $\boldsymbol{P}$ is an axial vector, while the unit vector along the electron momentum $\hat{p}_e$ is a vector, their scalar product is a pseudoscalar and hence a non-vanishing coefficient α would imply parity violation. The result was that electrons preferred to be emitted in the direction opposite to $\boldsymbol{P}$, and the measured value $\alpha \simeq -0.7$ had then profound implications for the physics of weak interactions.

The generalization by Lee and Yang of the Gamow Teller Hamiltonian was

$$H_{\rm LY} = \sum_i [\bar{p} O_i n][\bar{e} O^i (G_i + G_i' \gamma_5)\nu] + h.c..$$

Now the presence of terms such as $V \times A$ or $P \times S$ allows for parity violation, but clearly the situation became even more unpleasant since there are now 10 couplings (G_i and G_i') to determine, so that some order was really called for.

Soon the bright people in the field realized that there could be a simple explanation of why parity was violated in weak interactions, the only one involving neutrinos, and this had just to do with the nature of the neutrinos. Lee and Yang, Landau and Salam [7] realized that, if the neutrino was massless, there was no need to have both neutrino chirality states in the theory, and hence the handedness of the neutrino could be the origin for the parity violation. To see this, consider the chiral projections of a fermion

$$\Psi_{L,R} \equiv \frac{1 \mp \gamma_5}{2} \Psi.$$

We note that in the relativistic limit these two projections describe left and right handed helicity states (where the helicity, i.e. the spin projection in the direction of motion, is a constant of motion for a free particle), but in general an helicity eigenstate is a mixture of the two chiralities. For a massive particle, which has to move with a velocity smaller than the speed of light, it is always possible to make a boost to a system where the helicity is reversed, and hence the helicity is clearly not a Lorentz invariant while the chirality is (and hence has the desirable properties of a charge to which a gauge boson can be coupled). If we look now to the equation of motion for a Dirac particle as the one we are used to for the description of a charged massive particle such as an electron ($(i\partial\!\!\!/ - m)\Psi = 0$), in terms of the chiral projections this equation becomes

$$i\partial\!\!\!/\Psi_L = m\Psi_R$$

$$i\partial\!\!\!/\Psi_R = m\Psi_L$$

and hence clearly a mass term will mix the two chiralities. However, from these equations we see that for $m = 0$, as could be the case for the neutrinos, the two equations are decoupled, and one could write a consistent theory using only one of the two chiralities (which in this case would coincide with the helicity). If the Lee Yang Hamiltonian were just to depend on a single neutrino chirality, one would have then $G_i = \pm G'_i$ and parity violation would indeed be maximal. This situation has been described by saying that neutrinos are like vampires in Dracula's stories: when they were to look to themselves into a mirror they would be unable to see their reflected images.

The actual helicity of the neutrino was measured by Goldhaber et al. [8]. The experiment consisted in observing the K-electron capture in ^{152}Eu ($J = 0$) which produced ^{152}Sm* ($J = 1$) plus a neutrino. This excited nucleus then decayed into ^{152}Sm ($J = 0$) + γ. Hence the measurement of the polarization of the photon gave the required information on the helicity of the neutrino emitted initially. The conclusion was that '...Our results seem compatible with ... 100% negative helicity for the neutrinos', i.e. that the neutrinos are left handed particles.

This paved the road for the $V - A$ theory of weak interactions advanced by Feynman and Gell Mann, and Marshak and Soudarshan [9], which stated that weak interactions only involved vector and axial vector currents, in the combination $V - A$ which only allows the coupling to left handed fields, i.e.

$$J_\mu = \bar{e}_L\gamma_\mu\nu_L + \bar{n}_L\gamma_\mu p_L$$

with $H = (G_F/\sqrt{2})J^\dagger_\mu J^\mu$. This interaction also predicted the existence of purely leptonic weak charged currents, e.g. $\nu + e \to \nu + e$, to be experimentally observed much later[1].

[1] A curious fact was that the new theory predicted a cross section for the inverse beta decay a factor of two larger than the Bethe and Peierls original result, which had already been confirmed in 1958 to the 5% accuracy by Reines and Cowan. However, in a new experiment in 1969, Reines and Cowan found a new value consistent with the new prediction, what shows that many times when the experiment agrees with the theory of the moment the errors tend to be underestimated.

The current involving nucleons is actually not exactly $\propto \gamma_\mu(1-\gamma_5)$ (only the interaction at the quark level has this form), but is instead $\propto \gamma_\mu(g_V - g_A\gamma_5)$. The vector coupling remains however unrenormalized ($g_V = 1$) due to the so called conserved vector current hypothesis (CVC), which states that the vector part of the weak hadronic charged currents ($J^\pm_\mu \propto \bar{\Psi}\gamma_\mu\tau^\pm\Psi$, with $\tau^\pm$ the raising and lowering operators in the isospin space $\Psi^T = (p,n)$) together with the isovector part of the electromagnetic current (i.e. the term proportional to τ_3 in the decomposition $J^{\rm em}_\mu \propto \bar{\Psi}\gamma_\mu(1+\tau_3)\Psi$) form an isospin triplet of conserved currents. On the other hand, the axial vector hadronic current is not protected from strong interaction renormalization effects and hence g_A does not remain equal to unity. The measured value, using for instance the lifetime of the neutron, is $g_A = 1.26$, so that at the nucleonic level the charged current weak interactions are actually "$V - 1.26A$".

With the present understanding of weak interactions, we know that the clever idea to explain parity violation as due to the non-existence of one of the neutrino chiralities (the right handed one) was completely wrong, although it lead to major advances in the theory and ultimately to the correct interaction. Today we understand that the parity violation is a property of the gauge boson (the W) responsible for the gauge interaction, which couples only to the left handed fields, and not due to the absence of right handed fields. For instance, in the quark sector both left and right chiralities exists, but parity is violated because the right handed fields are singlets for the weak charged currents.

1.3 The Trilogy

In 1947 the muon was discovered in cosmic rays by Anderson and Neddermeyer. This particle was just a heavier copy of the electron, and as was suggested by Pontecorvo, it also had weak interactions $\mu + p \to n + \nu$ with the same universal strength $G_{\rm F}$. Hincks, Pontecorvo and Steimberger showed that the muon was decaying to three particles, $\mu \to e\nu\nu$, and the question arose whether the two emitted neutrinos were similar or not. It was then shown by Feinberg [10] that, assuming the two particles were of the same kind, weak interactions couldn't be mediated by gauge bosons (an hypothesis suggested in 1938 by Klein). The reasoning was that if the two neutrinos were equal, it would be possible to join the two neutrino lines and attach a photon to the virtual charged gauge boson (W) or to the external legs, so as to generate a diagram for the radiative decay $\mu \to e\gamma$. The resulting branching ratio would be larger than 10^{-5} and was hence already excluded at that time. This was probably the first use of 'rare decays' to constrain the properties of new particles.

The correct explanation for the absence of the radiative decay was put forward by Lee and Yang, who suggested that the two neutrinos emitted in the muon decay had different flavor, i.e. $\mu \to e + \nu_e + \nu_\mu$, and hence it was not possible to join the two neutrino lines to draw the radiative decay diagram. This was confirmed at Brookhaven in the first accelerator neutrino experiment[11]. They used an almost pure $\bar{\nu}_\mu$ beam, something which can be obtained from charged pion decays, since the $V - A$ theory implies that $\Gamma(\pi \to \ell + \bar{\nu}_\ell) \propto m_\ell^2$, i.e. this process requires a chirality flip in the final lepton line which strongly suppresses the decays $\pi \to e+\bar{\nu}_e$. Putting a detector in front of this beam they were able to observe the process $\bar{\nu}+p \to n+\mu^+$, but no production of positrons, what proved that the neutrinos produced in a weak decay in association with a muon were not the same as those produced in a beta decay (in association with an electron). Notice that although the neutrino fluxes are much smaller at accelerators than at reactors, their higher energies make their detection feasible due to the larger cross sections ($\sigma \propto E^2$ for $E \ll m_p$, and $\sigma \propto E$ for $E \gtrsim m_p$).

In 1975 the τhird charged lepton was discovered by Perl at SLAC, and being just a heavier copy of the electron and the muon, it was concluded that a third neutrino flavor had also to exist. Although the direct detection through e.g. $\bar{\nu}_\tau + p \to n + \tau^+$ has not yet been possible, due to the difficulty of producing a ν_τ beam and of detecting the very short τ track, there is little doubt about its existence, and we furthermore know today that the number of light weakly interacting neutrinos is precisely three (see below), so that the proliferation of neutrino species seems to be now under control.

1.4 The Gauge Theory

As was just mentioned, Klein had suggested that the short range charged current weak interaction could be due to the exchange of a heavy charged vector boson, the $W^\pm$. This boson exchange would look at small momentum transfers ($Q^2 \ll M_W^2$) as the non renormalizable four fermion interactions discussed before. If the gauge interaction is described by the Lagrangian $\mathcal{L} = -(g/\sqrt{2})J_\mu W^\mu + h.c.$, from the low energy limit one can identify the Fermi coupling as $G_{\rm F} = \sqrt{2}g^2/(8M_W^2)$. In the sixties, Glashow, Salam and Weinberg showed that it was possible to write down a unified description of electromagnetic and weak interactions with a gauge theory based in the group $SU(2)_L \times U(1)_Y$ (weak isospin × hypercharge, with the electric charge being $Q = T_3 + Y$), which was spontaneously broken at the weak scale down to the electromagnetic $U(1)_{em}$. This (nowadays standard) model involves the three gauge bosons in the adjoint of $SU(2)$, V_i^μ (with $i = 1,2,3$), and the hypercharge gauge field B^μ, so that the starting Lagrangian is

$$\mathcal{L} = -g\sum_{i=1}^{3} J_\mu^i V_i^\mu - g' J_\mu^Y B^\mu + h.c.,$$

with $J_\mu^i \equiv \sum_a \bar{\Psi}_{aL}\gamma_\mu(\tau_i/2)\Psi_{aL}$. The left handed leptonic and quark isospin doublets are $\Psi^T = (\nu_{eL}, e_L)$ and (u_L, d_L) for the first generation (and similar ones for the other two heavier generations) and the right handed fields are SU(2) singlets. The hypercharge current is obtained by summing over both left and right handed fermion chiralities and is $J_\mu^Y \equiv \sum_a Y_a \bar{\Psi}_a \gamma_\mu \Psi_a$.

After the electroweak breaking one can identify the weak charged currents with $J^\pm = J^1 \pm iJ^2$, which couple to the W boson $W^\pm = (V^1 \mp iV^2)/\sqrt{2}$, and the two neutral vector bosons V^3 and B will now combine through a rotation

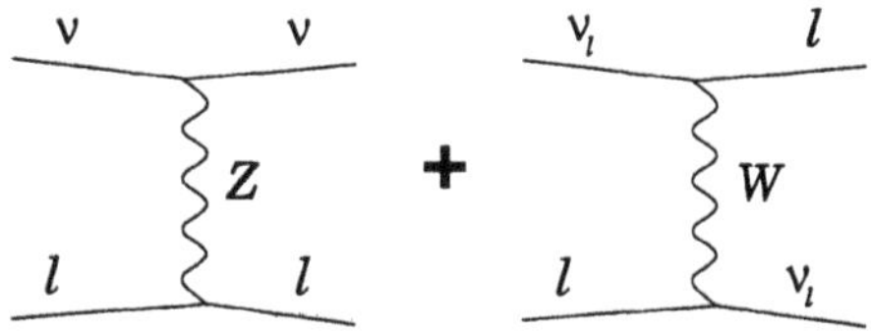

Fig. 1. Neutral and charged current contributions to neutrino lepton scattering

by the weak mixing angle θ_W (with $\text{tg}\theta_W = g'/g$), to give

$$\begin{pmatrix} A_\mu \\ Z_\mu \end{pmatrix} = \begin{pmatrix} \text{c}\,\theta_W & \text{s}\,\theta_W \\ -\text{s}\,\theta_W & \text{c}\,\theta_W \end{pmatrix} \begin{pmatrix} B_\mu \\ V_\mu^3 \end{pmatrix} . \tag{1}$$

We see that the broken theory has now, besides the massless photon field A_μ, an additional neutral vector boson, the heavy Z_μ, whose mass turns out to be related to the W boson mass through $s^2\theta_W = 1 - (M_W^2/M_Z^2)$. The electromagnetic and neutral weak currents are given by

$$J_\mu^{\text{em}} = J_\mu^Y + J_\mu^3$$

$$J_\mu^0 = J_\mu^3 - s^2\theta_W J_\mu^{\text{em}} ,$$

with the electromagnetic coupling being $e = g\ \text{s}\,\theta_W$.

The great success of this model came in 1973 with the experimental observation of the weak neutral currents using muon neutrino beams at CERN (Gargamelle) and Fermilab, using the elastic process $\nu_\mu e \to \nu_\mu e$. The semileptonic processes $\nu N \to \nu X$ where also studied and the comparison of neutral and charged current rates provided a measure of the weak mixing angle. From the theoretical side t'Hooft proved the renormalizability of the theory, so that the computation of radiative corrections became also meaningful.

The Hamiltonian for the leptonic weak interactions $\nu_\ell + \ell' \to \nu_\ell + \ell'$ can be obtained, using the Standard Model just presented, from the two diagrams in Fig. 1. In the low energy limit ($Q^2 \ll M_W^2,\ M_Z^2$), it is just given by

$$H_{\nu_\ell \ell'} = \frac{G_F}{\sqrt{2}} [\bar{\nu}_\ell \gamma_\mu (1-\gamma_5)\nu_\ell][\bar{\ell}'\gamma^\mu (c_L P_L + c_R P_R)\ell']$$

where the left and right couplings are $c_L = \delta_{\ell\ell'} + s^2\theta_W - 0.5$ and $c_R = s^2\theta_W$. The $\delta_{\ell\ell'}$ term in c_L is due to the charged current diagram, which clearly only appears when $\ell = \ell'$. On the other hand, one sees that due to the B component in the Z boson, the weak neutral currents also couple to the charged lepton right handed chiralities (i.e. $c_R \neq 0$). This interaction leads to the cross section (for $E_\nu \gg m_{\ell'}$)

$$\sigma(\nu + \ell \to \nu + \ell) = \frac{2G_F^2}{\pi} m_\ell E_\nu \left[c_L^2 + \frac{c_R^2}{3} \right] ,$$

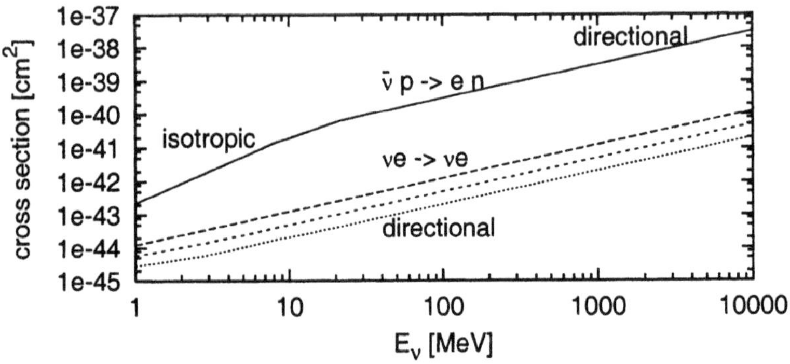

Fig. 2. Neutrino nucleon and neutrino lepton cross sections (the three lines correspond, from top to bottom, to the ν_e, $\bar{\nu}_e$ and $\nu_{\mu,\tau}$ lepton cross sections)

and a similar expression with $c_L \leftrightarrow c_R$ for antineutrinos. Hence, we have the following relations for the neutrino elastic scatterings off electrons

$$\sigma(\nu_e e) \simeq 9 \times 10^{-44}\mathrm{cm}^2 \left(\frac{E_\nu}{10\ \mathrm{MeV}}\right) \simeq 2.5\sigma(\bar{\nu}_e e) \simeq 6\sigma(\nu_{\mu,\tau} e) \simeq 7\sigma(\bar{\nu}_{\mu,\tau} e).$$

Regarding the angular distribution of the electron momentum with respect to the incident neutrino direction, in the center of mass system of the process $d\sigma(\nu_e e)/d\cos\theta \simeq 1{+}0.1[(1{+}\cos\theta)/2]^2$, and it is hence almost isotropic. However, due to the boost to the laboratory system, there will be a significant correlation between the neutrino and electron momenta for $E_\nu \gg$MeV, and this actually allows to do astronomy with neutrinos. For instance, water Cherenkov detectors such as Superkamiokande detect solar neutrinos using this process, and have been able to reconstruct a picture of the Sun with neutrinos. It will turn also to be relevant for the study of neutrino oscillations that these kind of detectors are six times more sensitive to electron type neutrinos than to the other two neutrino flavors.

Considering now the neutrino nucleon interactions, one has at low energies (1 MeV$< E_\nu <$ 50 MeV)

$$\sigma(\nu_e n \to pe) \simeq \sigma(\bar{\nu}_e p \to ne^+) \simeq \frac{G_\mathrm{F}^2}{\pi}\mathrm{c}^2\theta_C(g_V^2 + 3g_A^2)E_\nu^2,$$

where we have now introduced the Cabibbo mixing angle θ_C which relates, if we ignore the third family, the quark flavor eigenstates q^0 to the mass eigenstates q, i.e. $d^0 = \mathrm{c}\theta_C d + \mathrm{s}\theta_C s$ and $s^0 = -\mathrm{s}\theta_C d + \mathrm{c}\theta_C s$ (choosing a flavor basis so that the up type quark flavor and mass eigenstates coincide).

At $E_\nu \gtrsim$ 50 MeV, the nucleon no longer looks like a point-like object for the neutrinos, and hence the vector (v_μ) and axial (a_μ) hadronic currents involve now momentum dependent form factors, i.e.

$$\langle N(p')|v_\mu|N(p)\rangle = \bar{u}(p')\left[\gamma_\mu F_V + \frac{i}{2m_N}\sigma_{\mu\nu}q^\nu F_W\right]u(p)$$

$$\langle N(p')|a_\mu|N(p)\rangle = \bar{u}(p')\left[\gamma_\mu\gamma_5 F_A + \frac{\gamma_5}{2m_N}q_\mu F_P\right]u(p),$$

where $F_V(q^2)$ can be measured using electromagnetic processes and the CVC relation $F_V = F_V^{em,p} - F_V^{em,n}$ (i.e. as the difference between the proton and neutron electromagnetic vector form factors). Clearly $F_V(0) = 1$ and $F_A(0) = 1.26$, while F_W is related to the magnetic moments of the nucleons. The q^2 dependence has the effect of significantly flattening the cross section. In the deep inelastic regime, $E_\nu \gtrsim$ GeV, the neutrinos interact directly with the quark constituents. The cross section in this regime grows linearly with energy, and this provided an important test of the parton model. The main characteristics of the neutrino cross section just discussed are depicted in Fig. 2.

The final test of the standard model came with the direct production of the $W^\pm$ and Z gauge bosons at CERN in 1984, and with the precision test achieved with the Z factories LEP and SLC after 1989. These e^+e^- colliders working at and around the Z resonance ($s = M_Z^2 = (91\ \text{GeV})^2$) turned out to be also crucial for neutrino physics, since studying the shape of the $e^+e^- \to f\bar{f}$ cross section near the resonance, which has the Breit-Wigner form

$$\sigma \simeq \frac{12\pi\Gamma_e\Gamma_f}{M_Z^2}\frac{s}{(s-M_Z^2)^2 + M_Z^2\Gamma_Z^2},$$

it becomes possible to determine the total Z width Γ_Z. This width is just the sum of all possible partial widths, i.e.

$$\Gamma_Z = \sum_f \Gamma_{Z\to f\bar{f}} = \Gamma_{vis} + \Gamma_{inv}.$$

The visible (i.e. involving charged leptons and quarks) width Γ_{vis} can be measured directly, and hence one can infer a value for the invisible width Γ_{inv}. Since in the standard model this last arises from the decays $Z \to \nu_i\bar{\nu}_i$, whose expected rate for decays into a given neutrino flavour is $\Gamma^{th}_{Z\to\nu\bar{\nu}} = 167$ MeV, one can finally obtain the number of neutrinos coupling to the Z as $N_\nu = \Gamma_{inv}/\Gamma^{th}_{Z\to\nu\bar{\nu}}$. The present best value for this quantity is $N_\nu = 2.994 \pm 0.012$, giving then a strong support to the three generation standard model.

Going through the history of the neutrinos we have seen that they have been extremely useful to understand the standard model. On the contrary, the standard model is of little help to understand the neutrinos. Since in the standard model there is no need for ν_R, neutrinos are massless in this theory. There is however no deep principle behind this (unlike the masslessness of the photon which is protected by the electromagnetic gauge symmetry), and indeed in many extensions of the standard model neutrinos turn out to be massive. This makes the search for non-zero neutrino masses a very important issue, since it provides a window to look for physics beyond the standard model. There are many other important questions concerning the neutrinos which are not addressed by the standard model, such as whether they are Dirac or Majorana particles, whether lepton number is conserved, if the neutrino flavours are mixed (like the quarks through the Cabibbo Kobayashi Maskawa matrix) and hence oscillate when they propagate, as many hints suggest today, whether they have magnetic moments, if they decay, if they violate CP, and so on. In conclusion, although in the standard model neutrinos are a little bit boring, many of its extensions contemplate new possibilities which make the neutrino physics a very exciting field.

2 Neutrino Masses

2.1 Dirac or Majorana?

In the standard model, charged leptons (and also quarks) get their masses through their Yukawa couplings to the Higgs doublet field $\phi^T = (\phi_0, \phi_-)$

$$-\mathcal{L}_Y = \lambda \bar{L}\phi^* \ell_R + h.c. \ ,$$

where $L^T = (\nu, \ell)_L$ is a lepton doublet and ℓ_R an SU(2) singlet field. When the electroweak symmetry gets broken by the vacuum expectation value of the neutral component of the Higgs field $\langle \phi_0 \rangle = v/\sqrt{2}$ (with $v = 246$ GeV), the following 'Dirac' mass term results

$$-\mathcal{L}_m = m_\ell(\bar{\ell}_L \ell_R + \bar{\ell}_R \ell_L) = m_\ell \bar{\ell}\ell,$$

where $m_\ell = \lambda v/\sqrt{2}$ and $\ell = \ell_L + \ell_R$ is the Dirac spinor field. This mass term is clearly invariant under the $U(1)$ transformation $\ell \to \exp(i\alpha)\ell$, which corresponds to the lepton number (and actually in this case also to the electromagnetic gauge invariance). From the observed fermion masses, one concludes that the Yukawa couplings range from $\lambda_t \simeq 1$ for the top quark up to $\lambda_e \simeq 10^{-5}$ for the electron.

Notice that the mass terms always couple fields with opposite chiralities, i.e. requires a $L \leftrightarrow R$ transition. Since in the standard model the right handed neutrinos are not introduced, it is not possible to write a Dirac mass term, and hence the neutrino results massless. Clearly the simplest way to give the neutrino a mass would be to introduce the right handed fields just for this purpose (having no gauge interactions, these sterile states would be essentially undetectable and unproduceable). Although this is a logical possibility, it has the ugly feature that in order to get reasonable neutrino masses, below the eV, would require unnaturally small Yukawa couplings ($\lambda_\nu < 10^{-11}$). Fortunately it turns out that neutrinos are also very special particles in that, being neutral, there are other ways to provide them a mass. Furthermore, in some scenarios it becomes also possible to get a natural understanding of why neutrino masses are so much smaller than the charged fermion masses.

The new idea is that the left handed neutrino field actually involves two degrees of freedom, the left handed neutrino associated with the positive beta decay (i.e. emitted in association with a positron) and the other one being the right handed 'anti'-neutrino emitted in the negative beta decays (i.e. emitted in association with an electron). It may then be possible to write down a mass term using just these two degrees of freedom and involving the required $L \leftrightarrow R$ transition. This possibility was first suggested by Majorana in 1937, in a paper

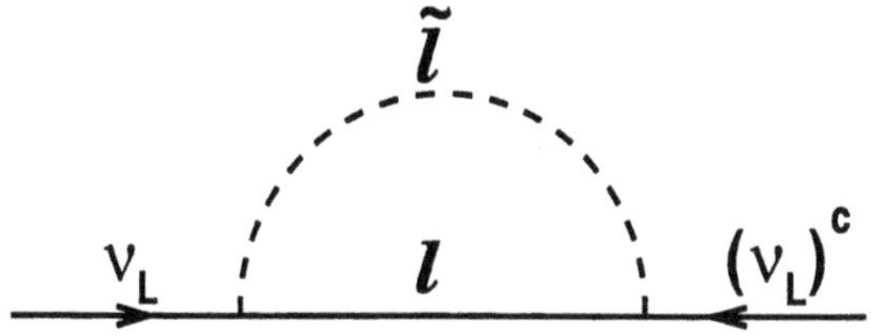

Fig. 3. Example of loop diagram leading to a Majorana mass term in supersymmetric models with broken R parity

named 'Symmetric theory of the electron and positron', and devoted mainly to the problem of getting rid of the negative energy sea of the Dirac equation[12]. As a side product, he found that for neutral particles there was 'no more any reason to presume the existence of antiparticles', and that 'it was possible to modify the theory of beta emission, both positive and negative, so that it came always associated with the emission of a neutrino'. The spinor field associated to this formalism was then named in his honor a Majorana spinor.

To see how this works it is necessary to introduce the so called antiparticle field, $\psi^c \equiv C\bar{\psi}^T = C\gamma_0^T\psi^*$. The charge conjugation matrix C has to satisfy $C\gamma_\mu C^{-1} = -\gamma_\mu^T$, so that for instance the Dirac equation for a charged fermion in the presence of an electromagnetic field, $(i\not{\partial} - e\not{A} - m)\psi = 0$ implies that $(i\not{\partial} + e\not{A} - m)\psi^c = 0$, i.e. that the antiparticle field has opposite charges as the particle field and the same mass. Since for a chiral projection one can show that $(\psi_L)^c = (P_L\psi)^c = P_R\psi^c = (\psi^c)_R$, i.e. this conjugation changes the chirality of the field, one has that ψ^c is related to the CP conjugate of ψ. Notice that $(\psi_L)^c$ describes exactly the same two degrees of freedom described by ψ_L, but somehow using a CP reflected formalism. For instance for the neutrinos, the ν_L operator annihilates the left handed neutrino and creates the right handed antineutrino, while the $(\nu_L)^c$ operator annihilates the right handed antineutrino and creates the left handed neutrino.

We can then now write the advertised Majorana mass term, as

$$-\mathcal{L}_M = \frac{1}{2}m\left[\overline{\nu_L}(\nu_L)^c + \overline{(\nu_L)^c}\nu_L\right] .$$

This mass term has the required Lorentz structure (i.e. the $L \leftrightarrow R$ transition) but one can see that it does not preserve any $U(1)$ phase symmetry, i.e. it violates the so called lepton number by two units. If we introduce the Majorana field $\nu \equiv \nu_L + (\nu_L)^c$, which under conjugation transforms into itself ($\nu^c = \nu$), the mass term becomes just $\mathcal{L}_M = -m\bar{\nu}\nu/2$.

Up to now we have introduced the Majorana mass by hand, contrary to the case of the charged fermions where it arose from a Yukawa coupling in a spontaneously broken theory. To follow the same procedure with the neutrinos presents however a difficulty, because the standard model neutrinos belong to $SU(2)$ doublets, and hence to write an electroweak singlet Yukawa coupling it is necessary to introduce an $SU(2)$ triplet Higgs field $\boldsymbol{\Delta}$ (something which is

Fig. 4. The see-saw model

not particularly attractive). The coupling $\mathcal{L} \propto \overline{L^c}\boldsymbol{\sigma}L \cdot \boldsymbol{\Delta})$ would then lead to the Majorana mass term after the neutral component of the scalar gets a VEV. Alternatively, the Majorana mass term could be a loop effect in models where the neutrinos have lepton number violating couplings to new scalars, as in the so-called Zee models or in the supersymmetric models with R parity violation (as illustrated in Fig. 3). These models have as interesting features that the masses are naturally suppressed by the loop, and they are attractive also if one looks for scenarios where the neutrinos have relatively large dipole moments, since a photon can be attached to the charged particles in the loop.

However, by far the nicest possibility to give neutrinos a mass is the so-called see-saw model introduced by Gell Man, Ramond and Slansky and by Yanagida in 1979[13]. In this scenario, which naturally occurs in grand unified models such as $SO(10)$, one introduces the $SU(2)$ singlet right handed neutrinos. One has now not only the ordinary Dirac mass term, but also a Majorana mass for the singlets which is generated by the VEV of an $SU(2)$ singlet Higgs, whose natural scale is the scale of breaking of the grand unified group, i.e. in the range 10^{12}–10^{16} GeV. Hence the Lagrangian will contain

$$\mathcal{L}_M = \frac{1}{2}\overline{(\nu_L, (N_R)^c)} \begin{pmatrix} 0 & m_D \\ m_D & M \end{pmatrix} \begin{pmatrix} (\nu_L)^c \\ N_R \end{pmatrix} + h.c..$$

The mass eigenstates are two Majorana fields with masses $m_{light} \simeq m_D^2/M$ and $m_{heavy} \simeq M$. Since $m_D/M \ll 1$, we see that $m_{light} \ll m_D$, and hence the lightness of the known neutrinos is here related to the heaviness of the sterile states N_R, as Fig. 4 illustrates.

If we actually introduce one singlet neutrino per family, the mass terms in eq. (2.1) are 3×3 matrices. Notice that if m_D is similar to the up-type quark masses, as happens in $SO(10)$, one would have $m_{\nu_\tau} \sim m_t^2/M \simeq 4\,\text{eV}(10^{13}\text{GeV}/M)$. It is clear then that in these scenarios the observation of neutrino masses below the eV would point out to new physics at about the GUT scale.

2.2 The Quest for the Neutrino Mass

Direct Searches. Already in his original paper on the theory of weak interactions Fermi had noticed that the observed shape of the electron spectrum was suggesting a small mass for the neutrino. The sensitivity to m_{ν_e} in the decay $X \to X' + e + \bar{\nu}_e$ arises clearly because the larger m_ν, the less available kinetic energy remains for the decay products, and hence the maximum electron energy is reduced. To see this consider the phase space factor of the decay, $d\Gamma \propto d^3p_e d^3p_\nu \propto p_e E_e dE_e p_\nu E_\nu dE_\nu \delta(E_e + E_\nu - Q)$, with the Q-value being the total available energy in the decay: $Q \simeq M_X - M_{X'} - m_e$. This leads to a differential electron spectrum proportional to $d\Gamma/dE_e \propto p_e E_e (Q - E_e)\sqrt{(Q-E_e)^2 - m_\nu^2}$, whose shape near the endpoint ($E_e \simeq Q - m_\nu$) depends on m_ν (actually the slope becomes infinite at the endpoint for $m_\nu \neq 0$, while it vanishes for $m_\nu = 0$).

Since the fraction of events in an interval ΔE_e around the endpoint is $\sim (\Delta E_e/Q)^3$, to enhance the sensitivity to the neutrino mass it is better to use processes with small Q-values, what makes the tritium the most sensitive nucleus ($Q = 18.6$ keV). Recent experiments at Mainz and Troitsk have allowed to set the bound $m_{\nu_e} \leq 3$ eV. To improve this bound is quite hard because the fraction of events within say 10 eV of the endpoint is already $\sim 10^{-10}$.

Regarding the muon neutrino, a direct bound on its mass can be set by looking to its effects on the available energy for the muon in the decay of a pion at rest, $\pi^+ \to \mu^+ + \nu_\mu$. From the knowledge of the π and μ masses, and measuring the momentum of the monochromatic muon, one can get the neutrino mass through the relation

$$m_{\nu_\mu}^2 = m_\pi^2 + m_\mu^2 - 2m_\pi\sqrt{p_\mu^2 + m_\mu^2}.$$

The best bounds at present are $m_{\nu_\mu} \leq 170$ keV from PSI, and again they are difficult to improve through this process since the neutrino mass comes from the difference of two large quantities. There is however a proposal to use the muon $(g-2)$ experiment at BNL to become sensitive down to $m_{\nu_\mu} \leq 8$ keV.

Finally, the bound on the ν_τ mass is $m_{\nu_\tau} \leq 17$ MeV and comes from the effects it has on the available phase space of the pions in the decay $\tau \to 5\pi + \nu_\tau$ measured at LEP.

To look for the electron neutrino mass, besides the endpoint of the ordinary beta decay there is another interesting process, but which is however only sensitive to a Majorana (lepton number violating) mass. This is the so called double beta decay. Some nuclei can undergo transitions in which two beta decays take place simultaneously, with the emission of two electrons and two antineutrinos ($2\beta 2\nu$ in Fig. 5). These transitions have been observed in a few isotopes (^{82}Se, ^{76}Ge, ^{100}Mo, ^{116}Cd, ^{150}Nd) in which the single beta decay is forbidden, and the associated lifetimes are huge (10^{19}–10^{24} yr). However, if the neutrino were a Majorana particle, the virtual antineutrino emitted in one vertex could flip chirality by a mass insertion and be absorbed in the second vertex as a neutrino, as exemplified in Fig. 5 ($2\beta 0\nu$). In this way only two electrons would be emitted and this could be observed as a monochromatic line in the added spectrum of the two electrons. The non observation of this effect has allowed to set the bound $m_{\nu_e}^{Maj} \leq 0.3$ eV (by the Heidelberg-Moscow collaboration at Gran Sasso). There are projects to improve the sensitivity of $2\beta 0\nu$ down to $m_{\nu_e} \sim 10^{-2}$ eV, and we note that this bound is quite relevant since as we have seen, if neutrinos are

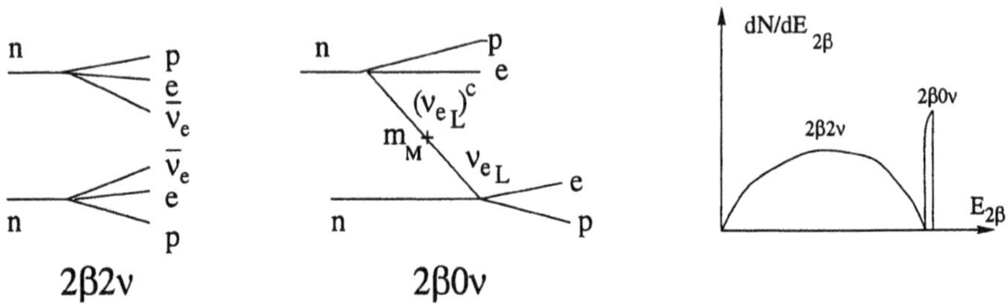

Fig. 5. Double beta decay with and without neutrino emission, and qualitative shape of the expected added spectrum of the two electrons

indeed massive it is somehow theoretically favored (e.g. in the see saw models) that they are Majorana particles.

At this point it is important to extend the discussion to take into account that there are three generations of neutrinos. If neutrinos turn out to be massive, there is no reason to expect that the mass eigenstates (ν_k, with $k = 1, 2, 3$) would coincide with the flavor (gauge) eigenstates (ν_α, with $\alpha = e, \mu, \tau$), and hence, in the same way that quark states are mixed through the Cabibbo, Kobayashi and Maskawa matrix, neutrinos would be related through the Maki, Nakagawa and Sakita mixing matrix [14], i.e. $\nu_\alpha = V_{\alpha k}\nu_k$. The MNS matrix can be parametrized as ($c_{12} \equiv \cos\theta_{12}$, etc.)

$$V = \begin{pmatrix} c_{12}c_{13} & c_{13}s_{12} & s_{13} \\ -c_{23}s_{12}e^{i\delta} - c_{12}s_{13}s_{23} & c_{12}c_{23}e^{i\delta} - s_{12}s_{13}s_{23} & c_{13}s_{23} \\ s_{23}s_{12}e^{i\delta} - c_{12}c_{23}s_{13} & -c_{12}s_{23}e^{i\delta} - c_{23}s_{12}s_{13} & c_{13}c_{23} \end{pmatrix} \begin{pmatrix} e^{i\alpha} & 0 & 0 \\ 0 & e^{i\beta} & 0 \\ 0 & 0 & 1 \end{pmatrix}$$

When the electron neutrino is a mixture of mass eigenstates, the $2\beta 0\nu$ decay amplitude will be proportional now to an 'effective electron neutrino mass' $\langle m_{\nu_e}\rangle = V_{ek}^2 m_k$, where here we adopted the Majorana neutrino fields as self-conjugates ($\chi_k^c = \chi_k$). If one allows for Majorana creation phases in the fields, $\chi_k^c = e^{i\alpha_k}\chi_k$, these phases will appear in the effective mass, $\langle m_{\nu_e}\rangle = {V'_{ek}}^2 e^{i\alpha_k} m_k$. Clearly $\langle m_{\nu_e}\rangle$ has to be independent of the unphysical phases α_k, so that the matrix diagonalizing the mass matrix in the new basis has to change accordingly, i. e. $V'_{ek} = e^{-i\alpha_k/2}V_{ek}$. In particular, α and β may be removed from V in this way, but they would anyhow reappear at the end in $\langle m\rangle$ through the propagators of the Majorana fields, which depend on the creation phases. When CP is conserved, it is sometimes considered convenient to choose basis so that V_{ek} is real (i.e. $\delta = 0$ from CP conservation and α and β are reabsorbed in the Majorana creation phases of the fields). In this case each contribution to $\langle m\rangle$ turns out to be multiplied by the intrinsic CP-parity of the mass eigenstate, $\langle m_{\nu_e}\rangle = |\sum_k |V_{ek}|^2 \eta_{CP}(\chi_k) m_k|$, with $\eta_{CP} = \pm i$. States with opposite CP parities can then induce cancellations in $2\beta 0\nu$ decays[2].

[2] In particular, Dirac neutrinos can be thought of as two degenerate Majorana neutrinos with opposite CP parities, and hence lead to a vanishing contribution to $2\beta 0\nu$, as would be expected from the conservation of lepton number in this case.

Double beta decay is the only process sensitive to the phases α and β. These phases can be just phased away for Dirac neutrinos, and hence in all experiments (such as oscillations) where it is not possible to distinguish between Majorana and Dirac neutrinos, it is not possible to measure them. However, oscillation experiments are the most sensitive way to measure small neutrino masses and their mixing angles, as we now turn to discuss[3].

2.3 Neutrino Oscillations

The possibility that neutrino flavor eigenstates be a superposition of mass eigenstates, as was just discussed, allows for the phenomenon of neutrino oscillations. This is a quantum mechanical interference effect (and as such it is sensitive to quite small masses) and arises because different mass eigenstates propagate differently, and hence the flavor composition of a state can change with time.

To see this consider a flavor eigenstate neutrino ν_α with momentum p produced at time $t = 0$ (e.g. a ν_μ produced in the decay $\pi^+ \to \mu^+ + \nu_\mu$). The initial state is then

$$|\nu_\alpha\rangle = \sum_k V_{\alpha k} |\nu_k\rangle .$$

We know that the mass eigenstates evolve with time according to $|\nu_k(t,x)\rangle = \exp[i(px - E_k t)]|\nu_k\rangle$. In the relativistic limit relevant for neutrinos, one has that $E_k = \sqrt{p^2 + m_k^2} \simeq p + m_k^2/2E$, and thus the different mass eigenstates will acquire different phases as they propagate. Hence, the probability of observing a flavor ν_β at time t is just

$$P(\nu_\alpha \to \nu_\beta) = |\langle \nu_\beta | \nu(t) \rangle|^2 = |\sum_k V_{\alpha k} \mathrm{e}^{-\mathrm{i}\frac{m_k^2}{2E}t} V^*_{\beta k}|^2 .$$

In the case of two generations, taking V just as a rotation with mixing angle θ, one has

$$P(\nu_\alpha \to \nu_\beta) = \sin^2 2\theta \, \sin^2 \left(\frac{\Delta m^2 x}{4E} \right),$$

which depends on the squared mass difference $\Delta m^2 = m_2^2 - m_1^2$, since this is what gives the phase difference in the propagation of the mass eigenstates. Hence, the amplitude of the flavor oscillations is given by $\sin^2 2\theta$ and the oscillation length of the modulation is $L_{osc} \equiv \frac{4\pi E}{\Delta m^2} \simeq 2.5$ m $E[\mathrm{MeV}]/\Delta m^2[\mathrm{eV}^2]$. We see then that neutrinos will typically oscillate with a macroscopic wavelength. For instance, putting a detector at ~ 100 m from a reactor allows to test oscillations of ν_e's to another flavor (or into a singlet neutrino) down to $\Delta m^2 \sim 10^{-2}$ eV^2 if $\sin^2 2\theta$ is not too small (≥ 0.1). The CHOOZ experiment has even reached $\Delta m^2 \sim 10^{-3}$ eV^2 putting a large detector at 1 km distance, and the proposed KAMLAND experiment will be sensitive to reactor neutrinos arriving from $\sim 10^2$ km, and hence will test $\Delta m^2 \sim 10^{-5}$ eV^2 in a few years (see Fig. 6).

[3] Oscillations may even allow to measure the CP violating phase δ, e.g. by comparing $\nu_\mu \to \nu_e$ amplitudes with the $\bar{\nu}_\mu \to \bar{\nu}_e$ ones, as is now being considered for future neutrino factories at muon colliders.

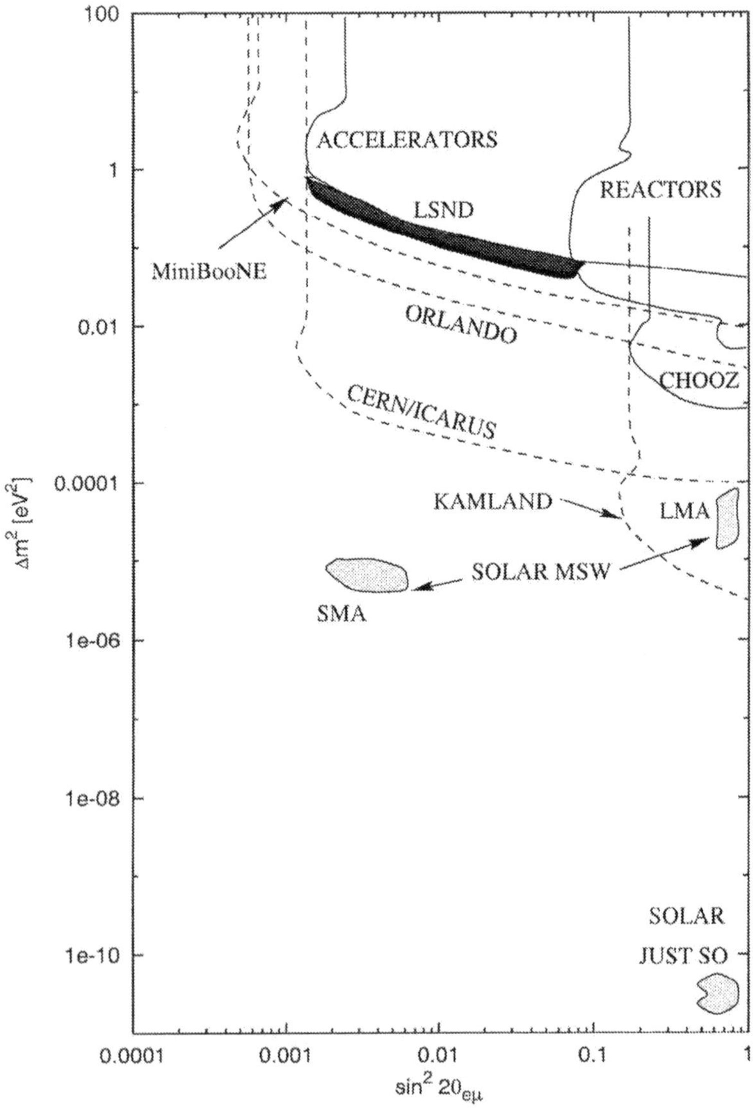

Fig. 6. Present bounds (solid lines), projected sensitivities of future experiments (dashed lines) and values suggested by LSND and solar neutrino experiments for $\nu_e \leftrightarrow \nu_\mu$ oscillations

These kind of experiments look essentially for the disappearance of the reactor ν_e's, i.e. to a reduction in the original ν_e flux. When one uses more energetic neutrinos from accelerators, it becomes possible also to study the appearance of a flavor different from the original one, with the advantage that one becomes

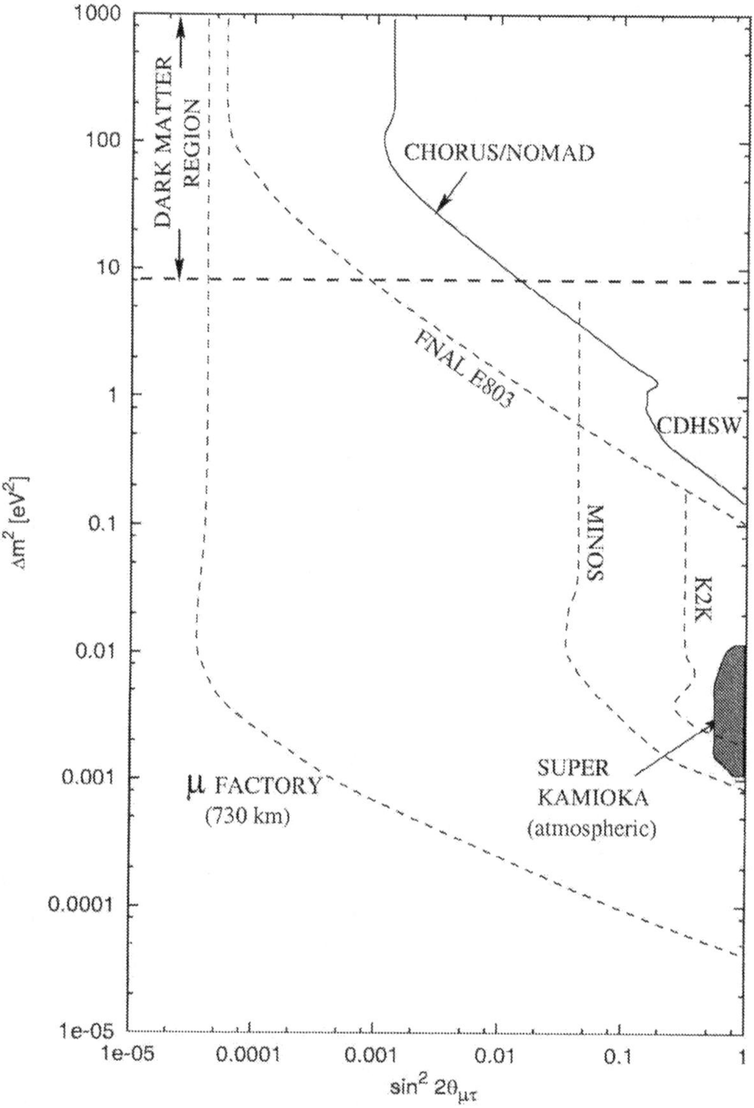

Fig. 7. Present bounds (solid lines), projected sensitivities of future experiments (dashed lines) and values suggested by the atmospheric neutrino anomaly for $\nu_\mu \leftrightarrow \nu_\tau$ oscillations. Also shown is the region where neutrinos would constitute a significant fraction of the dark matter ($\Omega_\nu > 0.1$)

sensitive to very small oscillation amplitudes (i.e. small $\sin^2 2\theta$ values), since the observation of only a few events is enough to establish a positive signal. At present there is one experiment (LSND) claiming a positive signal of $\nu_\mu \to \nu_e$

conversion, suggesting the neutrino parameters in the region indicated in Fig. 6, once the region excluded by other experiments is taken into account. The appearance of ν_τ's out of a ν_μ beam was searched at CHORUS and NOMAD at CERN without success, allowing to exclude the region indicated in Fig. 7, which is a region of relevance for cosmology since neutrinos heavier than $\sim$ eV would contribute to the dark matter in the Universe significantly.

In Figs. 6 and 7 we also display the sensitivity of various new experiments under construction or still at the proposal level, showing that significant improvements are to be expected in the near future (a useful web page with links to the experiments is the Neutrino Industry Homepage[4]). These new experiments will in particular allow to test some of the most clear hints we have at present in favor of massive neutrinos, which come from the two most important natural sources of neutrinos that we have: the atmospheric and the solar neutrinos.

3 Neutrinos in Astrophysics and Cosmology

We have seen that neutrinos made their shy appearance in physics just by steeling a little bit of the momentum of the electrons in a beta decay. In astrophysics however, neutrinos have a major (sometimes preponderant) role, being produced copiously in several environments.

3.1 Atmospheric Neutrinos

When a cosmic ray (proton or nucleus) hits the atmosphere and knocks a nucleus a few tens of km above ground, an hadronic (and electromagnetic) shower is initiated, in which pions in particular are copiously produced. The charged pion decays are the main source of atmospheric neutrinos through the chain $\pi \to \mu\nu_\mu \to e\nu_e\nu_\mu\nu_\mu$. One expects then twice as many ν_μ's than ν_e's (actually at very high energies, $E_\nu \gg$ GeV, the parent muons may reach the ground and hence be stopped before decaying, so that the expected ratio $R \equiv (\nu_\mu + \bar{\nu}_\mu)/(\nu_e + \bar{\nu}_e)$ should be even larger than two at high energies). However, the observation of the atmospheric neutrinos by IMB, Kamioka, Soudan, MACRO and Super Kamiokande indicates that there is a deficit of muon neutrinos, with $R_{obs}/R_{th} \simeq 0.6$ below $E_\nu \sim$ GeV. More remarkably, at multi-GeV energies (for which a neutrino oscillation length would increase) the Super Kamiokande experiment observes a zenith angle dependence indicating that neutrinos coming from above (with pathlengths $d \sim 20$ km) had not enough time to oscillate, while those from below ($d \sim 13000$ km) have already oscillated. The most plausible explanation for these effects is an oscillation $\nu_\mu \to \nu_\tau$ with maximal mixing and $\Delta m^2 \simeq$ few$\times 10^{-3}$ eV^2, as indicated in Fig. 7.

3.2 Solar Neutrinos

The sun gets its energy from the fusion reactions taking place in its interior, where essentially four protons form a He nucleus. By charge conservation this

[4] http://www.hep.anl.gov/NDK/Hypertext/nuindustry.html

has to be accompanied by the emission of two positrons and by lepton number conservation in the weak processes two ν_e's have to be produced. This fusion liberates 27 MeV of energy, which is eventually emitted mainly (97%) as photons and the rest (3%) as neutrinos. Knowing the energy flux of the solar radiation reaching the Earth ($k_\odot \simeq 1.5$ kW/m^2), it is then simple to estimate that the solar neutrino flux at Earth is $\Phi_\nu \simeq 2k_\odot/27$ MeV $\simeq 6 \times 10^{10}\nu_e/\text{cm}^2\text{s}$, which is a very large number indeed.

Many experiments have looked for these solar neutrinos and the puzzling result which has been with us for the last thirty years is that only between 1/2 to 1/3 of the expected fluxes are observed. Remarkably, Pontecorvo [15] noticed even before the first observation of solar neutrinos by Davies that neutrino oscillations could reduce the expected rates. We note that the oscillation length of solar neutrinos ($E \sim 0.1$–10 MeV) is of the order of 1 AU for $\Delta m^2 \sim 10^{-10}$ eV^2, and hence even those tiny neutrino masses can have observable effects if the mixing angles are large (this would be the 'just so' solution to the solar neutrino problem). Much more remarkable is the possibility of explaining the puzzle by resonantly enhanced oscillations of neutrinos as they propagate outwards through the Sun. Indeed, the solar medium affects ν_e's differently than $\nu_{\mu,\tau}$'s (since only the first interact through charged currents with the electrons present), and this modifies the oscillations in a beautiful way through an interplay of neutrino mixings and matter effects, in the so called MSW effect [16]. Two possible solutions using this mechanism require $\Delta m^2 \simeq 10^{-5}$ eV^2 and small mixings $s^2 2\theta \simeq \text{few}\times 10^{-3}$ eV^2 (SMA) or large mixing (LMA), as shown in Fig. 6.

Atmospheric and solar neutrinos are extremely fashionable nowadays. For instance more than a dozen review papers on the subject have appeared in the last year and hence I will avoid going with more details into them (see e.g. [17,18]), although the second lecture dealt exclusively with this subject.

3.3 Supernova Neutrinos

The most spectacular neutrino fireworks in the Universe are the supernova explosions, which correspond to the death of a very massive star. In this process the inner Fe core ($M_c \simeq 1.4\ M_\odot$), unable to get pressure support gives up to the pull of gravity and collapses down to nuclear densities (few$\times 10^{14}$g/cm^3), forming a very dense proto-neutron star. At this moment neutrinos become the main character on stage, and 99% of the gravitational binding energy gained (few$\times 10^{53}$ ergs) is released in a violent burst of neutrinos and antineutrinos of the three flavours, with typical energies of a few tens of MeV[5]. Being the density so high, even the weakly interacting neutrinos become trapped in the core, and they diffuse out in a few seconds to be emitted from the so called neutrinospheres (at $\rho \sim 10^{12}$ g/cm^3). These neutrino fluxes then last for ~ 10 s, after which the initially trapped lepton number is lost and the neutron star cools more slowly.

During those ~ 10 s the neutrino luminosity of the supernova ($\sim 10^{52}$ erg/s) is comparable to the total luminosity of the Universe (c.f. $L_\odot \simeq 4 \times 10^{33}$ erg/s),

[5] Actually there is first a brief (msec) ν_e burst from the neutronisation of the Fe core.

Fig. 8. Pulsar kicks from neutrino rockets ?

but unfortunately only a couple of such events occur in our galaxy per century, so that one has to be patient. Fortunately, on February 1987 a supernova exploded in the nearby ($d \sim 50$ kpc) Large Magellanic Cloud, producing a dozen neutrino events in the Kamiokande and IMB detectors. This started extra solar system neutrino astronomy and provided a very basic proof of the explosion mechanism. With the new larger detectors under operation at present (SuperKamioka and SNO) it is expected that a future galactic supernova ($d \sim 10$ kpc) would produce several thousand neutrino events and hence allow detailed studies of the supernova physics.

Also sensitive test of neutrino properties will be feasible if a galactic supernova is observed. The simplest example being the limits on the neutrino mass which would result from the measured burst duration as a function of the neutrino energy. Indeed, if neutrinos are massive, their velocity will be $v = c\sqrt{1-(m_\nu/E)^2}$, and hence the travel time from a SN at distance d would be $t \simeq \frac{d}{c}[1-\frac{1}{2}(m_\nu/E)^2]$, implying that lower energy neutrinos ($E \sim 10$ MeV) would arrive later than high energy ones by an amount $\Delta t \sim 0.5(d/10 \text{ kpc})$ $(m_\nu/10 \text{ eV})^2$s. Looking for this effect a sensitivity down to $m_{\nu_{\mu,\tau}} \sim 25$ eV would be achievable from a supernova at 10 kpc, and this is much better than the present direct bounds on the $\nu_{\mu,\tau}$ masses.

What remains after a (type II) supernova explosion is a pulsar, i.e. a fastly rotating magnetised neutron star. One of the mysteries related to pulsars is that they move much faster (few hundred km/s) than their progenitors (few tenths of km/s). There is no satisfactory standard explanation of how these initial kicks are imparted to the pulsar and here neutrinos may also have something to say. It has been suggested that these kicks could be due to a macroscopic manifestation of the parity violation of weak interactions, i.e. that in the same way as electrons preferred to be emitted in the direction opposite to the polarisation of the ^{60}Co nuclei in the experiment of Wu (and hence the neutrinos preferred to be emitted in the same direction), the neutrinos in the supernova explosions would be biased towards one side of the star because of the polarisation induced in the matter by the large magnetic fields present [19], leading to some kind of neutrino rocket effect, as shown in Fig. 8.

Although only a 1% asymmetry in the emission of the neutrinos would be enough to explain the observed velocities, the magnetic fields required are $\sim 10^{16}$ G, much larger than the ones inferred from observations ($\sim 10^{12}$–10^{13} G). An attempt has also been done [20] to exploit the fact that the neutrino oscillations in matter are affected by the magnetic field, and hence the resonant flavor conversion would take place in an off-centered surface. Since ν_τ's (or ν_μ's) interact less than ν_e's, an oscillation from $\nu_e \to \nu_\tau$ in the region where ν_e's are still trapped but ν_τ's can freely escape would generate a ν_τ flux from a deeper region of the star in one side than in the other. Hence if one assumes that the temperature profile is isotropic, neutrinos from the deeper side will be more energetic than those from the opposite side and can then be the source of the kick. This would require however $\Delta m^2 > 100$ eV^2, which is uncomfortably large, and $B > 10^{14}$ G. Moreover, it has been argued [21] that the assumption of an isotropic T profile near the neutrinospheres will not hold, since the side where the escaping neutrinos are more energetic will rapidly cool (the neutrinosphere region has negligible heat capacity compared to the core) adjusting the temperature gradient so that the isotropic energy flux generated in the core will manage ultimately to get out isotropically.

An asymmetric neutrino emission due to an asymmetric magnetic field affecting asymmetrically the ν_e opacities has also been proposed, but again the magnetic fields required are too large ($B \sim 10^{16}$ G) [22].

As a summary, to explain the pulsar kicks as due to an asymmetry in the neutrino emission is attractive theoretically, but unfortunately doesn't seem to work. Maybe when three dimensional simulations of the explosion would become available, possibly including the presence of a binary companion, larger asymmetries would be found just from standard hydrodynamical processes.

Supernovae are also helpful for us in that they throw away into the interstellar medium all the heavy elements produced during the star's life, which are then recycled into second generation stars like the Sun, planetary systems and so on. However, 25% of the baryonic mass of the Universe was already in the form of He nuclei well before the formation of the first stars, and as we understand now this He was formed a few seconds after the big bang in the so-called primordial nucleosynthesis. Remarkably, the production of this He also depends on the neutrinos, and the interplay between neutrino physics and primordial nucleosynthesis provided the first important astro-particle connection.

3.4 Cosmic Neutrino Background and Primordial Nucleosynthesis

In the same way as the big bang left over the 2.7°K cosmic background radiation, which decoupled from matter after the recombination epoch ($T \sim$ eV), there should also be a relic background of cosmic neutrinos (CνB) left over from an earlier epoch ($T \sim$ MeV), when weakly interacting neutrinos decoupled from the ν_i-e-γ primordial soup. Slightly after the neutrino decoupling, e^+e^- pairs annihilated and reheated the photons, so that the present temperature of the CνB is $T_\nu \simeq 1.9$°K, slightly smaller than the photon one. This means that there should be today a density of neutrinos (and antineutrinos) of each flavour $n_{\nu_i} \simeq 110$ cm^{-3}.

Primordial nucleosynthesis occurs between $T \sim 1$ MeV and 10^{-2} MeV, an epoch at which the density of the Universe was dominated by radiation (photons and neutrinos). This means that the expansion rate of the Universe depended on the number of neutrino species N_ν, becoming faster the bigger N_ν

($H \propto \sqrt{\rho} \propto \sqrt{\rho_\gamma + N_\nu \rho_\nu}$, with ρ_ν the density for one neutrino species). Helium production just occurred after deuterium photodissociation became inefficient at $T \sim 0.1$ MeV, with essentially all neutrons present at this time ending up into He. The crucial point is that the faster the expansion rate, the larger fraction of neutrons (w.r.t protons) would have survived to produce He nuclei. This implies that an observational upper bound on the primordial He abundance will translate into an upper bound on the number of neutrino species. Actually the predictions also depend in the total amount of baryons present in the Universe ($\eta = n_B/n_\gamma$), which can be determined studying the very small amounts of primordial D and ^{7}Li produced. The observational D measurements are somewhat unclear at presents, with determinations in the low side implying the strong constraint $N_\nu < 3.3$, while those in the high side implying $N_\nu < 4.8$ [23]. It is important that nucleosynthesis bounds on N_ν were established well before the LEP measurement of the number of standard neutrinos.

As a side product of primordial nucleosynthesis theory one can determine that the amount of baryonic matter in the Universe has to satisfy $\eta \simeq 1\text{–}6 \times 10^{-10}$. The explanation of this small number is one of the big challenges for particle physics and another remarkable fact of neutrinos is that they might be ultimately responsible for this matter-antimatter asymmetry.

3.5 Leptogenesis

The explanation of the observed baryon asymmetry as due to microphysical processes taking place in the early Universe is known to be possible provided the three Sakharov conditions are fulfilled: *i)* the existence of baryon number violating interactions ($\not{B}$), *ii)* the existence of C and CP violation ($\not{C}$ and $\not{CP}$) and *iii)* departure from chemical equilibrium ($\not{Eq}$). The simplest scenarios fulfilling these conditions appeared in the seventies with the advent of GUT theories, where heavy color triplet Higgs bosons can decay out of equilibrium in the rapidly expanding Universe (at $T \sim M_T \sim 10^{13}$ GeV) violating B, C and CP. In the middle of the eighties it was realized however that in the Standard Model non-perturbative $\not{B}$ and $\not{L}$ (but $B - L$ conserving) processes where in equilibrium at high temperatures ($T > 100$ GeV), and would lead to a transmutation between B and L numbers, with the final outcome that $n_B \simeq n_{B-L}/3$. This was a big problem for the simplest GUTs like SU(5), where $B - L$ is conserved (and hence $n_{B-L} = 0$), but it was rapidly turned into a virtue by Fukugita and Yanagida [24], who realised that it could be sufficient to generate initially a lepton number asymmetry and this will then be reprocessed into a baryon number asymmetry. The nice thing is that in see-saw models the generation of a lepton asymmetry (leptogenesis) is quite natural, since the heavy singlet Majorana neutrinos would decay out of equilibrium (at $T \lesssim M_N$) through $N_R \to \ell H^*$, $\bar{\ell} H$, i.e. into final states with different L, and the CP violation appearing at one loop through the diagrams in Fig. 9 would lead [25] to $\Gamma(N \to \ell H^*) \neq \Gamma(N \to \bar{\ell} H)$, so that a final L asymmetry will result. Reasonable parameter values lead naturally to the required asymmetries ($\eta \sim 10^{-10}$), making this scenario probably the simplest baryogenesis mechanism.

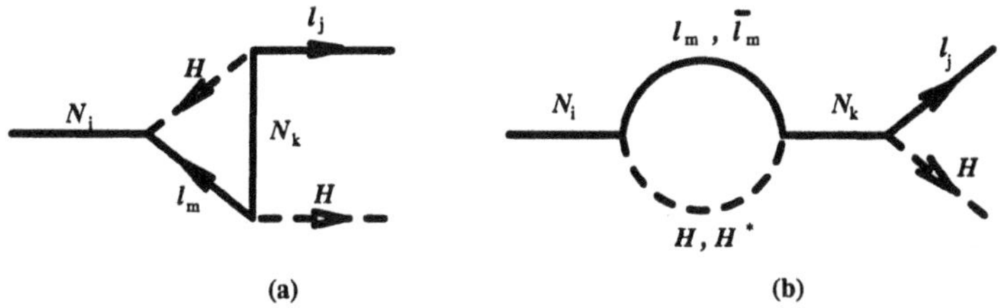

Fig. 9. One loop CP violating diagrams for leptogenesis

3.6 Neutrinos as Dark Matter

Neutrinos may not only give rise to the observed baryonic matter, but they could also themselves be the dark matter in the Universe. This possibility arises [26] because if the ordinary neutrinos are massive, the large number of them present in the CνB will significantly contribute to the mass density of the Universe, in an amount[6] $\Omega_\nu \simeq \sum_i m_{\nu_i}/(92h^2 \text{ eV})$. Hence, in order for neutrinos not to overclose the Universe it is necessary that $\sum_i m_{\nu_i} \lesssim 30$ eV, which is a bound much stronger than the direct ones for $m_{\nu_{\mu,\tau}}$. On the other hand, a neutrino mass ~ 0.1 eV (as suggested by the atmospheric neutrino anomaly) would imply that the mass density in neutrinos is already comparable to that in ordinary baryonic matter ($\Omega_B \sim 0.003$), and $m_\nu \gtrsim 1$ eV would lead to an important contribution of neutrinos to the dark matter.

The nice things of neutrinos as dark matter is that they are the only candidates that we know for sure that they exist, and that they are very helpful to generate the structures observed at large supercluster scales ($\sim$ 100 Mpc). However, they are unable to give rise to structures at galactic scales (they are 'hot' and hence free stream out of small inhomogeneities). Furthermore, even if those structures were formed, it would not be possible to pack the neutrinos sufficiently so as to account for the galactic dark halo densities, due to the lack of sufficient phase space [27], since to account for instance for the local halo density $\rho^0 \simeq 0.3$ GeV/cm^3 would require $n_\nu^0 \simeq 10^7(30 \text{ eV}/m_\nu)/\text{cm}^3$, which is a very large overdensity with respect to the average value 110/cm^3. The Tremaine Gunn phase-space constraint requires for instance that to be able to account for the dark matter in our galaxy one neutrino should be heavier than $\sim$ 50 eV, so that neutrinos can clearly account at most for a fraction of the galactic dark matter.

The direct detection of the dark matter neutrinos will be extremely difficult [28], because of their very small energies ($E \simeq m_\nu v^2/2 \simeq 10^{-6} m_\nu c^2$) which leads to very tiny cross sections with matter and with tiny momentum transfers. This has lead people to talk about kton detectors at mK temperatures in zero gravity environments ..., and hence this remains clearly as a challenge for the next millennium.

[6] The reduced Hubble constant is $h \equiv H/(100 \text{ km/s-Mpc}) \simeq 0.6$.

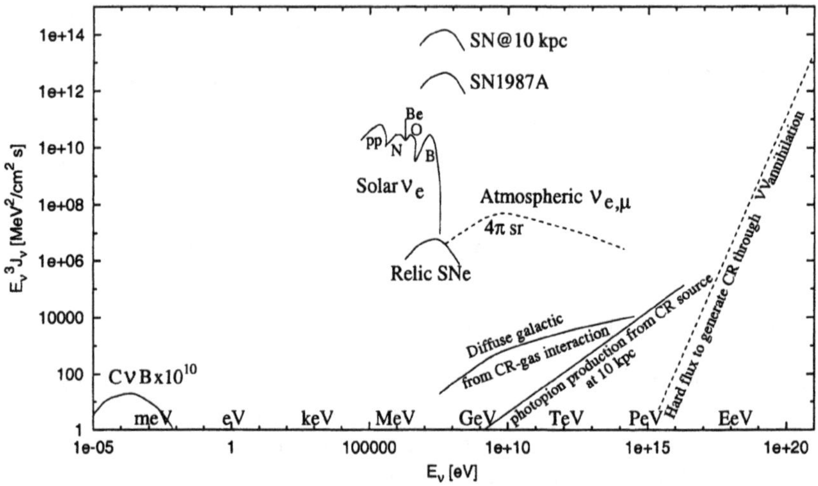

Fig. 10. Neutrino spectra. We show the cosmic neutrino background (CνB) multiplied by 10^{10}, solar and supernova neutrinos, the isotropic atmospheric neutrinos, those coming from the galactic plane due to cosmic ray gas interactions, an hypothetical galactic source at 10 kpc, whose detection at $E > 10$ TeV would require a good angular resolution to reject the atmospheric background (similar considerations hold for AGN neutrinos not displayed). Finally the required flux to produce the CR beyond the GZK cutoff by annihilations with the dark matter neutrinos from the other end of the spectrum

One speculative proposal to observe the dark matter neutrinos indirectly is through the observation of the annihilation of cosmic ray neutrinos of ultra high energies ($E_\nu \sim 10^{21}$ eV$/(m_\nu/4$ eV$)$) with dark matter ones at the Z-resonance pole where the cross section is enhanced [29]. Moreover, this process has been suggested as a possible way to generate the observed hadronic cosmic rays above the GZK cutoff [30], since neutrinos can travel essentially unattenuated for cosmological distances ($\gg 100$ Mpc) and induce hadronic cosmic rays locally through the annihilation with dark matter neutrinos. This proposal requires however extremely powerful neutrino sources.

In Fig. 10 we summarize qualitatively different fluxes which can appear in the neutrino sky and whose search and observation is opening new windows to understand the Universe.

References

1. E. Fermi: Z. Phys. **88**, 161 (1934)
2. H. Bethe and R. Peierls: Nature **133**, 532 (1934)
3. F. Reines and C. Cowan: Phys. Rev. **113**, 273 (1959)
4. G. Gamow and E. Teller: Phys. Rev. **49**, 895 (1936)
5. T. D. Lee and C. N. Yang: Phys. Rev. **104**, 254 (1956)

6. C. S. Wu et al.: Phys. Rev. **105**, 1413 (1957)
7. T. D. Lee and C. N. Yang: Phys. Rev. **105**, 1671 (1957); L. D. Landau: Nucl. Phys. **3**, 127 (1957); A. Salam: Nuovo Cimento **5**, 299 (1957)
8. M. Goldhaber, L. Grodzins and A. W. Sunyar: Phys. Rev. **109**, 1015 (1958)
9. R. Feynman and M. Gell-Mann: Phys. Rev. **109**, 193 (1958); E. Sudarshan and R. Marshak: Phys. Rev. **109**, 1860 (1958)
10. G. Feinberg: Phys. Rev. **110**, 1482 (1958)
11. G. Danby et al.: Phys. Rev. Lett. **9**, 36 (1962)
12. E. Majorana: Nuovo Cimento **14**, 170 (1937)
13. M. Gell-Mann, P. Ramond and R. Slansky, in *Supergravity*, p. 135, ed. by F. van Nieuwenhuizen and D. Freedman (1979); T. Yanagida, Proc. of the *Workshop on unified theory and baryon number in the universe*, KEK, Japan (1979)
14. Z. Maki, M. Nakagawa and S. Sakata: Prog. Theoret. Phys. **28**, 870 (1962)
15. B. Pontecorvo: Sov. Phys. JETP **26**, 984 (1968)
16. S. P. Mikheyev and A. Yu. Smirnov: Sov. J. Nucl. Phys. **42**, 913 (1985); L. Wolfenstein: Phys. Rev. **D17**, 2369 (1979)
17. G. Gelmini and E. Roulet: Rep. Prog. Phys. **58**, 1207 (1995)
18. P. Langacker: hep-ph/9811460; W. C. Haxton: nucl-th/9901076; A. Yu. Smirnov: hep-ph/9901208; G. G. Raffelt: hep-ph/9902271; E. Torrente-Lujan: hep-ph/9902339; J. Valle: hep-ph/9906378; R. D. Peccei: hep-ph/9906509; J. Ellis: hep-ph/9907458
19. N. Chugai: Pis'ma Astron. Zh. **10** 87 (1984); Dorofeev et al.: Sov. Astron. Lett. **11**, 123 (1985)
20. A. Kusenko and G. Segrè: Phys. Rev. Lett. **77**, 4972 (1996)
21. H-T. Janka and G. Raffelt: Phys. Rev. **D59**, 023005 (1999)
22. G. S. Bisnovatyi-Kogan: astro-ph/9707120; E. Roulet: JHEP **01**, 013 (1998)
23. K. Olive, G. Steigman and T. P. Walker: astro-ph/9905320
24. M. Fukugita and T. Yanagida: Phys. Lett. **B174**, 45 (1986)
25. L. Covi, E. Roulet and F. Vissani: Phys. Lett. **B384**, 169 (1996)
26. S. Gershtein and Ya. B. Zeldovich: JETP Lett. **4**, 120 (1966); R. Cowsik and J. Mc Clelland: Phys. Rev. Lett **29**, 669 (1972)
27. S. Tremaine and J. E. Gunn: Phys. Rev. Lett. **42**, 407 (1979)
28. P. Langacker, J. P. Leveille and J. Sheiman: Phys. Rev. **D27**, 1228 (1983)
29. T. Weiler: Phys. Rev. Lett. **49**, 234 (1982) and Astrophys. J. **285**, 495 (1984); E. Roulet: Phys. Rev. **D47**, 5247 (1993)
30. T. Weiler: Astropart. Phys. **11**, 303 (1999); D. Fargion, B. Mele and A. Salis: astro-ph/9710029; G. Gelmini and A. Kusenko: hep-ph/9908276; T. Weiler: hep-ph/9910316

Particle and Astrophysical Aspects of Ultra-high Energy Cosmic Rays

Günter Sigl

DARC, Observatoire de Paris-Meudon, F-92195 Meudon Cédex, France
and
Department of Astronomy & Astrophysics, Enrico Fermi Institute, The University of Chicago, Chicago, IL 60637-1433, USA

Abstract. The origin of cosmic rays is one of the major unresolved astrophysical questions. In particular, the highest energy cosmic rays observed possess macroscopic energies and their origin is likely to be associated with the most energetic processes in the Universe. Their existence triggered a flurry of theoretical explanations ranging from conventional shock acceleration to particle physics beyond the Standard Model and processes taking place at the earliest moments of our Universe. Furthermore, many new experimental activities promise a strong increase of statistics at the highest energies and a combination with $\gamma-$ray and neutrino astrophysics will put strong constraints on these theoretical models. Detailed Monte Carlo simulations indicate that charged ultra-high energy cosmic rays can also be used as probes of large scale magnetic fields whose origin may open another window into the very early Universe. We give an overview over this quickly evolving research field.

1 Introduction

After almost 90 years of research on cosmic rays (CRs), their origin is still an open question, for which the degree of uncertainty increases with energy: Only below 1 GeV, the modulation of the CR flux with solar activity proves that these particles must be solar in origin. The bulk of the CRs up to at least an energy of $E = 4 \times 10^{15}$ eV is believed to originate within our Galaxy. Above that energy, which is associated with the so called "knee", the flux of particles per area, time, solid angle, and energy, which can be well approximated by broken power laws $\propto E^{-\gamma}$, steepens from a power law index $\gamma \simeq 2.7$ to one of index $\simeq 3.2$. Above the so called "ankle" at $E \simeq 5 \times 10^{18}$ eV, the spectrum flattens again to a power law of index $\gamma \simeq 2.8$. This latter feature is often interpreted as a cross over from a steeper Galactic component to a harder component of extragalactic origin. Fig. 1 shows the measured CR spectrum above 100 MeV, up to 3×10^{20} eV, the highest energy measured so far for an individual CR.

The conventional scenario assumes that all high energy charged particles are accelerated in magnetized astrophysical shocks, whose size and typical magnetic field strength determines the maximal achievable energy, similar to the situation in man made particle accelerators. The most likely astrophysical accelerators for CR up to the knee, and possibly up to the ankle are the shocks associated with remnants of past Galactic supernova explosions, whereas for the presumed

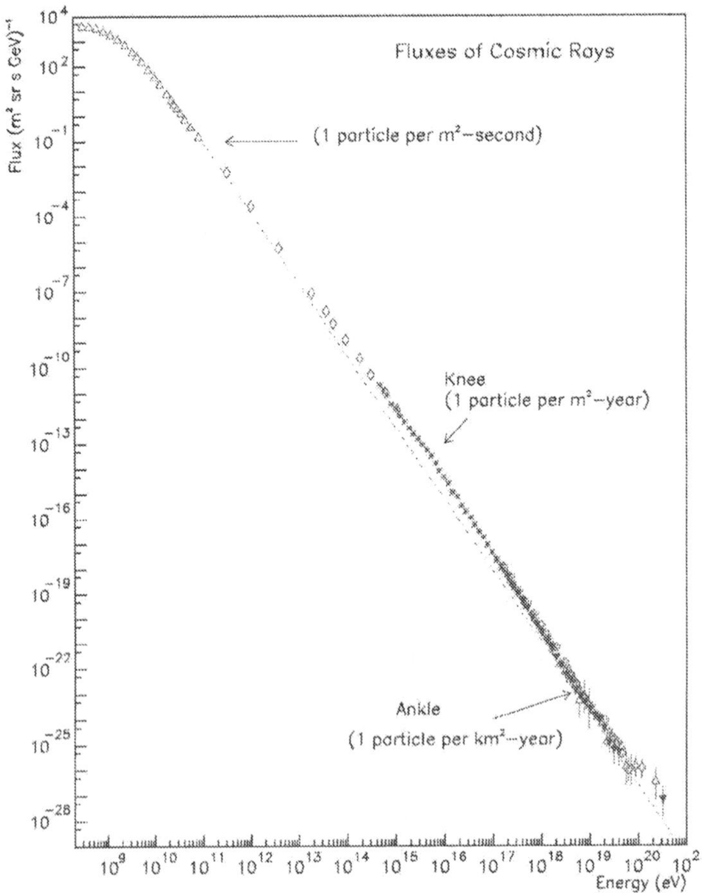

Fig. 1. The cosmic ray all particle spectrum [1]. Approximate integral fluxes are also shown

extragalactic component powerful objects such as active galactic nuclei are envisaged.

The main focus of this contribution will be on ultrahigh energy cosmic rays (UHECRs), those with energy $\gtrsim 10^{18}\,$eV [2–4,7–9]. For more details on CRs at lower energies up to a few hundred TeV see also the contribution by Trevor Weekes in this volume. In particular, extremely high energy (EHE)[1] cosmic

[1] We shall use the abbreviation EHE to specifically denote energies $E \gtrsim 10^{20}\,$eV, while the abbreviation UHE for "Ultra-High Energy" will sometimes be used to denote $E \gtrsim$ 1 EeV, where 1 EeV $= 10^{18}\,$eV. Clearly UHE includes EHE but not vice versa.

rays pose a serious challenge for conventional theories of CR origin based on acceleration of charged particles in powerful astrophysical objects. The question of the origin of these EHECRs is, therefore, currently a subject of much intense debate and discussions as well as experimental efforts; see [5,6,10], and [11] for a recent brief review, and [12] for a detailed review. In Sect. 2 we will summarize detection techniques and present and future experimental projects.

The current theories of origin of EHECRs can be broadly categorized into two distinct "scenarios": the "bottom-up" acceleration scenario, and the "top-down" decay scenario, with various different models within each scenario. As the names suggest, the two scenarios are in a sense exact opposite of each other. The bottom-up scenario is just an extension of the conventional shock acceleration scenario in which charged particles are accelerated from lower energies to the requisite high energies in certain special astrophysical environments. On the other hand, in the top-down scenario, the energetic particles arise simply from decay of certain sufficiently massive particles originating from physical processes in the early Universe, and no acceleration mechanism is needed.

The problems encountered in trying to explain EHECRs in terms of acceleration mechanisms have been well-documented in a number of studies; see, e.g., [13–15]. Even if it is possible, in principle, to accelerate particles to EHECR energies of order 100 EeV in some astrophysical sources, it is generally extremely difficult in most cases to get the particles come out of the dense regions in and/or around the sources without losing much energy. Currently, the most favorable sources in this regard are perhaps a class of powerful radio galaxies (see, e.g., [16,17] for recent reviews and references to the literature), although the values of the relevant parameters required for acceleration to energies $\gtrsim$ 100 EeV are somewhat on the extreme side [15]. However, even if the requirements of energetics are met, the main problem with radio galaxies as sources of EHECRs is that most of them seem to lie at large cosmological distances, $\gg$ 100 Mpc, from Earth. This is a major problem if EHECR particles are conventional particles such as nucleons or heavy nuclei. The reason is that nucleons above $\simeq$ 70 EeV lose energy drastically during their propagation from the source to Earth due to the Greisen-Zatsepin-Kuzmin (GZK) effect [18,19], namely, photo-production of pions when the nucleons collide with photons of the cosmic microwave background (CMB), the mean-free path for which is $\sim$ few Mpc [20]. This process limits the possible distance of any source of EHE nucleons to $\lesssim$ 100 Mpc. If the particles were heavy nuclei, they would be photo-disintegrated [21,22] in the CMB and infrared (IR) background within similar distances. Thus, nucleons or heavy nuclei originating in distant radio galaxies are unlikely to survive with EHECR energies at Earth with any significant flux, even if they were accelerated to energies of order 100 EeV at source. In addition, since EHECRs are not likely to be deflected strongly at least by the large scale intergalactic and/or Galactic magnetic fields, their arrival directions should point back to their sources in the sky (see Sect. 5 for details). Thus, EHECRs may offer us the unique opportunity of doing charged particle astronomy. Yet, for the observed EHECR events so far, no powerful sources close to the arrival directions of individual events are found within about 100 Mpc [23,14]. Very recently, it has been suggested by Boldt and Ghosh [24] that particles may be accelerated to energies $\sim 10^{21}$ eV near the event horizons of spinning supermassive black holes associated with presently *inactive* quasar remnants whose numbers within the local cosmological Universe (i.e., within a GZK distance of order 50 Mpc) may be sufficient to explain the observed EHECR flux. This would solve the problem of absence of suitable currently *active* sources associated with EHECRs. A detailed model incorporating this suggestion, however, remains to be worked out.

There are, of course, ways to avoid the distance restriction imposed by the GZK effect, provided the problem of energetics is somehow solved separately and provided one allows new physics beyond the Standard Model of particle physics; we shall discuss those suggestions in Sect. 3.

On the other hand, in the top-down scenario, which will be discussed in Sect. 4, the problem of energetics is trivially solved from the beginning. Here, the EHECR particles owe their origin to decay of some supermassive "X" particles of mass $m_X \gg 10^{20}$ eV, so that their decay products, envisaged as the EHECR particles, can have energies all the way up to $\sim m_X$. Thus, no acceleration mechanism is needed. The sources of the massive X particles could be topological defects such as cosmic strings or magnetic monopoles that could be produced in the early Universe during symmetry-breaking phase transitions envisaged in Grand Unified Theories (GUTs). In an inflationary early Universe, the relevant topological defects could be formed at a phase transition at the end of inflation. Alternatively, the X particles could be certain supermassive metastable relic particles of lifetime comparable to or larger than the age of the Universe, which could be produced in the early Universe through, for example, particle production processes associated with inflation. Absence of nearby powerful astrophysical objects such as AGNs or radio galaxies is not a problem in the top-down scenario because the X particles or their sources need not necessarily be associated with any specific active astrophysical objects. In certain models, the X particles themselves or their sources may be clustered in galactic halos, in which case the dominant contribution to the EHECRs observed at Earth would come from the X particles clustered within our Galactic Halo, for which the GZK restriction on source distance would be of no concern.

By focusing primarily on "non-conventional" scenarios involving new particle physics beyond the electroweak scale, we do not wish to give the wrong impression that these scenarios explain all aspects of EHECRs. In fact, as we shall see below, essentially each of the specific models that have been studied so far has its own peculiar set of problems. Indeed, the main problem of non-astrophysical solutions of the EHECR problem in general is that they are highly model dependent. On the other hand, it is precisely because of this reason that these scenarios are also attractive – they bring in ideas of new physics beyond the Standard Model of particle physics (such as Grand Unification and new interactions beyond the reach of terrestrial accelerators) as well as ideas of early Universe cosmology (such as topological defects and/or massive particle production in inflation) into the realms of EHECRs where these ideas have the potential to be tested by future EHECR experiments.

The physics and astrophysics of UHECRs are intimately linked with the emerging field of neutrino astronomy (for reviews see [25,26]) as well as with the already established field of $\gamma-$ray astronomy (for reviews see, e.g., [27] and the contribution by Trevor Weekes in this volume) which in turn are important subdisciplines of particle astrophysics (for a review see, e.g., [28]). Indeed, as we shall see, all scenarios of UHECR origin, including the top-down models, are severely constrained by neutrino and $\gamma-$ray observations and limits. In turn, this linkage has important consequences for theoretical predictions of fluxes of extragalactic neutrinos above a TeV or so whose detection is a major goal of next-generation neutrino telescopes (see Sect. 2): If these neutrinos are produced as secondaries of protons accelerated in astrophysical sources and if these protons are not absorbed in the sources, but rather contribute to the UHECR flux observed, then the energy content in the neutrino flux can not be higher than the one in UHECRs, leading to the so called Waxman Bahcall bound [29,30]. If one of these assumptions does not apply, such as for acceleration sources that are opaque to nucleons or in the TD scenarios where X particle decays produce much fewer nucleons than $\gamma-$rays and neutrinos, the Waxman Bahcall bound does not apply, but the neutrino flux is still constrained by the observed diffuse $\gamma-$ray flux in the GeV range (see Sect. 4.4).

Finally, in Sect. 5 we shall discuss how, apart from the unsolved problem of the source mechanism, EHECR observations have the potential to yield important information on Galactic and extragalactic magnetic fields.

2 Present and Future UHE CR and Neutrino Experiments

The CR primaries are shielded by the Earth's atmosphere and near the ground reveal their existence only by indirect effects such as ionization. Indeed, it was the height dependence of this latter effect which lead to the discovery of CRs by Hess in 1912. Direct observation of CR primaries is only possible from space by flying detectors with balloons or spacecraft. Naturally, such detectors are very limited in size and because the differential CR spectrum is a steeply falling function of energy (see Fig. 1), direct observations run out of statistics typically around a few 100 TeV.

Above $\sim$ 100 TeV, the showers of secondary particles created in the interactions of the primary CR with the atmosphere are extensive enough to be detectable from the ground. In the most traditional technique, charged hadronic particles, as well as electrons and muons in these Extensive Air Showers (EAS) are recorded on the ground [31] with standard instruments such as water Cherenkov detectors used in the old Volcano Ranch [2] and Haverah Park [4] experiments, and scintillation detectors which are used now-a-days. Currently operating ground arrays for UHECR EAS are the Yakutsk experiment in Russia [7] and the Akeno Giant Air Shower Array (AGASA) near Tokyo, Japan, which is the largest one, covering an area of roughly 100 km^2 with about 100 detectors mutually separated by about 1 km [9]. The Sydney University Giant Air

Shower Recorder (SUGAR) [3] operated until 1979 and was the largest array in the Southern hemisphere. The ground array technique allows one to measure a lateral cross section of the shower profile. The energy of the shower-initiating primary particle is estimated by appropriately parameterizing it in terms of a measurable parameter; traditionally this parameter is taken to be the particle density at 600 m from the shower core, which is found to be quite insensitive to the primary composition and the interaction model used to simulate air showers.

The detection of secondary photons from EAS represents a complementary technique. The experimentally most important light sources are the fluorescence of air nitrogen excited by the charged particles in the EAS and the Cherenkov radiation from the charged particles that travel faster than the speed of light in the atmospheric medium. The first source is practically isotropic whereas the second one produces light strongly concentrated on the surface of a cone around the propagation direction of the charged source. The fluorescence technique can be used equally well for both charged and neutral primaries and was first used by the Fly's Eye detector [8] and will be part of several future projects on UHECRs (see below). The primary energy can be estimated from the total fluorescence yield. Information on the primary composition is contained in the column depth $X_{\rm max}$ (measured in $\rm g\,cm^{-2}$) at which the shower reaches maximal particle density. The average of $X_{\rm max}$ is related to the primary energy E by

$$\langle X_{\rm max}\rangle = X_0' \ln\left(\frac{E}{E_0}\right) . \tag{1}$$

Here, X_0' is called the elongation rate and E_0 is a characteristic energy that depends on the primary composition. Therefore, if $X_{\rm max}$ and X_0' are determined from the longitudinal shower profile measured by the fluorescence detector, then E_0 and thus the composition, can be extracted after determining the energy E from the total fluorescence yield. Comparison of CR spectra measured with the ground array and the fluorescence technique indicate systematic errors in energy calibration that are generally smaller than $\sim 40\%$. For a more detailed discussion of experimental EAS analysis with the ground array and the fluorescence technique see, e.g., [32].

As an upscaled version of the old Fly's Eye Cosmic Ray experiment, the High Resolution Fly's Eye detector is currently under construction at Utah, USA [34]. Taking into account a duty cycle of about 10% (a fluorescence detector requires clear, moonless nights), the effective aperture of this instrument will be $\simeq 600\,\rm km^2\,sr$, about 10 times the AGASA aperture, with a threshold around $10^{17}\,$eV. Another project utilizing the fluorescence technique is the Japanese Telescope Array [35] which is currently in the proposal stage. Its effective aperture will be about 15–20 times that of AGASA above $10^{17}\,$eV, and it can also be used as a Cherenkov detector for TeV $\gamma-$ray astrophysics. Probably the largest up-coming project is the international Pierre Auger Giant Array Observatories [36] which will be a combination of a ground array of about 1700 particle detectors mutually separated from each other by about 1.5 km and covering about $3000\,\rm km^2$, and one or more fluorescence Fly's Eye type detectors. The ground array component will have a duty cycle of nearly 100%, leading to an effective aperture about 200 times as large as the AGASA array, and an event rate of 50–100 events per year above 10^{20} eV. About 10% of the events will be detected by both the ground array and the fluorescence component and can be used for cross calibration and detailed EAS studies. The energy threshold will be around 10^{19} eV. For maximal sky coverage it is furthermore planned to construct one site in each hemisphere. The southern site will be in Argentina, and the northern site probably in Utah, USA.

Recently NASA initiated a concept study for detecting EAS from space [38] by observing their fluorescence light from an Orbiting Wide-angle Light-collector (OWL). This would provide an increase by another factor ~ 50 in aperture compared to the Pierre Auger Project, corresponding to an event rate of up to a few thousand events per year above 10^{20} eV. Similar concepts such as the AIRWATCH [39] and Maximum-energy air-Shower Satellite (MASS) [40] missions are also being discussed. The energy threshold of such instruments would be between 10^{19} and 10^{20} eV. This technique would be especially suitable for detection of very small event rates such as those caused by UHE neutrinos which would produce deeply penetrating EAS (see Sect. 4.4). For more details on these recent experimental considerations see [10].

High energy neutrino astronomy is aiming towards a kilometer scale neutrino observatory. The major technique is the optical detection of Cherenkov light emitted by muons created in charged current reactions of neutrinos with nucleons either in water or in ice. The largest pilot experiments representing these two detector media are the now defunct Deep Undersea Muon and Neutrino Detection (DUMAND) experiment [41] in the deep sea near Hawaii and the Antarctic Muon And Neutrino Detector Array (AMANDA) experiment [42] in the South Pole ice. Another water based experiment is situated at Lake Baikal [43]. Next generation deep sea projects include the French Astronomy with a Neutrino Telescope and Abyss environmental RESearch (ANTARES) [45] and the underwater Neutrino Experiment SouthwesT Of GReece (NESTOR) project in the Mediterranean [46], whereas ICECUBE [47] represents the planned kilometer scale version of the AMANDA detector. Also under consideration are neutrino detectors utilizing techniques to detect the radio pulse from the electromagnetic showers created by neutrino interactions in ice. This technique could possibly be scaled up to an effective area of $10^4\,\mathrm{km}^2$ and a prototype is represented by the Radio Ice Cherenkov Experiment (RICE) experiment at the South Pole [48]. Neutrinos can also initiate horizontal EAS which can be detected by giant ground arrays such as the Pierre Auger Project [49]. Furthermore, as mentioned above, deeply penetrating EAS could be detected from space by instruments such as the proposed OWL detector [38]. More details and references on neutrino astronomy detectors are contained in [25,50], and some recent overviews on neutrino astronomy can be found in [26].

3 New Primary Particles and New Interactions

A possible way around the problem of missing counterparts within acceleration scenarios is to propose primary particles whose range is not limited by the GZK effect. Within the Standard Model the only candidate is the neutrino, whereas in supersymmetric extensions of the Standard Model, new neutral hadronic bound states of light gluinos with quarks and gluons, so-called R-hadrons that are heavier than nucleons, and therefore have a higher GZK threshold, have been suggested [51].

In both the neutrino and R-hadron scenario the particle propagating over extragalactic distances would have to be produced as a secondary in interactions of a primary proton that is accelerated in a powerful AGN which can, in contrast to the case of EAS induced by nucleons, nuclei, or γ−rays, be located at high redshift. Consequently, these scenarios predict a correlation between primary arrival directions and high redshift sources. In fact, possible evidence for an angular correlation of the five highest energy events with compact radio quasars at redshifts between 0.3 and 2.2 was recently reported [52]. Only a few more events could confirm or rule out the correlation hypothesis. Note, however, that these scenarios require the primary proton to be accelerated up to at least $10^{21}\,\mathrm{eV}$, demanding a very powerful astrophysical accelerator.

3.1 New Neutrino Interactions

Neutrino primaries have the advantage of being well established particles, however, within the Standard Model their interaction cross section with nucleons falls short by about five orders of magnitude to produce ordinary air showers. Interestingly, in theories with n additional large compact dimensions the exchange of bulk gravitons (Kaluza-Klein modes) leads to an extra contribution to any two-particle cross section. Such scenarios are motivated by string theory and, for an effective quantum gravity scale in $n+4$ dimensions, $M_{4+n} \sim \mathrm{TeV}$ provide a solution to the hierarchy problem in grand unifications of gauge interactions and therefore recently received much attention in the literature. The bulk graviton exchange cross section is given by [53]

$$\sigma_g \simeq \frac{4\pi s}{M_{4+n}^4} \simeq 10^{-27} \left(\frac{M_{4+n}}{\mathrm{TeV}}\right)^{-4} \left(\frac{E}{10^{20}\,\mathrm{eV}}\right) \mathrm{cm}^2 \,, \tag{2}$$

where in the last expression we specified to a neutrino of energy E hitting a nucleon at rest. Note that a neutrino would typically start to interact in the atmosphere for $\sigma_{\nu N} \gtrsim 10^{-27}\,\mathrm{cm}^2$, i.e. for $E \gtrsim 10^{20}\,\mathrm{eV}$, assuming $M_{4+n} \simeq 1\,\mathrm{TeV}$. The neutrino therefore becomes a primary candidate for the observed EHECR events. A specific signature of this scenario would be the absence of any events above the energy where σ_g grows beyond $\simeq 10^{-27}\,\mathrm{cm}^2$ in neutrino telescopes based on ice or water as detector medium [26], and a hardening of the spectrum above this energy in atmospheric detectors such as the Pierre Auger Project [36] and the Orbital Wide-angle Light Collector (OWL) [38]. Furthermore, according to (2), the average atmospheric column depth of the first interaction point of neutrino induced EAS in this scenario is predicted to depend linearly on energy. This should be easy to distinguish from the logarithmic scaling, (1), expected for nucleons, nuclei, and γ−rays.

3.2 Supersymmetric Particles

Light gluinos binding to quarks, anti-quarks and/or gluons can occur in supersymmetric theories involving gauge-mediated supersymmetry (SUSY) breaking [54] where the resulting gluino mass arises dominantly from radiative corrections and can vary between $\sim 1\,\mathrm{GeV}$ and $\sim 100\,\mathrm{GeV}$. In these scenarios, the gluino can be the lightest supersymmetric particle (LSP). There are also arguments against a light quasi-stable gluino [55], mainly based on constraints on the abundance of anomalous heavy isotopes of hydrogen and oxygen which could be formed as bound states of these nuclei and the gluino. Furthermore, accelerator constraints have become quite stringent [56] and seem to be inconsistent with the original scenario from [51]. However, the scenario with a "tunable" gluino mass [54] still seems possible and suggests either the gluino-gluon bound state $g\tilde{g}$, called glueballino R_0, or the isotriplet $\tilde{g}-(u\bar{u}-d\bar{d})_8$, called $\tilde{\rho}$, as the lightest quasi-stable R-hadron. For a summary of scenarios with light gluinos consistent with accelerator constraints see [57]. The case of a light quasi-stable gluino does not seem to be settled.

An astrophysical constraint on new neutral massive and strongly interacting EAS primaries results from the fact that the nucleon interactions producing these particles in the source also produce neutrinos and especially γ−rays. The resulting fluxes from powerful discrete acceleration sources may be easily detectable in the GeV range by space-borne γ−ray instruments such as EGRET and GLAST, and in the TeV range by ground based γ−ray detectors such as HEGRA and WHIPPLE and the planned VERITAS, HESS, and MAGIC projects (for reviews discussing these instruments see [27] and the contribution by Trevor Weekes in this volume). At least the latter three ground based instruments should have energy thresholds low enough to detect γ−rays from the postulated sources at redshift $z \sim 1$. Such observations in turn imply constraints on the required branching ratio of proton interactions into the R-hadron which, very roughly, should be larger than ~ 0.01. These constraints, however, will have to be investigated in more detail for specific sources. One could also search for heavy neutral baryons in the data from Cherenkov instruments in the TeV range in this context. To demonstrate these points, a schematic example of fluxes predicted for the new heavy particle and for γ−rays and neutrinos are shown in Fig. 2.

A further constraint on new EAS primary particles in general comes from the character of the air showers created by them: The observed EHECR air showers are consistent with nucleon primaries and limits the possible primary rest mass to less than $\simeq 50\,\mathrm{GeV}$ [58]. With the statistics expected from upcoming experiments such as the Pierre Auger Project, this upper limit is likely to be lowered down to $\simeq 10\,\mathrm{GeV}$.

It is interesting to note in this context that in case of a confirmation of the existence of new neutral particles in UHECRs, a combination of accelerator, air shower, and astrophysics data would be highly restrictive in terms of the underlying physics: In the above scenario, for example, the gluino would have to be in a narrow mass range, 1–10 GeV, and the newest accelerator constraints on the Higgs mass, $m_h \gtrsim 90\,\mathrm{GeV}$, would require the presence of a D term of an

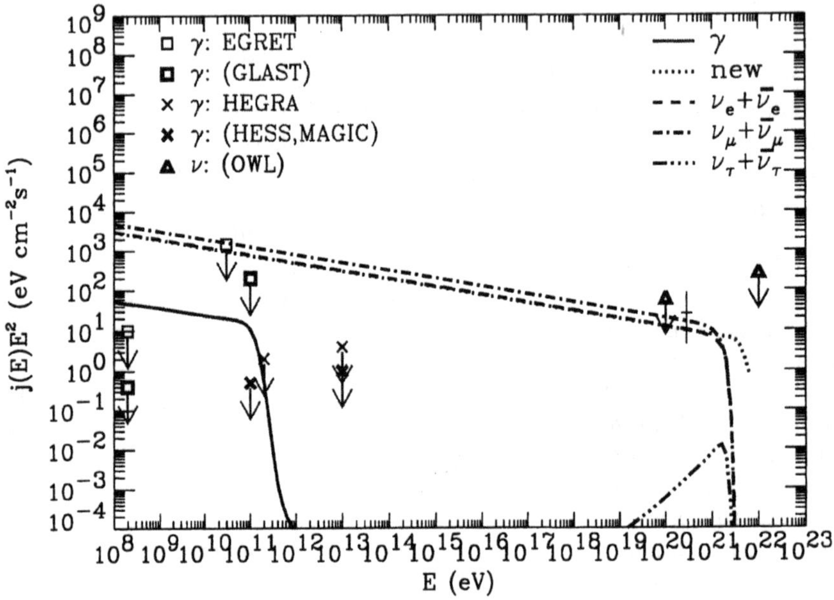

Fig. 2. Schematic predictions for the fluxes of the putative new neutral heavy particle (dotted line), electron, muon, and τ−neutrinos (dashed and dash-dotted lines, as indicated), and γ−rays (solid line) for a source at redshift $z = 1$. Assumed were a proton spectrum $\propto E^{-2.2}$ extending at least up to 10^{22} eV at the source, a branching ratio for production of the heavy neutral in nucleon interactions of 0.01, and a beaming factor of 10 for neutrinos and the heavy neutrals. The 1 sigma error bar at 3×10^{20} eV represents the point flux corresponding to the highest energy Fly's Eye event. The predicted fluxes were normalized such that this highest energy event is explained as a new heavy particle. The points with arrows on the right part represent projected approximate neutrino point source sensitivities for the OWL concept using the acceptance estimated in [38] for non-detection over a five year period. The points with arrows in the lower left part represent approximate γ−ray point source sensitivities of existing detectors such as EGRET and HEGRA, and of planned instruments such as the satellite detector GLAST, the Cherenkov telescope array HESS and the single dish instrument MAGIC, for 50 hours and 1 month observation time for the ground based and satellite detectors, respectively

anomalous $U(1)_X$ gauge symmetry, in addition to a gauge-mediated contribution to SUSY breaking at the messenger scale [54].

4 Top-Down Scenarios

4.1 The Main Idea

As mentioned in the introduction, all top-down scenarios involve the decay of X particles of mass close to the GUT scale which can basically be produced in two ways: If they are very short lived, as usually expected in many GUTs, they have to be produced continuously. The only way this can be achieved is by emission from topological defects left over from cosmological phase transitions that may have occurred in the early Universe at temperatures close to the GUT scale, possibly during reheating after inflation. Topological defects necessarily occur between regions that are causally disconnected, such that the orientation of the order parameter associated with the phase transition, can not be communicated between these regions and consequently will adopt different values. Examples are cosmic strings (similar to vortices in superfluid helium), magnetic monopoles, and domain walls (similar to Bloch walls separating regions of different magnetization in a ferromagnet). The defect density is consequently given by the particle horizon in the early Universe and their formation can even be studied in solid state experiments where the expansion rate of the Universe corresponds to the quenching speed with which the phase transition is induced [59]. The defects are topologically stable, but in the cosmological case time dependent motion leads to the emission of particles with a mass comparable to the temperature at which the phase transition took place. The associated phase transition can also occur during reheating after inflation.

Alternatively, instead of being released from topological defects, X particles may have been produced directly in the early Universe and, due to some unknown symmetries, have a very long lifetime comparable to the age of the Universe. In contrast to Weakly-Interacting Massive Particles (WIMPS) below a few hundred TeV which are the usual dark matter candidates motivated by, for example, supersymmetry and can be produced by thermal freeze out, such superheavy X particles have to be produced non-thermally. Several such mechanisms operating in the post-inflationary epoch in the early Universe have been studied. They include gravitational production through the effect of the expansion of the background metric on the vacuum quantum fluctuations of the X particle field, or creation during reheating at the end of inflation if the X particle field couples to the inflaton field. The latter case can be divided into three subcases, namely "incoherent" production with an abundance proportional to the X particle annihilation cross section, non-adiabatic production in broad parametric resonances with the oscillating inflaton field during preheating (analogous to energy transfer in a system of coupled pendula), and creation in bubble wall collisions if inflation is completed by a first order phase transition. In all these cases, such particles, also called "WIMPZILLAs", would contribute to the dark matter and their decays could still contribute to UHE CR fluxes today, with an anisotropy pattern that reflects the dark matter distribution in the halo of our Galaxy.

It is interesting to note that one of the prime motivations of the inflationary paradigm was to dilute excessive production of "dangerous relics" such as topological defects and superheavy stable particles. However, such objects can be produced right after inflation during reheating in cosmologically interesting abundances, and with a mass scale roughly given by the inflationary scale which in turn is fixed by the CMB anisotropies to $\sim 10^{13}\,\text{GeV}$ [60]. The reader will realize that this mass scale is somewhat above the highest energies observed in

CRs, which implies that the decay products of these primordial relics could well have something to do with EHECRs which in turn can probe such scenarios!

The X particle injection rate is assumed to be spatially uniform and for dimensional reasons can only depend on the mass scale m_X and on cosmic time t in the combination

$$\dot{n}_X(t) = \kappa\, m_X^p\, t^{-4+p}\,, \tag{3}$$

where κ and p are dimensionless constants whose value depend on the specific top-down scenario [61], For example, the case $p = 1$ is representative of scenarios involving release of X particles from topological defects, such as ordinary cosmic strings [62], necklaces [63] and magnetic monopoles [64]. This can be easily seen as follows: The energy density ρ_s in a network of defects has to scale roughly as the critical density, $\rho_s \propto \rho_{\rm crit} \propto t^{-2}$, where t is cosmic time, otherwise the defects would either start to overclose the Universe, or end up having a negligible contribution to the total energy density. In order to maintain this scaling, the defect network has to release energy with a rate given by $\dot{\rho}_s = -a\rho_s/t \propto t^{-3}$, where $a = 1$ in the radiation dominated area, and $a = 2/3$ during matter domination. If most of this energy goes into emission of X particles, then typically $\kappa \sim \mathcal{O}(1)$. In the numerical simulations presented below, it was assumed that the X particles are nonrelativistic at decay.

The X particles could be gauge bosons, Higgs bosons, superheavy fermions, etc. depending on the specific GUT. They would have a mass m_X comparable to the symmetry breaking scale and would decay into leptons and/or quarks of roughly comparable energy. The quarks interact strongly and hadronize into nucleons (Ns) and pions, the latter decaying in turn into γ-rays, electrons, and neutrinos. Given the X particle production rate, dn_X/dt, the effective injection spectrum of particle species a ($a = \gamma, N, e^\pm, \nu$) via the hadronic channel can be written as $(\mathrm{d}n_X/\mathrm{d}t)(2/m_X)(\mathrm{d}N_a/\mathrm{d}x)$, where $x \equiv 2E/m_X$, and $\mathrm{d}N_a/\mathrm{d}x$ is the relevant fragmentation function (FF).

We adopt the Local Parton Hadron Duality (LPHD) approximation [65] according to which the total hadronic FF, dN_h/dx, is taken to be proportional to the spectrum of the partons (quarks/gluons) in the parton cascade (which is initiated by the quark through perturbative QCD processes) after evolving the parton cascade to a stage where the typical transverse momentum transfer in the QCD cascading processes has come down to $\sim R^{-1} \sim$ few hundred MeV, where R is a typical hadron size. The parton spectrum is obtained from solutions of the standard QCD evolution equations in modified leading logarithmic approximation (MLLA) which provides good fits to accelerator data at LEP energies [65]. We will specifically use a recently suggested generalization of the MLLA spectrum that includes the effects of supersymmetry [66]. Within the LPHD hypothesis, the pions and nucleons after hadronization have essentially the same spectrum. The LPHD does not, however, fix the relative abundance of pions and nucleons after hadronization. Motivated by accelerator data, we assume the nucleon content f_N of the hadrons to be in the range 3 to 10%, and the rest pions distributed equally among the three charge states. According to recent Monte Carlo simulations [67], the nucleon-to-pion ratio may be significantly higher in certain ranges of x values at the extremely high energies of interest here. Unfortunately, however, due to the very nature of these Monte Carlo calculations, it is difficult to understand the precise physical reason for the unexpectedly high baryon yield relative to mesons. While more of these Monte Carlo calculations of the relevant FFs in the future will hopefully clarify the situation, we will use here the range of $f_N \sim 3$ to 10% mentioned above. The standard pion decay spectra then give the injection spectra of γ-rays, electrons, and neutrinos. For more details concerning uncertainties in the X particle decay spectra see [68].

4.2 Numerical Simulations

The γ-rays and electrons produced by X particle decay initiate electromagnetic (EM) cascades on low energy radiation fields such as the CMB. The high energy photons undergo electron-positron pair production (PP; $\gamma\gamma_b \rightarrow e^-e^+$), and at energies below $\sim 10^{14}$ eV they interact mainly with the universal infrared and optical (IR/O) backgrounds, while above ~ 100 EeV they interact mainly with the universal radio background (URB). In the Klein-Nishina regime, where the center of mass energy is large compared to the electron mass, one of the outgoing particles usually carries most of the initial energy. This "leading" electron (positron) in turn can transfer almost all of its energy to a background photon via inverse Compton scattering (ICS; $e\gamma_b \rightarrow e'\gamma$). EM cascades are driven by this cycle of PP and ICS. The energy degradation of the "leading" particle in this cycle is slow, whereas the total number of particles grows exponentially with time. This makes a standard Monte Carlo treatment difficult. Implicit numerical schemes have therefore been used to solve the relevant kinetic equations. A detailed account of the transport equation approach used in the calculations whose results are presented in this contribution can be found in [69]. All EM interactions that influence the γ-ray spectrum in the energy range $10^8\,\text{eV} < E < 10^{25}\,\text{eV}$, namely PP, ICS, triplet pair production (TPP; $e\gamma_b \rightarrow ee^-e^+$), and double pair production (DPP, $\gamma\gamma_b \rightarrow e^-e^+e^-e^+$), as well as synchrotron losses of electrons in the large scale extragalactic magnetic field (EGMF), are included.

Similarly to photons, UHE neutrinos give rise to neutrino cascades in the primordial neutrino background via exchange of W and Z bosons [70,71]. Besides the secondary neutrinos which drive the neutrino cascade, the W and Z decay products include charged leptons and quarks which in turn feed into the EM and hadronic channels. Neutrino interactions become especially significant if the relic neutrinos have masses m_ν in the eV range and thus constitute hot dark matter, because the Z boson resonance then occurs at an UHE neutrino energy $E_{\text{res}} = 4 \times 10^{21}(\text{eV}/m_\nu)$ eV. In fact, this has been proposed as a significant source of EHECRs [72,73]. Motivated by recent experimental evidence for neutrino mass we assumed a mass of 1 eV for all three neutrino flavors (for simplicity) and implemented the relevant W boson interactions in the t-channel and the Z boson exchange via t- and s-channel. Hot dark matter is also expected to cluster, potentially increasing secondary γ-ray and nucleon production [72,73]. This influences mostly scenarios where X decays into neutrinos only. We parameterize massive neutrino clustering by a length scale l_ν and an overdensity f_ν over the average density $\bar{n_\nu}$. The Fermi distribution with a velocity dispersion v yields $f_\nu \lesssim v^3 m_\nu^3/(2\pi)^{3/2}/\bar{n_\nu} \simeq 330\,(v/500\,\mathrm{km\,sec}^{-1})^3\,(m_\nu/\mathrm{eV})^3$ [74]. Therefore, values of $l_\nu \simeq$ few Mpc and $f_\nu \simeq 20$ are conceivable on the local Supercluster scale [73].

The relevant nucleon interactions implemented are pair production by protons ($p\gamma_b \to pe^-e^+$), photoproduction of single or multiple pions ($N\gamma_b \to N\,n\pi$, $n \geq 1$), and neutron decay. In TD scenarios, the particle injection spectrum is generally dominated by the "primary" γ-rays and neutrinos over nucleons. These primary γ-rays and neutrinos are produced by the decay of the primary pions resulting from the hadronization of quarks that come from the decay of the X particles. The contribution of secondary γ-rays, electrons, and neutrinos from decaying pions that are subsequently produced by the interactions of nucleons with the CMB, is in general negligible compared to that of the primary particles; we nevertheless include the contribution of the secondary particles in our code.

We assume a flat Universe with no cosmological constant, and a Hubble constant of $h = 0.65$ in units of 100 km sec^{-1}Mpc^{-1} throughout. The numerical calculations follow *all* produced particles in the EM, hadronic, and neutrino channel, whereas the often-used continuous energy loss (CEL) approximation (e.g., [75]) follows only the leading cascade particles. The CEL approximation can significantly underestimate the cascade flux at lower energies.

The two major uncertainties in the particle transport are the intensity and spectrum of the URB for which there exists only an estimate above a few MHz frequency [76], and the average value of the EGMF. To bracket these uncertainties, simulations have been performed for the observational URB estimate from [76] that has a low-frequency cutoff at 2 MHz ("minimal"), and the medium and maximal theoretical estimates from [77], as well as for EGMFs between zero and 10^{-9} G, the latter motivated by limits from Faraday rotation measurements, see Sect. 5.2 below. A strong URB tends to suppress the UHE γ-ray flux by direct absorption whereas a strong EGMF blocks EM cascading (which otherwise develops efficiently especially in a low URB) by synchrotron cooling of the electrons. For the IR/O background we used the most recent data [78].

4.3 Results: $\gamma-$ray and Nucleon Fluxes

Figure 3 shows results from [68] for the time averaged $\gamma-$ray and nucleon fluxes in a typical TD scenario, assuming no EGMF, along with current observational constraints on the $\gamma-$ray flux. The spectrum was optimally normalized to allow for an explanation of the observed EHECR events, assuming their consistency with a nucleon or $\gamma-$ray primary. The flux below $\lesssim 2 \times 10^{19}$ eV is presumably due to conventional acceleration in astrophysical sources and was not fit. Similar spectral shapes have been obtained in [80], where the normalization was chosen to match the observed differential flux at 3×10^{20} eV. This normalization, however, leads to an overproduction of the integral flux at higher energies, whereas above 10^{20} eV, the fits shown in Figs. 3 and 4 have likelihood significances above 50% (see [81] for details) and are consistent with the integral flux above 3×10^{20} eV

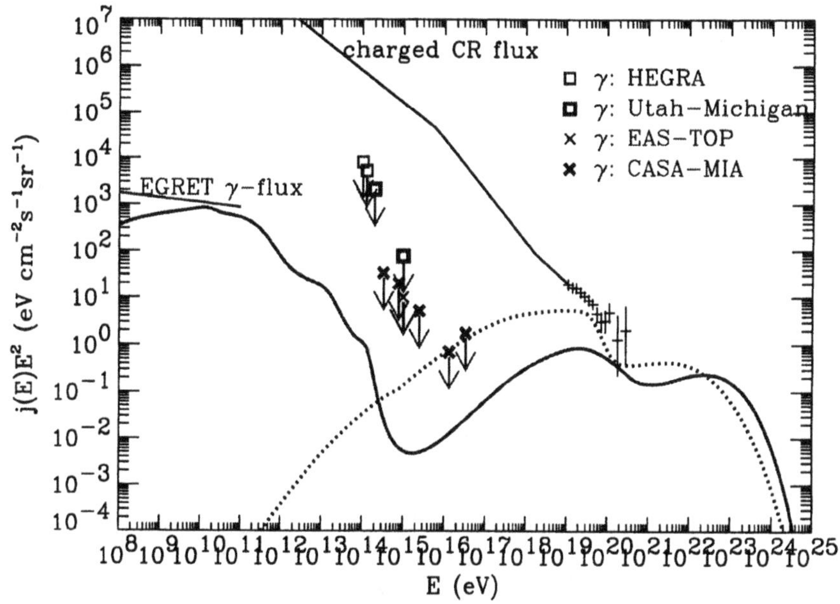

Fig. 3. Predictions for the differential fluxes of $\gamma-$rays (solid line) and protons and neutrons (dotted line) in a TD model characterized by $p = 1$, $m_X = 10^{16}$ GeV, and the decay mode $X \to q+q$, assuming the supersymmetric modification of the fragmentation function [66], with a fraction of about 10% nucleons. The calculation used the code described in [68] and assumed the strongest URB version from [77] and an EGMF $\ll 10^{-11}$ G. 1 sigma error bars are the combined data from the Haverah Park [4], the Fly's Eye [8], and the AGASA [9] experiments above 10^{19} eV. Also shown are piecewise power law fits to the observed charged CR flux (thick solid line) and the EGRET measurement of the diffuse $\gamma-$ray flux between 30 MeV and 100 GeV [79] (solid line on left margin). Points with arrows represent upper limits on the $\gamma-$ray flux from the HEGRA, the Utah-Michigan, the EAS-TOP, and the CASA-MIA experiments, as indicated

estimated in [8,9]. The PP process on the CMB depletes the photon flux above 100 TeV, and the same process on the IR/O background causes depletion of the photon flux in the range 100 GeV–100 TeV, recycling the absorbed energies to energies below 100 GeV through EM cascading (see Fig. 3). The predicted background is *not* very sensitive to the specific IR/O background model, however [82]. The scenario in Fig. 3 obviously obeys all current constraints within the normalization ambiguities and is therefore quite viable. Note that the diffuse $\gamma-$ray background measured by EGRET [79] up to 10 GeV puts a strong constraint on these scenarios, especially if there is already a significant contribution to this background from conventional sources such as unresolved $\gamma-$ray blazars [83]. However, the $\gamma-$ray background constraint can be circumvented by assuming that TDs or the decaying long lived X particles do not have a uniform

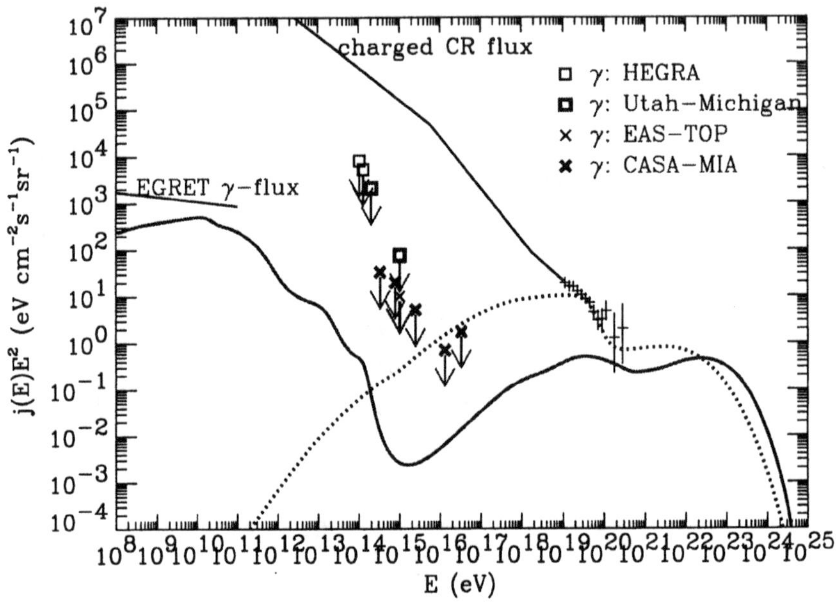

Fig. 4. Same as Fig. 3, but for an EGMF of 10^{-9} G

density throughout the Universe but cluster within galaxies [84]. As can also be seen, at energies above 100 GeV, TD models are not significantly constrained by observed γ–ray fluxes yet (see [12] for more details on these measurements).

Figure 4 shows results for the same TD scenario as in Fig. 3, but for a high EGMF $\sim 10^{-9}$ G, somewhat below the current upper limit, see (10) below. In this case, rapid synchrotron cooling of the initial cascade pairs quickly transfers energy out of the UHE range. The UHE γ–ray flux then depends mainly on the absorption length due to pair production and is typically much lower [75,85]. (Note, though, that for $m_X \gtrsim 10^{25}$ eV, the synchrotron radiation from these pairs can be above 10^{20} eV, and the UHE flux is then not as low as one might expect.) We note, however, that the constraints from the EGRET measurements do not change significantly with the EGMF strength as long as the nucleon flux is comparable to the γ–ray flux at the highest energies, as is the case in Figs. 3 and 4. The results of [68] differ from those of [80] which obtained more stringent constraints on TD models because of the use of an older fragmentation function from [86], and a stronger dependence on the EGMF because of the use of a weaker EGMF which lead to a dominance of γ–rays above $\simeq 10^{20}$ eV.

The energy loss and absorption lengths for UHE nucleons and photons are short ($\lesssim$ 100 Mpc). Thus, their predicted UHE fluxes are independent of cosmological evolution. The γ–ray flux below $\simeq 10^{11}$ eV, however, scales as the total X particle energy release integrated over all redshifts and increases with decreasing p [87]. For $m_X = 2 \times 10^{16}$ GeV, scenarios with $p < 1$ are therefore ruled out (as can be inferred from Figs. 3 and 4), whereas constant comoving injection models ($p = 2$) are well within the limits.

We now turn to signatures of TD models at UHE. The full cascade calculations predict γ–ray fluxes below 100 EeV that are a factor $\simeq 3$ and $\simeq 10$ higher than those obtained using the CEL or absorption approximation often used in the literature, in the case of strong and weak URB, respectively. Again, this shows the importance of non-leading particles in the development of unsaturated EM cascades at energies below $\sim 10^{22}$ eV. Our numerical simulations give a γ/CR flux ratio at 10^{19} eV of $\simeq 0.1$. The experimental exposure required to detect a γ–ray flux at that level is $\simeq 4 \times 10^{19}\,\mathrm{cm}^2\,\mathrm{sec\,sr}$, about a factor 10 smaller than the current total experimental exposure. These exposures are well within reach of the Pierre Auger Cosmic Ray Observatories [36], which may be able to detect a neutral CR component down to a level of 1% of the total flux. In contrast, if the EGMF exceeds $\sim 10^{-11}$ G, then UHE cascading is inhibited, resulting in a lower UHE γ–ray spectrum. In the 10^{-9} G scenario of Fig. 4, the γ/CR flux ratio at 10^{19} eV is 0.02, significantly lower than for no EGMF.

It is clear from the above discussions that the predicted particle fluxes in the TD scenario are currently uncertain to a large extent due to particle physics uncertainties (e.g., mass and decay modes of the X particles, the quark fragmentation function, the nucleon fraction f_N, and so on) as well as astrophysical uncertainties (e.g., strengths of the radio and infrared backgrounds, extragalactic magnetic fields, etc.). More details on the dependence of the predicted UHE particle spectra and composition on these particle physics and astrophysical uncertainties are contained in [68]. We stress here that there are viable TD scenarios which predict nucleon fluxes that are comparable to or even higher than the γ–ray flux at all energies, even though γ–rays dominate at production. This occurs, e.g., in the case of high URB and/or for a strong EGMF, and a nucleon fragmentation fraction of $\simeq 10\%$; see, for example, Fig. 4. Some of these TD scenarios would therefore remain viable even if EHECR induced EAS should be proven inconsistent with photon primaries (see, e.g., [88]).

The normalization procedure to the EHECR flux described above imposes the constraint $Q^0_{\mathrm{EHECR}} \lesssim 10^{-22}\,\mathrm{eV\,cm^{-3}\,sec^{-1}}$ within a factor of a few [80,68,89] for the total energy release rate Q_0 from TDs at the current epoch. In most TD models, because of the unknown values of the parameters involved, it is currently not possible to calculate the exact value of Q_0 from first principles, although it has been shown that the required values of Q_0 (in order to explain the EHECR flux) mentioned above are quite possible for certain kinds of TDs. Some cosmic string simulations suggest that strings may lose most of their energy in the form of X particles and estimates of this rate have been given [90]. If that is the case, the constraint on Q^0_{EHECR} translates via (3) into a limit on the symmetry breaking scale η and hence on the mass m_X of the X particle: $\eta \sim m_X \lesssim 10^{13}$ GeV [91]. Independently of whether or not this scenario explains EHECR, the EGRET measurement of the diffuse GeV γ–ray background leads to a similar bound, $Q^0_{\mathrm{EM}} \lesssim 2.2 \times 10^{-23}\,h(3p-1)\,\mathrm{eV\,cm^{-3}\,sec^{-1}}$, which leaves the bound on η and m_X practically unchanged. Furthermore, constraints from limits

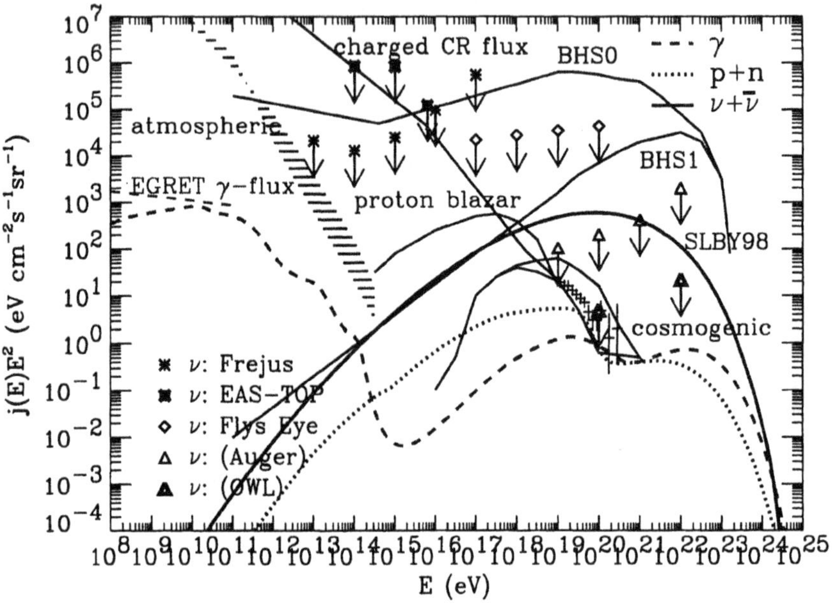

Fig. 5. Predictions for the summed differential fluxes of all neutrino flavors (solid lines) from the atmospheric background for different zenith angles [95] (hatched region marked "atmospheric"), from proton blazars that are photon optically thick to nucleons but contribute to the diffuse γ–ray flux [92] ("proton blazar"), from UHECR interactions with the CMB [93] ("cosmogenic"), for the TD model from [61] with $p = 0$ ("BHS0") and $p = 1$ ("BHS1"), and for the TD model from Fig. 3, assuming an EGMF of $\lesssim 10^{-12}$ G ("SLBY98", from [68]). Also shown are the fluxes of γ–rays (dashed line), and nucleons (dotted lines) for this latter TD model. The data shown for the CR flux and the diffuse γ–ray flux from EGRET are as in Figs. 3 and 4. Points with arrows represent approximate upper limits on the diffuse neutrino flux from the Frejus [96], the EAS-TOP [97], and the Fly's Eye [98] experiments, as indicated. The projected sensitivity for the Pierre Auger project is using the acceptance estimated in [49], and the one for the OWL concept study is based on [38], both assuming observations over a few years period

on CMB distortions and light element abundances from ^{4}He-photodisintegration are comparable to the bound from the directly observed diffuse GeV γ-rays [87].

4.4 Results: Neutrino Fluxes

As discussed in Sect. 4.1, in TD scenarios most of the energy is released in the form of EM particles and neutrinos. If the X particles decay into a quark and a lepton, the quark hadronizes mostly into pions and the ratio of energy release into the neutrino versus EM channel is $r \simeq 0.3$.

Figure 5 shows predictions of the total neutrino flux for the same TD model on which Fig. 3 is based, as well as some of the older estimates from [61]. In the absence of neutrino oscillations the electron neutrino and anti-neutrino fluxes that are not shown are about a factor of 2 smaller than the muon neutrino and anti-neutrino fluxes, whereas the $\tau-$neutrino flux is in general negligible. In contrast, if the interpretation of the atmospheric neutrino deficit in terms of nearly maximal mixing of muon and $\tau-$neutrinos proves correct, the muon neutrino fluxes shown in Fig. 5 would be maximally mixed with the $\tau-$neutrino fluxes. To put the TD component of the neutrino flux in perspective with contributions from other sources, Fig. 5 also shows the atmospheric neutrino flux, a typical prediction for the diffuse flux from photon optically thick proton blazars [92] that are not subject to the Waxman Bahcall bound and were normalized to recent estimates of the blazar contribution to the diffuse $\gamma-$ray background [83], and the flux range expected for "cosmogenic" neutrinos created as secondaries from the decay of charged pions produced by UHE nucleons [93]. The TD flux component clearly dominates above $\sim 10^{19}\,$eV.

In order to translate neutrino fluxes into event rates, one has to fold in the interaction cross sections with matter. At UHEs these cross sections are not directly accessible to laboratory measurements. Resulting uncertainties therefore translate directly to bounds on neutrino fluxes derived from, for example, the non-detection of UHE muons produced in charged-current interactions. In the following, we will assume the estimate [94]

$$\sigma_{\nu N}(E) \simeq 2.36 \times 10^{-32}(E/10^{19}\,\mathrm{eV})^{0.363}\,\mathrm{cm}^2 \quad (10^{16}\,\mathrm{eV} \lesssim E \lesssim 10^{21}\,\mathrm{eV})\,. \quad (4)$$

based on the Standard Model for the charged-current muon-neutrino-nucleon cross section $\sigma_{\nu N}$ if not indicated otherwise.

For an (energy dependent) ice or water equivalent acceptance $A(E)$ (in units of volume times solid angle), one can obtain an approximate expected rate of UHE muons produced by neutrinos with energy $> E$, $R(E)$, by multiplying $A(E)\sigma_{\nu N}(E)n_{\mathrm{H_2O}}$ (where $n_{\mathrm{H_2O}}$ is the nucleon density in water) with the integral muon neutrino flux $\simeq Ej_{\nu_\mu}$. This can be used to derive upper limits on diffuse neutrino fluxes from a non-detection of muon induced events. Figure 5 shows bounds obtained from several experiments: The Frejus experiment derived upper bounds for $E \gtrsim 10^{12}\,$eV from their non-detection of almost horizontal muons with an energy loss inside the detector of more than 140 MeV per radiation length [96]. The EAS-TOP collaboration published two limits from horizontal showers, one in the regime $10^{14} - 10^{15}\,$eV, where non-resonant neutrino-nucleon processes dominate, and one at the Glashow resonance which actually only applies to $\bar{\nu}_e$ [97]. The Fly's Eye experiment derived upper bounds for the energy range between $\sim 10^{17}\,$eV and $\sim 10^{20}\,$eV [98] from the non-observation of deeply penetrating particles. The AKENO group has published an upper bound on the rate of near-horizontal, muon-poor air showers [99]. Horizontal air showers created by electrons or muons that are in turn produced by charged-current reactions of electron and muon neutrinos within the atmosphere have recently also been pointed out as an important method to constrain or measure UHE neutrino fluxes [49] with next generation detectors.

The $p = 0$ TD model BHS0 from the early work of [61] is not only ruled out by the constraints from Sect. 4.3, but also by some of the experimental limits on the UHE neutrino flux, as can be seen in Fig. 5. Further, although both the BHS1 and the SLBY98 models correspond to $p = 1$, the UHE neutrino flux above $\simeq 10^{20}$ eV in the latter is almost two orders of magnitude smaller than in the former. The main reason for this is the different flux normalization adopted in the two papers: First, the BHS1 model was obtained by normalizing the predicted *proton* flux to the observed UHECR flux at $\simeq 4\times10^{19}$ eV, whereas in the SLBY98 model the actually "visible" sum of the nucleon and γ−ray fluxes was normalized in an optimal way. Second, the BHS1 assumed a nucleon fraction about a factor 3 smaller [61]. Third, the BHS1 scenario used an older fragmentation function from [86] which has more power at larger energies. Clearly, the SLBY98 model is not only consistent with the constraints discussed in Sect. 4.3, but also with all existing neutrino flux limits within 2–3 orders of magnitude.

What, then, are the prospects of detecting UHE neutrino fluxes predicted by TD models? In a $1\,\mathrm{km}^3\,2\pi\,\mathrm{sr}$ size detector, the SLBY98 scenario from Fig. 5, for example, predicts a muon-neutrino event rate of $\simeq 0.15\,\mathrm{yr}^{-1}$, and an electron neutrino event rate of $\simeq 0.089\,\mathrm{yr}^{-1}$ above 10^{19} eV, where "backgrounds" from conventional sources should be negligible. Further, the muon-neutrino event rate above 1 PeV should be $\simeq 1.2\,\mathrm{yr}^{-1}$, which could be interesting if conventional sources produce neutrinos at a much smaller flux level. Of course, above $\simeq 100$ TeV, instruments using ice or water as detector medium, have to look at downward going muon and electron events due to neutrino absorption in the Earth. However, τ−neutrinos obliterate this Earth shadowing effect due to their regeneration from τ decays [100]. The presence of τ−neutrinos, for example, due to mixing with muon neutrinos, as suggested by recent experimental results from Super-Kamiokande, can therefore lead to an increased upward going event rate [101]. For recent compilations of UHE neutrino flux predictions from astrophysical and TD sources see [102] and references therein.

For detectors based on the fluorescence technique such as the HiRes [34] and the Telescope Array [35] (see Sect. 2), the sensitivity to UHE neutrinos is often expressed in terms of an effective aperture $a(E)$ which is related to $A(E)$ by $a(E) = A(E)\sigma_{\nu N}(E)n_{\mathrm{H_2O}}$. For the cross section of (4), the apertures given in [34] for the HiRes correspond to $A(E) \simeq 3\,\mathrm{km}^3 \times 2\pi\,\mathrm{sr}$ for $E \gtrsim 10^{19}$ eV for muon neutrinos. The expected acceptance of the ground array component of the Pierre Auger project for horizontal UHE neutrino induced events is $A(10^{19}\,\mathrm{eV}) \simeq 20\,\mathrm{km}^3\,\mathrm{sr}$ and $A(10^{23}\,\mathrm{eV}) \simeq 200\,\mathrm{km}^3\,\mathrm{sr}$ [49], with a duty cycle close to 100%. We conclude that detection of neutrino fluxes predicted by scenarios such as the SLBY98 scenario shown in Fig. 5 requires running a detector of acceptance $\gtrsim 10\,\mathrm{km}^3 \times 2\pi\,\mathrm{sr}$ over a period of a few years. Apart from optical detection in air, water, or ice, other methods such as acoustical and radio detection [25] (see, e.g., the RICE project [48] for the latter) or even detection from space [38] appear to be interesting possibilities for detection concepts operating at such scales (see Sect. 2). For example, the OWL satellite concept, which aims to detect EAS from space, would have an aperture of $\simeq 3 \times 10^6\,\mathrm{km}^2\,\mathrm{sr}$ in the atmosphere, corresponding to $A(E) \simeq 6 \times 10^4\,\mathrm{km}^3\,\mathrm{sr}$ for $E \gtrsim 10^{20}$ eV, with a duty cycle of $\simeq 0.08$ [38]. The backgrounds seem to be in general negligible [71,103]. As indicated by the numbers above and by the projected sensitivities shown in Fig. 5, the Pierre Auger Project and especially the OWL project should be capable of detecting typical TD neutrino fluxes. This applies to any detector of acceptance $\gtrsim 100\,\mathrm{km}^3\,\mathrm{sr}$. Furthermore, a 100 day search with a radio telescope of the NASA Goldstone type for pulsed radio emission from cascades induced by neutrinos or cosmic rays in the lunar regolith could reach a sensitivity comparable or better to the Pierre Auger sensitivity above $\sim 10^{19}\,\mathrm{eV}$ [105].

A more model independent estimate [89] for the average event rate $R(E)$ can be made if the underlying scenario is consistent with observational nucleon and γ−ray fluxes and the bulk of the energy is released above the PP threshold on the CMB. Let us assume that the ratio of energy injected into the neutrino versus EM channel is a constant r. As discussed in Sect. 4.3, cascading effectively reprocesses most of the injected EM energy into low energy photons whose spectrum peaks at $\simeq 10\,\mathrm{GeV}$ [82]. Since the ratio r remains roughly unchanged during propagation, the height of the corresponding peak in the neutrino spectrum should roughly be r times the height of the low-energy γ−ray peak, i.e., we have the condition $\max_E\left[E^2 j_{\nu_\mu}(E)\right] \simeq r \max_E\left[E^2 j_\gamma(E)\right]$. Imposing the observational upper limit on the diffuse γ−ray flux around 10 GeV shown in Fig. 5, $\max_E\left[E^2 j_{\nu_\mu}(E)\right] \lesssim 2\times 10^3 r\,\mathrm{eV cm^{-2} sec^{-1} sr^{-1}}$, then bounds the average diffuse neutrino rate above PP threshold on the CMB, giving

$$R(E) \lesssim 0.34\,r\left[\frac{A(E)}{1\,\mathrm{km}^3\times 2\pi\,\mathrm{sr}}\right]\left(\frac{E}{10^{19}\,\mathrm{eV}}\right)^{-0.6}\mathrm{yr}^{-1}\quad (E\gtrsim 10^{15}\,\mathrm{eV})\,. \tag{5}$$

For $r \lesssim 20(E/10^{19}\,\mathrm{eV})^{0.1}$ this bound is consistent with the flux bounds shown in Fig. 5 that are dominated by the Fly's Eye constraint at UHE. We stress again that TD models are not subject to the Waxman Bahcall bound because the nucleons produced are considerably less abundant than and are not the primaries of produced γ−rays and neutrinos.

In typical TD models such as the one discussed above where primary neutrinos are produced by pion decay, $r \simeq 0.3$. However, in TD scenarios with $r \gg 1$ neutrino fluxes are only limited by the condition that the *secondary* γ−ray flux produced by neutrino interactions with the relic neutrino background be below the experimental limits. An example for such a scenario is given by X particles exclusively decaying into neutrinos (although this is not very likely in most particle physics models, but see [68] and Fig. 6 for a scenario involving topological defects and [106] for a scenario involving decaying superheavy relic particles, both of which explain the observed EHECR events as secondaries of neutrinos interacting with the primordial neutrino background). Such scenarios could induce appreciable event rates above $\sim 10^{19}\,\mathrm{eV}$ in a km^3 scale detector. A detection would thus open the exciting possibility to establish an experimental lower limit on r. Being based solely on energy conservation, (5) holds regardless of whether or not the underlying TD mechanism explains the observed EHECR events.

The transient neutrino event rate could be much higher than (5) in the direction to discrete sources which emit particles in bursts. Corresponding pulses in the EHE nucleon and γ−ray fluxes would only occur for sources nearer than $\simeq 100\,\mathrm{Mpc}$ and, in case of protons, would be delayed and dispersed by deflection

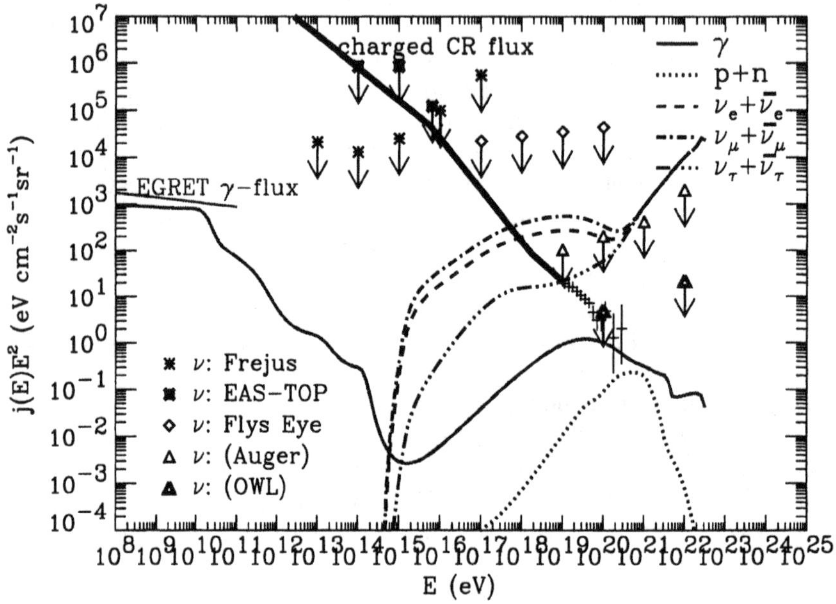

Fig. 6. Flux predictions for a TD model characterized by $p = 1$, $m_X = 10^{14}$ GeV, with X particles exclusively decaying into neutrino-antineutrino pairs of all flavors (with equal branching ratio), assuming a neutrino mass $m_\nu = 1$ eV. For neutrino clustering, an overdensity of $\simeq 30$ over a scale of $l_\nu \simeq 5$ Mpc was assumed. The calculation assumed the strongest URB version from [77] and an EGMF $\ll 10^{-11}$ G. The line key is as in Figs. 3 and 5

in Galactic and extragalactic magnetic fields [107,108]. The recent observation of a possible clustering of CRs above $\simeq 4 \times 10^{19}$ eV by the AGASA experiment [109] might suggest sources which burst on a time scale $t_b \ll 1$ yr. A burst fluence of $\simeq r \left[A(E)/1\,\text{km}^3 \times 2\pi\,\text{sr}\right] (E/10^{19}\,\text{eV})^{-0.6}$ neutrino induced events within a time t_b could then be expected. Associated pulses could also be observable in the GeV – TeV γ−ray flux if the EGMF is smaller than $\simeq 10^{-15}$ G in a significant fraction of extragalactic space [110].

In contrast, the neutrino flux is comparable to (not significantly larger than) the UHE photon plus nucleon fluxes in the models involving metastable superheavy relic particles discussed above. This can be understood because the neutrino flux is dominated by the extragalactic contribution which scales with the extragalactic nucleon and γ−ray contribution in exactly the same way as in the unclustered case, whereas the extragalactic contribution to the "visible" flux to be normalized to the UHECR data is much smaller in the clustered case. The resulting neutrino fluxes would be hardly detectable even with next generation experiments.

5 UHE Cosmic Rays and Cosmological Large Scale Magnetic Fields

5.1 Deflection and Delay of Charged Hadrons

Whereas for UHE electrons the dominant influence of large scale magnetic fields is synchrotron loss rather than deflection, for charged hadrons the opposite is the case. A relativistic particle of charge qe and energy E has a gyroradius $r_g \simeq E/(qeB_\perp)$ where $B_\perp$ is the field component perpendicular to the particle momentum. If this field is constant over a distance d, this leads to a deflection angle

$$\theta(E,d) \simeq \frac{d}{r_g} \simeq 0.52^\circ q \left(\frac{E}{10^{20}\,\mathrm{eV}}\right)^{-1} \left(\frac{d}{1\,\mathrm{Mpc}}\right) \left(\frac{B_\perp}{10^{-9}\,\mathrm{G}}\right) . \tag{6}$$

Magnetic fields beyond the Galactic disk are poorly known and include a possible extended field in the halo of our Galaxy and a large scale EGMF. In both cases, the magnetic field is often characterized by an r.m.s. strength B and a correlation length l_c, i.e. it is assumed that its power spectrum has a cut-off in wavenumber space at $k = 2\pi/l_c$ and in real space it is smooth on scales below l_c. If we neglect energy loss processes for the moment, then the r.m.s. deflection angle over a distance d in such a field is $\theta(E,d) \simeq (2dl_c/9)^{1/2}/r_g$, or

$$\theta(E,d) \simeq 0.8^\circ\, q \left(\frac{E}{10^{20}\,\mathrm{eV}}\right)^{-1} \left(\frac{d}{10\,\mathrm{Mpc}}\right)^{1/2} \left(\frac{l_c}{1\,\mathrm{Mpc}}\right)^{1/2} \left(\frac{B}{10^{-9}\,\mathrm{G}}\right) , \tag{7}$$

for $d \gtrsim l_c$, where the numerical prefactors were calculated from the analytical treatment in [107]. There it was also pointed out that there are two different limits to distinguish: For $\mathrm{d}\theta(E,d) \ll l_c$, particles of all energies "see" the same magnetic field realization during their propagation from a discrete source to the observer. In this case, (7) gives the typical coherent deflection from the line-of-sight source direction, and the spread in arrival directions of particles of different energies is much smaller. In contrast, for $\mathrm{d}\theta(E,d) \gg l_c$, the image of the source is washed out over a typical angular extent again given by (7), but in this case it is centered on the true source direction. If $\mathrm{d}\theta(E,d) \simeq l_c$, the source may even have several images, similar to the case of gravitational lensing. Therefore, observing images of UHECR sources and identifying counterparts in other wavelengths would allow one to distinguish these limits and thus obtain information on cosmic magnetic fields. If d is comparable to or larger than the interaction length for stochastic energy loss due to photo-pion production or photodisintegration, the spread in deflection angles is always comparable to the average deflection angle.

Deflection also implies an average time delay of $\tau(E,d) \simeq \mathrm{d}\theta(E,d)^2/4$, or

$$\tau(E,d) \simeq 1.5 \times 10^3\, q^2 \left(\frac{E}{10^{20}\,\mathrm{eV}}\right)^{-2} \left(\frac{d}{10\,\mathrm{Mpc}}\right)^{2} \left(\frac{l_c}{1\,\mathrm{Mpc}}\right) \left(\frac{B}{10^{-9}\,\mathrm{G}}\right)^{2} \mathrm{yr} \tag{8}$$

relative to rectilinear propagation with the speed of light. It was pointed out in [111] that, as a consequence, the observed UHECR spectrum of a bursting source at a given time can be different from its long-time average and would typically peak around an energy E_0, given by equating $\tau(E,d)$ with the time of observation relative to the time of arrival for vanishing time delay. Higher energy particles would have passed the observer already, whereas lower energy particles would not have arrived yet. Similarly to the behavior of deflection angles, the width of the spectrum around E_0 would be much smaller than E_0 if both d is smaller than the interaction length for stochastic energy loss and $d\theta(E,d) \ll l_c$. In all other cases the width would be comparable to E_0.

Constraints on magnetic fields from deflection and time delay cannot be studied separately from the characteristics of the "probes", namely the UHECR sources, at least as long as their nature is unknown. An approach to the general case is discussed in Sect. 5.3.

5.2 Constraints on EHECR Source Locations

As pointed out in Sect. 1, nucleons, nuclei, and γ–rays above a few 10^{19} eV cannot have originated much further away than $\simeq 50$ Mpc. Together with (7) this implies that above a few 10^{19} eV the arrival direction of such particles should in general point back to their source within a few degrees [14]. This argument is often made in the literature and follows from the Faraday rotation bound on the EGMF and a possible extended field in the halo of our Galaxy, which in its historical form reads $Bl_c^{1/2} \lesssim 10^{-9}\,\mathrm{G\,Mpc}^{1/2}$ [112], as well as from the known strength and scale height of the field in the disk of our Galaxy, $B_g \simeq 3\times10^{-6}$ G, $l_g \lesssim 1$ kpc. Furthermore, the deflection in the disk of our Galaxy can be corrected for in order to reconstruct the extragalactic arrival direction: Maps of such corrections as a function of arrival direction have been calculated in [113] for plausible models of the Galactic magnetic field. The deflection of UHECR trajectories in the Galactic magnetic field may, however, also give rise to several other important effects [114] such as (de)magnification of the UHECR fluxes due to the magnetic lensing effect mentioned in the previous section (which can modify the UHECR spectrum from individual sources), formation of multiple images of a source, and apparent "blindness" of the Earth towards certain regions of the sky with regard to UHECRs. These effects may in turn have important implications for UHECR source locations.

However, important modifications of the Faraday rotation bound on the EGMF have recently been discussed in the literature: The average electron density which enters estimates of the EGMF from rotation measures, can now be more reliably estimated from the baryon density $\Omega_b h^2 \simeq 0.02$, whereas in the original bound the closure density was used. Assuming an unstructured Universe and $\Omega_0 = 1$ results in the much weaker bound [115]

$$B \lesssim 3\times10^{-7}\left(\frac{\Omega_b h^2}{0.02}\right)^{-1}\left(\frac{h}{0.65}\right)\left(\frac{l_c}{\mathrm{Mpc}}\right)^{-1/2}\,\mathrm{G}\,, \quad (9)$$

which suggests much stronger deflection. However, taking into account the large scale structure of the Universe in the form of voids, sheets, filaments etc., and assuming flux freezing of the magnetic fields whose strength then approximately scales with the 2/3 power of the local density, leads to more stringent bounds: Using the Lyman α forest to model the density distribution yields [115]

$$B \lesssim 10^{-9} - 10^{-8}\,\mathrm{G} \tag{10}$$

for the large scale EGMF for coherence scales between the Hubble scale and 1 Mpc. This estimate is closer to the original Faraday rotation limit. However, in this scenario the maximal fields in the sheets and voids can be as high as a μG [116,115].

Therefore, according to (7) and (10), deflection of UHECR nucleons is still expected to be on the degree scale if the local large scale structure around the Earth is not strongly magnetized. However, rather strong deflection can occur if the Supergalactic Plane is strongly magnetized, for particles originating in nearby galaxy clusters where magnetic fields can be as high as 10^{-6} G [112] (see Sect. 5.3 below) and/or for heavy nuclei such as iron [23]. In this case, magnetic lensing in the EGMF can also play an important role in determining UHECR source locations [117,118].

5.3 Angle-Energy-Time Images of UHECR Sources

Small Deflection

For small deflection angles and if photo-pion production is important, one has to resort to numerical Monte Carlo simulations in 3 dimensions. Such simulations have been performed in [119] for the case $d\theta(E,d) \gg l_c$ and in [108,120,121] for the general case.

In [108,120,121] the Monte Carlo simulations were performed in the following way: The magnetic field was represented as a Gaussian random field with zero mean and a power spectrum with $\langle B^2(k)\rangle \propto k^{n_H}$ for $k < k_c$ and $\langle B^2(k)\rangle = 0$ otherwise, where $k_c = 2\pi/l_c$ characterizes the numerical cut-off scale and the r.m.s. strength is $B^2 = \int_0^\infty \mathrm{d}k\, k^2 \langle B^2(k)\rangle$. The field is then calculated on a grid in real space via Fourier transformation. For a given magnetic field realization and source, nucleons with a uniform logarithmic distribution of injection energies are propagated between two given points (source and observer) on the grid. This is done by solving the equations of motion in the magnetic field interpolated between the grid points, and subjecting nucleons to stochastic production of pions and (in case of protons) continuous loss of energy due to PP. Upon arrival, injection and detection energy, and time and direction of arrival are recorded. From many (typically 40000) propagated particles, a histogram of average number of particles detected as a function of time and energy of arrival is constructed for any given injection spectrum by weighting the injection energies correspondingly. This histogram can be scaled to any desired total fluence at the detector and, by convolution in time, can be constructed for arbitrary emission time scales of the source. An example for the distribution of arrival times and energies of UHECRs from a bursting source is given in Fig. 7.

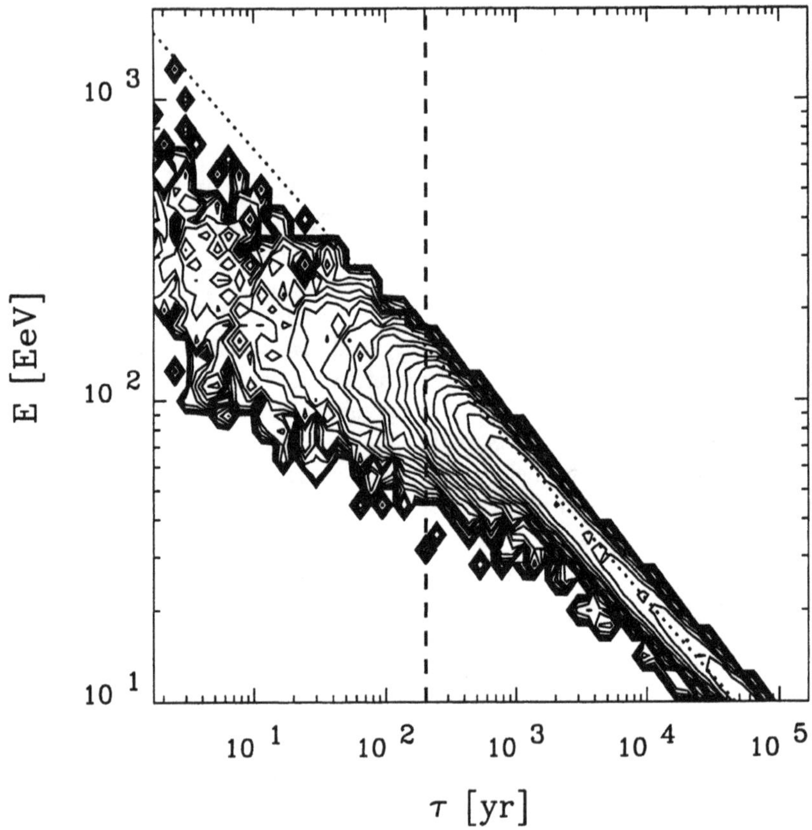

Fig. 7. Contour plot of the UHECR image of a bursting source at $d = 30\,\mathrm{Mpc}$, projected onto the time-energy plane, with $B = 2\times10^{-10}\,\mathrm{G}$, $l_c = 1\,\mathrm{Mpc}$, from [108]. The contours decrease in steps of 0.2 in the logarithm to base 10. The dotted line indicates the energy-time delay correlation $\tau(E,d) \propto E^{-2}$ as would be obtained in the absence of pion production losses. Clearly, $\mathrm{d}\theta(E,d) \ll l_c$ in this example, since for $E < 4\times10^{19}\,\mathrm{eV}$, the width of the energy distribution at any given time is much smaller than the average (see Sect. 5.1). The dashed lines, which are not resolved here, indicate the location (arbitrarily chosen) of the observational window, of length $T_{obs} = 5\,\mathrm{yr}$

We adopt the following notation for the parameters: τ_{100} denotes the time delay due to magnetic deflection at $E = 100\,\mathrm{EeV}$ and is given by (8) in terms of the magnetic field parameters; T_S denotes the emission time scale of the source; $T_S \ll 1\mathrm{yr}$ corresponds to a burst, and $T_S \gg 1\mathrm{yr}$ (roughly speaking) to a continuous source; γ is the differential index of the injection energy spectrum; N_0 denotes the fluence of the source with respect to the detector, *i.e.*, the total

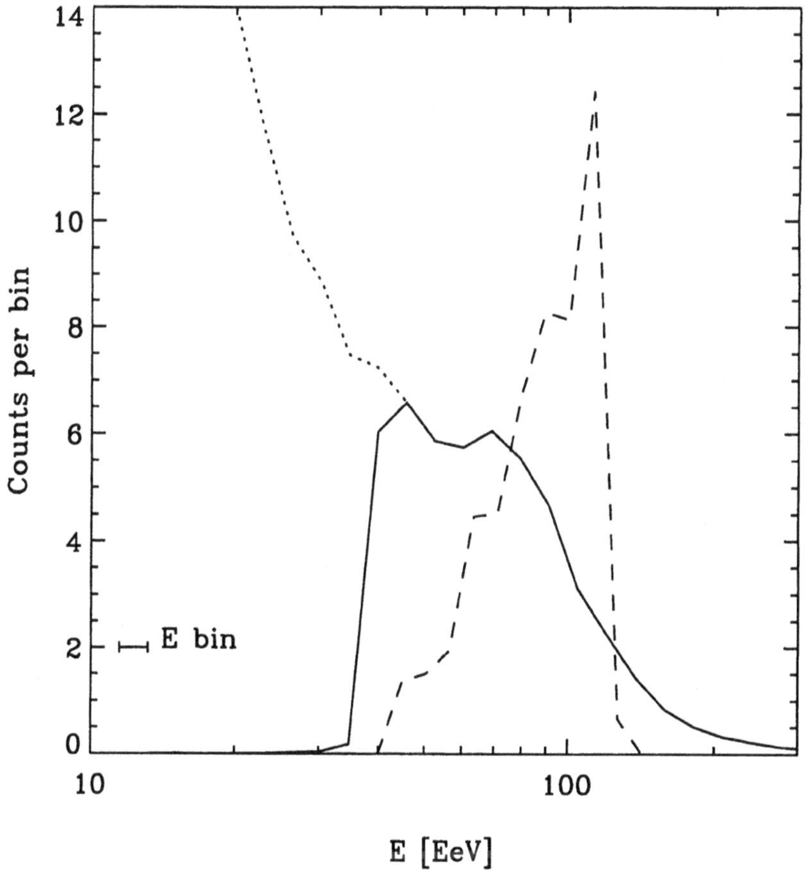

Fig. 8. Energy spectra for a continuous source (solid line), and for a burst (dashed line), from [108]. Both spectra are normalized to a total of 50 particles detected. The parameters corresponding to the continuous source case are: $T_S = 10^4$ yr, $\tau_{100} = 1.3 \times 10^3$ yr, and the time of observation is $t = 9 \times 10^3$ yr, relative to rectilinear propagation with the speed of light. A low energy cutoff results at the energy $E_S = 4 \times 10^{19}$ eV where $\tau_{E_S} = t$. The dotted line shows how the spectrum would continue if $T_S \ll 10^4$ yr. The case of a bursting source corresponds to a slice of the image in the $\tau_E - E$ plane, as indicated in Fig. 7 by dashed lines. For both spectra, $d = 30$ Mpc, and $\gamma = 2$

number of particles that the detector would detect from the source on an infinite time scale; finally, $\mathcal{L}$ is the likelihood function of the above parameters.

By putting windows of width equal to the time scale of observation over these histograms one obtains expected distributions of events in energy and time and direction of arrival for a given magnetic field realization, source distance and

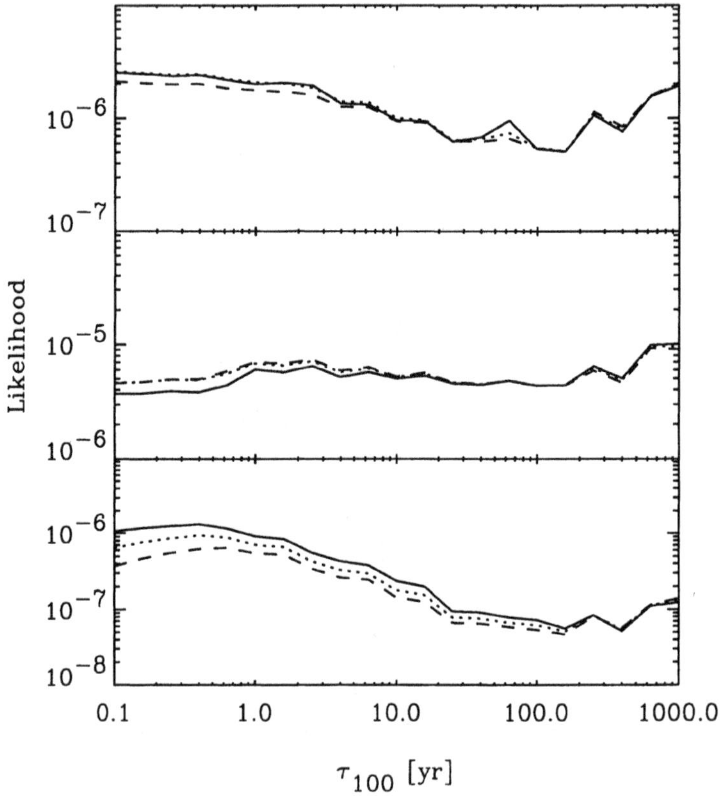

Fig. 9. The likelihood, $\mathcal{L}$, marginalized over T_S and N_0 as a function of the average time delay at 10^{20} eV, τ_{100}, assuming a source distance $d = 30$ Mpc. The panels are for pair # 3 through # 1, from top to bottom, of the AGASA pairs [109]. Solid lines are for $\gamma = 1.5$, dotted lines for $\gamma = 2.0$, and dashed lines for $\gamma = 2.5$

position, emission time scale, total fluence, and injection spectrum. Examples of the resulting energy spectrum are shown in Fig. 8. By dialing Poisson statistics on such distributions, one can simulate corresponding observable event clusters.

Conversely, for any given real or simulated event cluster, one can construct a likelihood of the observation as a function of the time delay, the emission time scale, the differential injection spectrum index, the fluence, and the distance. In order to do so, and to obtain the maximum of the likelihood, one constructs histograms for many different parameter combinations as described above, randomly puts observing time windows over the histograms, calculates the likelihood function from the part of the histogram within the window and the cluster events, and averages over different window locations and magnetic field realizations.

In [120] this approach has been applied to and discussed in detail for the three pairs observed by the AGASA experiment [109], under the assumption that all events within a pair were produced by the same discrete source. Although the inferred angle between the momenta of the paired events acquired in the EGMF is several degrees [122], this is not necessarily evidence against a common source, given the uncertainties in the Galactic field and the angular resolution of AGASA which is $\simeq 2.5°$. As a result of the likelihood analysis, these pairs do not seem to follow a common characteristic; one of them seems to favor a burst, another one seems to be more consistent with a continuously emitting source. The current data, therefore, does not allow one to rule out any of the models of UHECR sources. Furthermore, two of the three pairs are insensitive to the time delay. However, the pair which contains the 200 EeV event seems to significantly favor a comparatively small average time delay, $\tau_{100} \lesssim 10\,\mathrm{yr}$, as can be seen from the likelihood function marginalized over T_S and N_0 (see Fig. 9). According to (8) this translates into a tentative bound for the r.m.s. magnetic field, namely,

$$B \lesssim 2 \times 10^{-11} \left(\frac{l_c}{1\,\mathrm{Mpc}}\right)^{-1/2} \left(\frac{d}{30\,\mathrm{Mpc}}\right)^{-1} \mathrm{G}\,, \tag{11}$$

which also applies to magnetic fields in the halo of our Galaxy if d is replaced by the lesser of the source distance and the linear halo extent. If confirmed by future data, this bound would be at least two orders of magnitude more restrictive than the best existing bounds which come from Faraday rotation measurements [see (10)] and, for a homogeneous EGMF, from CMB anisotropies [123]. UHECRs are therefore at least as sensitive a probe of cosmic magnetic fields as other measures in the range near existing limits such as the polarization [124] and the small scale anisotropy [125] of the CMB.

More generally, confirmation of a clustering of EHECRs would provide significant information on both the nature of the sources and on large-scale magnetic fields [126]. This has been shown quantitatively [121] by applying the hybrid Monte Carlo likelihood analysis discussed above to simulated clusters of a few tens of events as they would be expected from next generation experiments [6] such as the High Resolution Fly's Eye [34], the Telescope Array [35], and most notably, the Pierre Auger Project [36] (see Sect. 2), provided the clustering recently suggested by the AGASA experiment [109,127] is real. The proposed OWL satellite observatory concept [38] might even allow one to detect clusters of hundreds of such events.

Five generic situations of UHECR time-energy images were discussed in [121], classified according to the values of the time delay τ_E induced by the magnetic field, the emission timescale of the source T_{S}, as compared to the lifetime of the experiment. The likelihood calculated for the simulated clusters in these cases presents different degeneracies between different parameters, which complicates the analysis. As an example, the likelihood is degenerate in the ratios N_0/T_{S}, or $N_0/\Delta\tau_{100}$, where N_0 is the total fluence, and $\Delta\tau_{100}$ is the spread in arrival time; these ratios represent rates of detection. Another example is given by the degeneracy between the distance d and the injection energy spectrum index γ.

Yet another is the ratio $(d\tau_E)^{1/2}/l_c$, that controls the size of the scatter around the mean of the $\tau_E - E$ correlation. Therefore, in most general cases, values for the different parameters cannot be pinned down, and generally, only domains of validity are found. In the following the reconstruction quality of the main parameters considered is summarized.

The distance to the source can be obtained from the pion production signature, above the GZK cut-off, when the emission timescale of the source dominates over the time delay. Since the time delay decreases with increasing energy, the lower the energy $E_{\rm C}$, defined by $\tau_{E_{\rm C}} \simeq T_{\rm S}$, the higher the accuracy on the distance d. The error on d is, in the best case, typically a factor 2, for one cluster of $\simeq$ 40 events. In this case, where the emission timescale dominates over the time delay at all observable energies, information on the magnetic field is only contained in the angular image, which was not systematically included in the likelihood analysis of [121] due to computational limits. Qualitatively, the size of the angular image is proportional to $B(dl_c)^{1/2}/E$, whereas the structure of the image, *i.e.*, the number of separate images, is controlled by the ratio $d^{3/2}B/l_c^{1/2}/E$. Finally, in the case when the time delay dominates over the emission timescale, with a time delay shorter than the lifetime of the experiment, one can also estimate the distance with reasonable accuracy.

Some sensitivity to the injection spectrum index γ exists whenever events are recorded over a sufficiently broad energy range. At least if the distance d is known, it is in general comparatively easy to rule out a hard injection spectrum if the actual $\gamma \gtrsim 2.0$, but much harder to distinguish between $\gamma = 2.0$ and 2.5.

If the lifetime of the experiment is the largest time scale involved, the strength of the magnetic field can only be obtained from the time-energy image because the angular image will not be resolvable. When the time delay dominates over the emission timescale, and is, at the same time, larger than the lifetime of the experiment, only a lower limit corresponding to this latter timescale, can be placed on the time delay and hence on the strength of the magnetic field. When combined with the Faraday rotation upper limit (10), this would nonetheless allow one to bracket the r.m.s. magnetic field strength within a few orders of magnitude. In this case also, significant information is contained in the angular image. If the emission time scale is larger then the delay time, the angular image is obviously the only source of information on the magnetic field strength.

The coherence length l_c enters in the ratio $(d\tau_E)^{1/2}/l_c$ that controls the scatter around the mean of the $\tau_E - E$ correlation in the time-energy image. It can therefore be estimated from the width of this image, provided the emission timescale is much less than τ_E (otherwise the correlation would not be seen), and some prior information on d and τ_E is available.

An emission timescale much larger than the experimental lifetime may be estimated if a lower cut-off in the spectrum is observable at an energy $E_{\rm C}$, indicating that $T_{\rm S} \simeq \tau_{E_{\rm C}}$. The latter may, in turn, be estimated from the angular image size via (8), where the distance can be estimated from the spectrum visible above the GZK cut-off, as discussed above. An example of this scenario is shown

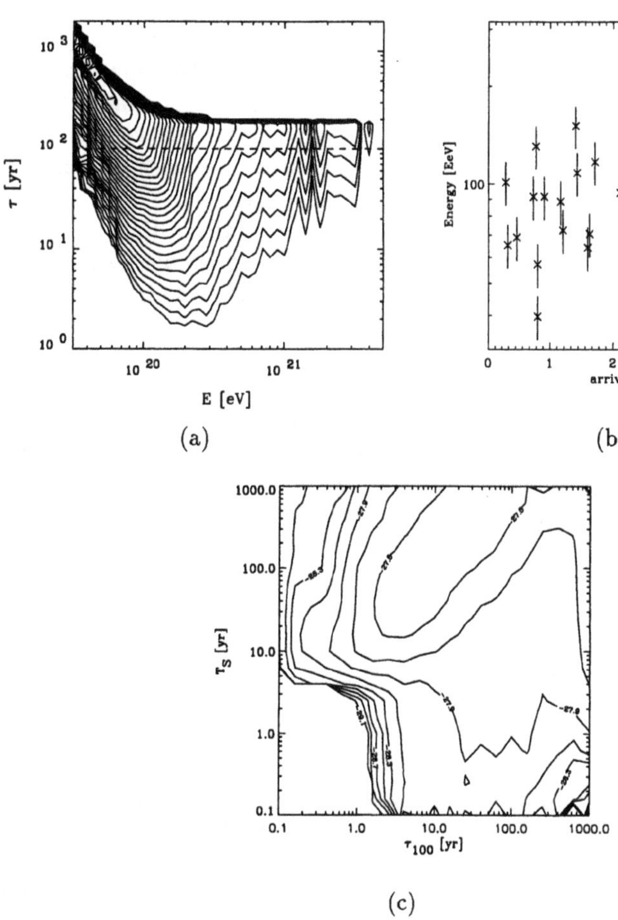

Fig. 10. (a) Arrival time-energy histogram for $\gamma = 2.0$, $\tau_{100} = 50\,\mathrm{yr}$, $T_{\mathrm{S}} = 200\,\mathrm{yr}$, $l_c \simeq 1\,\mathrm{Mpc}$, $d = 50\,\mathrm{Mpc}$, corresponding to $B \simeq 3 \times 10^{-11}\,\mathrm{G}$. Contours are in steps of a factor $10^{0.4} = 2.51$; (b) Example of a cluster in the arrival time-energy plane resulting from the cut indicated in (a) by the dashed line at $\tau \simeq 100\,\mathrm{yr}$; (c) The likelihood function, marginalized over N_0 and γ, for $d = 50\,\mathrm{Mpc}$, $l_c \simeq \mathrm{Mpc}$, for the cluster shown in (b), in the $T_{\mathrm{S}} - \tau_{100}$ plane. The contours shown go from the maximum down to about 0.01 of the maximum in steps of a factor $10^{0.2} = 1.58$. Note that the likelihood clearly favors $T_{\mathrm{S}} \simeq 4\tau_{100}$. For τ_{100} large enough to be estimated from the angular image size, $T_{\mathrm{S}} \gg T_{\mathrm{obs}}$ can, therefore, be estimated as well

in Fig. 10. For angular resolutions $\Delta\theta$, timescales in the range

$$3 \times 10^3 \left(\frac{\Delta\theta}{1^\circ}\right)^2 \left(\frac{d}{10\,\mathrm{Mpc}}\right) \mathrm{yr} \lesssim T_{\mathrm{S}} \simeq \tau_E \lesssim 10^4 - 10^7 \left(\frac{E}{100\,\mathrm{EeV}}\right)^{-2} \mathrm{yr} \quad (12)$$

could be probed. The lower limit follows from the requirement that it should be possible to estimate τ_E from θ_E, using (8), otherwise only an upper limit on T_S, corresponding to this same number, would apply. The upper bound in (12) comes from constraints on maximal time delays in cosmic magnetic fields, such as the Faraday rotation limit in the case of cosmological large-scale field (smaller number) and knowledge on stronger fields associated with the large-scale galaxy structure (larger number). Equation (12) constitutes an interesting range of emission timescales for many conceivable scenarios of UHECRs. For example, the hot spots in certain powerful radio galaxies that have been suggested as UHECR sources [128], have a size of only several kpc and could have an episodic activity on timescales of $\sim 10^6$ yr.

A detailed comparison of analytical estimates for the distributions of time delays, energies, and deflection angles of nucleons in weak random magnetic fields with the results of Monte Carlo simulations has been presented in [129]. In this work, deflection was simulated by solving a stochastic differential equation and observational consequences for the two major classes of source scenarios, namely continuous and impulsive UHECR production, were discussed. In agreement with earlier work [111] it was pointed out that at least in the impulsive production scenario and for an EGMF in the range $0.1\text{–}1\times10^{-9}$ G, as required for cosmological GRB sources, there is a typical energy scale $E_b \sim 10^{20.5} - 10^{21.5}$ eV below which the flux is quasi-steady due to the spread in arrival times, whereas above which the flux is intermittent with only a few sources contributing.

General Case

Unfortunately, neither the diffusive limit nor the limit of nearly rectilinear propagation is likely to be applicable to the propagation of UHECRs around 10^{20} eV in general. This is because in magnetic fields in the range of a few 10^{-8} G, values that are realistic for the Supergalactic Plane [116,115], the gyro radii of charged particles is of the order of a few Mpc which is comparable to the distance to the sources. An accurate, reliable treatment in this regime can only be achieved by numerical simulation.

To this end, the Monte Carlo simulation approach of individual trajectories developed in [120,121] has recently been generalized to arbitrary deflections [117]. The Supergalactic Plane was modeled as a sheet with a thickness of a few Mpc and a Gaussian density profile. The same statistical description for the magnetic field was adopted as in [120,121], but with a field power law index $n_H = -11/3$, representing a turbulent Kolmogorov type spectrum, and weighted with the sheet density profile. It should be mentioned, however, that other spectra, such as the Kraichnan spectrum, corresponding to $n_H = -7/2$, are also possible. The largest mode with non-zero power was taken to be the largest turbulent eddy whose size is roughly the sheet thickness. In addition, a coherent field component B_c is allowed that is parallel to the sheet and varies proportional to the density profile.

When CR backreaction on the weakly turbulent magnetic field is neglected, the diffusion coefficient of CR of energy E is determined by the magnetic field

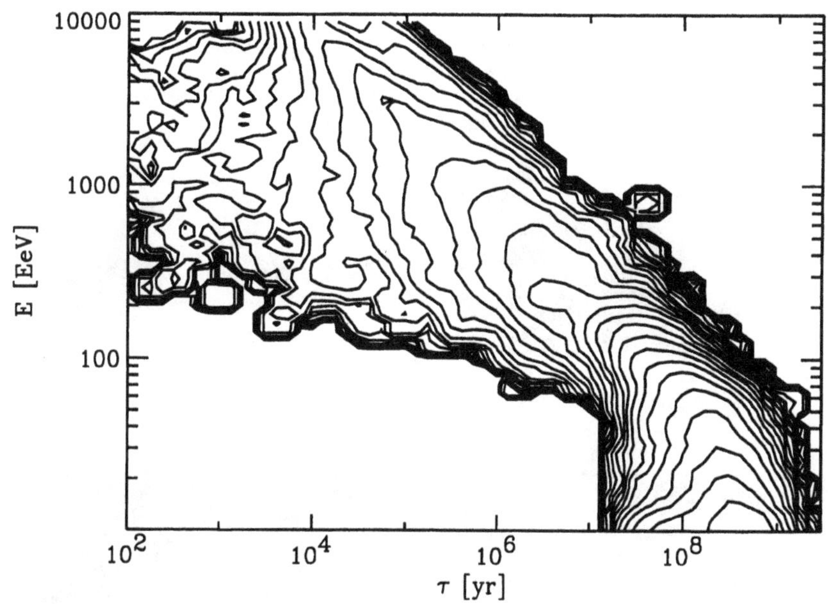

Fig. 11. The distribution of time delays τ_E and energies E for a burst with spectral index $\gamma = 2.4$ at a distance $d = 10\,\text{Mpc}$, similar to Fig. 7, but for the Supergalactic Plane scenario discussed in the text. The turbulent magnetic field component in the sheet center is $B = 3 \times 10^{-7}\,\text{G}$. Furthermore, a vanishing coherent field component is assumed. The inter-contour interval is 0.25 in the logarithm to base 10 of the distribution per logarithmic energy and time interval. The three regimes discussed in the text, $\tau_E \propto E^{-2}$ in the rectilinear regime $E \gtrsim 200\,\text{EeV}$, $\tau_E \propto E^{-1}$ in the Bohm diffusion regime $60\,\text{EeV} \lesssim E \lesssim 200\,\text{EeV}$, and $\tau_E \propto E^{-1/3}$ for $E \lesssim 60\,\text{EeV}$ are clearly visible

power on wavelengths comparable to the particle Larmor radius, and can be approximated by

$$D(E) \simeq \frac{1}{3} r_g(E) \frac{B}{\int_{1/r_g(E)}^{\infty} \mathrm{d}k\, k^2 \langle B^2(k) \rangle} . \tag{13}$$

As a consequence, for the Kolmogorov spectrum, in the diffusive regime, where $\tau_E \gtrsim d$, the diffusion coefficient should scale with energy as $D(E) \propto E^{1/3}$ for $r_g \lesssim L/(2\pi)$, and as $D(E) \propto E$ in the so called Bohm diffusion regime,$r_g \gtrsim L/(2\pi)$. This should be reflected in the dependence of the time delay τ_E on energy E: From the rectilinear regime, $\tau_E \lesssim d$, hence at the largest energies, where $\tau_E \propto E^{-2}$, this should switch to $\tau_E \propto E^{-1}$ in the regime of Bohm diffusion, and eventually to $\tau_E \propto E^{-1/3}$ at the smallest energies, or largest time delays. Indeed, all three regimes can be seen in Fig. 11 which shows an example of the distribution of arrival times and energies of UHECRs from a bursting source.

The numerical results indicate an effective gyroradius that is roughly a factor 10 higher than the analytical estimate, with a correspondingly larger diffusion

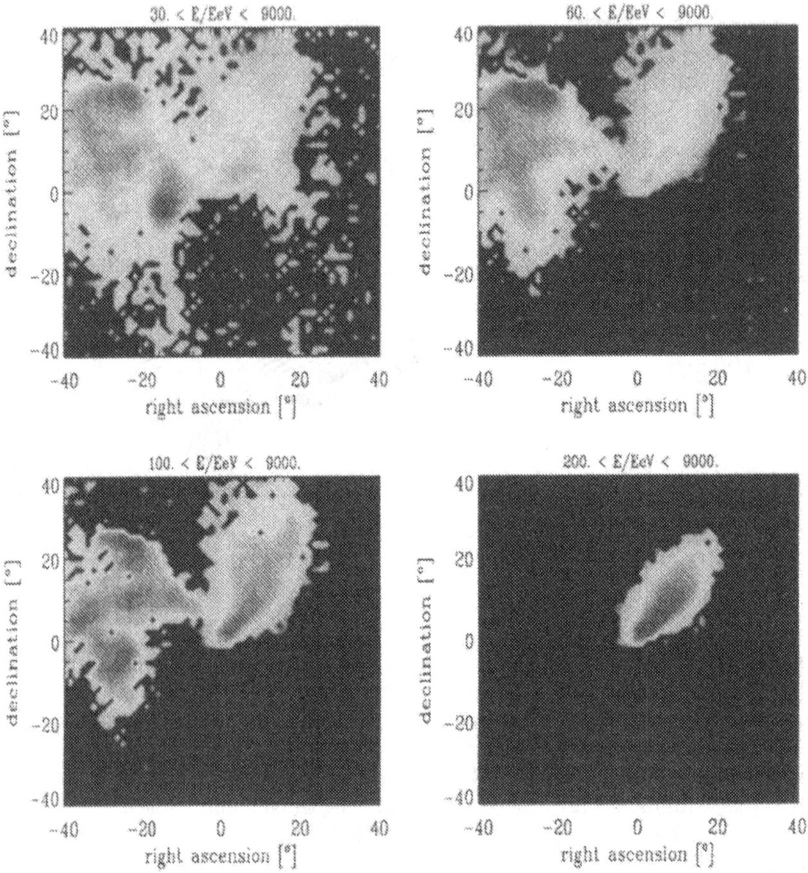

Fig. 12. Angular image of a point-like source in a magnetized Supergalactic Plane, corresponding to one particular magnetic field realization with a maximal magnetic field in the plane center, $B_{\rm max} = 5 \times 10^{-8}$ G, all other parameters being the same as in Fig. 11. The image is shown in different energy ranges, as indicated, as seen by a detector of $\simeq 1°$ angular resolution. A transition from several images at lower energies to only one image at the highest energies occurs where the linear deflection becomes comparable to the effective field coherence length. The difference between neighboring shade levels is 0.1 in the logarithm to base 10 of the integral flux per solid angle

coefficient compared to (13). In addition, the fluctuations of the resulting spectra between different magnetic field realizations can be substantial. This is a result of the fact that most of the magnetic field power is on the largest scales where there are the fewest modes. These considerations mean that the applicability of analytical flux estimates of discrete sources in specific magnetic field configurations is rather limited.

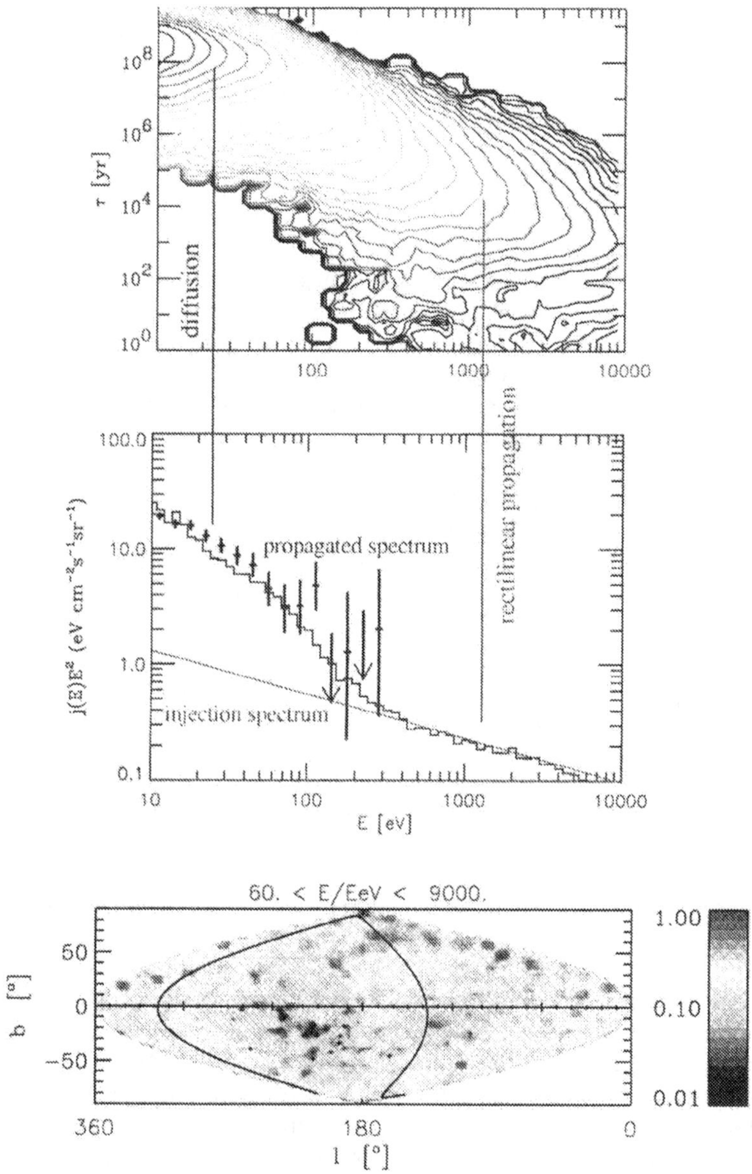

Fig. 13. The distribution of arrival times and energies (top), the solid angle integrated spectrum (middle, with 1 sigma error bars showing combined data from the Haverah Park [4], the Fly's Eye [8], and the AGASA [9] experiments above 10^{19} eV), and the angular distribution of arrival directions in Galactic coordinates (bottom, with color scale showing the intensity per solid angle) in the Supercluster scenario with continuous source distribution explained in the text, averaged over 4 magnetic field realizations with 20000 particles each

In a steady state situation, diffusion leads to a modification of the injection spectrum by roughly a factor τ_E, at least in the absence of significant energy loss and for a homogeneous, infinitely extended medium that can be described by a spatially constant diffusion coefficient. Since in the non-diffusive regime the observed spectrum repeats the shape of the injection spectrum, a change to a flatter observed spectrum at high energies is expected in the transition region [130]. From the spectral point of view this suggests the possibility of explaining the observed UHECR flux above $\simeq 10\,\mathrm{EeV}$ including the highest energy events with only one discrete source [131].

Angular images of discrete sources in a magnetized Supercluster in principle contain information on the magnetic field structure. For the recently suggested field strengths between $\sim 10^{-8}\,\mathrm{G}$ and $\simeq 1\mu\,\mathrm{G}$ the angular images are large enough to exploit that information with instruments of angular resolution in the degree range. An example where a transition from several images at low energies to one image at high energies allows one to estimate the magnetic field coherence scale is shown in Fig. 12.

The newest AGASA data [127], however, indicate an isotropic distribution of EHECR. To explain this with only one discrete source would require the magnetic fields to be so strong that the flux beyond $10^{20}\,\mathrm{eV}$ would most likely be too strongly suppressed by pion production, as discussed above. This suggests a more continuous source distribution which may also still reproduce the observed UHECR flux above $\simeq 10^{19}\,\mathrm{eV}$ with only one spectral component [132]. A more systematic parameter study of sky maps and spectra in UHECR in different scenarios is therefore now being pursued [133,118].

Intriguingly, scenarios in which a diffuse source distribution follows the density in the Supergalactic Plane within a certain radius, can accommodate both the large scale isotropy (by diffusion) and the small scale clustering (by magnetic lensing) revealed by AGASA if a magnetic field of strength $B \gtrsim 0.05\mu\,\mathrm{G}$ permeates the Supercluster [118].

Figure 13 shows the distribution of arrival times and energies, the solid angle integrated spectrum, and the angular distribution of arrival directions in Galactic coordinates in such a scenario where the UHECR sources with spectral index $\gamma = 2.4$ are distributed according to the matter density in the Local Supercluster, following a pancake profile with scale height of 5 Mpc and scale length 20 Mpc. The r.m.s. magnetic field has a Kolmogorov spectrum with a maximal field strength $B_{\mathrm{max}} = 5 \times 10^{-7}\,\mathrm{G}$ in the plane center, and also follows the matter density. The observer is within 2 Mpc of the Supergalactic Plane whose location is indicated by the solid line in the lower panel and at a distance $d = 20\,\mathrm{Mpc}$ from the plane center. The absence of sources within 2 Mpc from the observer was assumed. The transition discussed above from the diffusive regime below $\simeq 2 \times 10^{20}\,\mathrm{eV}$ to the regime of almost rectilinear propagation above that energy is clearly visible.

Detailed Monte Carlo simulations performed on these distributions reveal that the anisotropy decreases with increasing magnetic field strength due to diffusion and that small scale clustering increases with coherence and strength of the magnetic field due to magnetic lensing. Both anisotropy and clustering also increase with the (unknown) source distribution radius. Furthermore, the discriminatory power between models with respect to anisotropy and clustering strongly increases with exposure [118].

As a result, a diffuse source distribution associated with the Supergalactic Plane can explain most of the currently observed features of ultra-high energy cosmic rays at least for field strengths close to $0.5\,\mu$ G. The large-scale anisotropy and the clustering predicted by this scenario will allow strong discrimination against other models with next generation experiments such as the Pierre Auger Project.

6 Conclusions

Ultra-high energy cosmic rays have the potential to open a window to and act as probes of new particle physics beyond the Standard Model as well as processes occuring in the early Universe at energies close to the Grand Unification scale. Even if their origin will turn out to be attributable to astrophysical shock acceleration with no new physics involved, they will still be witnesses of one of the most energetic processes in the Universe. Furthermore, complementary to other methods such as Faraday rotation measurements, ultra-high energy cosmic rays can be used as probes of the poorly known large scale cosmic magnetic fields. The future appears promising and exciting due to the anticipated arrival of several large scale experiments.

References

1. S. Swordy: private communication. The data represent published results of the LEAP, Proton, Akeno, AGASA, Fly's Eye, Haverah Park, and Yakutsk experiments
2. J. Linsley: Phys. Rev. Lett. **10**, 146 (1963); Proc. 8th *International Cosmic Ray Conference* **4** 295 (1963)
3. R. G. Brownlee et al.: Can. J. Phys. **46**, S259 (1968); M. M. Winn et al.: J. Phys. G **12** (1986) 653; see also http://www.physics.usyd.edu.au/hienergy/sugar.html
4. See, e.g., M. A. Lawrence, R. J. O. Reid, and A. A. Watson: J. Phys. G **17**, 733 (1991), and references therein; see also http://ast.leeds.ac.uk/haverah/havhome.html
5. Proc. International Symposium on *Astrophysical Aspects of the Most Energetic Cosmic Rays*, ed. by M. Nagano and F. Takahara (World Scientific, Singapore 1991)
6. Proc. of International Symposium on *Extremely High Energy Cosmic Rays: Astrophysics and Future Observatories*, ed. by M. Nagano (Institute for Cosmic Ray Research, Tokyo 1996)
7. N. N. Efimov et al.: in [5], p. 20; B. N. Afnasiev: in [6], p. 32
8. D. J. Bird et al.: Phys. Rev. Lett. **71**, 3401 (1993); Astrophys. J. **424**, 491 (1994); ibid. **441**, 144 (1995)
9. N. Hayashida et al.: Phys. Rev. Lett. **73**, 3491 (1994); S. Yoshida et al.: Astropart. Phys. **3**, 105 (1995); M. Takeda et al.: Phys. Rev. Lett. **81**, 1163 (1998); see also http://icrsun.icrr.u-tokyo.ac.jp/as/project/agasa.html

10. Proc. of *Workshop on Observing Giant Cosmic Ray Air Showers from* $> 10^{20}$ *eV Particles from Space*, ed. by J. F. Krizmanic, J. F. Ormes, and R. E. Streitmatter (AIP Conference Proceedings 433, 1997)
11. J. W. Cronin: Rev. Mod. Phys. **71**, S165 (1999)
12. P. Bhattacharjee and G. Sigl: e-print astro-ph/9811011, to appear in Phys. Rept
13. A. M. Hillas: Ann. Rev. Astron. Astrophys. **22**, 425 (1984)
14. G. Sigl, D. N. Schramm, and P. Bhattacharjee: Astropart. Phys. **2**, 401 (1994)
15. C. A. Norman, D. B. Melrose, and A. Achterberg: Astrophys. J. **454**, 60 (1995)
16. P. L. Biermann: J. Phys. G **23**, 1 (1997)
17. J. G. Kirk and P. Duffy: J. Phys. G **25**, R163 (1999)
18. K. Greisen: Phys. Rev. Lett. **16**, 748 (1966)
19. G. T. Zatsepin and V. A. Kuzmin: Pis'ma Zh. Eksp. Teor. Fiz. **4**, 114 (1966) [JETP. Lett. **4**, 78 (1966)]
20. F. W. Stecker: Phys. Rev. Lett. **21**, 1016 (1968)
21. J. L. Puget, F. W. Stecker, and J. H. Bredekamp: Astrophys. J. **205**, 638 (1976)
22. L. N. Epele and E. Roulet: Phys. Rev. Lett. **81**, 3295 (1998); J. High Energy Phys. **9810**, 009 (1998); F. W. Stecker: Phys. Rev. Lett. **81**, 3296 (1998); F. W. Stecker and M. H. Salamon: Astrophys. J. **512**, 521 (1999)
23. J. W. Elbert, and P. Sommers: Astrophys. J. **441**, 151 (1995)
24. E. Boldt and P. Ghosh: e-print astro-ph/9902342, to appear in Mon. Not. R. Astron. Soc. (1999)
25. T. K. Gaisser, F. Halzen, and T. Stanev: Phys. Rept. **258**, 173 (1995)
26. F. Halzen: e-print astro-ph/9810368, lectures presented at the TASI School, July 1998; e-print astro-ph/9904216, talk presented at the *17th International Workshop on Weak Interactions and Neutrinos*, Cape Town, South Africa, January 1999
27. R. A. Ong: Phys. Rept. **305**, 95 (1998); M. Catanese and T. C. Weekes: e-print astro-ph/9906501, invited review to appear in Publ. Astron. Soc. of the Pacific
28. K. Mannheim: Rev. Mod. Astron. **12**, 101 (1999)
29. E. Waxman and J. Bahcall: Phys. Rev. D. **59**, 023002 (1999); J. Bahcall and E. Waxman: e-print hep-ph/9902383
30. K. Mannheim, R. J. Protheroe, and J. P. Rachen: e-print astro-ph/9812398, submitted to Phys. Rev. D.; J. P. Rachen, R. J. Protheroe, and K. Mannheim: e-print astro-ph/9908031
31. See, e.g., S. Petrera: Nuovo Cimento **19C**, 737 (1996)
32. S. Yoshida and H. Dai: J. Phys. G **24**, 905 (1998); P. Sokolsky, *Introduction to Ultrahigh Energy Cosmic Ray Physics* (Addison Wesley, Redwood City, California, 1989); P. Sokolsky, P. Sommers, and B. R. Dawson: Phys. Rept. **217**, 225 (1992); M. V. S. Rao and B. V. Sreekantan, *Extensive Air Showers* (World Scientific, Singapore, 1998)
33. Proc. 24th *International Cosmic Ray Conference* (Istituto Nazionale Fisica Nucleare, Rome, Italy, 1995)
34. S. C. Corbató et al.: Nucl. Phys. B (Proc. Suppl.) **28B**, 36 (1992); D. J. Bird et al.: in [33], Vol. **2**, 504; Vol. **1** 750; M. Al-Seady et al.: in [6], p. 191; see also http://bragg.physics.adelaide.edu.au/astrophysics/HiRes.html
35. M. Teshima et al.: Nucl. Phys. B (Proc. Suppl.) **28B**, 169 (1992); M. Hayashida et al.: in [6], p. 205; see also http://www-ta.icrr.u-tokyo.ac.jp/
36. J. W. Cronin: Nucl. Phys. B (Proc. Suppl.) **28B**, 213 (1992); The Pierre Auger Observatory Design Report (2nd edition), March 1997; see also http://http://www.auger.org/ and http://www-lpnhep.in2p3.fr/auger/welcome.html

37. Proc. 25th *International Cosmic Ray Conference*, ed. by M. S. Potgieter et al. (Durban, 1997)
38. J. F. Ormes et al.: in [37], Vol. **5**, 273; Y. Takahashi et al.: in [6], p. 310; see also http://lheawww.gsfc.nasa.gov/docs/gamcosray/hecr/OWL/.
39. J. Linsley: in [37], Vol. **5**, 381
40. J. Linsley et al.: in [37], Vol. **5**, 385; P. Attinà et al.: ibid., 389; J. Forbes et al.: ibid., 273
41. See http://dumand.phys.washington.edu/ dumand/
42. For general information see http://amanda.berkeley.edu/; see also F. Halzen: New Astron. Rev **42**, 289 (1999)
43. For general information see http://www.ifh.de/baikal/baikalhome.html; also see Baikal Collaboration: e-print astro-ph/9906255, talk given at the *8th Int. Workshop on Neutrino Telescopes*, Venice, Feb 1999
44. Proceedings of the *19th Texas Symposium on Relativistic Astrophysics* (Paris, France, 1998)
45. For general information see http://antares.in2p3.fr/antares/antares.html; see also S. Basa: in [44] (e-print astro-ph/9904213); ANTARES Collaboration: e-print astro-ph/9907432
46. For general information see http://www.roma1.infn.it/nestor/nestor.html
47. For general information see http://www.ps.uci.edu/ icecube/workshop.html; see also F. Halzen: Am. Astron. Soc. Meeting 192, **62** 28 (1998); AMANDA collaboration: e-print astro-ph/9906205, talk presented at the *8th Int. Workshop on Neutrino Telescopes*, Venice, Feb 1999
48. For general information see http://kuhep4.phsx.ukans.edu/ iceman/index.html
49. J. J. Blanco-Pillado, R. A. Vázquez, and E. Zas: Phys. Rev. Lett. **78**, 3614 (1997); K. S. Capelle, J. W. Cronin, G. Parente, and E. Zas: Astropart. Phys. **8**, 321 (1998)
50. J. G. Learned: Phil. Trans. Roy. Soc. London **A 346**, 99 (1994)
51. G. R. Farrar: Phys. Rev. Lett. **76**, 4111 (1996); D. J. H. Chung, G. R. Farrar, and E. W. Kolb: Phys. Rev. D **57**, 4696 (1998)
52. G. R. Farrar and P. L. Biermann: Phys. Rev. Lett. **81**, 3579 (1998)
53. S. Nussinov and R. Shrock: Phys. Rev. **D 59**, 105002 (1999)
54. S. Raby: Phys. Lett. **B 422**, 158 (1998); S. Raby and K. Tobe: Nucl. Phys. **B 539**, 3 (1999)
55. M. B. Voloshin and L. B. Okun: Sov. J. Nucl. Phys. **43**, 495 (1986)
56. I. F. Albuquerque et al. (E761 collaboration): Phys. Rev. Lett. **78**, 3252 (1997); J. Adams et al. (KTeV Collaboration): Phys. Rev. Lett. **79**, 4083 (1997); A. Alavi-Harati et al. (KTeV collaboration): e-print hep-ex/9903048
57. L. Clavelli: e-print hep-ph/9908342
58. I. F. M. Albuquerque, G. F. Farrar, and E. W. Kolb: Phys. Rev. **D 59**, 015021 (1999)
59. see, e.g., T. Vachaspati: Contemp. Phys. **39**, 225 (1998)
60. for a brief review see V. Kuzmin and I. Tkachev: e-print hep-ph/9903542, submitted to Phys. Rept
61. P. Bhattacharjee, C. T. Hill, and D. N. Schramm: Phys. Rev. Lett. **69**, 567 (1992)
62. see, e.g., P. Bhattacharjee and N. C. Rana: Phys. Lett. **B 246**, 365 (1990)
63. V. Berezinsky and A. Vilenkin: Phys. Rev. Lett. **79**, 5202 (1997)
64. P. Bhattacharjee and G. Sigl: Phys. Rev. **D 51**, 4079 (1995)
65. Yu. L. Dokshitzer, V. A. Khoze, A. H. Müller, and S. I. Troyan: *Basics of Perturbative QCD* (Editions Frontieres, Singapore, 1991)

66. V. Berezinsky and M. Kachelrieß : Phys. Lett. **B 434**, 61 (1998)
67. M. Birkel and S. Sarkar: Astropart. Phys. **9**, 297 (1998)
68. G. Sigl, S. Lee, P. Bhattacharjee, and S. Yoshida: Phys. Rev. **D 59**, 043504 (1999)
69. S. Lee: Phys. Rev. **D 58**, 043004 (1998)
70. T. J. Weiler: Phys. Rev. Lett. **49**, 234 (1982); Astrophys. J. **285**, 495 (1984); E. Roulet: Phys. Rev. **D 47**, 5247 (1993); S. Yoshida: Astropart. Phys. **2**, 187 (1994)
71. S. Yoshida, H. Dai, C. C. H. Jui, and P. Sommers: Astrophys. J. **479**, 547 (1997)
72. T. J. Weiler: Astropart. Phys. **11**, 317 (1999)
73. S. Yoshida, G. Sigl, and S. Lee: Phys. Rev. Lett. **81**, 5505 (1998)
74. see, e.g., P. J. E. Peebles: *Principles of Physical Cosmology*, Princeton University Press, New Jersey, 1993
75. F. A. Aharonian, P. Bhattacharjee, and D. N. Schramm: Phys. Rev. **D 46**, 4188 (1992)
76. T. A. Clark, L. W. Brown, and J. K. Alexander: Nature **228**, 847 (1970)
77. R. J. Protheroe and P. L. Biermann: Astropart. Phys. **6**, 45 (1996)
78. See, e.g., S. Biller et al.: Phys. Rev. Lett. **80**, 2992 (1998); T. Stanev and A. Franceschini: Astrophys. J. **494**, L159; (1998) F. W. Stecker and O. C. de Jager: Astron. Astrophys. **334**, L85 (1998)
79. P. Sreekumar et al.: Astrophys. J. **494**, 523 (1998)
80. R. J. Protheroe and T. Stanev: Phys. Rev. Lett. **77**, 3708; (1996) erratum, ibid. **78**, 3420 (1997)
81. G. Sigl, S. Lee, D. N. Schramm, and P. Bhattacharjee: Science **270**, 1977 (1995)
82. P. S. Coppi and F. A. Aharonian: Astrophys. J. **487**, L9 (1997)
83. R. Mukherjee and J. Chiang: Astropart. Phys. **11**, 213 (1999)
84. V. Berezinsky, M. Kachelrieß , and A. Vilenkin: Phys. Rev. Lett. **79**, 4302 (1997)
85. S. Lee, A. V. Olinto, and G. Sigl: Astrophys. J. **455**, L21 (1995)
86. C. T. Hill: Nucl. Phys. **B 224**, 469 (1983)
87. G. Sigl, K. Jedamzik, D. N. Schramm, and V. Berezinsky: Phys. Rev. **D 52**, 6682 (1995)
88. F. Halzen, R. V'azques, T. Stanev, and H. P. Vankov: Astropart. Phys. **3**, 151 (1995)
89. G. Sigl, S. Lee, D. N. Schramm, and P. S. Coppi: Phys. Lett. **B 392**, 129 (1997)
90. G. R. Vincent, N. D. Antunes, and M. Hindmarsh: Phys. Rev. Lett. **80**, 2277 (1998); G. R. Vincent, M. Hindmarsh, and M. Sakellariadou: Phys. Rev. **D 56**, 637 (1997)
91. U. F. Wichoski, J. H. MacGibbon, and R. H. Brandenberger: e-print hep-ph/9805419, submitted to Phys. Rev. D
92. R. Protheroe: in *Accretion Phenomena and Related Outflows*, Vol. 163 of IAU Colloquium, ed. by D. Wickramasinghe, G. Bicknell, and L. Ferrario (Astron. Soc. of the Pacific, 1997), p. 585
93. R. J. Protheroe and P. A. Johnson: Astropart. Phys. **4**, 253 (1996), and erratum ibid. **5**, 215 (1996)
94. R. Gandhi, C. Quigg, M. H. Reno, and I. Sarcevic: Astropart. Phys. **5**, 81 (1996); Phys. Rev. **D 58**, 093009 (1998)
95. P. Lipari: Astropart. Phys. **1**, 195 (1993)
96. W. Rhode et al.: Astropart. Phys. **4**, 217 (1996)
97. M. Aglietta et al. (EAS-TOP collaboration): in [33], Vol. 1, 638
98. R. M. Baltrusaitis et al.: Astrophys. J. **281**, L9 (1984); Phys. Rev. **D 31**, 2192 (1985)

99. M. Nagano et al.: J. Phys. G **12**, 69 (1986)
100. F. Halzen and D. Saltzberg: Phys. Rev. Lett. **81**, 4305 (1998)
101. K. Mannheim: Astropart. Phys. **11**, 49 (1999)
102. R. Protheroe: e-print astro-ph/9809144, invited talk at *Neutrino 98*, Takayama 4-9 June 1998; M. Roy and H. J. Crawford, e-print astro-ph/9808170, submitted to Astropart. Phys
103. P. B. Price: Astropart. Phys. **5**, 43 (1996)
104. Proc. 26th *International Cosmic Ray Conference*, (Utah, 1999)
105. P. W. Gorham, K. M. Liewer, and C. J. Naudet: e-print astro-ph/9906504, to appear in [104]
106. G. Gelmini and A. Kusenko: e-print hep-ph/9908276
107. E. Waxman and J. Miralda-Escudé: Astrophys. J. **472**, L89 (1996)
108. M. Lemoine, G. Sigl, A. V. Olinto, and D.N. Schramm: Astrophys. J., **486**, L115 (1997)
109. N. Hayashida et al.: Phys. Rev. Lett. **77**, 1000 (1996)
110. E. Waxman and P. S. Coppi: Astrophys. J. **464**, L75 (1996)
111. J. Miralda-Escudé and E. Waxman: Astrophys. J. **462**, L59 (1996)
112. P. P. Kronberg: Rep. Prog. Phys. **57**, 325; (1994) J. P. Vallée: Fundamentals of Cosmic Physics, Vol. **19**, 1 (1997)
113. T. Stanev: Astrophys. J. **479** 290 (1997); G. A. Medina Tanco, E. M. de Gouveia Dal Pino, and J. E. Horvath: Astrophys. J. **492**, 200 (1998)
114. D. Harari, S. Mollerach, and E. Roulet: e-print astro-ph/9906309
115. P. Blasi, S. Burles, and A. V. Olinto: Astrophys. J. **514**, L79 (1999)
116. D. Ryu, H. Kang, and P. L. Biermann: Astron. Astrophys. **335**, 19 (1998)
117. G. Sigl, M. Lemoine, and P. Biermann: Astropart. Phys. **10**, 141 (1999)
118. M. Lemoine, G. Sigl, and P. Biermann: e-print astro-ph/9903124
119. G. A. Medina Tanco, E. M. de Gouveia Dal Pino, and J. E. Horvath: Astropart. Phys. **6**, 337 (1997)
120. G. Sigl, M. Lemoine, and A. Olinto: Phys. Rev. **D. 56**, 4470 (1997)
121. G. Sigl and M. Lemoine: Astropart. Phys. **9**, 65 (1998)
122. G. A. Medina Tanco: Astrophys. J. **495**, L71 (1998)
123. J. D. Barrow, P. G. Ferreira, and J. Silk: Phys. Rev. Lett. **78**,
124. A. Loeb and A. Kosowsky: Astrophys. J. **469**, 1 (1996)
125. K. Subramanian and J. D. Barrow: Phys. Rev. Lett. **81**, 3575 (1998)
126. G. Sigl, D. N. Schramm, S. Lee, and C. T. Hill: Proc. Natl. Acad. Sci. USA, **94**, 10501 (1997)
127. M. Takeda et al.: e-print astro-ph/9902239, submitted to Astrophys. J.
128. J. P. Rachen and P. L. Biermann: Astron. Astrophys. **272**, 161 (1993)
129. A. Achterberg, Y. Gallant, C. A. Norman, and D. B. Melrose: e-print astro-ph/9907060, submitted to Mon. Not. R. Astron. Soc.
130. J. Wdowczyk and A. W. Wolfendale: Nature **281**, 356 (1979); M. Giler, J. Wdowczyk, and A. W. Wolfendale: J. Phys. G. **6**, 1561 (1980); V. S. Berezinskii, S. I. Grigo'eva, and V. A. Dogiel: Sov. Phys. JETP **69**, 453 (1989)
131. P. Blasi and A. V. Olinto: Phys. Rev. **D. 59**, 023001 (1999)
132. G. Medina Tanco: Astrophys. J. **510**, L91 (1999); e-print astro-ph/9905239, to appear in [104]
133. G. Medina Tanco: e-print astro-ph/9809219, to appear in *Topics in cosmic-ray astrophysics*, ed. by M. A. DuVernois (Nova Scientific, New York 1999)

Galaxies Behind the Milky Way and the Great Attractor

Renée C. Kraan-Korteweg

Departamento de Astronomía, Universidad de Guanajuato, Apartado Postal 144, 36000 Guanajuato GTO, Mexico

Abstract. Dust and stars in the plane of the Milky Way create a "Zone of Avoidance" in the extragalactic sky. Galaxies are distributed in gigantic labyrinth formations, filaments and great walls with occasional dense clusters. They can be traced all over the sky, except where the dust within our own galaxy becomes too thick – leaving about 25% of the extragalactic sky unaccounted for. Our Galaxy is a natural barrier which constrains the studies of large-scale structures in the Universe, the peculiar motion of our Local Group of galaxies and other streaming motions (cosmic flows) which are important for understanding formation processes in the Early Universe and for cosmological models.

Only in recent years have astronomers developed the techniques to peer through the disk and uncover the galaxy distribution in the Zone of Avoidance. I present the various observational multi-wavelength procedures (optical, far infrared, near infrared, radio and X-ray) that are currently being pursued to map the galaxy distribution behind our Milky Way, including a discussion of the (different) limitations and selection effects of these (partly) complementary approaches. The newly unveiled large-scale structures are discussed and compared to predictions from theoretical reconstructions of the mass density field. Particular emphasis is given to discoveries in the Great Attractor region – a from streaming motions predicted huge overdensity centered behind the Galactic Plane. The recently unveiled massive rich cluster A3627 seems to constitute the previously unidentified core of the Great Attractor.

1 The Zone of Avoidance

A first reference to the Zone of Avoidance (ZOA), or the "Zone of few Nebulae" was made in 1878 by Proctor [1], based on the distribution of nebulae in the "General Catalogue of Nebulae" by Sir John Herschel [2]. This zone becomes considerably more prominent in the distribution of nebulae presented by Charlier [3] using data from the "New General Catalogue" by Dreyer [4,5]. These data also reveal first indications of large-scale structure: the nebulae display a very clumpy distribution. Currently well-known galaxy clusters such as Virgo, Fornax, Perseus, Pisces and Coma are easily recognizable even though Dreyer's catalog contains both Galactic and extragalactic objects as it was not known then that the majority of the nebulae actually are external stellar systems similar to the Milky Way. Even more obvious in this distribution, though, is the absence of galaxies around the Galactic Equator. As extinction was poorly known at that time, no connection was made between the Milky Way and the "Zone of few Nebulae".

A first definition of the ZOA was proposed by Shapley [6], as the region delimited by "the isopleth of five galaxies per square degree from the Lick and Harvard surveys" (compared to a mean of 54 gal./sq.deg. found in unobscured regions by Shane & Wirtanen [7]). This "Zone of Avoidance" used to be "avoided" by astronomers interested in the extragalactic sky because of the inherent difficulties in analyzing the few obscured galaxies known there.

Merging data from more recent galaxy catalogs, i.e. the Uppsala General Catalog UGC [8] for the north ($\delta \geq -2\overset{\circ}{.}5$), the ESO Uppsala Catalog [9] for the south ($\delta \leq -17\overset{\circ}{.}5$), and the Morphological Catalog of Galaxies MCG [10] for the strip inbetween ($-17\overset{\circ}{.}5 < \delta < -2\overset{\circ}{.}5$), a whole-sky galaxy catalog can be defined. To homogenize the data determined by different groups from different survey material, the following adjustments have to be applied to the diameters: $D = 1.15 \cdot D_{\rm UGC}, D = 0.96 \cdot D_{\rm ESO}$ and $D = 1.29 \cdot D_{\rm MCG}$ [11]. According to Hudson & Lynden-Bell [12] this "whole-sky" catalog then is complete for galaxies larger than $D = 1\overset{\prime}{.}3$.

The distribution of these galaxies is displayed in Galactic coordinates in Fig. 1 in an equal-area Aitoff projection centered on the Galactic Bulge ($\ell = 0^\circ, b = 0^\circ$). The galaxies are diameter-coded, so that structures relevant for the dynamics in the local Universe stand out accordingly. Most conspicuous in this distribution is, however, the very broad, nearly empty band of about 20°. Why this Zone of Avoidance? Optical galaxy catalogs are limited to the largest galaxies. They therefore become increasingly incomplete close to the Galactic Equator where the dust thickens. This diminishes the light emission of the galaxies and reduces their visible extent. Such obscured galaxies are not included in diameter- or magnitude-limited catalogs because they appear small and faint – even though they might be intrinsically large and bright. A further complication is the growing number of foreground stars close to the Galactic Plane (GP) which fully or partially block the view of galaxy images.

Comparing this "band of few galaxies" with the currently available dust extinction maps of the DIRBE experiment [13], we can see that the ZOA – the area where the galaxy counts become severely incomplete – is described almost perfectly by the absorption contour in the blue A_B of $1\overset{\rm m}{.}0$ (where A_B is 4.14 times the extinction $E(B-V)$ [14]). This contour matches the ZOA defined by Shapley [6] closely.

1.1 Constraints Due to the Milky Way

Why is the distribution of galaxies behind the Milky Way important, and why is it not sufficient to study galaxies and their large-scale distribution away from the foreground "pollution" of the Milky Way?

In the last 20 years, enormous effort and observation time has been devoted to map the galaxy distribution in space. It was found that galaxies are located predominantly in clusters, sheets and filaments, leaving large areas devoid of luminous matter (see [15] for a detailed observational description of "Large-Scale Structures in the Universe").

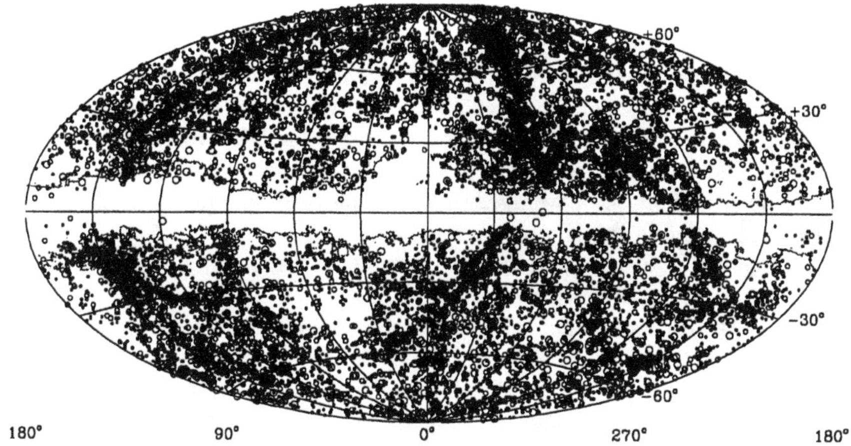

Fig. 1. Aitoff equal-area projection in Galactic coordinates of galaxies with $D \geq 1\overset{'}{.}3$. The galaxies are diameter-coded: small circles represent galaxies with $1\overset{'}{.}3 \leq D < 2'$, larger circles $2' \leq D < 3'$, and big circles $D \geq 3'$. The contour marks absorption in the blue of $A_B = 1\overset{\mathrm{m}}{.}0$ as determined from the Schlegel et al. [13] dust extinction maps. The displayed contour surrounds the area where the galaxy distribution becomes incomplete (the ZOA) remarkably well

Our Galaxy is part of the Local Group (LG) of galaxies, a small, gravitationally bound group of galaxies consisting of a few bright spiral galaxies and about 2 dozen dwarf galaxies. Our LG lies in the outskirts of the Local Supercluster, a flattened structure of about 30 Mpc, centered on the Virgo galaxy cluster with a few thousand galaxies (including its numerous dwarfs). Many such superclusters have meanwhile been charted. The nearby ones can actually be identified in the 2-dimensional galaxy distribution of Fig. 1: the Local Supercluster is visible as a great circle (the Supergalactic Plane) centered on the Virgo cluster at $\ell = 284°, b = 74°$, the Perseus-Pisces supercluster which bends into the ZOA at $\ell = 95°$ and $\ell = 165°$, and the general galaxy overdensity in the Great Attractor (GA) region ($280 \lesssim \ell \lesssim 360°, |b| \lesssim 30°$). Most of these superclusters and wall-like structures have massive clusters at their centers.

The lack of data in the ZOA severely constrains the studies of these structures in the nearby Universe, the origin of the peculiar velocity of the Local Group, and other streaming motions. Such studies are dependent on an accurate description of the whole sky distribution of galaxies, as described in the following sections.

Peculiar Motion of the Local Group of Galaxies. The Cosmic Microwave Background radiation (CMB) of 2.7° K – the relic radiation of the hot early Universe – shows a dipole of about 0.1%. This dipole is explained by a peculiar motion of the LG on top of the uniform Hubble expansion of 630 km s^{-1} towards the Galactic coordinates $\ell = 268°, b = 27°$ [16] induced by the gravitational attraction of the irregular mass distribution in the nearby Universe (see Fig. 1). Part of this motion can be explained by the acceleration of the LG towards Virgo, the center of the Local Supercluster (~ 220 km s^{-1} towards $\ell = 284°, b = 75°$). The remaining component of ~ 495 km s^{-1} towards $\ell = 274°, b = 12°$ [17,18] hence must arise from other mass concentrations and/or voids in the nearby Universe. The determination of the peculiar motion on the LG, i.e. its net gravity field, requires whole-sky coverage. Here, the lack of data in about 25% of the optical extragalactic sky is a severe handicap.

Various dipole determinations have assumed a uniformly filled ZOA or have used cloning methods which transplant the fairly well-mapped adjacent regions into the ZOA. Both procedures are unsatisfactory, because inhomogeneous data coverage will introduce non-existing flow fields. The derived results on the apex of the LG motion, as well as the distance at which convergence is attained, still are controversial. Kolatt et al. [19], for instance, have shown that the mass distribution within the inner $\pm 20°$ of the ZOA – as derived from theoretical reconstructions of the density field (see Sect. 7) – is crucial to the derivation of the gravitational acceleration of the LG: the direction of the motion measured within a volume of 6000 km s^{-1} will change by 31° when the (reconstructed) mass within the ZOA is included. Care should therefore be taken on how to extrapolate the galaxy density field across the ZOA. Obviously, a reliable consensus on the galaxy distribution in the ZOA is important to minimize these uncertainties.

Nearby Galaxies. In this context, not only the identification of unknown and suspected clusters, filaments and voids are relevant, but also the detection of nearby smaller entities. The peculiar velocity of the LG, $\boldsymbol{v}_p$, is proportional to the net gravity field $\boldsymbol{G}$, which can be determined by summing up the masses $\mathcal{M}_i$ of the individual galaxies at their distances $\boldsymbol{r}_i$:

$$\boldsymbol{v}_p \propto f(\boldsymbol{G}) \propto \frac{\Omega_0^{0.6}}{b} \sum \frac{\mathcal{M}_i}{r_i^2} \hat{\mathbf{r}}_{\mathrm{i}},$$

where Ω_0 is the density parameter and b the bias parameter. The gravity field as well as the light flux of a galaxy decreases with r^{-2}. The direction and amplitude of the peculiar velocity therefore is directly related to the sum of the *apparent magnitudes* of the galaxies in the sky through

$$\boldsymbol{v}_p \propto \sum_{\mathrm{i}} 10^{-0.4\mathrm{m}} \hat{\mathbf{r}}_{\mathrm{i}},$$

for a constant mass-to-light ratio. This has important implications and suggests, for instance, that the galaxy Cen A with an absorption-corrected magnitude of $B^o = 6\overset{\mathrm{m}}{.}1$ exerts a stronger luminosity-indicated gravitational attraction on the Local Group than the whole Virgo cluster. However, in this context, the question whether the mass-to-light ratio is constant, i.e. no biasing occurs, is doubtful, a problem inherent to all cumulative dipole determinations. These calculations also predict that the 8 apparently brightest galaxies – which are all nearby ($v < 300$ km s^{-1}) – are responsible for 20% of the total dipole as determined from optically known galaxies within $v \lesssim 6000\ \mathrm{km\,s^{-1}}$. Hence, a major part of the peculiar motion of the LG is generated by a few average, but nearby galaxies.

In this sense, the detection of other nearby galaxies hidden by the obscuration of the Galaxy can be as important as the detection of entire clusters at larger distances. The expectation of finding additional nearby galaxies in the ZOA is not unrealistic. Six of the nine apparently brightest galaxies are located in the ZOA: IC342, Maffei 1 and 2, NGC4945, CenA and the recently discovered galaxy Dwingeloo 1 (see Sect. 5.1). Moreover, the presence of an unknown Andromeda-like galaxy behind the Milky Way would have implications for the internal dynamics of the LG, the mass determination of the LG, and the present density of the Universe from timing arguments [20].

Cosmic Flow Fields such as in the Great Attractor Region. Density enhancements locally decelerate the uniform expansion field, as has been observed within our own Local Supercluster. Vice versa, systematic streaming motions over and above the uniform expansion field usually indicate mass overdensities (accelerations) or voids (decelerations). Knowing (a) the observed recessional velocity v_{obs} of a galaxy through its redshift z

$$v_{\mathrm{obs}} = cz = c\,\frac{\lambda(t) - \lambda_0}{\lambda_0},$$

where λ_0 is the rest wavelength, and $\lambda(t)$ is the observed wavelength, and (b) a redshift-independent distance estimate r, the peculiar motion of a galaxy $\boldsymbol{v}_p$ due to the underlying mass density field can be determined:

$$\boldsymbol{v}_p = \boldsymbol{v}_{\mathrm{obs}} - \boldsymbol{v}_{\mathrm{Hub}},$$

where v_{Hub} is the recession velocity a galaxy would have in an unperturbed expansion field ($v_{\mathrm{Hub}} = H_0 \cdot r$). In this manner, the mass density field can be determined independent of the galaxy distribution and/or an assumption on the mass-to-light ratio.

Based on these considerations, Dressler et al. [21] identified a systematic infall pattern from peculiar velocities of about 400 elliptical galaxies which was interpreted as being due to a hypothetical Great Attractor with a mass of $\sim 5 \times 10^{16}\mathcal{M}_{\odot}$, at a position in redshift space of $(\ell, b, v) = (307°, 9°, \sim 4400\ \mathrm{km\,s^{-1}})$ [22]. A more recent study by Kolatt et al. [19], based on a larger data set (elliptical *and* spiral galaxies) and the potential reconstruction method POTENT (see Sect. 7 and Fig. 17) place the center of the GA right behind the Milky Way. Recent consensus is that the GA is an extended region ($\sim$ 40°x40°) of moderately enhanced galaxy density centered behind the Galactic Plane. Although there is a considerable excess of optical galaxies and IRAS-selected galaxies in this region (see Fig. 1 and Fig. 9), no dominant cluster or central peak can been seen. However, a major part of the GA is hidden by the Milky Way.

Connectivity of Superclusters Across the ZOA. Various large-scale structures are 'bisected' by the Milky Way. What is their true extent? These large-scale structures, their sizes, and the distribution of the various galaxy types within these structures, carry information on the conditions and formation processes of the early Universe, providing important constraints which must be reproduced in cosmological models. It is therefore valuable to fully outline these superclusters across the ZOA.

It is curious, that the two major superclusters in the local Universe, i.e. Perseus-Pisces and the Great Attractor overdensity, lie at similar distances on opposite sides of the LG, and that both are partially obscured by the ZOA. It is therefore of particular interest to map these structures in detail, determine their extent and masses, in order to find out which one of the two is dominant in the tug-of-war on the Local Group.

1.2 Unveiling Large-Scale Structures Behind the Milky Way

For all of the above reasons, the unveiling of galaxies behind the Milky Way has turned into a research field of its own in the last ten years. In the following, I discuss all the various observational multi-wavelength techniques that are currently being employed to uncover the galaxy distribution in the ZOA such as deep optical searches, far-infrared and near-infrared surveys, systematic blind radio surveys and searches for hidden massive X-ray clusters. I will describe the different limitations and selection effects inherent to each method and present results obtained with these various methods – describing the results and discoveries in detail for the Great Attractor region. Predictions from reconstructions of the density field in the ZOA are also presented and compared with observational evidence. The comparison between reconstructed density fields and the observed galaxy distribution are important as they allow derivations of the density and biasing parameters Ω_0 and b.

2 Optical Galaxy Searches

Systematic optical galaxy catalogs are generally limited to the largest galaxies (typically with diameters $D \gtrsim 1'$, e.g. [9]). These catalogs become, however, increasingly incomplete for galaxies the closer they are to the Galactic Plane. With the thickening of the dust layer, the absorption increases and reduces the brightness of the galaxies and their 'visible' extension. Obviously such galaxies are not intrinsically faint; they only appear faint because of the dimming by the dust. Systematical deeper searches for partially obscured galaxies – down to fainter magnitudes and smaller dimensions compared to existing catalogs – have been performed on sky surveys with the aim of reducing this ZOA.

2.1 Early Searches and Results

One of the first attempts to detect galaxies in the ZOA was carried out by Böhm-Vitense in 1956 [23]. She did follow-up observations in selected fields in the GP in which Shane & Wirtanen [24] found objects that "looked like extragalactic nebulae" but were not believed to be galaxies because they were so close to the dust equator. She confirmed many galaxies and concluded that the obscuring matter in the plane must be extremely thin and full of holes between $\ell = 125°$-$130°$.

Because extinction was known to be low in Puppis, Fitzgerald [25] performed a galaxy search on a field there ($\ell \sim 245°$) and discovered 18 small and faint galaxies. Two years later, Dodd & Brand [26] examined 3 fields adjacent to this area ($\ell \sim 243°$) and detected another 29 galaxies. Kraan-Korteweg & Huchtmeier [27] observed these galaxies at radio wavelengths with the 100 m radio telescope at Effelsberg in Germany. This method was chosen because extinction is unimportant at these long wavelengths and the neutral gas of spiral galaxies can easily be observed at 21 cm (see Sect. 5). With these observations, a previously unknown nearby cluster at $(\ell, b, v) = (245°, 0°, \sim 1500\ \mathrm{km\,s^{-1}})$ could be identified. Adding far-infrared data (see Sect. 3), it was shown that this Puppis cluster is comparable to the Virgo cluster and that it contributes a significant component to the peculiar motion of the LG [28].

During a search for infrared objects Weinberger et al. [29], detected two galaxy candidates near the Galactic Plane ($\ell \sim 88°$) which Huchra et al. [30] confirmed in 1977 to be the brightest members of a galaxy cluster at 4200 $\mathrm{km\,s^{-1}}$. This discovery led Weinberger [31] to start the first *systematic* galaxy search. Using the red prints of the Palomar Sky Survey, he covered the whole northern GP ($\ell = 33°$-$213°$) in a thin strip ($|b| \leq 2°$). He found 207 galaxies, the distribution of which is highly irregular: large areas disclose no galaxies, the "hole" pointed out by Böhm-Vitense was verified, but most conspicuous was a huge excess of galaxies around $\ell = 160°$-$165°$. In 1984, Focardi et al. [32] made the connection with large-scale structures: they interpreted the excess as the possible continuation of the Perseus-Pisces cluster [PP] across the plane to the cluster A569. Radio-redshift measurements by Hauschildt [33] established that the PP cluster at a mean redshift of $v = 5500\ \mathrm{km\,s^{-1}}$ extends to the cluster 3C129 in the GP ($\ell = 160°, b = 0.°1$). Additional H I and optical redshift measurements of Zwicky galaxies by Chamaraux et al. [34] indicate that this chain can be followed even further to the A569 cloud at $v \sim 6000\ \mathrm{km\,s^{-1}}$ on the other side of the ZOA.

These early searches proved that large-scale structure can be traced to very low Galactic latitudes despite the foreground obscuration and its patchy nature which shows clumpiness and clustering in the galaxy distribution independent of large-scale structure. The above investigations did confirm suspected large-scale features across the plane through searches in selected regions and follow-up redshift observations. To study large-scale structure, systematically broader latitude strips covering the whole Milky Way, respectively the whole ZOA (see Fig. 1) are required.

2.2 Status of Systematic Optical Searches

Using existing sky surveys such as the first and second generation Palomar Observatory Sky Surveys POSS I and POSS II in the north, and the ESO/SRC (United Kingdom Science Research Council) Southern Sky Atlas, various groups have performed systematic deep searches for "partially obscured" galaxies. They catalogued galaxies down to fainter magnitudes and smaller dimensions ($D \gtrsim 0\farcm1$) than previous catalogs. Here, examination by eye remains the best technique. A separation of galaxy and star images can as yet not be done on a viable basis below $|b| \lesssim 10°$-$15°$ by automated measuring machines such as e.g. COSMOS [35] or APM [36] and sophisticated extraction algorithms, nor with the application of Artificial Neural Networks. Thus, although surveys by eye clearly are both very trying and time consuming – and maybe not as objective – they currently still provide the best technique to identify partially obscured galaxies in crowded star fields.

Meanwhile, through the efforts of various collaborations, nearly the whole ZOA has been surveyed and over 50000 previously unknown galaxies could be discovered in this way. These surveys are not biased with respect to any particular morphological type. The various surveyed regions are displayed in Fig. 2. Details and results on the uncovered galaxy distributions can be found in the respective references listed below:

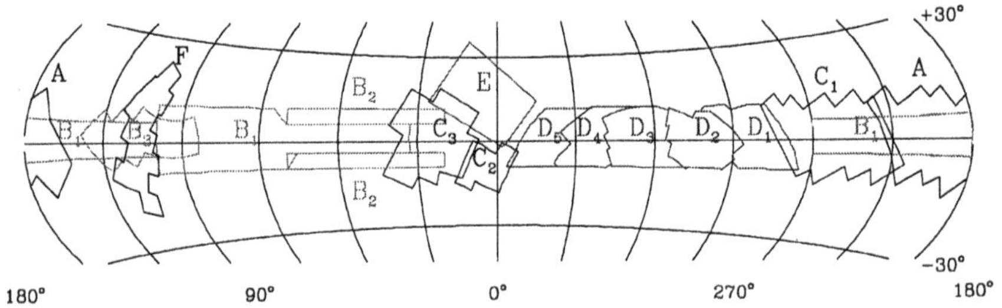

Fig. 2. An overview of the different optical galaxy surveys in the ZOA centered on the Galaxy. The labels identifying the search areas are explained in the text. Note that the surveyed regions cover the entire ZOA as defined by the foreground extinction level of $A_B = 1\fm0$ displayed in Fig. 1

A: the Perseus-Pisces Supercluster by Pantoja [37]; $\mathbf{B}_{1-3}$: the northern Milky Way (B_1 by Seeberger et al. [38–40], Lercher et al. [41], and Saurer et al. [42], from POSS I; B_2 by Marchiotto et al. [43] also from POSS II; B_3 by Weinberger et al. [44] from POSS II);
$\mathbf{C}_{1-3}$: the Puppis region by Saito et al. [45,46] [C_1], the Sagittarius/Galactic region by Roman et al. [47] [C_2], and the Aquila and Sagittarius region by Roman et al. [48] [C_3];
$\mathbf{D}_{1-5}$: the southern Milky Way (the Hydra to Puppis region [D_1] by Salem & Kraan-Korteweg [49], the Hydra/Antlia Supercluster region [D_2] by Kraan-Korteweg [50], the Crux region [D_3] by Woudt [51], Woudt & Kraan-Korteweg [52], the GA region [D_4] by Woudt [51], Woudt & Kraan-Korteweg [53], and the Scorpius region [D_5] by Fairall & Kraan-Korteweg [54]; **E**: the Ophiuchus Supercluster by Wakamatsu et al. [55],Hasegawa et al. [56];**F**: the northern GP/SGP crossing by Hau et al. [57].

Comparing the surveyed regions (Fig. 2) with the ZOA as outlined in Fig. 1 clearly demonstrates that nearly the whole ZOA has been covered by systematic deep optical galaxy searches.

2.3 The Galaxy Distribution in the Great Attractor Region

Most of these searches have quite similar characteristics. As an example, I discuss in the following the optical galaxy search performed by our group in the Great Attractor region (D_{1-5}).

The tools for this galaxy search were simple. It comprised a viewer with the ability to magnify 50 times and the IIIaJ film copies of the ESO/SRC survey. The viewer projects an area of $3\farcm5 \times 4\farcm0$ on a screen, making the visual, systematic scanning of these plates quite straightforward and comfortable.

Even though Galactic extinction effects are stronger in the blue, the IIIaJ films were searched rather than their red counterparts. Comparison between the various surveys demonstrated that the hypersensitized and fine grained emulsion of the IIIaJ films go deeper and show higher resolution. Even in the deepest extinction layers of the ZOA, the red films were found to have no advantage over the IIIaJ films.

A diameter limit of $D \gtrsim 0\farcm2$ was imposed. Below this diameter the reflection crosses of the stars disappear, making it hard to differentiate consistently between stars or blended stars and faint galaxies. The positions of all the galaxies are measured with the Optronics, a high precision measuring machine, at ESO (European Southern Observatories) in Garching, Germany. The accuracy of these positions is about $1''$. For every galaxy we recorded the major and minor diameter, an estimate of the average surface brightness and the morphological type of the galaxy. From the diameters and the average surface brightness a magnitude estimate was derived. A surprisingly good relation was found for the estimated magnitudes, with no deviations from linearity even for the faintest galaxies, and a scatter of only $\sigma = 0\fm5$ [50]. In this manner over 17 000 galaxies in about 1800 sq. deg. could be identified, of which $\sim$ 97% were previously unknown. Their distribution is displayed in Fig. 3 together with all the Lauberts galaxies larger than $D \geq 1\farcm3$ (diameter-coded as in Fig. 1) as well as the DIRBE foreground extinction contours of $A_B = 1\fm0$, $3\fm0$ and $5\fm0$.

The distribution reveals that galaxies can easily be traced through obscuration layers of 3 magnitudes, thereby narrowing the ZOA considerably. A few galaxies are still recognizable up to extinction levels of $A_B = 5\fm0$ and a handful of very small galaxy candidates have been found at even higher extinction levels. The latter most likely indicate holes in the dust layer. Overall, the mean number density follows the dust distribution remarkably well at low Galactic latitudes. The contour level of $A_B = 5\fm0$, for instance, is nearly indistinguishable from the galaxy density contour at 0.5 galaxies per square degree.

At intermediate extinction levels (between the outer and second extinction contour $1\fm0 \leq A_B \leq 3\fm0$), distinct under- and overdensities are noticeable in the unveiled galaxy distribution that are uncorrelated with the foreground obscuration. They must be the signature of large-scale structures.

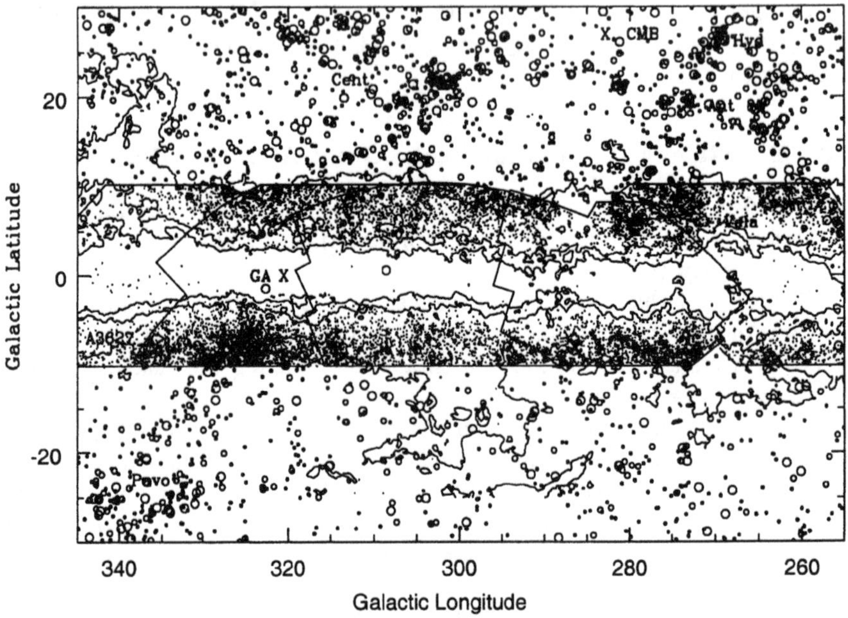

Fig. 3. Distribution of Lauberts galaxies with $D \geq 1'.3$ (open circles – coded as in Fig. 1) and galaxies with $D \geq 12''$ (small dots) identified in the deep optical galaxy searches D_1-D_5. The contours represent extinction levels of $A_B = 1^m.0$, $3^m.0$ and $5^m.0$. Note how the ZOA could be filled to $A_B = 3^m.0$ and that galaxy over- and underdensities uncorrelated with extinction can be recognized in this distribution

The most extreme overdensity is found at $(\ell, b) \sim (325°, -7°)$. It is at least a factor 10 denser compared to regions at similar extinction levels. This galaxy excess is centered on the cluster A3627. It is the only cluster out of 4076 clusters in the Abell cluster catalog [58].Although it is (a) classified as a rich, nearby cluster, (b) the only Abell cluster identified below $|b| < 10°$, and (c) within a few degrees of the predicted center of the GA [19], this cluster had not received any attention. This is mainly due to the foreground obscuration. A3627 is hardly discernable in, for instance, the distribution of Lauberts galaxies: the observed diameters of the galaxies in this density peak are just *below* the Lauberts diameter limit (due to the obscuration). This cluster is *not evident* in the far infrared (see Sect. 3). This can be explained by the predominance of early-type galaxies (50% in the core of this cluster, 25% within its Abell radius) which do not radiate in the far infrared but are a clear signature of rich clusters. The new data support the classification of A3627 as a rich cluster: over 600 likely new cluster members were identified compared to the 50 larger galaxies noted by Abell.

The galaxies detected in these searches are quite small ($< D >= 0'.4$) and faint ($< B_J >= 18^m.0$) on average. So the question arises whether these new galaxies and the newly uncovered over- and underdensities are relevant at all to our understanding of the dynamics in the local Unverse. To assess this, we have to understand the effects of extinction: galaxies are diminished by at least 1^m of foreground extinction at the highest latitudes ($|b| \sim 10°$) of the search areas. These effects increase considerably closer to the Galactic Equator. The effects of the absorption on the observed parameters of these low-latitude galaxies is reflected clearly in Fig. 4. Here, the magnitudes and major diameters of galaxies in the Hydra/Antlia search region (D_2) are plotted against the Galactic extinction $E(B-V)$ derived from the 100 micron DIRBE dust maps [13]. The top panels show the observed magnitudes (left) and diameters (right).

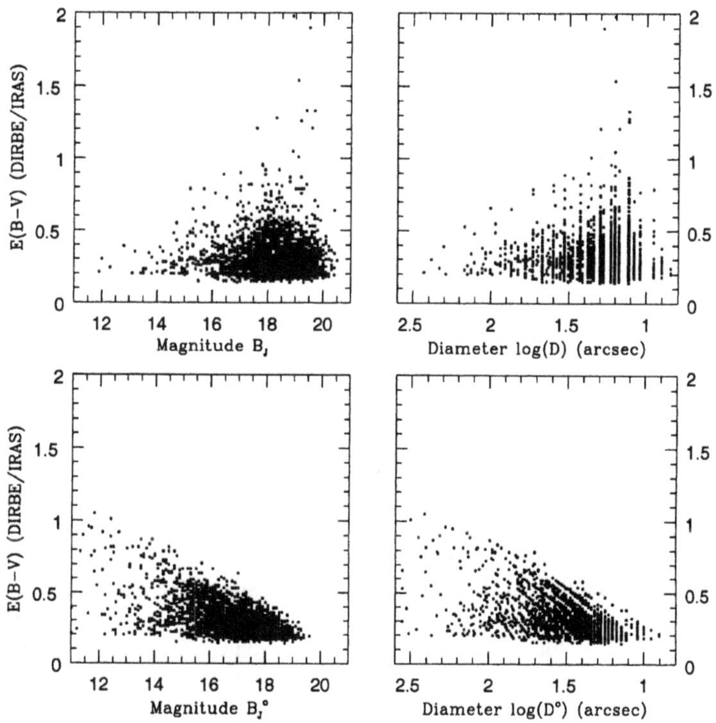

Fig. 4. The observed (top panels) and extinction-corrected (bottom) magnitudes (left) and diameters (right) of galaxy candidates in the Hydra/Antlia region as a function of the foreground extinction $E(B-V)$

The distribution of both the observed magnitudes and diameters show a distinct cut-off as a function of extinction – all the galaxies lie in the lower right triangle of the diagram, leaving the upper left triangle empty. At low extinction values, bright and faint galaxies can be identified, whereas apparently faint and small galaxies remain visible only at higher extinction values. The division in the diagram defines an upper envelope of the intrinsically brightest and largest galaxies. This fiducial line, i.e. the shift Δm to fainter apparent magnitudes of the intrinisically brightest galaxies, is a direct measure of the absorption A_B. In fact, this shift in magnitude is tightly correlated with the absorption in the blue $A_B = 4.14 \cdot E(B-V)$. The galaxies at these extinction levels are not intrinsically faint. They must in fact be intrinsically very bright galaxies to still be visible through the murk of the Milky Way.

The obscuration effects on the parameters of galaxies have been studied in detail by Cameron [59] who simulated the effects of absorption on the brightness profiles of various Virgo galaxies. This led to analytical descriptions of the diameter and isophotal magnitude corrections given in Table 1 for early-type and spiral galaxies:

Table 1. Obscurational effects on the diameter and isophotal magnitude.

	Reduction factor	Additional Δm
ellipticals/lenticulars	$10^{0.13 A_B{}^{1.3}}$	$0.08 A_B{}^{1.8}$
spirals	$10^{0.10 A_B{}^{1.7}}$	$0.07 A_B{}^{2.5}$

For example, a spiral galaxy, seen through an extinction of $A_B = 1^{\rm m}$, is reduced to $\sim 80\%$ of its unobscured size. Only $\sim 22\%$ of a (spiral) galaxy's original dimension is seen when it is observed through $A_B = 3^{\rm m}$, and its isophotal magnitude will be diminished by $4\overset{\rm m}{.}1$. Applying these corrections to the optical ZOA galaxy samples invert the trends in the magnitude and diameter distributions. This can be verified in the lower panels of Fig. 4 where the extinction-corrected magnitudes and diameters are plotted. At high extinction only the intrinsically bright galaxies can be identified. These deep optical galaxy searches hence do uncover intrinsically bright galaxies at lower latitudes.

Correcting the galaxies identified in deep optical searches for absorption partially lifts the veil of the Milky Way. Without the extinction layer, the Lauberts catalog would have, for instance, found 139 galaxies with $D \geq 1\overset{\prime}{.}0$ within the Abell radius $R_A = 3\,h_{50}^{-1}$ Mpc for A3627 compared to the previously identified 31 galaxies, where h_{50}, the dimensionless Hubble parameter is 1 for a Hubble constant of $H_0 = 50$ km s^{-1} Mpc^{-1} ($H_0 = 50\,h$ km s^{-1} Mpc^{-1}). This makes this cluster *the most prominent overdensity in the southern sky*. Were it not for the obscuration, it most likely would have been the best-studied cluster in the Universe.

2.4 Redshift Follow-ups and the Cluster A3627

Analazing the galaxy density as a function of the galaxy size, magnitude and/or morphology in combination with the foreground extinction has led to the identification of various important large-scale structures in the ZOA and their approximate distances. Redshift observations must be obtained to map the large-scale structures in redshift space. So far, this has been pursued extensively in the

Perseus-Pisces supercluster [37], the Puppis region [60], the Ophiuchus supercluster behind the Galactic Bulge area [56] and the southern ZOA. Here again, I concentrate on the results from various observing programs in the Great Attractor region. For a listing of the mapping of other large-scale structures and references see Kraan-Korteweg & Woudt [61].

For the survey regions D_{1-5} we use complementary observing approaches to obtain the redshifts (see [62] for a more detailed description):

– multifiber spectroscopy with the MEFOS instrument [63] at the 3.6m telescope of ESO. This instrument has the ability to obtain 29 spectra simultaneously within a one-degree circular field; ideally suited to probe the densest regions in the uncovered galaxy distribution,

– individual spectroscopy of all the brighter galaxies ($B_J \sim 17\overset{m}{.}0 - 17\overset{m}{.}5$, depending on the central surface brightness of a galaxy) with the 1.9m telescope of the South African Astronomical Observatory (SAAO) [64–66]. This method allows homogeneous coverage over the whole search area, – 21cm observations of extended, low surface-brightness spiral galaxies with the 64m radio telescope in Parkes, Australia [67]. The radio observations are an important addition as it is impossible to obtain good signal-to-noise optical spectra for highly obscured low-surface brightness galaxies whereas the 21cm radiation is not influenced by the dust.

With the above observations, we typically obtain redshifts of $\gtrsim 10\%$ of the galaxies and can trace large-scale structures out to recession velocities of $\sim$ 25000 $\mathrm{km\,s^{-1}}$. To focus again on the GA region, a redshift "slice" (the distribution of a certain region on the sky as a function of redshift) out to 10000 $\mathrm{km\,s^{-1}}$ is shown in the left-hand panel of Fig. 5 for our optical survey region ($260° \lesssim \ell \lesssim 350°$, $|b| \lesssim 10°$): a region that previously was largely blank now reveals clusters, superclusters and voids. In this illustration, the ZOA is now comparable to other unobscured regions of the sky. The radially very extended feature at $\ell = 325°$ – the location of the cluster A3627 – is the signature of a galaxy cluster: the "finger of God" feature due to the velocity dispersion of a virially bound cluster.

On the right-hand panel, all structures within the general GA region ($300° \leq \ell \leq 340°$) are displayed with structures adjacent to the Milky Way ($-45° \leq b \leq 45°$). Here we can clearly discern the Hydra ($b = 27°$), Antlia ($b = 19°$) and bimodal Centaurus clusters on the northern side of the Galactic Plane and the Pavo cluster ($-24°$) on the southern side. It is impressive to note that the new redshifts in the A3627 cluster area prove this cluster to be the dominant structure within the general GA overdensity. While this cluster includes the well-researched radio galaxy PKS1610−601, relatively few redshifts of other cluster members were known beforehand. Adding, however, the new ZOA redshift data, we find a near Gaussian distribution of the velocities, resulting in a mean observed velocity of $< v >= 4848$ $\mathrm{km\,s^{-1}}$ and a velocity dispersion of $\sigma = 896$ $\mathrm{km\,s^{-1}}$. This is displayed in Fig. 6 where the dark shaded histogram identifies previously known galaxies and the light shaded histogram the redshift data from our ZOA program.

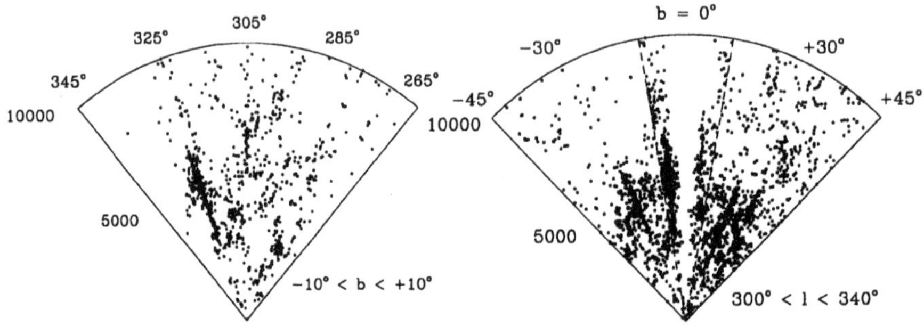

Fig. 5. Redshift slices out to 10000 $\mathrm{km\,s^{-1}}$. The left panel shows the distribution "in" the ZOA ($|b| \lesssim 10°$) along Galactic longitudes, the right panel the distribution in the GA region ($300° < \ell < 340°$) for the latitude range $|b| \leq 45°$

The large dispersion suggests A3627 to be a massive cluster. The dynamical mass within a radius R [68] is given by

$$\mathcal{M}(< R) = \frac{9\sigma^2 R_c}{\mathrm{G}}(\ln(x + (1+x^2)^{1/2}) - x(1+x^2)^{-1/2})$$

where σ is the measured line-of-sight velocity dispersion (corrected for the errors in the velocity measurements), R_c is the core radius [69], G is the gravitational constant, and $x = R/R_c$.

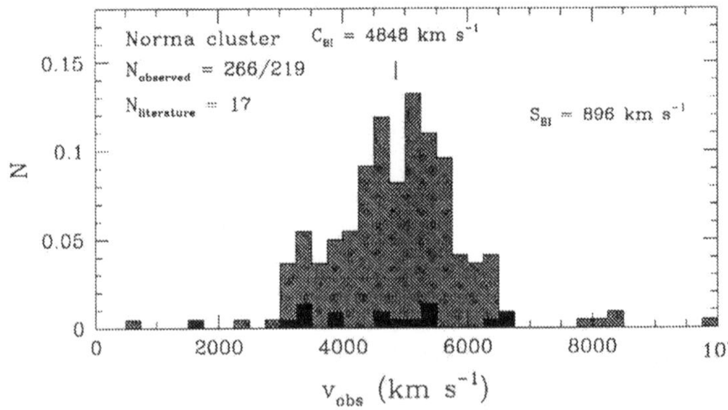

Fig. 6. The velocity histogram of galaxies within the Abell radius ($R_A = 3\,h_{50}^{-1}$ Mpc) of the Norma cluster. Galaxies with redshift information available in the literature before the ZOA redshift survey are indicated by the dark shaded histogram. A total of 219 likely cluster members are identified

With a core radius of 0.29 h_{50}^{-1} Mpc, a virial mass within the Abell radius $R_A = 3h_{50}^{-1}$ Mpc of

$$\mathcal{M}_{A3627} = 0.9 \cdot 10^{14} h_{50}^{-1} \mathcal{M}_{\odot}$$

is found for A3627. This mass is typical of rich clusters, and comparable, for instance, to the well-studied Coma cluster [70,71]. The latter was already identified in 1906 by Wolf [72] in the distribution of nebulae (galactic and extragalactic). With a mean redshift of 6960 $\mathrm{km\,s^{-1}}$, the Coma cluster counted as the nearest rich cluster. At a mean redshift of 4848 $\mathrm{km\,s^{-1}}$, this place is now being usurped by the A3627 cluster, also called Norma cluster for the constellation it lies in.

Rich massive clusters generally are strong X-ray emitters (see Sect. 6) and were identified early on with X-ray satellites (Einstein, HEAO, Uhuru) – except for A3627. However, A3627 was detected in a whole-sky survey by the X-ray satellite ROSAT, in which the Norma cluster ranks as the 6th brightest X-ray cluster in the sky compared to Coma, which ranks 4 [73].

The mean velocity of the Norma cluster puts it well within the predicted velocity range of the GA. Including the new results from the deep optical galaxy search, the Norma cluster now is the most massive galaxy cluster in the GA region known to date. It most likely marks the previously unidentified but predicted density-peak at the bottom of the potential well of the GA overdensity.

The mass excess of the GA is presumed to arise within an area of radius of about 20 Mpc [74]. These extended potential wells generally have a rich cluster at their center. This actually matches the emerging picture quite well: A3627 appears to lie at the center of an apparent "great wall"-like structure, similar to Coma in the (northern) Great Wall. The right-hand redshift slice of Fig. 5 suggests a very large-scale coherent structure, starting at Pavo $(332°, -24°)$ and moving towards the density peak of A3627 at slightly larger velocities. This supercluster then seems to bend towards or merge with the Vela supercluster at $(l, b, v) \sim (280°, 6°, \sim 6000\ \mathrm{km\,s^{-1}})$ postulated by Kraan-Korteweg et al. [62].

One can, however, not exclude the possibility that other unknown rich clusters reside in the GA region, as the ZOA has not been fully mapped with the optical galaxy searches (see Fig. 3 and right panel of Fig. 5). Finding a further uncharted, rich cluster of galaxies at the heart of the GA would have serious implications for our current understanding of this massive overdensity in the local Universe. Various indications suggest, for instance, that PKS1343−601, the second brightest extragalactic radio source in the southern sky, might form the center of yet another highly obscured rich cluster [61], particularly as it also shows significant X-ray emission. At $(\ell, b) \sim (310°, 2°)$, this radio galaxy lies behind an obscuration layer of about 12 magnitudes of extinction in the B-band, hence optical surveys are ineffective. Still, West & Tarenghi observed this source in 1989 [75]: with an extinction-corrected diameter of $D^o \sim 4'$ and a recession velocity of $v = 3872\ \mathrm{km\,s^{-1}}$ this galaxy appears to be a giant elliptical galaxy and giant ellipticals are mainl found at the cores of clusters.

Since PKS1343−601 is so heavily obscured, little data are available to substantiate the existence of this prospective cluster. In Fig. 7 the A3627 cluster at a mean extinction $A_B = 1\overset{\mathrm{m}}{.}5$ as seen in deep optical searches is compared to the

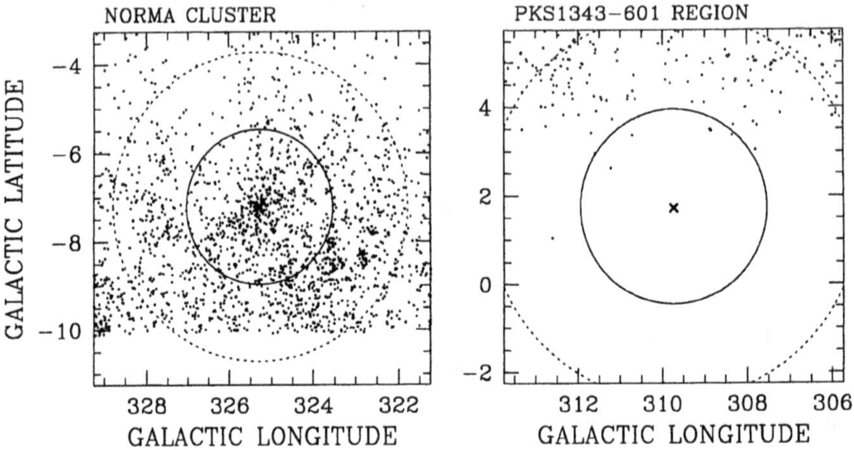

Fig. 7. Sky distribution of galaxies identified in the deep optical galaxy search around the rich A3627 cluster ($A_B \sim 1\overset{m}{.}5$) and around the suspected cluster centered on PKS1343−601 ($A_B \sim 12^m$), both in the GA region. The inner circle marks the Abell radius $R_A = 3\,h_{50}^{-1}$ Mpc

prospective PKS1343 cluster at ($309\overset{\circ}{.}7, +1\overset{\circ}{.}7$, 3872 km s^{-1}) with an extinction of 12^m. One can clearly see, that at the low Galactic latitude of the suspected cluster PKS1343, the optical galaxy survey could not retrieve the underlying galaxy distribution, especially not within the Abell radius of the suspected cluster (the inner circle in the right panel of Fig. 7). To verify this cluster, other observational approaches are necessary. Interestingly enough, deep H I observations did uncover a significant excess of galaxies at this position in velocity space (see Sect. 5.3) although a "finger of God", the characteristic signature of a cluster in redshift space, is not seen. Hence, the Norma cluster A3627 remains the best candidate for the center of the extended GA overdensity.

2.5 Completeness of Optical Galaxy Searches

In order to merge the various deep optical ZOA surveys with existing galaxy catalogs, Kraan-Korteweg [50] and Woudt [51] have analyzed the completeness of their ZOA galaxy catalogs as a function of the foreground extinction. By studying the apparent diameter distribution as a function of the extinction, as shown in Fig. 4, as well as the location of the flattening in the slope of the cumulative observed and extinction-corrected diameter curves $(\log D) - (\log N)$ and $(\log D^o) - (\log N)$ for various extinction intervals (cf. Fig. 6 in [50]), they concluded that the optical ZOA surveys are complete to an apparent diameter of $D = 14''$ – where the diameters correspond to an isphote of 24.5 mag/arcsec^2 – for extinction levels less than $A_B = 3\overset{m}{.}0$ (see also Fig. 4).

What about the intrinsic diameters, i.e. the diameters galaxies would have if they were unobscured? Applying the Cameron corrections, it was found that at $A_B = 3\overset{m}{.}0$, an obscured spiral or an elliptical galaxy at the completeness limit

$D = 14''$ would have an intrinsic diameter of $D^o \sim 60''$, respectively $D^o \sim 50''$. At extinction levels higher than $A_B = 3\overset{m}{.}0$, an elliptical galaxy with $D^o = 60''$ would appear smaller than the completeness limit $D = 14''$ and might have gone unnoticed. These optical galaxy catalogs should therefore be complete to $D^o \geq 60''$ for all galaxy types down to extinction levels of $A_B \leq 3\overset{m}{.}0$, with the possible exception of extremely low-surface brightness galaxies. Only intrinsically very large and bright galaxies – particularly galaxies with high surface brightness – will be recovered in deeper extinction layers. This completeness limit could be confirmed by independently analyzing the diameter vs. extinction and the cumulative diameter diagrams for extinction-corrected diameters.

We can thus supplement the ESO, UGC and MCG catalogs (see Fig. 1), which are complete to $D = 1\overset{'}{.}3$, with galaxies from optical ZOA galaxy searches that have $D^o \geq 1\overset{'}{.}3$ and $A_B \leq 3\overset{m}{.}0$. As our completeness limit lies well above the ESO, UGC and MCG catalogs, we can assume that the other similarly performed optical galaxy searches in the ZOA should also be complete to $D^o = 1\overset{'}{.}3$ for extinction levels of $A_B \leq 3\overset{m}{.}0$.

With Fig. 8, the first attempt has been made to arrive at an improved whole-sky galaxy distribution with a reduced ZOA. In this Aitoff projection all the UGC, ESO, MCG galaxies that have *extinction-corrected* diameters $D^o \geq 1\overset{'}{.}3$ are plotted [remember that galaxies adjacent to the optical galaxy search regions are also affected by absorption though to a lesser extent ($A_B \leq 1\overset{m}{.}0$)], including the galaxies other optical surveys for which positions and diameters were available. The regions for which these data are not yet available are marked in Fig. 8. As some searches were performed on older generation POSS I plates, which are less deep compared to the second generation POSS II and ESO/SRC plates, an additional correction was applied to those diameters, i.e. the same correction as for the UGC galaxies which also are based on POSS I survey material ($D_{25} = 1.15 \cdot D_{\mathrm{POSS\,I}}$).

A comparison of Fig. 1 with Fig. 8 demonstrates convincingly how the deep optical galaxy searches realize a considerable reduction of the ZOA; we can now trace the large-scale structures in the nearby Universe to extinction levels of $A_B = 3\overset{m}{.}0$. Inspection of Fig. 8 reveals that the galaxy density enhancement in the GA region is even more pronounced and a connection of the Perseus-Pisces chain across the Milky Way at $\ell = 165°$ more likely. Hence, these supplemented whole-sky maps certainly should improve our understanding of the velocity flow fields and the total gravitational attraction on the Local Group.

Optical galaxy searches, however, fail in the most opaque part of the Milky Way, the region encompassed by the $A_B = 3\overset{m}{.}0$ contour in Fig. 8 – a sufficiently large region to hide further dynamically important galaxy densities. Here, other systematic surveys in other wavebands can be applied to reduce the current ZOA even further. The success and status of these approaches are discussed in the following sections.

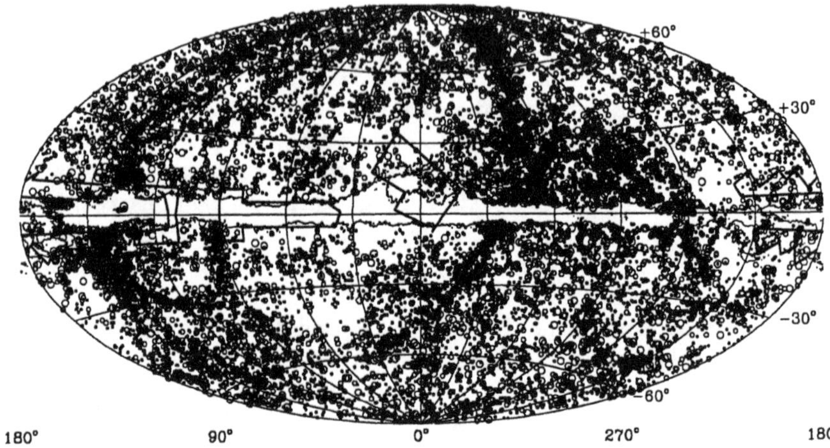

Fig. 8. Aitoff equal-area distribution in Galactic coordinates of ESO, UGC, MCG galaxies with extinction-corrected diameters $D^o \geq 1\farcm3$, including galaxies identified in the optical ZOA galaxy searches for extinction-levels of $A_B \leq 3\fm0$ (contour). The diameters are coded as in Fig. 1. With the exception of the areas for which either the positions of the galaxies or their diameters are not yet available (demarcated areas), the ZOA could be reduced considerably compared to Fig. 1

3 Far Infrared Surveys and the ZOA

In 1983, the Infrared Astronomical Satellite IRAS surveyed 96% of the whole sky in the far infrared bands at 12, 25, 60 and 100 μm, resulting in a catalog of 250 000 point sources, i.e. the IRAS Point Source Catalogue [76]. The latter has been used extensively to quantify extragalactic large-scale structures. The identification of the galaxies from the IRAS data base is quite different compared to the optical: only the fluxes at the 4 far infrared (FIR) IRAS passbands are available but no images. The identification of galaxies is strictly based on the relation of the fluxes. For instance, Yamada et al. [77] used the criteria: **1.** $f_{60} > 0.6$Jy, **2.** $f_{60}^2 > f_{12} f_{25}$, **3.** $0.8 < f_{100}/f_{60} < 5.0$, to select galaxy candidates from the IRAS PSC.

With these flux and color criteria mainly normal spiral galaxies and starburst galaxies are identified. Hardly any dwarf galaxies enter the IRAS galaxy sample, nor the dustless elliptical galaxies, as they do not radiate in the far infrared. The upper cut-off in the third criterion is imposed to minimize the contamination with cool cirrus sources and young stellar object within our Galaxy. This, however, also makes the IRAS surveys less complete for nearby galaxies [51,50].

The advantage of using IRAS data for large-scale structure studies is its homogeneous sky coverage (all data from one instrument) and the negligible effect of the extinction on the flux at these long wavelengths. Even so, it remains difficult to probe the inner part of the ZOA with IRAS data because of cirrus, high source counts of Galactic objects in the Galaxy, and confusion with these objects – most of them have the same IRAS characteristics as external galaxies. The difficulty in obtaining unambiguous galaxy identifications at these latitudes was demonstrated by Lu et al. [78], who found that the detection rate of IRAS galaxy candidates decreases strongly as a function of Galactic latitude (from $|b| = 16°$ to $|b| = 2°$). This can only be explained by the increase in faulty IRAS galaxy identifications. Yamada et al. [77] also found a dramatic and unrealistic increase in possible galaxies close to the Galactic Plane in their systematic IRAS galaxy survey of the southern Milky Way ($|b| \leq 15°$).

So, despite the various advantages given with IRAS data, the sky coverage in which reliable IRAS galaxy identifications can be made (84%) provides only a slight improvement over optical galaxy catalogs (compare e.g. the light-grey mask in Fig. 9 with the optical ZOA-contour as displayed in Fig. 1). In addition to that, the density enhancements are very weak in IRAS galaxy samples because (a) the IRAS luminosity function is very broad, which results in a more diluted distribution since a larger fraction of distant galaxies will enter a flux-limited sample compared to an optical galaxy sample, and (b) IRAS is insensitive to elliptical galaxies, which reside mainly in galaxy clusters, and mark the peaks in the mass density distribution of the Universe. This is quite apparent in a comparison of the IRAS galaxy distribution (Fig. 9) with the optical galaxy distribution (Fig. 1 and Fig. 8).

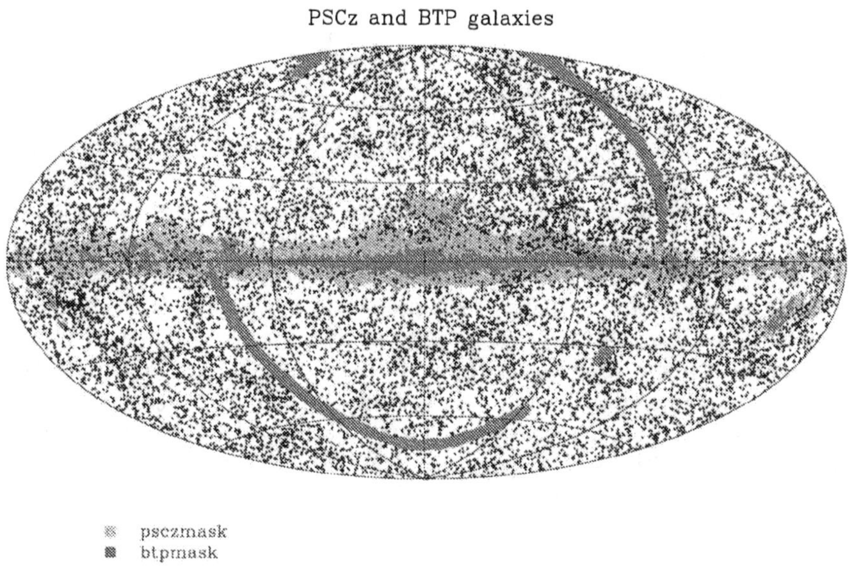

Fig. 9. The PSCz and BTP IRAS galaxy catalogs centered on the Galaxy with the PSCz incompleteness mask (light-grey mask) and the BTP mask (dark-grey). Note the dramatic reduction of the incompleteness around the Galactic Equator due to the BTP survey

Nevertheless, dedicated searches for large-scale clustering within the whole ZOA ($|b| \leq 15°$) have been made by various Japanese collaborations (see [79] for a summary). They used IRAS color criteria to select galaxy candidates which were subsequently verified through visual examination on sky surveys, such as the POSS of the northern hemisphere and the ESO/SRC for the southern sky. Because of their verification procedure, this data-set suffers, however, from the same limitations in highly obscured regions as optical surveys.

Based on redshift follow-ups of these ZOA IRAS galaxy samples, they established various filamentary features and connections across the ZOA. Most coincide with the structures uncovered in optical work. In the northern Milky Way both crossings of the Perseus-Pisces arms into the ZOA are very prominent – considerably stronger in the FIR than at optical wavelengths – and they furthermore identified a new structure: the Cygnus-Lyra filament at ($60° - 90°, 0°, 4000\,\mathrm{km\,s^{-1}}$). Across the southern Milky Way they confirmed the three general concentrations of galaxies around Puppis ($\ell = 245°$), the Hydra-Antlia extension ($\ell = 280°$, [64]) and the Centauraus Wall ($\ell = 315°$). However, the cluster A3627 is not seen, nor is the Great Attractor very prominent compared to the optical or to the POTENT reconstructions described in Sect. 7.

Besides the search for the continuity of structures across the Galactic Plane, the IRAS galaxy samples have been widely used for the determination of the peculiar motion of the Local Group, as well as the reconstructions of large-scale structure across the Galactic Plane (see Sect. 7). This has been performed on two-dimensional IRAS galaxy distribution and, in recent years, as well as on their distribution in redshift space with the availability of redshift surveys for progressively deeper IRAS galaxy samples, i.e. 2658 galaxies to $f_{60\mu m} = 1.9$ Jy [80], 5321 galaxies to $f_{60\mu m} = 1.2$ Jy [81], and lately the PSCz catalog of 15411 galaxies complete to $f_{60\mu m} = 0.6$ Jy with 84% sky coverage and a depth of 20000 $\mathrm{km\,s^{-1}}$ [82].

The PSCz is in principal deep enough to see convergence of the dipole. Saunders and collaborators realized, however, that the 16% of the sky missing from the survey causes significant uncertainty, particularly because of the location behind the Milky Way of many of the prominent large-scale structures (superclusters as well as voids). In 1994, they therefore started a longterm program to increase the sky coverage of the PSCz. Optimizing their color criteria to minimize contamination by Galactic sources ($f_{60}/f_{25} > 2$, $f_{60}/f_{12} > 4$, and $1.0 < f_{100}/f_{60} < 5.0$), they extracted a further 3500 IRAS galaxy candidates at lower Galactic latitudes (light-grey area of Fig. 9), reducing the coverage gap to a mere 7% (dark-grey area). Taking K' band snapshots of all the galaxy candidates of their 'Behind The Plane' [BTP] survey, they could add a thousand galaxies to the PSCz sample.

The resulting sky map of 16,400 galaxies (PSCz plus BTP) is shown in Fig. 9 (from [83]). The BTP survey has reduced the "IRAS ZOA" dramatically. Some incompleteness remains towards the Galactic Center, but large-scale structures can easily be identified across most of the Galactic Plane. In the Great Attractor region, the galaxies can be traced (for the first time with IRAS data) to the rich cluster A3627 – the suspected core of the GA [84]. The IRAS galaxies overall seem to align well with the Norma supercluster [85]. The BTP collaboration is currently working hard on obtaining redshifts for these new and heavily obscured galaxies and exciting new results on large-scale structure across the Milky Way and dipole determinations can be expected in the near future.

4 Near Infrared Surveys and the ZOA

Observations in the near infrared (NIR) can provide important complementary data to other surveys. With extinction decreasing as a function of wavelength, NIR photons are up to 10 times less affected by absorption compared to optical surveys – an important aspect in the search and study of galaxies behind the obscuration layer of the Milky Way. The NIR is sensitive to early-type galaxies – tracers of massive groups and clusters – which are missed in IRAS and H I surveys (Sect. 3 and 5). In addition, confusion with Galactic objects is considerably lower compared to the FIR surveys. Furthermore, because recent star formation contributes only little to the NIR flux of galaxies (in contrast to optical and FIR emission), NIR data give a better estimation of the stellar mass content of galaxies.

4.1 The NIR Surveys DENIS and 2MASS

Two systematic near infrared surveys are currently being performed. DENIS, the DEep Near Infrared Southern Sky Survey, is imaging the southern sky from $-88° < \delta < +2°$ in the I_c (0.8μm), J (1.25μm) and K_s (2.15μm) bands. 2MASS, the 2 Micron All Sky Survey, is covering the whole sky in the J (1.25μm), H (1.65μm) and K_s (2.17μm) bands. The mapping of the sky is performed in declination strips, which are 30° in length and 12 arcmin wide for DENIS, and $6° \times 8\overset{'}{.}5$ for 2MASS. Both the DENIS and 2MASS surveys are expected to complete their observations by the end of 2000. The main characteristics of the 2 surveys and their respective completeness limits for extended sources are given in Table 2 [86–89].

Details and updates on completeness, data releases and data access for DENIS and 2MASS can be found on the websites http://www-denis.iap.fr, and http://www.ipac.caltech.edu/2mass, respectively.

The DENIS completeness limits (total magnitudes) for highly reliable automated galaxy extraction (determined away from the ZOA, i.e. $|b| > 10°$) are $I = 16\overset{m}{.}5$, $J = 14\overset{m}{.}8$, $K_s = 12\overset{m}{.}0$ [90]. The number counts per square degrees for these completeness limits are 50, 28 and 3 respectively. For 2MASS, the completeness limits are $J = 15\overset{m}{.}0$, $H = 14\overset{m}{.}2$, $K_s = 13\overset{m}{.}5$ (isophotal magnitudes), with number counts of 48, ~40 and 24. In all wavebands, except I_c, the number counts are quite imprecise due to the low number statistics and the strong dependence on the star crowding in the analyzed fields. Still, they suffice to reveal the promise of NIR surveys at very low Galactic latitudes. As illustrated in Fig. 10, the galaxy density in the B band in unobscured regions is 110 galaxies per square degree for the completeness limit of $B_J \leq 19\overset{m}{.}0$ [91]. These counts drop rapidly with increasing obscuration: $N(A_B) \simeq 110 \times \mathrm{dex}(0.6\,[-A_B])\ \mathrm{deg}^{-2}$. The decrease in detectable galaxies due to extinction is much slower in the NIR, i.e. 45%, 21%, 14% and 9% compared to the optical for the I_c, J, H and K_s bands. This dependence makes NIR surveys very powerful at low Galactic latitudes even though they are not as deep as the POSS and ESO/SRC sky surveys: the NIR counts of the shallower NIR surveys overtake the optical counts at extinction levels of $A_B \gtrsim 2\text{-}3^{\mathrm{m}}$. The location of the reversal in efficiency is particularly opportune because the NIR surveys become more efficient where deep optical galaxy searches become incomplete, i.e. at $A_B \gtrsim 3\overset{\mathrm{m}}{.}0$ (see Sect. 2.5).

Table 2. Main characteristics of the DENIS and 2MASS surveys

	DENIS			2MASS		
Channel	I_c	J	K_s	J	H	K_s
Central wavelength	0.8μm	1.25μm	2.15μm	1.25μm	1.65μm	2.15μm
Arrays	1024x1024	256x256	256x256	256x256	256x256	256x256
Pixel size	$1\overset{\prime\prime}{.}0$	$3\overset{\prime\prime}{.}0$	$3\overset{\prime\prime}{.}0$	$2\overset{\prime\prime}{.}0$	$2\overset{\prime\prime}{.}0$	$2\overset{\prime\prime}{.}0$
Integration time	9s	10s	10s	7.8s	7.8s	7.8s
Completeness limit for extended sources	$16\overset{\mathrm{m}}{.}5$	$14\overset{\mathrm{m}}{.}8$	$12\overset{\mathrm{m}}{.}0$	$15\overset{\mathrm{m}}{.}0$	$14\overset{\mathrm{m}}{.}2$	$13\overset{\mathrm{m}}{.}5$
Number counts for the completeness limits	50	28	3	48	~40	24
Extinction compared to the optical A_B	0.45	0.21	0.09	0.21	0.14	0.09

The above predictions do not take into account any dependence on morphological type, surface brightness, intrinsic color, orientation and crowding, which may lower the counts of actually detectable galaxies counts.

4.2 Pilot Studies with DENIS Data in the Great Attractor Region

To compare the above predictions with real data, Schröder et al. [92,93] and Kraan-Korteweg et al. [94] examined the efficiency of uncovering galaxies at high extinctions using DENIS images. The analyzed regions include the rich cluster A3627 $(\ell, b) = (325\overset{\circ}{.}3, -7\overset{\circ}{.}2)$ at the heart of the GA (Norma) supercluster as well as its suspected extension across the Galactic Plane.

Three high-quality DENIS strips cross the cluster A3627. The 66 images on these strips that lie within the Abell-radius were inspected by eye. This covers about one-eighth of the cluster area. The extinction over the regarded cluster area varies as $1\overset{\mathrm{m}}{.}2 \leq A_B \leq 2\overset{\mathrm{m}}{.}0$.

On these 66 images, 151 galaxies had previously been identified in the deep optical ZOA galaxy search [53]. Of these, 122 were recovered in the I_c, 100 in

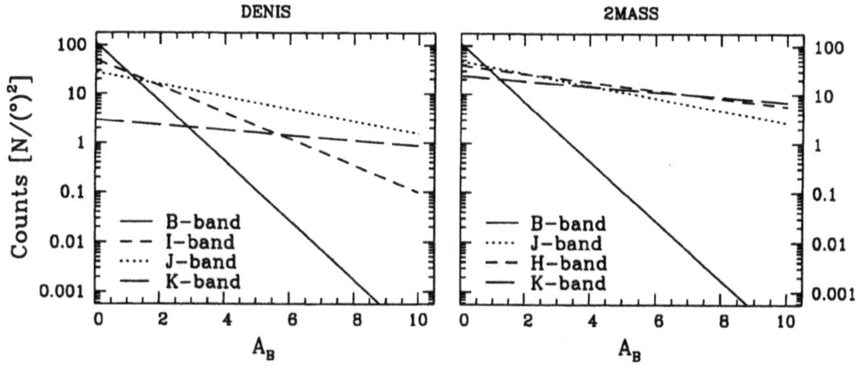

Fig. 10. Predicted I_c, J and K_s galaxy counts for DENIS (left panel), and J, H and K_s counts for 2MASS (right panel) for their respective galaxy completeness limits as a function of the absorption in the B band. For comparison both panels also show the B counts of an optical galaxy sample extracted from sky surveys

the J, and 74 in the K_s band. Most of the galaxies not re-discovered in K_s are low surface brightness spiral galaxies.

Surprisingly, the J band provided better galaxy detection than the I_c band. In the latter, the severe star crowding makes identification of faint galaxies very difficult. At these extinction levels, the optical survey does remain the most efficient in *identifying* obscured galaxies.

The search for more obscured galaxies was made in the region $320° \leq \ell \leq 325°$ and $|b| \leq 5°$, i.e. the suspected crossing of the GA. Of the 1800 images in that area, 385 of the then available DENIS images were inspected by eye (308 in K_s). 37 galaxies at higher latitudes were known from the optical survey. 28 of these could be re-identified in I_c, 26 in J, and 14 in the K_s band. In addition, 15 new galaxies were found in I_c and J, 11 of which also appear in the K_s band. The ratios of galaxies found in I_c compared to B, and of K_s compared to I_c are higher than in the A3627 cluster. This is due to the higher obscuration level (starting with $A_B \simeq 2\overset{m}{.}3 - 3\overset{m}{.}1$ at the high-latitude border).

On average, about 3.5 galaxies per square degree were found in the I_c band. This roughly agrees with the predictions of Fig. 10. Because of star crowding, one does not expect to find galaxies below latitudes of $b \simeq 1°$-$2°$ in this longitude range [95]. Low-latitude images substantiate this – the images are nearly fully covered with stars. Indeed, the lowest Galactic latitude galaxies were found at $b \simeq 1\overset{\circ}{.}2$ and $A_B \simeq 11^{\rm m}$ (in J and K_s only).

Figure 11 shows a few characteristic examples of highly obscured galaxies found in the DENIS blind search. I_c band images are at the top, J in the middle and K_s at the bottom. The first galaxy located at $(l, b) = (324\overset{\circ}{.}6, -4\overset{\circ}{.}5)$ is viewed through an extinction layer of $A_B = 2\overset{m}{.}0$ according to the DIRBE extinction maps [13]. It is barely visible in the J band. The next galaxy at $(l, b) = (324\overset{\circ}{.}7, -3\overset{\circ}{.}5)$ is subject to heavier extinction ($A_B = 2\overset{m}{.}7$), and indeed easier to recognize in the NIR. It is most distinct in the J band. The third galaxy at even higher extinction $(l, b, A_B) = (320\overset{\circ}{.}1, +2\overset{\circ}{.}5, 5\overset{m}{.}7)$ is – in agreement with the prediction of Fig. 10 – not visible in the B band. Neither is the fourth galaxy at $b = +1\overset{\circ}{.}9$ and $A_B = 9\overset{m}{.}6$: this galaxy can not be seen in I_c band either and is very faint only in J and K_s.

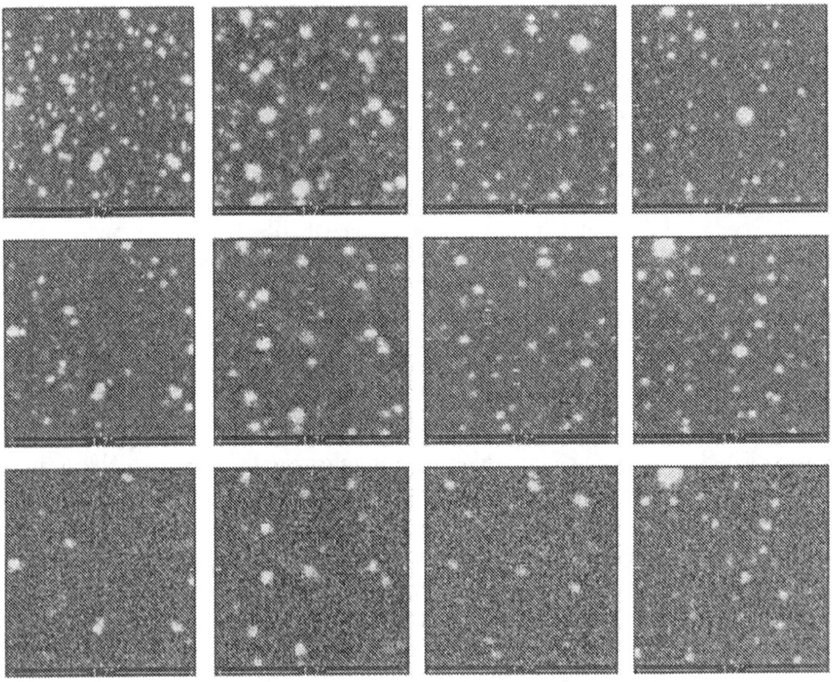

Fig. 11. DENIS survey images (before bad pixel filtering) of four galaxies found in the deepest extinction layer of the Milky Way; the I_c band image is at the top, J in the middle and K_s at the bottom

4.3 Conclusions

The conclusions from this pilot study are that at *intermediate latitudes and extinction* ($|b| \gtrsim 5°$, $A_B \lesssim 4$-5^m) optical surveys are superior for identifying galaxies. But despite the extinction and the star crowding at these latitudes, I_c, J and K_s photometry from the survey data could be performed successfully at these low latitudes. The NIR data (magnitudes, colors) of these galaxies can therefore add important data in the analysis of these obscured galaxies. They led, for instance, to the preliminary I_c^o, J^o and K_s^o galaxy luminosity functions in A3627 (Fig. 2 in [94]).

At *lowest latitudes and high extinction* ($|b| \lesssim 5°$ and $A_B \gtrsim 4$-5^m), the search for 'invisible' obscured galaxies on existing DENIS-images implicate that NIR-surveys can trace galaxies down to about $|b| \gtrsim 1°$-$1\overset{\circ}{.}5$. The J band was found to be optimal for identifying galaxies up to $A_B \simeq 7^{\mathrm{m}}$. NIR surveys can hence further reduce the width of the ZOA.

The NIR surveys are particularly useful for the mapping of massive early-type galaxies – tracers of density peaks in the mass distribution – as these can not be detected with any of the techniques that are efficient in tracing the spiral population in more opaque regions (Sect. 3 and 5).

Nevertheless, NIR surveys are also important with regard to the blue and low surface-brightness spiral galaxies because a significant fraction of them are also detectable in the near infrared. This is confirmed, for instance, with the serendipitous discovery in the ZOA of a large, nearby ($v = 750$ km s^{-1}) edge-on spiral galaxy by 2MASS [96]: with an extension in the K_s band of 5 arcmin, this large galaxy is – not unexpectedly for its extinction of $A_B = 6\overset{\mathrm{m}}{.}6$ at the position of $(\ell, b) = (236\overset{\circ}{.}8, -1\overset{\circ}{.}8)$ – not seen in the optical [46]. Furthermore, the overlap of galaxies found in NIR and H I surveys allows the determination of redshift independent distances via the NIR Tully–Fisher relation [97], and therewith the peculiar velocity field. This will provide important new input on the mass density field "in the ZOA" (Sect. 7).

5 Blind HI Surveys in the ZOA

In the regions of the highest obscuration and infrared confusion, the Galaxy is fully transparent to the 21cm line radiation of neutral hydrogen. H I-rich galaxies can readily be found at lowest latitudes through the detection of their redshifted 21cm emission, though early-type galaxies – tracers of massive groups and clusters – are gas-poor and will not be identified in these surveys. Also very low-velocity extragalactic sources might be missed due to the strong Galactic H I emission, and galaxies close to radio continuum sources.

An advantage of blind H I surveys is the immediate availability of rotational properties of a detected galaxy, next to its redshift, providing insight on the intrinsic properties of these obscured galaxies. The rotational velocity can furthermore be used (in combination with e.g. NIR photometry) to determine the distance in real space from the Tully–Fisher relation, leading to determinations of the mass density field from the peculiar velocities.

Until recently, radio receivers were not sensitive and efficient enough to attempt systematic surveys of the ZOA. Kerr & Henning [98] demonstrated, however, the effectiveness of this approach: they pointed the late 300-ft telescope of Green Bank to 1900 locations in the ZOA (1.5% coverage) and detected 19 previously unknown spiral galaxies.

Since then two systematic blind H I searches for galaxies behind the Milky Way were initiated. The first – the Dwingeloo Obscured Galaxies Survey (DOGS) – used the 25 m Dwingeloo radio to survey the whole northern Galactic Plane for galaxies out to 4000 km s^{-1} [99–101]. A more sensitive survey, probing a considerably larger volume (out to 12700 km s^{-1}), is being performed for the southern Milky Way at the 64 m radiotelescope of Parkes [102–105].

In the following, the observing techniques of these two surveys as well as the first results will be discussed.

5.1 The Dwingeloo Obscured Galaxies Survey

Since 1994, the Dwingeloo 25 m radio telescope has been dedicated to a systematic search for galaxies in the northern Zone of Avoidance ($30° \leq \ell \leq 220°$, $|b| \leq 5\overset{\circ}{.}25$). The last few patches of the survey were completed early 1999, using the Westerbork array in total power mode. The 20 MHz bandwidth was tuned to cover the velocity range $0 \leq v \leq 4000$ km s^{-1}.

The 25 m Dwingeloo telescope has a half-power-beamwidth (HPBW) of 36 arcmin. The 15000 survey points required for the survey coverage are ordered in a honeycomb pattern with a grid spacing of $0\overset{\circ}{.}4$. Galaxies are generally detected in various adjacent pointings, facilitating a more accurate determination of their positions through interpolations. The rms noise per channel typically was $\sigma_{ch} = 40$ mJy for a 1 hr integration (12 x 5min).

Because of the duration of the project (15000 hours not including overhead and downtime) the strategy was to first conduct a fast search of 5min integrations (rms = 175 mJy) to uncover possible massive nearby galaxies whose effect might yield important clues to the dynamics of the Local Group.

The shallow Dwingeloo search (rms = 175 mJy) has been completed in 1996 yielding five objects (cf. [100] for details), three of which were known previously. The most exciting discovery was the barred spiral galaxy Dwingeloo 1 [99].

This galaxy candidate was detected early on in the survey through a strong signal (peak intensity of 1.4 Jy) at the very low redshift of $v = 110$ km s^{-1} in the spectra of four neighboring pointings, suggestive of a galaxy of large angular extent. The optimized position of $(\ell, b) = (138\overset{\circ}{.}5, -0\overset{\circ}{.}1)$ coincided with a very low surface brightness feature on the Palomar Sky Survey plate of $2\overset{\prime}{.}2$, detected earlier by Hau et al. [57] in his optical galaxy search of the northern Galactic/SuperGalactic Plane crossing (cf. Sect. 2.2). Despite foreground obscuration of about 6^{m} in the optical, follow-up observations in the V, R and I band at the INT (La Palma) confirmed this galaxy candidate as a barred, possibly grand-design spiral galaxy of type SBb of 4.2 x 4.2 arcmin (cf. Fig. 12).

Dwingeloo 1 has been the subject of much follow-up observations (optical: Loan et al. [106], Buta & McCall [107]; HI-synthesis: Burtonet al. [108];CO observations: Kuno et al. [109], Li et al. [110], Tilanus & Burton [111]; X-ray: Reynolds et al. [112]). To summarize, it is a massive barred spiral, with rotation velocity of 130 km s^{-1}, implying a dynamical mass of roughly one-third the mass of the Milky Way. Its approximate distance of ~ 3 Mpc and angular location place it within the IC342/Maffei group of galaxies. The follow-up HI synthesis observations [108] furthermore revealed a counterrotating dwarf companion, Dwingeloo 2. Since then various further dwarf galaxies in this nearby galaxy group have been discovered.

60% of the deeper Dwingeloo survey (rms = 40 mJy) has been analyzed [101]. 36 galaxies were detected, 23 of which were previously unknown. Five of the 36 sources were originally identified by the shallow survey. Based on the survey sensitivity, the registered number of galaxies is in agreement with the Zwaan et al. [113] HI mass function which predicts 50 to 100 detections for the full survey.

Fig. 12. Composite V, R, I-image of the Dwingeloo 1 galaxy at $\ell = 138^\circ\!.5, b = -0^\circ\!.1$. The displayed 484 x 484 pixels of $0''\!.6$ cover an area of $4'\!.8$ x $4'\!.8$. The large diameter visible on this image is about $4'\!.2$. Dwingeloo 1 has a distinct bar, with 2 spiral arms that can be traced over nearly 180°. The morphology in this figure agrees with that of an SBb galaxy

Surprisingly, three dwarf galaxies were detected close to the nearby isolated galaxy NGC 6946 at $(\ell, b, v) = (95^\circ\!.7, 11^\circ\!.7, 46\ \mathrm{km\,s^{-1}})$. One of these had earlier been catalogued as a compact High Velocity Cloud [114]. Burton et al. [115], in their search for compact isolated high-velocity clouds in the Dwingeloo/Leiden Galactic H I survey [116,117], discovered a further member of this galaxy concentration. Now, seven galaxies with recessional velocities $v_{\mathrm{LSR}} \leq 250\ \mathrm{km\,s^{-1}}$ have been identified within 15° of the galaxy NGC 6946. More might be discovered as the DOGS data in this region have not yet been fully analyzed. The agglomeration of these various galaxies might indicate a new group or cloud of galaxies in the nearby Universe. As such it would be the only galaxy group in the nearby Universe that is strongly offset (by 40°) from the Supergalactic Plane [118,119].

5.2 The Parkes Multibeam ZOA Blind HI Survey

In March 1997, the systematic blind H I survey in the southern Milky Way ($212° \leq \ell \leq 36°$; $|b| \leq 5\overset{\circ}{.}5$) began with the Multibeam receiver at the 64 m Parkes telescope. The instrument has 13 beams each with a beamwidth of $14\overset{\prime}{.}4$. The beams are arranged in a hexagonal grid in the focal plane array [120], allowing rapid sampling of large areas.

The observations are being performed in driftscan mode. 23 contiguous fields of length $\Delta\ell = 8°$ have been defined. Each field is being surveyed along constant Galactic latitudes with latitude offsets 35 arcmin until the final width of $|b| \leq 5\overset{\circ}{.}5$ has been attained (17 passages back and forth). The ultimate goal is 25 repetitions per field. With an effective integration time of 25 min/beam a $3\,\sigma$ detection limit of 25 mJy is obtained. The survey covers the velocity range $-1200 \lesssim v \lesssim 12700$ km s^{-1} and will be sensitive to normal spiral galaxies well beyond the Great Attractor region.

So far, a shallow survey covering the whole southern Milky Way based on 2 out of the foreseen 25 driftscan passages has been analyzed (cf. [102,104,105]). A detailed study of the Great Attractor region ($308° \leq \ell \leq 332°$) based on 4 scans has been made by Juraszek et al. [121,122]. The first four full-sensitivity cubes are available for that region as well (Sect. 5.3).

In the shallow survey, 110 galaxies were catalogued with peak H I-flux densities of $\gtrsim 80$ mJy (rms = 15 mJy after Hanning smoothing). The detections show no dependence on Galactic latitude, nor the amount of foreground obscuration through which they have been detected. Though galaxies up to 6500 km s^{-1} were identified, most of the detected galaxies (80%) are quite local ($v < 3500$ km s^{-1}) due to the (yet) low sensitivity. About one third of the detected galaxies have a counterpart either in NED (NASA/IPAC Extragalactic Database) or in the deep optical surveys.

The distribution of the 110 H I-detected galaxies is displayed in the lower panel of Fig. 13. It demonstrates convincingly that galaxies can be traced through the thickest extinction layers of the Galactic Plane. The fact that hardly any galaxies are found behind the Galactic bulge ($\ell = 350°$ to $\ell = 30°$) is due to local structure: this is the region of the Local Void.

For comparative purposes, the top panel of Fig. 13 shows the distribution of all known galaxies with $v \leq 10000$ km s^{-1} (extracted from the Lyon-Meudon Extragalactic Database (LEDA). Although this constitutes an uncontrolled sample, it traces the main structures in the nearby Universe in a representative way. Note the increasing incompleteness for extinction levels of $A_B \gtrsim 1\overset{m}{.}0$ (outer contour) – reflecting the growing incompleteness of optical galaxy catalogs – and the near full lack of galaxy data for extinction levels $A_B \gtrsim 3\overset{m}{.}0$ (inner contour). The middle panel shows galaxies with $v <$10000 km s^{-1} from the follow-up observations of the deep optical galaxy search by Kraan-Korteweg and collaborators

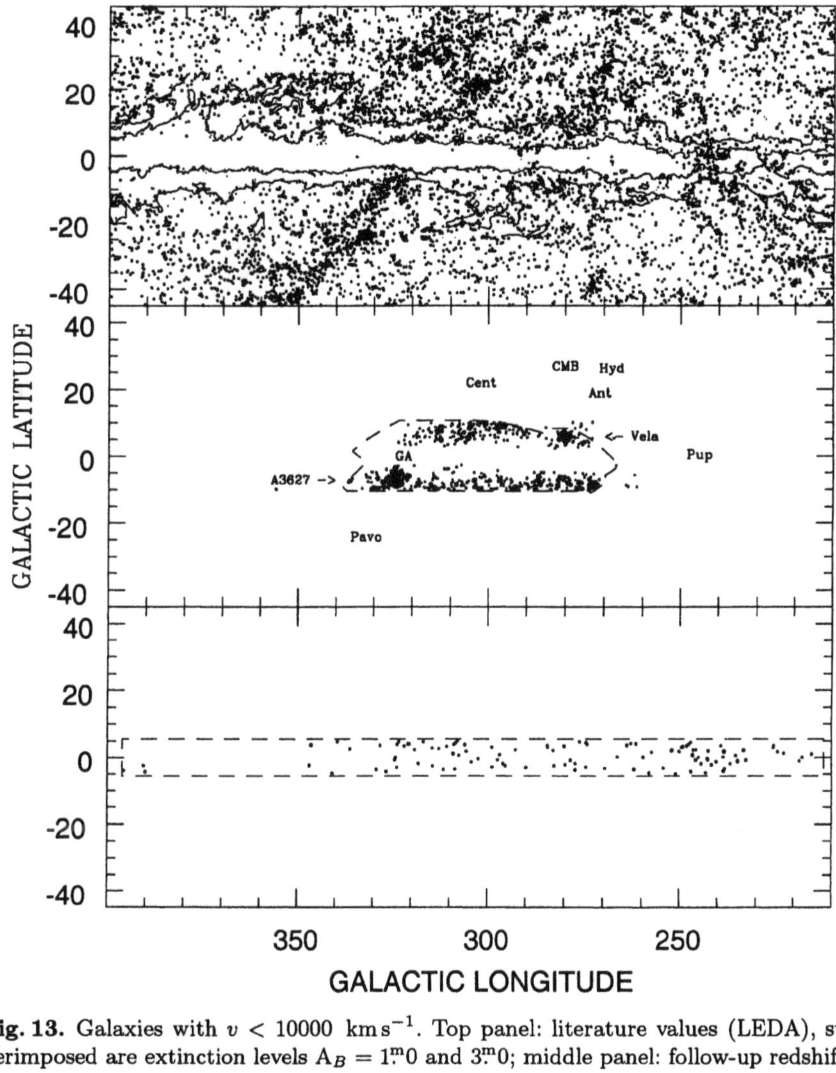

Fig. 13. Galaxies with $v < 10000$ km s^{-1}. Top panel: literature values (LEDA), superimposed are extinction levels $A_B = 1\overset{m}{.}0$ and $3\overset{m}{.}0$; middle panel: follow-up redshifts (ESO, SAAO and Parkes) from deep optical ZOA survey with locations of clusters and dynamically important structures; bottom panel: galaxies detected with the shallow Multibeam ZOA survey

(Sect. 2.4). Various new overdensities are apparent at low latitudes but the innermost part of our Galaxy remains obscured with this approach. Here, the blind H I data (lower panel) finally can provide the missing link for large-scale structure studies.

In Fig. 14, the data of Fig. 13 are combined in redshift slices. The achieved sensitivity of the shallow MB H I-survey fills in structures all the way across the

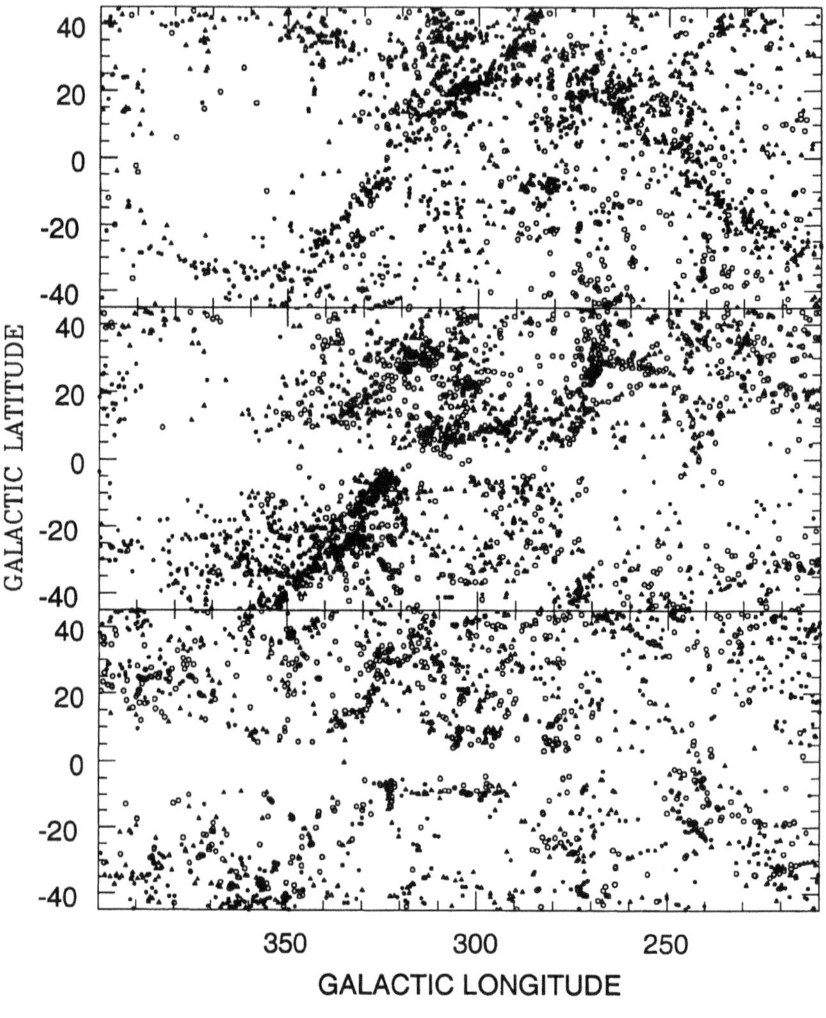

Fig. 14. Redshift slices from the data in Fig. 13: $500 < v < 3500$ (top), $3500 < v < 6500$ (middle), $6500 < v < 9500$ $\mathrm{km\,s^{-1}}$ (bottom). The open circles mark the nearest $\Delta v = 1000$ $\mathrm{km\,s^{-1}}$ slice in a panel, then triangles, then the filled dots the 2 more distant ones

ZOA for $v < 3500$ $\mathrm{km\,s^{-1}}$(upper panel) for the first time. Note the continuity of the thin filamentary sine-wave-like structure that dominates the whole southern sky and crosses the Galactic Equator twice. This structure snakes over $\sim 180°$ through the southern sky. Taking a mean distance of $30h^{-1}$ Mpc, this implies a linear size of $\sim 100h^{-1}$ Mpc, with a thickness of 'only' $\sim 5h^{-1}$ Mpc or less. Various other filaments spring forth from this dominant filament, always from a rich group or small cluster at the junction of these interleaving structures. This feature is very different from the thick, foamy Great Wall-like structure, the GA, in the middle panel.

Also note the prominence of the Local Void which is very well delineated in this presentation. No galaxies were found within the Local Void, but the three newly identified galaxies at $\ell \sim 30°$ help to define the boundary of the Void.

The full sensitivity ZOA MB-survey will fill in the large-scale structures in the more distant panels of Fig. 14. First results of the full sensitivity survey have been obtained in the Great Attractor region (Sect. 5.3).

Three nearby, very extended ($20'$ to $\gtrsim 1°$) galaxies were discovered with the shallow survey. Being likely candidates of dynamically important galaxies, immediate follow-up observations were initiated at the Australian Telscope Compact Array (ATCA). These objects did not turn out to be massive perturbing monsters, however. Two were seen to break up into H I complexes and both have unprecedented low H I column densities [103]. Systematic synthesis observations are being performed to investigate the frequency of these interacting and/or low H I column density systems in this purely H I-selected sample.

5.3 The Parkes ZOA MB Deep Survey and the Great Attractor

Four cubes centered on the Great Attractor region ($300° \geq \ell \geq 332°$, $|b| \leq 5\overset{\circ}{.}5$) of the full-sensitivity survey have been analyzed [122]. 236 galaxies above the 3σ detection level of 25 mJy have been uncovered. 70% of the detections had no previous identification.

In the left panel of Fig. 15, a sky distribution centered on the GA region displays all galaxies with redshifts $v \leq 10000$ km s^{-1}. Next to redshifts from the literature, redshifts from the follow-up observations of Kraan-Korteweg and collaborators in the Hy/Ant-Crux-GA ZOA surveys (dashed area) are plotted. They clearly reveal the prominence of the cluster A3627 at $(\ell, b, v) = (325°, -7°, 4882$ km s$^{-1})$ close to the core of the GA region at $(\ell, b, v) = (320°, 0°, 4500$ km s$^{-1})$. Adding now the new detections from the systematic blind H I MB-ZOA survey (box), structures can be traced all the way across the Milky Way. The new picture seems to support that the GA overdensity is a "great-wall" like structure starting close to the Pavo cluster, having its core at the A3627 cluster and then bending over towards shorter longitudes across the ZOA.

This becomes even clearer in the right panel of Fig. 15 (compare with right hand panel of Fig. 5) where the galaxies are displayed in a redshift cone out to $v \leq 10000$ km s^{-1} for the longitude range $300° \leq \ell \leq 332°$. The combined surveys in the GA region clearly substantiate that A3627 is the most massive galaxy cluster uncovered in this region and therefore the most likely candidate for the predicted density-peak at the bottom of the potential well of the GA overdensity. The new data do not unambigously confirm the existence of the suspected further cluster around the bright elliptical radio galaxy PKS1343−601 (Sect. 2.4). Although the MB data reveal an excess of galaxies at this position in velocity space ($b = +2°, v = 4000$ km s^{-1}) a "finger of God" is not seen. It could be that many central cluster galaxies are missed by the H I observations because spiral galaxies generally avoid the cores of clusters. The reality of this possible

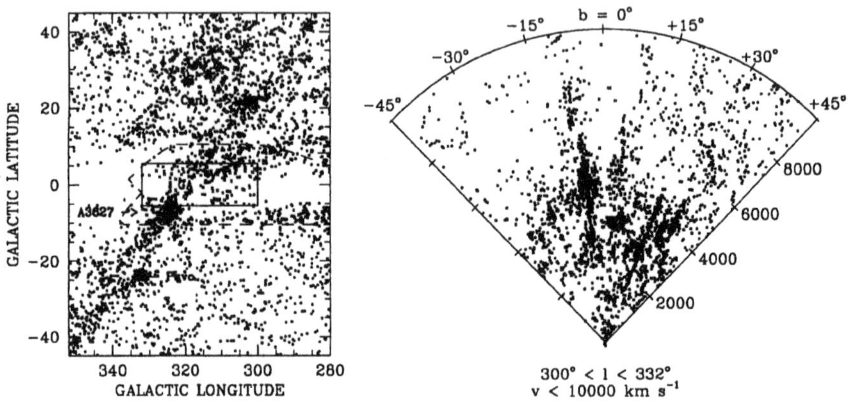

Fig. 15. A sky distribution (left) and redshift cone (right) for galaxies with $v < 10000$ km s^{-1} in the GA region. Circles mark redshifts from the literature (LEDA), squares redshifts from the optical galaxy search in the Hy/Ant-Crux-GA regions (outlined on left panel) and crosses detections in the full-sensitivity HI MB-ZOA survey (box)

cluster still remains a mystery. This prospective cluster has meanwhile been imaged in the *I*-band [123], where extinction effects are less severe compared to the optical (see Sect. 4). A first glimpse of the images do reveal various early-type galaxies. The forthcoming analysis should then unambiguously settle the question whether another cluster forms part of the GA overdensity.

5.4 Conclusions

The systematic probing of the galaxy distribution in the most opaque parts of the ZOA with H I surveys have proven very powerful. For the first time large-scale structure could be mapped without hindrance across the Milky Way (Figs. 14 and 15). This is the only approach that easily uncovers the galaxy distribution in the ZOA, allows the confirmation of implied connections and uncovers new connections behind the Milky Way.

¿From the analysis of the Dwingeloo survey and the shallow Parkes MB ZOA survey, it can be maintained that no Andromeda or other H I-rich Circinus-like galaxy is lurking undetected behind the deepest extinction layers of the Milky Way (although gas-poor, early-type galaxies might, of course, still remain hidden). The census of dynamically important, H I-rich nearby galaxies whose gravitational influence could significantly impact peculiar motion of the Local Group or its internal dynamics is now complete – at least for objects whose signal is not drowned within the strong Galactic H I emission.

6 X-ray Surveys

The X-ray band potentially is an excellent window for studies of large-scale structure in the Zone of Avoidance, because the Milky Way is transparent to the hard X-ray emission above a few keV, and because rich clusters are strong X-ray emitters. Since the X-ray luminosity is roughly proportional to the cluster mass as $L_X \propto M^{3/2}$ or M^2, depending on the still uncertain scaling law between the X-ray luminosity and temperature, massive clusters hidden by the Milky Way should be easily detectable through their X-ray emission.

This method is particularly attractive, because clusters are primarily composed of early-type galaxies which are not recovered by IRAS galaxy surveys (Sect. 3) or by systematic H I surveys (Sect. 5). Even in the NIR, the identification of early-type galaxies becomes difficult or impossible at the lowest Galactic latitudes because of the increasing extinction and crowding problems (Sect. 4). Rich clusters, however, play an important role in tracing large-scale structures because they generally are located at the center of superclusters and Great Wall-like structures. They mark the density peaks in the galaxy distribution and – with the very high mass-to-light ratios of clusters – the deepest potential wells within these structures. Their location within these overdensities will help us understand the observed velocity flow fields induced by these overdensities.

The X-ray all-sky surveys carried out by Uhuru, Ariel V, HEAO-1 (in the 2-10 keV band) and ROSAT (0.1-2.4 keV) provide an optimal tool to search for clusters of galaxies at low Galactic latitude. However, confusion with Galactic sources such as X-ray binaries and Cataclysmic Variables may cause serious problems, especially in the earlier surveys Uhuru, Ariel V and HEAO-1 which had quite low angular resolution. And although dust extinction and stellar confusion are unimportant in the X-ray band, photoelectric absorption by the Galactic hydrogen atoms – the X-ray absorbing equivalent hydrogen column density – does limit detections close to the Galactic Plane. The latter effect is particularly severe for the softest X-ray emission, as e.g. observed by ROSAT (0.1-2.4 keV) compared to the earlier 2-10 keV missions. On the other hand, the better resolution of the ROSAT All Sky Survey (RASS), compared to the HEAO-1 survey, will reduce confusion problems with Galactic sources as happened, for example, in the case of the cluster A3627 (see below).

Until recently, the possibility of searching for galaxy clusters behind the Milky Way through their X-ray emission has not been pursued in a systematic way, even though a large number of X-ray bright clusters are located at low Galactic latitudes [124]: for instance, four of the seven most X-ray luminous clusters in the 2-10 keV range, the Perseus, Ophiuchus, Triangulum Australis, and PKS0745−191 clusters ($L_X > 10^{45}$ erg s^{-1}) lie at latitudes below $|b| < 20°$ [125].

A first attempt to identify galaxy clusters in the ZOA through their X-ray emission had been made by Jahoda and Mushotzky in 1989 [126]. They used the HEAO-1 all-sky data to search for X-ray-emission of a concentration of clusters or one enormous cluster that might help explain the shortly before discovered large-scale deviations from the Hubble flow that were associated with the Great

Attractor. Unfortunately, this search missed the 6th brightest cluster A3627 in the ROSAT X-ray All Sky Survey [73,127] which had been identified as the most likely candidate for the predicted but unidentified core of the Great Attractor. A3627 was not seen in the HEAO-1 data because of the low angular resolution and the confusion with the neighbouring X-ray bright, Galactic X-ray binary 1H1556-605 (cf. Fig. 8 and 9 in [73]).

6.1 CIZA: Clusters in the Zone of Avoidance

Since 1997, a group led by Ebeling [128,129] have systematically searched for bright X-ray clusters of galaxies at $|b| < 20°$. Starting from the ROSAT Bright Source Catalog (BSC, [130]) which lists the 18811 X-ray brightest sources detected in the RASS, they apply the following criteria to search for clusters: (a) $|b| < 20°$, (b) a X-ray flux above $S > 5 \times 10^{-12}$ erg cm^{-2} s^{-1} (the flux limit of completeness of the ROSAT BCS), and (c) a spectral hardness ratio. Ebeling et al. demonstrated in 1998 that the X-ray hardness ratio is very effective in discriminating against softer, non-cluster X-ray sources. With these criteria, they select a candidate cluster sample which, although at this point still highly contaminated by non-cluster sources, contains the final CIZA cluster sample.

They first cross-identified their 520 cluster candidates against NED and SIMBAD, and checked unknown ones on the Digitized Sky Survey. The new cluster candidates, including known Abell clusters without photometric and spectroscopic data, were imaged in the R band, respectively in the K' band at high extinctions. With the subsequent spectroscopy of galaxies around the X-ray position, the real clusters could be confirmed.

Time and funding permitting, the CIZA team plans to extend their cluster survey to lower X-ray fluxes (2-3 $\times 10^{-12}$ erg cm^{-2} s^{-1}), the aim being a total sample of 200 X-ray selected clusters below $|b| < 20°$.

So far, 76 galaxy clusters were identified within $|b| < 20°$ of which 80% were not known before. Their distribution (reproduced from Ebeling et al. [129]) is displayed in Fig. 16. 14 of these clusters are relatively nearby ($z \leq 0.04$), and one was uncovered at a latitude of only $b = 0\overset{\circ}{.}3$ within the Perseus-Pisces chain.

6.2 Conclusions

With the discovery of so far 76 clusters of which only 20% were known before, Ebeling et al. [129] have proven the strength of the method to use X-ray criteria to search for galaxy clusters in the ZOA. As mentioned in the introduction to this section, this approach is complementary to the other wavelengths searches which all fail to uncover galaxy clusters at very low Galactic latitudes.

Having used the ROSAT BSC to select their galaxy cluster candidates, the CIZA collaboration can combine their final cluster sample with other X-ray selected cluster samples from the RASS, such as the ROSAT Brightest Cluster Sample at $|b| \geq 20°$ and $\delta \geq 0°$ [131] and the REFLEX sample at $|b| \geq 20°$ and $\delta \leq 2.5°$ (Böhringer et al. in prep.). The resulting, all-sky cluster list will be ideally suited to study large-scale structure and the connectivity of superclusters across the Galactic Plane.

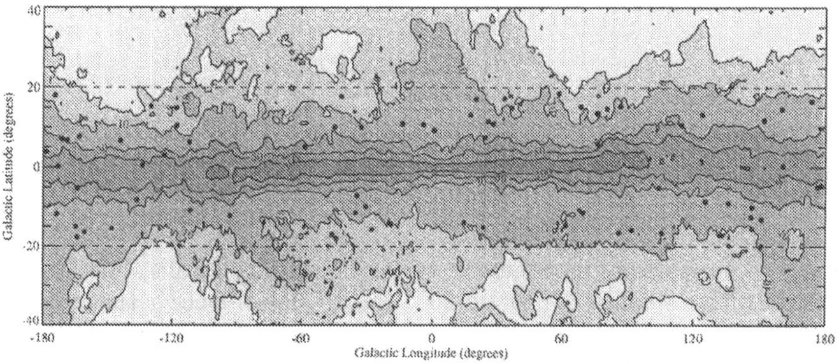

Fig. 16. Distribution in Galactic coordinates of the 76 by Ebeling et al. [129] so far spectroscopically confirmed X-ray clusters (solid dots) of which 80% were previously unknown. Superimposed are Galactic HI column densities in units of 10^{20} cm^{-2} (Dickey & Lockman 1990). Note that the region of relatively high absorption ($N_{\rm HI} > 5 \times 10^{21}$ cm^{-2}) actually is very narrow and that clusters could be identified to very low latitudes

7 Theoretical Reconstructions

Various mathematical methods exist to reconstruct the galaxy distribution in the ZOA without having access to direct observations.

One possibility is the expansion of galaxy distributions adjacent to the ZOA into spherical harmonics to recover the structures in the ZOA, either with 2-dimensional catalogs (sky positions) or 3-dimensional data sets (redshift catalogs).

A statistical method to reconstruct structures behind the Milky Way is the Wiener Filter (WF), developed explicitly for reconstructions of corrupt or incomplete data [132,133]. Using the WF in combination with linear theory allows the determination of the real-space density of galaxies, as well as their velocity and potential fields.

The POTENT analysis developed by [134] can reconstruct the potential field (mass distribution) from peculiar velocity fields in the ZOA [19]. The reconstruction of the potential fields versus density fields have the advantage that they can locate hidden overdensities (their signature) even if "unseen".

Because of the sparsity of data and the heavy smoothing applied in all these methods, only structures on large scales (superclusters) can be mapped. Individual (massive) nearby galaxies that can perturb the dynamics of the Universe quite locally (the vicinity of the Local Group or its barycenter) will not be uncovered in this manner. But even if theoretical methods can outline LSS accurately, the observational efforts do not become superfluous. The comparison of the real galaxy distribution δ_g ($\mathbf{r}$), from e.g. complete redshift surveys, with the peculiar velocity field $\mathbf{v}(\mathbf{r})$ will lead to an estimate of the density and biasing parameter

($\Omega^{0.6}/b$) through the equation

$$\nabla \cdot \mathbf{v}(\mathbf{r}) = -\frac{\Omega^{0.6}}{b}\, \delta_g(\mathbf{r}), \tag{1}$$

cf. Strauss & Willick [135] for a detailed review.

7.1 Early Predictions

Early reconstructions on relatively sparse data galaxy catalogs have been performed within volumes out to $v \leq 5000\ \mathrm{km\,s^{-1}}$. Despite heavy smoothing, they have been quite successful in pinpointing a number of important features:

• Scharf et al. [136] applied spherical harmonics to the 2-dimensional IRAS PSC and noted a prominent cluster behind the ZOA in Puppis ($\ell \sim 245°$) which was simultaneously discovered as a nearby cluster through H I-observations of obscured galaxies in that region by Kraan-Korteweg & Huchtmeier [27].

• Hoffman [133] predicted the Vela supercluster at $(280°, 6°, 6000\ \mathrm{km\,s^{-1}})$ using 3-dimensional WF reconstructions on the IRAS 1.9 Jy redshift catalog [80], which was observationally discovered just a bit earlier by Kraan-Korteweg & Woudt [137].

• Using POTENT analysis, Kolatt et al. [19] predicted the center of the Great Attractor overdensity – its density peak – to lie behind the ZOA at $(320°, 0°, 4500\ \mathrm{km\,s^{-1}}$, see Fig. 17). Shortly thereafter, Kraan-Korteweg et al. [84] unveiled the cluster A3627 as being very rich and massive and at the correct distance. It hence is the most likely candidate for the central density peak of the GA.

7.2 Deeper Reconstructions

Recent reconstructions have been applied to denser galaxy samples covering larger volumes ($v \lesssim 10000\ \mathrm{km\,s^{-1}}$) with smoothing scales of the order of 500 $\mathrm{km\,s^{-1}}$ (compared to 1200 $\mathrm{km\,s^{-1}}$in the earlier reconstructions). It therefore seemed of interest to see whether these reconstructions find evidence for unknown major galaxy structures at higher redshifts.

The currently most densely-sampled, well-defined galaxy redshift catalog is the Optical Redshift Survey [138]. However, this catalog is limited to $|b| \geq 20°$ and the reconstructions [139] within the ZOA are strongly influenced by 1.2 Jy IRAS Redshift Survey data and a mock galaxy distribution in the inner ZOA. I therefore concentrate on reconstructions based on the 1.2 Jy IRAS Redshift Survey only. In the following, the structures identified in the ZOA by (a) Webster et al. [140] using WF plus spherical harmonics and linear theory and (b) Bistolas [141] who applied a WF plus linear theory and non-constrained realizations on the 1.2 Jy IRAS Redshift Survey are discussed and compared to observational data. Fig. 2 in Webster et al. displays the reconstructed density fields on shells of 2000, 4000, 6000 and 8000 $\mathrm{km\,s^{-1}}$; Fig. 5.2 in Bistolas displays the density fields in the ZOA from 1500 to 8000 $\mathrm{km\,s^{-1}}$ in steps of 500 $\mathrm{km\,s^{-1}}$.

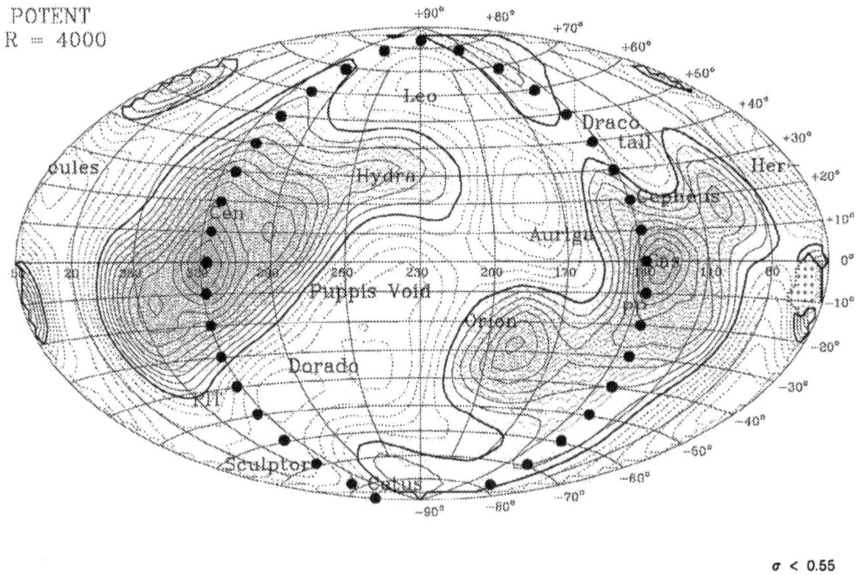

Fig. 17. The mass-density fluctuation field in a shell at 4000 $\mathrm{km\,s^{-1}}$ as determined with POTENT from peculiar velocity data. The density is smoothed by a three-dimensional Gaussian of radius 1200 $\mathrm{km\,s^{-1}}$. Density contour spacings are $\Delta\delta = 0.1$ with $\delta = 0$ as a heavy contour. Compared to Fig. 1 and 8 this Aitoff projection is displaced by $\Delta\ell = 50°$. The Supergalactic Plane is indicated (solid dots). (Figure 1b from [19])

The WLF reconstructions clearly find the recently by Roman et al. [47] identified nearby cluster at (33°, 5°-15°, 1500 $\mathrm{km\,s^{-1}}$), whereas Bistolas reveals no clustering in the region of the Local Void out to 4000 $\mathrm{km\,s^{-1}}$. At the same longitudes, clustering is indicated at 7500 $\mathrm{km\,s^{-1}}$ by Bistolas, but not by Webster et al. The Perseus-Pisces chain is strong in both reconstructions, and the 2nd Perseus-Pisces arm – which folds back at $\ell \sim 195°$ – is clearly confirmed. Both reconstructions find the Perseus-Pisces complex to be very extended in space, i.e. from 3500 $\mathrm{km\,s^{-1}}$ out to 9000 $\mathrm{km\,s^{-1}}$. Whereas the GA region is more prominent compared to Perseus-Pisces in the Webster et al. reconstructions, the signal of the Perseus-Pisces complex is considerably stronger than the GA in Bistolas, where it does not even reveal a well-defined central density peak. Both reconstructions find no evidence for the suspected PKS1343 cluster but its signal could be hidden in the central (A3627) density peak due to the smoothing. While the Cygnus-Lyra complex $(60°\text{-}90°, 0°, 4000\ \mathrm{km\,s^{-1}})$ discovered by Takata et al. [79] stands out clearly in Bistolas, it is not evident in Webster et al. Both reconstructions find a strong signal for the Vela supercluster $(285°, 6°, 6000\ \mathrm{km\,s^{-1}})$ identified by Kraan-Korteweg & Woudt [137] and Hoffman [133]. The Cen-Crux cluster identified by Woudt [51]is evident in Bistolas though less distinct in Webster et al. A suspected connection at $(\ell, v) \sim (345°, 6000\ \mathrm{km\,s^{-1}})$ – cf. Fig. 2 in

[102] – is supported by both methods. The Ophiuchus cluster [56] just becomes visible in the most distant reconstruction shells (8000 $\mathrm{km\,s^{-1}}$).

7.3 Conclusions

Not all reconstructions find the same features, and when they do, the prominence of the density peaks as well as their locations in space do vary considerably. At velocities of $\sim 4000\ \mathrm{km\,s^{-1}}$ most of the dominant structures lie close to the ZOA while at larger distances, clusters and voids seem to be more homogeneously distributed over the whole sky. Out to 8000 $\mathrm{km\,s^{-1}}$, none of the reconstructions predict any major structures which are not mapped or suggested from observational data. So, no major surprises seem to remain hidden in the ZOA. The various multi-wavelength explorations of the Milky Way will soon be able to verify this. Still, the combination of both the reconstructed potential fields and the observationally mapped galaxy distribution will lead to estimates of the cosmological parameters Ω_0 and b.

8 Conclusions

In the last decade, enormous progress has been made in unveiling the extragalactic sky behind the Milky Way. At optical wavebands, the entire ZOA has been systematically surveyed. It has been shown that these surveys are complete for galaxies larger than $D^o = 1\farcm3$ (corrected for absorption) down to extinction levels of $A_B = 3\fm0$. Combining these data with previous "whole-sky" maps results in a reduction of the "optical ZOA" of a factor of about 2-2.5 which allow an improved understanding of the velocity flow fields and the total gravitational attraction on the Local Group. Various previously unknown structures in the nearby Universe could be mapped in this way.

At higher extinction levels, other windows to the ZOA become more efficient in tracing the large-scale structures. Very promising in this respect are the current near-infrared surveys which find galaxies down to latitudes of $|b| \sim 1\fdg5$ and systematic H I surveys which detect gas-rich spiral galaxies all the way across the Galactic Plane – hampered slightly only at very low latitudes ($|b| \lesssim 1\fdg0$) because of the numerous continuum sources. The "Behind the Plane" Survey resulted in a reduction from 16% to 7% of the "FIR ZOA" and new indications of possible hidden massive clusters behind the Milky Way are now forthcoming from the CIZA project – although again an "X-ray ZOA" will remain due to the absorption of X-ray radiation by the thickening gas layer close to the Galactic Plane.

A difficult task is still awaiting us, i.e. to obtain a detailed understanding of the selection effects inherent to the various methods in order to merge the different data sets in a uniform, well-defined way. This is extremely important if we want to use this data for quantitative cosmography. Moreover, we need a better understanding of the obscurational effects on the observed properties of galaxies identified through the dust layer (at all wavelengths), in addition to an accurate high-resolution, well-calibrated map of the Galactic extinction.

Despite the fact that our knowledge of the above questions is as yet limited, a lot can and has been learned from ZOA research. This is evident, for instance, from the detailed and varied investigations of the Great Attractor region. Mapping the GA and understanding the from peculiar velocity fields inferred massive overdensity had remained an enigma due the fact that the major and central part of this extended density enhancement is largely hidden by the obscuring veil of the Milky Way. Does light trace mass in this region and where is the rich cluster which biasing predicts at the center of large-scale potential wells?

The results from the various ZOA surveys now clearly imply that the Great Attractor is, in fact, a nearby "great-wall" like supercluster, starting at the nearby Pavo cluster below the GP, moving across the massive galaxy cluster A3627 toward the shallow overdensity in Vela at 6000 $\mathrm{km\,s^{-1}}$. The cluster A3627 is the dominant central component of this structure, similar to the Coma cluster in the (northern) Great Wall. Whether a second massive cluster around PKS1343−601 is part of the core of the GA remains uncertain.

Acknowledgements

The enthusiastic collaborations of my colleagues in the exploration of the galaxy distribution behind the Milky Way is greatly appreciated. These are P.A. Woudt, C. Salem and A.P. Fairall with deep optical searches, C. Balkowski, V. Cayatte, A.P. Fairall, P.A. Henning with the redshift follow-ups of the optically identified galaxies, A. Schröder and G.A. Mamon in the exploration of the DENIS images at low Galactic latitude, W.B. Burton, P.A. Henning, O. Lahav and A. Rivers in the northern ZOA HI-survey (DOGS) and the HIPASS ZOA team members L. Staveley-Smith , R.D. Ekers, A.J. Green, R.F. Haynes, P.A. Henning, S. Juraszek, M. J. Kesteven, B. Koribalski, R.M. Price, E. Sadler and A. Schröder in the southern ZOA survey.

Particular thanks go to P.A. Woudt for his careful reading of the manuscript and his valuable suggestions, to W. Saunders for preparing Fig. 9, to A. Schröder and G. Mamon for their comments on the NIR section, and to H. Ebeling for his input with regard to the X-ray section and Fig. 16.

This research has made use of the NASA/IPAC Extragalactic Database (NED) which is operated by the Jet Propulsion Laboratory, Caltech, under contract with the National Aeronautics and Space Administration, as well as the Lyon-Meudon Extragalactic Database (LEDA), supplied by the LEDA team at the Centre de Recherche Astronomique de Lyon, Observatoire de Lyon.

References

1. Proctor, R.: *The Universe of Stars*, (Longmans, Green and Co. London 1878) pp. 41
2. Herschel, J.: Philosophical Transactions (1864)
3. Charlier C.V.L.: Arkiv för Mat. Astron. Fys. **16**, 1 (1922)
4. Dreyer J.L.E.: *A New General Catalogue of Nebulae and Clusters of Stars, being the catalogue of the late Sir John F.W. Herschel, revised, corrected and enlarged* Mem.R.A.S. XLIX, Part 1 (1888)

5. Dreyer J.L.E.: *Index Catalogue of Nebulae found in the Years 1888 to 1894, with Notes and Corrections* Mem.R.A.S. LI (1895)
6. Shapley, H.: in *Galaxies*, (Cambridge: Harvard University Press, 1961) pp. 159
7. Shane C.D., Wirtanen C.A.: Publ. Lick Obs. XXII, Pt. I (1967)
8. Nilson P.: *Uppsala General Catalog of Galaxies*, (Uppsala, University of Uppsala 1973)
9. Lauberts A.: *The ESO/Uppsala Survey of the ESO (B)* (Atlas, ESO, Garching 1982)
10. Vorontsov-Velyaminov B., Archipova V.: *Morphological Catalog of Galaxies*, Parts 2 to 5, (Moscow, Moscow University 1963-74)
11. Fouqué P., Paturel G.: A&A **150**, 192 (1985)
12. Hudson M.J., Lynden-Bell D.: MNRAS **252**, 219 (1991)
13. Schlegel D.J., Finkbeiner D.P., Davis M.: ApJ **500**, 525 (1998)
14. Cardelli J.A., Clayton G.C., Mathis J.S: ApJ **345**, 245 (1989)
15. Fairall A.P.: *Large-Scale Structures in the Universe* (Wiley Praxis Series in Astronomy and Astrophysics, Praxis Publishing, Chichester 1998)
16. Kogut A., Lineweaver C., Smoot G.F., Bennett C. L., Banday A., Boggess N.W., Cheng E.S., de Amici, G., Fixsen D.J., Hinshaw G., Jackson P. D., Janssen M., Keegstra P., Loewenstein K., Lubin P., Mather J.C., Tenorio L., Weiss R., Wilkinson D.T., Wright E.: ApJ **419**, 1 (1993)
17. Sandage A., Tammann G.A.: in *Large Scale Structures of the Universe, Cosmology and Fundamental Physics*, ed. by G. Setti & L. van Hove, (Garching: ESO 1984) pp. 127
18. Shaya E.J.: ApJ **280**, 470 (1984)
19. Kolatt T., Dekel A., Lahav O.: MNRAS **275**, 797 (1995)
20. Peebles, P.J.E.: ApJ **429**, 43 (1994)
21. Dressler A., Faber S.M., Burstein D., Davies, R.L., Lynden-Bell D., Terlevich R.J., Wegner G.: ApJ **313**, 37 (1987)
22. Lynden-Bell D., Faber S.M., Burstein D., Davies R.L., Dressler A., Terlevich R.J., Wegner G.: ApJ **326**, 19 (1988)
23. Böhm-Vitense E.: 1956, PASP 68, 430
24. Shane C.D., Wirtanen C.A.: AJ **59**, 285 (1954)
25. Fitzgerald M.P.: A&A **31**, 467 (1974)
26. Dodd R.J., Brand P.W.J.L.: A&AS **25**, 519 (1976)
27. Kraan-Korteweg R.C., Huchtmeier W.K.: A&A **266**, 150 (1992)
28. Lahav O., Yamada T., Scharf C.A., Kraan-Korteweg R.C.: MNRAS **262**, 711 (1993)
29. Weinberger R., Elsässer H., Beetz M., Birkle K.: A&A **48**, 327 (1976)
30. Huchra J., Hoessel J., Elias J.: AJ **82**, 674 (1977)
31. Weinberger R., A&AS **40**, 123 (1980)
32. Focardi P., Marano B., Vettolani G.: A&A **136**, 178 (1984)
33. Hauschildt M.: A&A **184**, 43 (1987)
34. Chamaraux P., Cayatte V., Balkowski C., Fontanelli P.: A&A **229**, 340 (1990)
35. Drinkwater M.J., Barnes D.G., Ellison S.L.: PASA **12**, 248 (1995)
36. Lewis G., Irwin M.: Spectrum, Newsletter of the Royal Observatories **12**, 22 (1996)
37. Pantoja C.A., Altschuler D.R., Giovanardi C., Giovanelli R.: AJ **113**, 905 (1997)
38. Seeberger R., Saurer W., Weinberger R., Lercher G.: in *Unveiling Large-Scale Structures Behind the Milky Way*, ed. by C. Balkowski, R.C. Kraan-Korteweg, ASP Conf. Ser. **67**, 81 (1994)

39. Seeberger R., Saurer W., Weinberger R.: A&AS **117**, 1 (1996)
40. Seeberger R., Saurer W.: A&AS **127**, 101 (1998)
41. Lercher G., Kerber F., Weinberger R.: A&AS **117**, 369 (1996)
42. Saurer W., Seeberger R., Weinberger R.: A&AS **126**, 247 (1997)
43. Marchiotto W., Wildauer H., Weinberger R.: (1999) in progress
44. Weinberger R., Gajdosik M., Zanin C.: A&AS **137**, 293 (1999)
45. Saito M., Ohtani A., Asomuna A., Kashikawa N., Maki T., Nishida S., Watanabe T.: PASJ **42**, 603 (1990)
46. Saito M., Ohtani A., Baba A., Hotta N., Kameno S., Kurosu S., Nakada K., Takata T.: PASJ **43**, 449 (1991)
47. Roman A.T., Takeuchi T.T., Nakanishi K., Saito M.: PASJ **50**, 47 (1998)
48. Roman A.T., Nakanishi K., Tomita A., Saito M.: PASJ **48**, 679 (1996)
49. Salem C., Kraan-Korteweg R.C.: in prep.,
50. Kraan-Korteweg R.C.: A&ASS **141**, 123 (2000)
51. Woudt P.A.: Ph.D. thesis, Univ. of Cape (Town 1998)
52. Woudt P.A., Kraan-Korteweg R.C.: A&AS (2000), in prep.
53. Woudt P.A., Kraan-Korteweg R.C.: A&AS (2000), in prep.
54. Fairall A.P., Kraan-Korteweg R.C.: in *Mapping the Hidden Universe*, ed. by R.C. Kraan-Korteweg, P.A. Henning & H. Andernach, ASP Conf. Ser. (2000) in press
55. Wakamatsu K., Hasegawa T., Karoji H., Sekiguchi K., Menzies J.W., Malkan M.: in *Unveiling Large-Scale Structures Behind the Milky Way*, ed. by C. Balkowski, R.C. Kraan-Korteweg, ASP Conf. Ser. **67**, 131 (1994)
56. Hasegawa T., Wakamatsu K., Malkan M., Sekiguchi K., Menzies J.W., Parker Q.A., Jugaku J., Karoji H., Okamura S.: MNRAS (2000), in press
57. Hau G.K.T., Ferguson H.C., Lahav O., Lynden-Bell D.: MNRAS **277**, 125 (1995)
58. Abell G.O., Corwin H.G., Olowin R.P.: ApJS **70**, 1 (1989)
59. Cameron L.M.: A&A **233**, 16 (1990)
60. Chamaraux P., Masnou J.-L., Kazés I., Saito M., Takata T., Yamada T.: MNRAS **307**, 263 (1999)
61. Kraan-Korteweg R.C., Woudt P.A.: PASA **16**, 53 (1999)
62. Kraan-Korteweg R.C., Cayatte V., Fairall A.P., Balkowksi C., Fairall A.P., Henning P.A.: x in *Unveiling Large-Scale Structures behind the Milky Way*, ed. by C. Balkowski and R.C. Kraan-Korteweg, ASP Conf. Ser. **67** (1999) pp. 99
63. Felenbok P., Guérin J., Fernandez A., Cayatte V., Balkowski C., Kraan-Korteweg R.C.: Experimental Astronomy **7**, 65 (1997)
64. Kraan-Korteweg R.C., Fairall A.P., Balkowski C.: A&A **297**, 617 (1995)
65. Fairall A.P., Woudt P.A., Kraan-Korteweg R.C.: A&AS **127**, 463 (1998)
66. Woudt P.A., Kraan-Korteweg R.C., Fairall A.P.: A&A **352**, 39 (1999)
67. Kraan-Korteweg R.C., Woudt P.A., Henning P.A.: PASA **14**, 15 (1997)
68. Sarazin C.L.: Rev. Mod. Phys. **58**, 1 (1986)
69. King I.: AJ **67**, 471 (1962)
70. Hughes J.P.: AJ **337**, 21 (1990)
71. White S.D.M., Briel U.G., Henry J.P.: MNRAS **261**, L8 (1993)
72. Wolf M.: Heidelberg Publ. **1**, 125 (1906)
73. Böhringer H., Neumann D.M., Schindler S., Kraan-Korteweg R.C.: ApJ **467**, 168 (1996)
74. Lynden-Bell D.: in *Observational Tests of Cosmological Inflation*, ed. by Shanks, T. et al. (1991) pp. 337
75. West R.M., Tarenghi M.: A&A **223**, 61 (1989)
76. Joint *IRAS* Science Working Group: *IRAS Point Source Catalog* Version 2 (Washington: US Govt. Printing Office 1988) (IRAS PSC)

77. Yamada T., Takata T., Djamaluddin T., Tomita A., Kentaro T., Saito M.: ApJS **89**, 57 (1993)
78. Lu N.Y., Dow M.W., Houck J.R., Salpeter E.E., Lewis B.M.: ApJ **357**, 388 (1990)
79. Takata T., Yamada T., Saito M.: ApJ **457**, 693 (1996)
80. Strauss M.A., Huchra J.P., Davis M., Yahil A., Fisher K.B., Tonry J.: ApJS **82**, 29 (1992)
81. Fisher K.B., Huchra J., Davis M., Strauss M.A., Yahil A., Schlegel D.: ApJS **100**, 69 (1995)
82. Saunders W., Sutherland W.J., Maddox S.J., Keeble O., Oliver S.J., Rowan-Robinson M., Efstathiou G.P., Tadros H., White S.D.M., Frenk C.S., Carramiñana A., Hawkins M.R.S.: MNRAS (2000) in press (astro-ph/0001117)
83. Saunders W., D'Mellow K.J., Tully R.B., Mobasher B., Maddox S.J., Sutherland W.J., Carrasco B.E., Hau G., Clements D.L., Staveley-Smith L.: in *Towards an Understanding of Cosmic Flows of Large-Scale Structure*, ed. by Courteau S., Strauss M., Willick J., ASP Conf. Ser. (2000) in press (astro-ph/9909174)
84. Kraan-Korteweg R.C., Woudt P.A., Cayatte V., Fairall A.P., Balkowski C., Henning P.A.: Nature **379** 519 (1996)
85. Woudt P.A., Kraan-Korteweg R.C., Fairall A.P.: in *Towards an Understanding of Cosmic Flows of Large-Scale Structure*, ed. by Courteau S., Strauss M., Willick J., ASP Conf. Ser. (2000) in press (astro-ph/9909094)
86. Epchtein N.: in *The Impact of Large Scale Near-Infrared Surveys.* ed. by. F. Garzón et al. (Kluwer, Dordrecht 1997) pp. 15
87. Epchtein N.: in *The Impact of Near-Infrared Sky Surveys on Galactic and Extragalactic Astronomy* ed. by N. Epchtein (Kluwer, Dordrecht 1998) pp. 3
88. Skrutskie M.F., Schneider S.E., Stiening R., Strom S.E., Weinberg M.D., Beichmann C., Chester T., Cutri R., Lonsdale C., Elias J., Elston R., Capps R., Carpentie J., Juchra J., Liebert J., Monet D., Price S., Seitzer P.: in *The Impact of Large Scale Near-Infrared Surveys*, ed. by F. Garzón et al. , (Kluwer, Dordrecht 1997) pp. 25
89. Skrutskie M.F.: in *The Impact of Near-Infrared Sky Surveys on Galactic and Extragalactic Astronomy*, ed. by N. Epchtein, (Kluwer, Dordrecht 1998) pp. 11
90. Mamon G.A.: in *Wide Field Surveys in Cosmology*, eds. Y. Mellier & S. Colombi, (Editions Frontières: Gif-sur-Yvette 1998) pp. 323
91. Gardner J.P., Sharples R.M., Carrasco B.E., Frenk C.S.: MNRAS **282**, L1 (1996)
92. Schröder A., Kraan-Korteweg R.C., Mamon G.A. Ruphy S.: in *Extragalactic Astronomy in the Infrared*, ed. by G. A. Mamon et al. (Editions Frontières: Gif-sur-Yvette 1997) pp. 381
93. Schröder A., Kraan-Korteweg R.C., Mamon G.A.: PASA **16**, 42 (1999)
94. Kraan-Korteweg R.C., Schröder A., Mamon G., Ruphy S.: in *The Impact of Near-Infrared Surveys on Galactic and Extragalactic Astronomy*, ed. by N. Epchtein (Kluwer, Dordrecht 1998) pp. 205
95. Mamon G.A.: in *Unveiling Large-Scale Structures Behind the Milky Way*, ed. by C. Balkowski, R.C. Kraan-Korteweg, ASP Conf. Ser. **67**, 53 (1994)
96. Hurt R.L., Jarrett T., Cutri R., Skrutskie M., Schneider S., van Driel W.: BAAS **194**, 832 (1999)
97. Tully R.B., Fisher J.R.: A&A **54**, 661 (1977)
98. Kerr F.J., Henning P.A.: ApJ **320**, L99 (1987)
99. Kraan-Korteweg R.C., Loan A.J., Burton W.B., Lahav O., Ferguson, H.C., Henning P.A., Lynden-Bell D.: Nature **372**, 77 (1994)
100. Henning P.A., Kraan-Korteweg R.C., Rivers A.J., Loan, A.J., Lahav O., Burton W.B.: AJ **115**, 584 (1998)

101. Rivers A.J., Henning P.A., Kraan-Korteweg R.C.: PASA **16**, 48 (1999)
102. Kraan-Korteweg R.C., Koribalski B., Juraszek S.: in *Looking Deep in the Southern Sky*, ed. by R. Morganti, W. Couch (Springer 1998) pp. 23
103. Staveley-Smith L., Juraszek S., Koribalski B.S. Ekers, R.D., Green, A.J., Haynes, R.F., Henning, P.A., Kesteven, M.J., Kraan-Korteweg, R.C., Price, R.M., Sadler, E.M.: AJ **116**, 2717 (1998)
104. Henning P.A., Staveley-Smith L., Kraan-Korteweg R.C., Sadler E.M.: PASA **16**, 35 (1999)
105. Henning P.A., Staveley-Smith L., Ekers R.D., Green A.J., Haynes R.F., Juraszek S., Kesteven M.J., Koribalski B., Kraan-Korteweg R.C., Price R.M. Sadler E.M., Schröder A.: Astron. J. (2000), in press
106. Loan A.J., Maddox S.J., Lahav O., Balcells M., Kraan-Korteweg R.C., Assendorp R., Almoznino E., Brosch N., Goldberg E., Ofek E.O.: MNRAS **280**, 537 (1996)
107. Buta R.J., McCall M.L.: ApJS **124**, 33 (1999)
108. Burton W.B., Verheijen M.A.W., Kraan-Korteweg R.C., Henning P.A.: A&A **309**, 687 (1996)
109. Kuno N., Vila-Vilaro B., Nishiyama K.: PASJ **48**, 19 (1996)
110. Li J.G., Zhao J.H., Ho P.T.P., Sage L.J.: A&A **307**, 424 (1996)
111. Tilanus R.P.J., Burton W.B.: A&A **324**, 899 (1997)
112. Reynolds C.S., Loan A.J., Fabian A.C., Makishima K., Brandt W.N., Mizuno T.: MNRAS **286**, 349 (1997)
113. Zwaan M., Briggs F., Sprayberry D.: PASA **14**, 126 (1997)
114. Wakker B.P.: Ph.D. thesis, Univ. of Groningen (1990)
115. Burton W.B., Braun R., Walterbos R.A.M., Hoopes C.G.: AJ **117**, 194 (1999)
116. Hartmann D.: Ph.D. thesis, Univ. of Leiden (1994)
117. Hartmann D., Burton W.B.: *Atlas of Galactic Neutral Hydrogen* (Cambridge University Press 1997)
118. Tammann G.A., Kraan-Korteweg R.C.: in *The Large Scale Structure of the Universe*; IAU Symp. 79, ed. by M.S. Longair and J. Einasto, (Reidel: Dordrecht 1978), pp. 71
119. Kraan-Korteweg R.C.: AN **300**, 181 (1979)
120. Staveley-Smith, L., Wilson, W.E., Bird, T.S., Disney, M.J., Ekers, R.D., Freeman, K.C., Haynes, R.F., Sinclair, M.W., Vaile, R.A., Webster, R.L., Wright, A.E.: PASA **13**, 243 (1996)
121. Juraszek S.: PASA **16**, 38 (1999)
122. Juraszek S., Staveley-Smith L., Kraan-Korteweg R.C., Green A.J., Ekers R.D., Henning P.A., Kesteven M.J., Koribalksi B., Sadler E.M., Schröder A.C.: AJ (2000), in press
123. Woudt P.A., Kraan-Korteweg R.C.: in progress.
124. Fabian A.C.: in *Unveiling Large-Scale Structures behind the Milky Way*. ed. by C. Balkowski and R.C. Kraan-Korteweg, ASP Conf. Ser. **67**, 76 (1994)
125. Edge A.C., Stewart G.C., Fabian A.C., Arnaud, K.A.: MNRAS **245**, 559 (1990)
126. Jahoda K., Mushotzky R.F.: ApJ **346**, 638 (1989)
127. Tamura T., Fukazawa Y., Kaneda H., Makishima K., Tashiro M., Tanaka Y., Böhringer H.: PASJ **50**, 195 (1998)
128. Ebeling H., Mullis C.R., Tully B.R.: BAAS **31** (HEAD meeting), 699 (1999)
129. Ebeling H., Mullis C.R., Tully B.R.: (1999) submitted to ApJ
130. Voges W., Aschenbach B., Boller T., Bäuninger H., Briel U., Burkert W., Dennerl K., Englhauser J., Gruber R., Haberl F., Hartner G., Hasinger G., Kürster M., Pfeffermann E., Pietsch W., Predehl P., Rosso, C., Schmitt J.H.M.M., Trümper J., Zimmermann H.-U.: A&A **349**, 389 (1999)

131. Ebeling H., Edge A.C., Böhringer H., Allen S.W., Crawford C.S., Fabian A.C., Voges W., Huchra J.P.: MNRAS **301**, 881 (1998)
132. Lahav O.: in *Unveiling Large-Scale Structures behind the Milky Way*, ed. by C. Balkowski and R.C. Kraan-Korteweg, ASP Conf. Ser. **67**, 171 (1994)
133. Hoffman Y.: in *Unveiling Large-Scale Structures behind the Milky Way*, ed. by C. Balkowski and R.C. Kraan-Korteweg, ASP Conf. Ser.**67**, 185 (1994)
134. Bertschinger E., Dekel A.: ApJ **336**, 5 (1989)
135. Strauss M.A., Willick J.A.: Phys. Rep. **26**, 27 (1995)
136. Scharf C., Hoffman Y., Lahav O., Lynden-Bell D.: MNRAS **256**, 229 (1992)
137. Kraan-Korteweg R.C., Woudt P.A.: in *Cosmic Velocity Fields*, ed. by F. Bouchet and M. Lachièze-Rey, (Editions Frontières: Gif-sur-Yvette 1993) pp. 557
138. Santiago B.X., Strauss M.A., Lahav O., Davis M., Dressler A., Huchra J.: ApJ **446**, 457 (1995)
139. Baker J.E., Davis M., Strauss M.A., Lahav O., Santiago B.X.: ApJ **508**, 6 (1998)
140. Webster M., Lahav O., Fisher K.: MNRAS **287**, 425 (1997)
141. Bistolas V.: Ph.D. thesis, Hebrew University, Jerusalem (1998)

List of Participants

INSTITUTO DE ASTRONOMIA, UNAM, Mexico

Vladimir Avila	avila@astroscu.unam.mx
Fernando Becerra	vinance@astroscu.unam.mx
Irene Cruz Gonzalez	irene@astroscu.unam.mx
Wilder Chicana	wilder@astroscu.unam.mx
Hector Hernandez	hector@astroscu.unam.mx
William Lee	wlee@astroscu.unam.mx
Dany Page	page@astroscu.unam.mx
Carlos Perez	capeto@astroscu.unam.mx
Leopoldo Pineda	pineda@astroscu.unam.mx
Victor Robledo	vico@astroscu.unam.mx
Armando Rojas	rojas@astroscu.unam.mx
Juan Segura	sosa@astroscu.unam.mx

INSTITUTO DE CIENCIAS NUCLEARES, UNAM, Mexico

Alexis Aguilar	alexis@nuclecu.unam.mx
Juan Carlos D'Olivo	dolivo@nuclecu.unam.mx
Peter Hess	hess@nuclecu.unam.mx
Jorge Hirsch	hirsch@nuclecu.unam.mx
Juan Carlos Lopez	vieyra@nuclecu.unam.mx
Lukas Nellen	lukas@nuclecu.unam.mx
Sarira Sahu	sarira@nuclecu.unam.mx
Luis Urrutia	urrutia@nuclecu.unam.mx

INSTITUTO DE FISICA, UNAM, Mexico

Armando Perdomo	perdomo@feynmann.ifisicacu.unam.mx

INSTITUTO DE GEOFISICA, UNAM, Mexico

Rogelio Caballero	
Guadalupe Cordero	
Rosa Diaz	
Juan Ramirez	
Jose Francisco Valdez	jfvaldes@tonatiuh.igeofcu.unam.mx

INAOE, Mexico

Alberto Carraminana	alberto@inaoep.mx
Luis Carrasco	carrasco@inaoep.mx
Lino Rodriguez	linorome@inaoep.mx
Daniel Rosa	danrosa@inaoep.mx
Juan Pablo Torres	papaqui@inaoep.mx
Olga Vega	ovega@inaoep.mx

DEPARTAMENTO DE FISICA, CINVESTAV-IPN, Mexico

Juan Arteaga	jarteaga@fis.cinvestav.mx
Alejandro Castillo	acastillo@fis.cinvestav.mx
Joaquin Esteves	joaquin@rosa.fis.cinvestav.mx
Julio Flores	julio@fis.cinvestav.mx
Francisco Guzman	siddahartha@fis.cinvestav.mx
Tonatiuh Matos	tmatosfis.cinvestav.mx
Rodrigo Pelayo	rpelayo@fis.cinvestav.mx
Luis Urena	lurena@fis.cinvestav.mx
Carlos Vargas	cvargas@fis.cinvestav.mx
Victor Velazquez	vvelaz@fis.cinvestav.mx
Arnulfo Zepeda	zepeda@fis.cinvestav.mx

INSTITUTO DE FISICA Y MATEMATICAS, UNIVERSIDAD MICHOACANA DE SNH, Mexico

Mariano Alarcon	mach@ginette.ifm.umich.mx
Elvira Garcia	elvira@itzel.ifm.umich.mx
Mauricio Gonzalez	aviles@itzel.ifm.umich.mx
Gerardo Leon	gleon@itzel.ifm.umich.mx
Martin Medina	
Luis Manuel Villasenior	villasen@zeus.ccu.umich.mx
Thomas Zannias	zannias@ginette.ifm.umich.mx

DEPARTAMENTO DE ASTRONOMIA, UNIVERSIDAD DE GUANAJUATO, Mexico

Victor Migenes	vmigenes@cibeles.astro.ugto.mx

UNIVERSIDAD NACIONAL DE COSTA RICA, Costa Rica

Ludmila Semionova	lsemiono@samara.una

CENTRO BRASILEIRO DE PESQUISAS FISICAS, Brasil

Santiago Perez	santiago@lafex.cbpf.br

UNIVERSIDAD DE LOS ANDES, Venezuela

Ingrid Inciarte	ingridi@ciens.ula.ve

UNIVERSIDAD DE GUATEMALA, Guatemala

Enrique Pazos
Hector Perez

Lecture Notes in Physics

For information about Vols. 1–524
please contact your bookseller or Springer-Verlag

Vol. 525: A. Ceresole, C. Kounnas, D. Lüst, S. Theisen (Eds.), Quantum Aspects of Gauge Theories, Supersymmetry and Unification. Proceedings, 1998. X, 511 pages. 1999.

Vol. 526: H.-P. Breuer, F. Petruccione (Eds.), Open Systems and Measurement in Relativistic Quantum Theory. Proceedings, 1998. VIII, 240 pages. 1999.

Vol. 527: D. Reguera, J. M. G. Vilar, J. M. Rubí (Eds.), Statistical Mechanics of Biocomplexity. Proceedings, 1998. XI, 318 pages. 1999.

Vol. 528: I. Peschel, X. Wang, M. Kaulke, K. Hallberg (Eds.), Density-Matrix Renormalization. Proceedings, 1998. XVI, 355 pages. 1999.

Vol. 529: S. Biringen, H. Örs, A. Tezel, J.H. Ferziger (Eds.), Industrial and Environmental Applications of Direct and Large-Eddy Simulation. Proceedings, 1998. XVI, 301 pages. 1999.

Vol. 530: H.-J. Röser, K. Meisenheimer (Eds.), The Radio Galaxy Messier 87. Proceedings, 1997. XIII, 342 pages. 1999.

Vol. 531: H. Benisty, J.-M. Gérard, R. Houdré, J. Rarity, C. Weisbuch (Eds.), Confined Photon Systems. Proceedings, 1998. X, 496 pages. 1999.

Vol. 532: S. C. Müller, J. Parisi, W. Zimmermann (Eds.), Transport and Structure. Their Competitive Roles in Biophysics and Chemistry. XII, 400 pages. 1999.

Vol. 533: K. Hutter, Y. Wang, H. Beer (Eds.), Advances in Cold-Region Thermal Engineering and Sciences. Proceedings, 1999. XIV, 608 pages. 1999.

Vol. 534: F. Moreno, F. González (Eds.), Light Scattering from Microstructures. Proceedings, 1998. XII, 300 pages. 2000

Vol. 535: H. Dreyssé (Ed.), Electronic Structure and Physical Properties of Solids: The Uses of the LMTO Method. Proceedings, 1998. XIV, 458 pages. 2000.

Vol. 536: T. Passot, P.-L. Sulem (Eds.), Nonlinear MHD Waves and Turbulence. Proceedings, 1998. X, 385 pages. 1999.

Vol. 537: S. Cotsakis, G. W. Gibbons (Eds.), Mathematical and Quantum Aspects of Relativity and Cosmology. Proceedings, 1998. XII, 251 pages. 1999.

Vol. 538: Ph. Blanchard, D. Giulini, E. Joos, C. Kiefer, I.-O. Stamatescu (Eds.), Decoherence: Theoretical, Experimental, and Conceptual Problems. Proceedings, 1998. XII, 345 pages. 2000.

Vol. 539: A. Borowiec, W. Cegła, B. Jancewicz, W. Karwowski (Eds.), Theoretical Physics. Fin de Siècle. Proceedings, 1998. XX, 319 pages. 2000.

Vol. 540: B. G. Schmidt (Ed.), Einstein's Field Equations and Their Physical Implications. Selected Essays. 1999. XIII, 429 pages. 2000.

Vol. 541: J. Kowalski-Glikman (Ed.), Towards Quantum Gravity. Proceedings, 1999. XII, 376 pages. 2000.

Vol. 542: P. L. Christiansen, M. P. Sørensen, A. C. Scott (Eds.), Nonlinear Science at the Dawn of the 21st Century. Proceedings, 1998. XXVI, 458 pages. 2000.

Vol. 543: H. Gausterer, H. Grosse, L. Pittner (Eds.), Geometry and Quantum Physics. Proceedings, 1999. VIII, 408 pages. 2000.

Vol. 544: T. Brandes (Ed.), Low-Dimensional Systems. Interactions and Transport Properties. Proceedings, 1999. VIII, 219 pages. 2000

Vol. 545: J. Klamut, B. W. Veal, B. M. Dabrowski, P. W. Klamut, M. Kazimierski (Eds.), New Developments in High-Temperature Superconductivity. Proceedings, 1998. VIII, 275 pages. 2000.

Vol. 546: G. Grindhammer, B. A. Kniehl, G. Kramer (Eds.), New Trends in HERA Physics 1999. Proceedings, 1999. XIV, 460 pages. 2000.

Vol. 547: D. Reguera, G. Platero, L.L. Bonilla, J.M. Rubí(Eds.), Statistical and Dynamical Aspects of Mesoscopic Systems. Proceedings, 1999. XII, 357 pages. 2000.

Vol. 548: D. Lemke, M. Stickel, K. Wilke (Eds.), ISO Surveys of a Dusty Universe. Proceedings, 1999. XIV, 432 pages. 2000.

Vol. 549: C. Egbers, G. Pfister (Eds.), Physics of Rotating Fluids. Selected Topics, 1999. XVIII, 437 pages. 2000.

Vol. 550: M. Planat (Ed.), Noise, Oscillators and Algebraic Randomness. Proceedings, 1999. VIII, 417 pages. 2000.

Vol. 551: B. Brogliato (Ed.), Impacts in Mechanical Systems. Analysis and Modelling. Lectures, 1999. IX, 273 pages. 2000.

Vol. 552: Z. Chen, R. E. Ewing, Z.-C. Shi (Eds.), Numerical Treatment of Multiphase Flows in Porous Media. Proceedings, 1999. XXI, 445 pages. 2000.

Vol. 553: J.-P. Rozelot, L. Klein, J.-C. Vial Eds.), Transport of Energy Conversion in the Heliosphere. Proceedings, 1998. IX, 214 pages. 2000.

Vol. 554: K. R. Mecke, D. Stoyan (Eds.), Statistical Physics and Spatial Statistics. The Art of Analyzing and Modeling Spatial Structures and Pattern Formation. Proceedings, 1999. XII, 415 pages. 2000.

Vol. 555: A. Maurel, P. Petitjeans (Eds.), Vortex Structure and Dynamics. Proceedings, 1999. XII, 319 pages. 2000.

Vol. 556: D. Page, J. G. Hirsch (Eds.), From the Sun to the Great Attractor. X, 330 pages. 2000.

Vol. 557: J. A. Freund, T. Pöschel (Eds.), Stochastic Processes in Physics, Chemistry, and Biology. X, 330 pages. 2000.

Vol. 558: P. Breitenlohner, D. Maison (Eds.), Quantum Field Theory. Proceedings, 1998. VIII, 323 pages. 2000

Vol. 559: H.-P. Breuer, F. Petruccione (Eds.), Relativistic Quantum Measurement and Decoherence. Proceedings, 1999. X, 140 pages. 2000.

Vol. 561: H. J. Carmichael, R. J. Glauber, M. O. Scully (Eds.), Directions in Quantum Optics. Proceedings, 1999. XVII, 369 pages. 2001.

Monographs

For information about Vols. 1–24
please contact your bookseller or Springer-Verlag

Vol. m 25: A. V. Bogdanov, G. V. Dubrovskiy, M. P. Krutikov, D. V. Kulginov, V. M. Strelchenya, Interaction of Gases with Surfaces. XIV, 132 pages. 1995.

Vol. m 26: M. Dineykhan, G. V. Efimov, G. Ganbold, S. N. Nedelko, Oscillator Representation in Quantum Physics. IX, 279 pages. 1995.

Vol. m 27: J. T. Ottesen, Infinite Dimensional Groups and Algebras in Quantum Physics. IX, 218 pages. 1995.

Vol. m 28: O. Piguet, S. P. Sorella, Algebraic Renormalization. IX, 134 pages. 1995.

Vol. m 29: C. Bendjaballah, Introduction to Photon Communication. VII, 193 pages. 1995.

Vol. m 30: A. J. Greer, W. J. Kossler, Low Magnetic Fields in Anisotropic Superconductors. VII, 161 pages. 1995.

Vol. m 31 (Corr. Second Printing): P. Busch, M. Grabowski, P.J. Lahti, Operational Quantum Physics. XII, 230 pages. 1997.

Vol. m 32: L. de Broglie, Diverses questions de mécanique et de thermodynamique classiques et relativistes. XII, 198 pages. 1995.

Vol. m 33: R. Alkofer, H. Reinhardt, Chiral Quark Dynamics. VIII, 115 pages. 1995.

Vol. m 34: R. Jost, Das Märchen vom Elfenbeinernen Turm. VIII, 286 pages. 1995.

Vol. m 35: E. Elizalde, Ten Physical Applications of Spectral Zeta Functions. XIV, 224 pages. 1995.

Vol. m 36: G. Dunne, Self-Dual Chern-Simons Theories. X, 217 pages. 1995.

Vol. m 37: S. Childress, A.D. Gilbert, Stretch, Twist, Fold: The Fast Dynamo. XI, 406 pages. 1995.

Vol. m 38: J. González, M. A. Martín-Delgado, G. Sierra, A. H. Vozmediano, Quantum Electron Liquids and High-Tc Superconductivity. X, 299 pages. 1995.

Vol. m 39: L. Pittner, Algebraic Foundations of Non-Com-mutative Differential Geometry and Quantum Groups. XII, 469 pages. 1996.

Vol. m 40: H.-J. Borchers, Translation Group and Particle Representations in Quantum Field Theory. VII, 131 pages. 1996.

Vol. m 41: B. K. Chakrabarti, A. Dutta, P. Sen, Quantum Ising Phases and Transitions in Transverse Ising Models. X, 204 pages. 1996.

Vol. m 42: P. Bouwknegt, J. McCarthy, K. Pilch, The W3 Algebra. Modules, Semi-infinite Cohomology and BV Algebras. XI, 204 pages. 1996.

Vol. m 43: M. Schottenloher, A Mathematical Introduction to Conformal Field Theory. VIII, 142 pages. 1997.

Vol. m 44: A. Bach, Indistinguishable Classical Particles. VIII, 157 pages. 1997.

Vol. m 45: M. Ferrari, V. T. Granik, A. Imam, J. C. Nadeau (Eds.), Advances in Doublet Mechanics. XVI, 214 pages. 1997.

Vol. m 46: M. Camenzind, Les noyaux actifs de galaxies. XVIII, 218 pages. 1997.

Vol. m 47: L. M. Zubov, Nonlinear Theory of Dislocations and Disclinations in Elastic Body. VI, 205 pages. 1997.

Vol. m 48: P. Kopietz, Bosonization of Interacting Fermions in Arbitrary Dimensions. XII, 259 pages. 1997.

Vol. m 49: M. Zak, J. B. Zbilut, R. E. Meyers, From Instability to Intelligence. Complexity and Predictability in Nonlinear Dynamics. XIV, 552 pages. 1997.

Vol. m 50: J. Ambjørn, M. Carfora, A. Marzuoli, The Geometry of Dynamical Triangulations. VI, 197 pages. 1997.

Vol. m 51: G. Landi, An Introduction to Noncommutative Spaces and Their Geometries. XI, 200 pages. 1997.

Vol. m 52: M. Hénon, Generating Families in the Restricted Three-Body Problem. XI, 278 pages. 1997.

Vol. m 53: M. Gad-el-Hak, A. Pollard, J.-P. Bonnet (Eds.), Flow Control. Fundamentals and Practices. XII, 527 pages. 1998.

Vol. m 54: Y. Suzuki, K. Varga, Stochastic Variational Approach to Quantum-Mechanical Few-Body Problems. XIV, 324 pages. 1998.

Vol. m 55: F. Busse, S. C. Müller, Evolution of Spontaneous Structures in Dissipative Continuous Systems. X, 559 pages. 1998.

Vol. m 56: R. Haussmann, Self-consistent Quantum Field Theory and Bosonization for Strongly Correlated Electron Systems. VIII, 173 pages. 1999.

Vol. m 57: G. Cicogna, G. Gaeta, Symmetry and Perturbation Theory in Nonlinear Dynamics. XI, 208 pages. 1999.

Vol. m 58: J. Daillant, A. Gibaud (Eds.), X-Ray and Neutron Reflectivity: Principles and Applications. XVIII, 331 pages. 1999.

Vol. m 59: M. Kriele, Spacetime. Foundations of General Relativity and Differential Geometry. XV, 432 pages. 1999.

Vol. m 60: J. T. Londergan, J. P. Carini, D. P. Murdock, Binding and Scattering in Two-Dimensional Systems. Applications to Quantum Wires, Waveguides and Photonic Crystals. X, 222 pages. 1999.

Vol. m 61: V. Perlick, Ray Optics, Fermat's Principle, and Applications to General Relativity. X, 220 pages. 2000.

Vol. m 62: J. Berger, J. Rubinstein, Connectivity and Superconductivity. XI, 246 pages. 2000.

Vol. m 63: R. J. Szabo, Ray Optics, Equivariant Cohomology and Localization of Path Integrals. XII, 315 pages. 2000.

Vol. m 64: I. G. Avramidi, Heat Kernel and Quantum Gravity. X, 143 pages. 2000.

www.ingramcontent.com/pod-product-compliance
Lightning Source LLC
Chambersburg PA
CBHW071238230926
51849CB00001B/151
9783540410645